AVIATION WEATHER

by Peter F. Lester

55 Inverness Drive East, Englewood, CO 80112-5498
JS319007A
International Standard Book Number 0-88487-178-9

ACKNOWLEDGEMENTS

This project could not have been accomplished in such a short time without the help of many individuals and groups. I thank my University, San Jose State, for giving me the opportunity to teach, to carry out my research, and to write. I thank my students, past and present. Their interest and energy is what makes it worthwhile. I am grateful to those who read the first draft of the manuscript: Jerry Steffens, who found time to help clarify some convoluted text, despite a heavy and continuously changing teaching load; Marc Burton, the ultimate aviation meteorologist, whose candor and sense of humor are unsurpassed; and John Rieger III, M.D., pilot, and philosopher. Thanks also to Art Rangno for some outstanding pictures and for his input on the icing chapter. The cooperation of Jeppesen DataPlan is greatly appreciated, in particular, Mike Cetinich and Cris Taylor.

I applaud the entire team at Jeppesen Sanderson: Matt Adams, Judi Glenn, Rich Hahn, Dean McBournie, Dave Schoeman, and all the others. Their amazing talents with art and manuscript, their patience with me, and their dedication and willingness to put in long hours were an example that kept me going.

I especially want to thank Liz Kailey who combined her talents as writer and experienced pilot/CFI with my sketchy outlines and notes to produce Chapters 16 and 17.

I gratefully acknowledge the editorial talent, management ability, and high professional standards of Senior Editor, Dick Snyder. His knowledge of the audience to whom the book is directed and his skill in bringing together writers, artists, and editors to solve some difficult problems in a short time were essential for the success of the effort.

Although they didn't see me much, my family, close and extended, provided a solid connection to what is really important in life. Thanks for putting up with my "outta sight . . . must be writing" behavior for the last year.

This book is dedicated to the loves of my life . . . Heather Marie and Rachael Phyllis . . . and to the memory of my Mom, Marie Burns Lester.

TABLE OF CONTENTS

Part I
Aviation Weather Basics

Part II
Atmospheric Circulation Systems

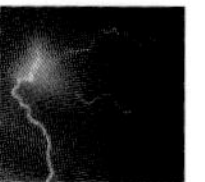

Part III
Aviation Weather Hazards

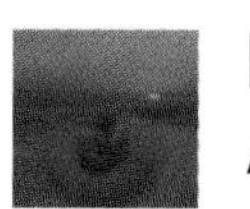

Part IV
Applying Weather Knowledge

Appendix

HOW THE SYSTEM WORKS

The Jeppesen Sanderson study/review concept of learning presents information in an uncomplicated way with coordinated text and illustrations. Key terms, FAA question material, and insight readings are presented in color to allow you to review important points and concepts. The major features of the book are presented in the following pages.

Learning Objectives

Learning objectives at the beginning of each Part help you focus on important concepts.

Learning Objectives

Learning objectives at the beginning of each Chapter help you focus on important concepts.

Chapter Outlines

Chapter outlines help you identify the major topics of the chapter.

CHAPTER 1

The Atmosphere

Introduction

The formal study of any physical system, such as an engine or an airplane, usually begins with a description of that system. Information about component parts, their location and dimensions, and terminology is necessary background for later examination and understanding of the system design and operation. Our study of aviation weather begins in a similar way. The "system" in this case, is the atmosphere.

When you complete this chapter, you should be able to describe the composition, dimensions, and average vertical structure of the atmosphere using proper technical vocabulary. Furthermore, you will have been introduced to a valuable reference tool, the standard atmosphere.

ATMOSPHERIC COMPOSITION
ATMOSPHERIC PROPERTIES
- *Temperature*
- *Density*
- *Pressure*
- *The Gas Law*

ATMOSPHERIC STRUCTURE
- *Dimensions*
- *Atmospheric Layers*
  - *Temperature Layers*
  - *Other Layers*
  - *Standard Atmosphere*

PRESSURE, ALTITUDE, AND DENSITY

peratures are standard. If the atmosphere is colder than standard, your indicated altitude will be higher than true altitude. If the atmosphere is warmer than standard, the indicated altitude will be lower than the true altitude. (Figure 3-12) Temperature errors are generally smaller than those associated with variations in sea level pressure. For example, if the actual temperature was 10C° warmer than standard, the true altitude would be about 4% higher. This is only 40 feet at 1,000 feet MSL. But the error increases with height. At 12,000 feet MSL, it is about 500 feet. Flight over high mountains in bad weather, requires close attention to possible temperature errors.

Remember, the pressure altimeter will not automatically show exact altitude in flight. It is pilot's responsibility to ensure terrain avoidance.(USAF, 1990)

On warm days pressure surfaces are raised and the indicated altitude is lower than true altitude.

The third pressure altimeter error that arises because of nonstandard atmospheric conditions is caused by rapid changes in vertical movements of the air. These changes upset the balance of forces that allows atmospheric pressure to be related directly to altitude. Significant errors may be expected in the extreme conditions of thunderstorms and in strong mountain waves. More on this problem is presented in Part III on aviation weather hazards.

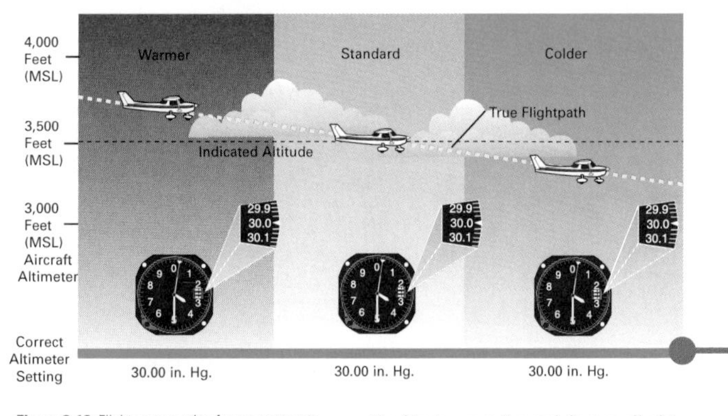

Figure 3-12. Flight cross section from a warmer to a colder airmass. Notice that standard atmospheric conditions only occur near the center of the diagram. The aircraft true altitude is higher than indicated in the warmer air and lower than indicated in the colder air.

The altimeter errors discussed above are all related to atmospheric conditions. Other errors may arise due to instrument problems. These include improper calibration, friction, lag, improper instrument location, and temperature changes of the instrument. These are beyond the scope of

3-14

Insight Readings

Insight readings are contained in green boxes throughout the text. These readings add insight to the information covered and help you relate to and understand the main topics in the chapter.

FAA Questions

FAA question material appears in blue boxes. This material summarizes important concepts behind the weather-related questions on the Private, Instrument, and Commercial airmen knowledge exams. You can review the blue boxes prior to taking your FAA exam.

Full Color Graphics

Full-color graphics make learning fun and meaningful. Detailed material is presented in an uncomplicated way. Graphics are carefully planned to complement and expand the concepts in the text.

Chapter Summaries

Chapter summaries are provided at the end of each chapter to help you review the important concepts contained in the chapter.

ATMOSPHERIC ENERGY AND TEMPERATURE

SUMMARY

The observed temperatures of the earth and the atmosphere are produced primarily by solar energy that is transferred across space via radiation. Solar energy is redistributed in the earth-atmosphere system by terrestrial radiation, conduction, and convection/advection. Solar and terrestrial radiation have distinctly different characteristics. A variety of global temperature patterns result when these differences are combined with changes in sun-earth geometry, variations in the nature of the earth's surface, the presence certain variable gases, and atmospheric motions. In subsequent chapters we will see that these temperature patterns and their changes are the root causes of atmospheric winds and weather.

Key Terms

Key terms are set off in blue type when they are first used and defined. Key terms are listed at the end of each chapter and included in a glossary at the end of the book.

KEY TERMS

Absolute Zero
Advection
Albedo
Boiling Point
Celsius
Conduction
Convection
Diurnal Variation
Energy Spectrum
Equinox
Fahrenheit
Frequency
Greenhouse Effect
Heat Capacity
Indicated Air Temperature (IAT)
Infrared (IR)
Melting Point
Outside Air Temperature (OAT)
Radiation
Room Temperature
Solar Declination
Solar Elevation Angle
Solar Radiation
Solstice
Sounding
Spectrum
Speed of Light
Standard Sea Level Temperature
Surface Air Temperature
Temperature Gradient
Terrestrial Radiation
True Air Temperature (TAT)
Ultraviolet (UV)
Upper Air Temperature
Wavelength

Chapter Questions

Chapter questions help you evaluate your understanding of the material presented in the chapter. A variety of question formats are included, such as open-ended, matching, true/false, and essay. Answers to the questions are provided by your instructor.

CHAPTER QUESTIONS

1. Convert the following temperatures from °C to°F.
   1. -60
   2. -40
   3. -15
   4. 5
   5. 35
2. Convert the following from °F to °C.
   1. -453
   2. -100
   3. 0
   4. 25
   5. 113

2-16

Color Photographs

Color photographs enhance learning and spark interest by showing actual conditions.

off to the side, the path is still straight. Note that if the direction of rotation is clockwise (cw), the deflection of the ball is to the left.

When we attempt to describe the motion of the atmosphere (or anything else) relative to the rotating earth, we must also consider Coriolis force. However, things become a little more involved because the earth is a rotating sphere, rather than a rotating disk. In the case of the merry-go-round, it did not matter where the thrower and the catcher were located on the rotating platform. For a fixed rotation rate and a constant speed of the ball, Coriolis force was the same everywhere on the platform. This is because the axis of rotation of the merry-go-round was vertical; that is, perpendicular to the platform across which the ball was moving. This is not the case with the earth. For a fixed rotation rate and speed of the ball, Coriolis force is different at different latitudes. The variation is illustrated in figure 4-8.

If, as shown in figure 4-8, our merry-go-round is attached to the earth at the North Pole, it rotates counterclockwise. Note that the axis of the earth and the axis of rotation of the merry-go-round are parallel at the pole. If we let the earth's rotation rate and the speed of the ball be the same as in the previous example, then (considering only Coriolis force) we would see the same effect on the ball. At the equator, the situation is different. In that location, the axis of the earth is perpendicular to the axis of the merry-go-round, so the merry-go-round does not rotate

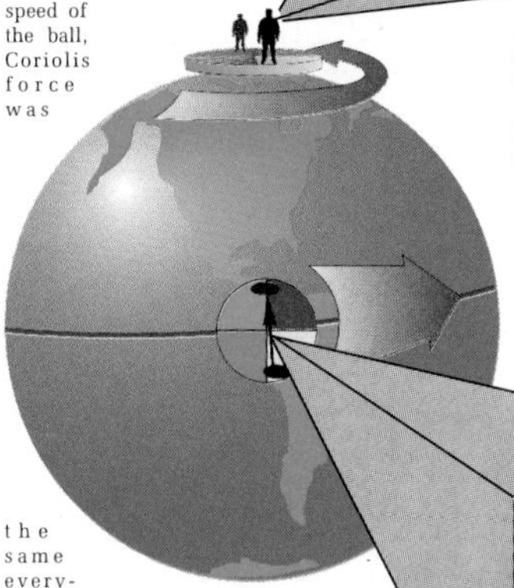

Figure 4-8. For a given speed of the ball and a constant rotation rate, Coriolis force is a maximum at the poles and zero at the equator.

References and Recommended Readings

References and recommended readings are included in the appendix. This helps you locate additional information on the topics and concepts presented in the book.

REFERENCES AND RECOMMENDED READING

Ahrens, C.D., 1994: *Meteorology Today: an introduction to weather, climate, and the environment. Fifth Edition.* West Publishing Co., St. Paul, MN. 592pp.

Anderson, J.D., 1985: *Introduction to Flight.* Second Edition. McGraw-Hill. New York. 560pp.

Atkinson, B.W., 1981: *Mesoscale Atmospheric Circulations.* Academic Press. 495pp.

Atlas, D. (Ed.), 1990: *Radar in Meteorology.* American Meteorological Society, Boston. 806pp.

Bradbury, T.A., and J.P. Kuettner (Eds.), 1976: *Forecasters Manual for Soaring Flight.* Organisation Scientifique et Technique International du Vol a Voile (OSTIV), Geneva. 119pp.

Buck, R.N.,1988: *Weather Flying.* Third Edition. MacMillan, New York. 311pp.

Byers, H.R., 1974: *General Meteorology.* McGraw-Hill, New York.

Byers, H.R., and R.R.Braham, 1949: *The Thunderstorm.* U.S. Weather Bureau, Washington. 287pp.

Caracena, F., R.L. Holle, C.A. Doswell III, 1990: *Microbursts, A Handbook for Visual Identification.* Second Edition. NOAA, ERL, NSSL. 35pp.

Collins, R.L., 1982: *Thunderstorms and Airplanes.* Delacorte Press/Eleanor Friede. New York. 280pp.

Crossley, A.F. and A.G. Forsdyke, 1960: *Handbook of Aviation Meteorology. M.O. 630 (A.P.3340).* Her Majesty's Stationery Office. London.

Elsberry, R.L. (Ed.), W.M. Frank, G.J. Holland, J.D. Jarrel, R.L. Southern, 1987: *A Global View of Tropical Cyclones.* USNPGS, Monterey. Publication sponsored by Office of Naval Research, Marine Meteorological Program. 185pp.

Evans, J., and M.L. Stone, 1993: *Role of the Aviation Weather System in providing a real-time ATC Volcanic Ash Advisory System. Preprints, Fifth Annual Conference on Aviation Weather Systems,* Vienna, VA. American Meteorological Society. Boston, MA.

FAA, 1975: *Aviation Weather. AC 00-6A.* U.S. Department of Transportation, Federal Aviation Administration and U.S. Department of Commerce, National Oceanic and Atmospheric Administration, National Weather Service, Washington, D.C. 219pp.

FAA, 1983: *Thunderstorms. AC 00-24B.* U.S. Department of Transportation, Federal Aviation Administration, Washington, D.C. 7pp.

FAA, 1988: *Pilot Wind Shear Guide. AC 00-54.* U.S. Department of Transportation, Federal Aviation Administration, Washington, D.C. 56pp.

FAA, 1991: *Wake Turbulence. AC 90-23D.* U.S. Department of Transportation, Federal Aviation Administration, Washington, D.C.

FAA, 1995: *Airman's Information Manual.* U.S. Department of Transportation, Federal Aviation Administration.

FAA, 1995: *Aviation Weather Services. AC 00-45.* U.S. Department of Transportation, Federal Aviation Administration, and U.S. Department of Commerce, National Oceanic and Atmospheric Administration, National Weather Service, Washington, D.C.

This textbook is designed to facilitate self-study. To get the most out of this textbook in the shortest amount of time, the following self-study procedure is recommended.

1. Review the Chapter Outline at the beginning of each chapter, noting major topics and subtopics.
2. Review the list of Key Terms and Chapter Questions at the end of the chapter.
3. Read the chapter Introduction and Summary Sections.
4. Skim through the chapter, reading the FAA Question Material, Insight Readings, and Illustrations.
5. Read the chapter.
6. Answer the Chapter Questions.
7. For a general review, repeat steps 1 through 4.

PREFACE

Meteorology is the study of the atmosphere and its phenomena; in many texts, it is simply referred to as atmospheric science. In contrast, weather is technically defined as the state of the atmosphere at an instant in time. Although the study of atmospheric impacts on aviation deals both with meteorology and with weather, it is traditionally referred to as aviation weather. We will use the latter terminology, clarifying differences where necessary.

Meteorology is a relatively "young" science. The vast majority of important developments in the field have only taken place in the 20th century. Driven by hot and cold war technological breakthroughs and, more recently, by environmental concerns, our understanding of the atmosphere and our ability to predict its behavior have improved dramatically in the last 50 years.

In the middle of this rapid growth has been the airplane. Much of the progress in modern meteorology has been driven by, and for, aviation. As aircraft designs improved and more and more aircraft were able to fly higher, faster, and farther, previously unobserved details of fronts, jet streams, turbulence, thunderstorms, mountain waves, hurricanes, and many other atmospheric characteristics were encountered.

The aviation industry turned to formal atmospheric research for the practical reason that aircraft are extremely vulnerable to certain atmospheric conditions. Aircraft designers needed careful measurements of those conditions; subsequent studies by specially equipped weather research aircraft produced even more details of the weather environment of flight.

In the early days of aviation, it became obvious that a regular supply of weather information was necessary to serve day-to-day operational needs. In the 1920's, many weather stations and the first weather data communication networks were established in the U.S. to serve the growing aviation industry. These were the forerunners of the modern weather data communication systems that, today, serve a wide variety of public and private users across the entire world.

The strong interdependence of aviation and meteorology should be apparent from the beginning of your aviation experience. Whether your connection to flying is as a pilot, controller, dispatcher, scientist, engineer, or as an interested passenger, you will quickly discover that it is nearly impossible to discuss any aspect of aviation without some reference to the meteorological environment in which the aircraft operates.

The objective of this text is to help the new student of aviation understand the atmosphere for the purpose of maximizing aircraft performance while minimizing exposure to weather hazards.

The book is also meant to provide a review of meteorology basics in preparation for the FAA examinations. It brings together information from a variety of sources and should serve as an up-to-date reference text. It is written with a minimum of mathematics and a maximum of practical information.

The text is divided into four Parts:

Part I (Chapters 1-6) addresses the "basics." This is important background in elementary meteorology that provides concepts and vocabulary necessary to understand aviation weather applications.

Part II (Chapters 7-10) deals with the wide variety of atmospheric circulation systems, their causes, behavior, and their related aviation weather.

Part III (Chapters 11-15) focuses specifically on the flight hazards produced by the circulation systems described in Part II.

Part IV (Chapters 16 and 17) considers the weather forecast process and the task of interpreting weather information. These final chapters provide a framework to put the information presented in previous chapters to practical use.

As you begin your study of aviation weather, a brief "pretest" is useful to emphasize the importance of the study of aviation meteorology. Given the following meteorological phenomena:

> Rain, gusts, whiteout, drizzle, high density altitude, mountain wave, low ceiling, downdrafts, haze, lightning, obscuration, microburst, high winds, snow, thunderstorm

1. Can you define/describe the meteorological conditions under which the phenomena occur?
2. Can you explain why, when, and where the conditions are likely to occur?
3. Can you describe the conditions' specific hazards to flight, and explain how to minimize those hazards?

If you cannot, then consider that all of the items listed above are cited as causes or contributing factors in the more than 400 General Aviation accidents that occurred in 1991 (NTSB 1994). These weather-related accidents accounted for 19% of the total General Aviation accidents that occurred that year. Another sobering statistic is that, if only accidents involving fatalities are considered, weather was the cause or a contributing factor in more than 23% of the total accident cases . . . nearly one in four!

When you complete this study of aviation meteorology, you should be able to return to this page and answer these questions with confidence and with respect for the atmosphere and its vagaries.

Part I

Basic Aviation Weather

PART I
Aviation Weather Basics

Part I provides you with the fundamentals of meteorology. These "basics" are the foundation of the entire study of aviation weather. The time you spend reading and understanding the basics will pay off in later parts of the text when you turn your attention to more complex topics such as the behavior and prediction of weather systems, and weather related flight hazards.

When you complete Part I, you will have developed a vocabulary of aviation weather terms and a knowledge of the essential properties and weather-producing processes of the atmosphere. As a pilot, you must be fully aware of weather and its influences on flight. Your task of understanding these concepts is made much easier with a solid foundation in Aviation Weather Basics.

CHAPTER 1

The Atmosphere

Introduction

The formal study of any physical system, such as an engine or an airplane, usually begins with a description of that system. Information about component parts, their location and dimensions, and terminology is necessary background for later examination and understanding of the system design and operation. Our study of aviation weather begins in a similar way. The "system" in this case, is the atmosphere.

When you complete this chapter, you should be able to describe the composition, dimensions, and average vertical structure of the atmosphere using proper technical vocabulary. Furthermore, you will have been introduced to a valuable reference tool, the standard atmosphere.

ATMOSPHERIC COMPOSITION

ATMOSPHERIC PROPERTIES

- *Temperature*
- *Density*
- *Pressure*
- *The Gas Law*

ATMOSPHERIC STRUCTURE

- *Dimensions*
- *Atmospheric Layers*
  - *Temperature Layers*
  - *Other Layers*
  - *Standard Atmosphere*

Section A

ATMOSPHERIC COMPOSITION

Each planet in our solar system is different, a product of the planet's original composition as well as its size and distance from the sun. Most planets, including Earth, have an atmosphere; that is, an envelope of gases surrounding the planet.

The earth's atmosphere is a unique mixture of gases along with small amounts of water, ice, and other particulates. The gases are mainly nitrogen and oxygen with only small amounts of a variety of other gases. (Figure 1-1)

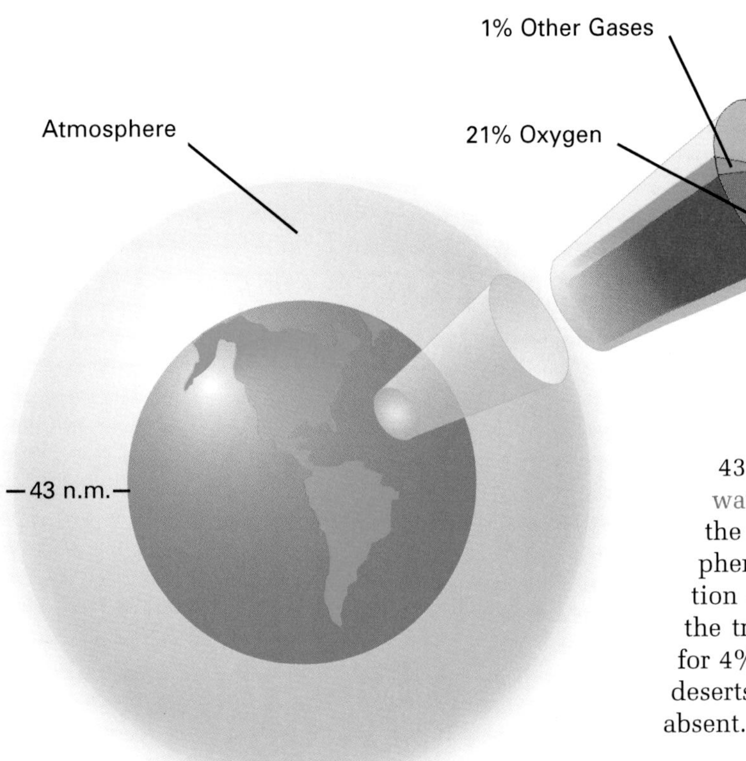

Figure 1-1. Primary permanent components of the mixture of gases in the lower atmosphere. Below an altitude of about 43 n.m. (260,000 feet), the ratios of these gases (78:21:1) remain relatively constant. Above this altitude, energy from the sun is strong enough to break down molecular structures and the ratios.

Although nitrogen (N_2) takes up most of the volume of the atmosphere, it doesn't contribute to weather-producing processes under ordinary atmospheric conditions. Exceptions occur when N_2 is subjected to very high temperatures, for example, when air passes through an internal combustion engine. In that case, nitrogen combines with oxygen to form air pollutants known as "oxides of nitrogen" (NO_x).

By and large, the most important role of atmospheric oxygen (O_2) is the support of life as we know it. Because oxygen concentration decreases with altitude, all pilots must be aware of the serious effects of oxygen deprivation on aircrews and passengers. Oxygen supports combustion and contributes to both the formation and the destruction of air pollutants through chemical combinations with other gases.

In figure 1-1, N_2 and O_2 and many of the "other" atmospheric gases are "permanent," which means that their proportions remain about the same, at least in the lower 260,000 feet (about 43 n.m) of the atmosphere. In contrast, water vapor (H_2O) is a "variable" gas; that is, the percentage of water vapor in the atmosphere can vary greatly, depending on the location and source of the air. For example, over the tropical oceans, water vapor may account for 4% of the total volume of gases, while over deserts or at high altitudes, it may be nearly absent.

Water vapor is an important gas for weather production, even though it exists in very small amounts compared to O_2 and N_2. This is because it can also exist as a liquid (water) and as a solid (ice). These contribute to the formation of fog, clouds, precipitation, and icing, well-known aviation weather problems.

Water vapor also absorbs radiant energy from the earth (terrestrial radiation). This reduces cooling, causing temperatures at the surface to be warmer than would otherwise be expected. Detailed information about this and other characteristics of water vapor, water, and ice are given in later chapters.

Gases that occupy a very small part of the total volume of the atmosphere are generally referred to as "trace gases." Two of the more important of these are also variable gases: carbon dioxide and ozone. Although their concentrations are extremely small, their impact on atmospheric processes may be very large. For example, carbon dioxide (CO_2) also absorbs terrestrial radiation. The concentration of this trace gas has been increasing over the last century due to the worldwide burning of fossil fuels and rain forest depletion, which may contribute to the long term warming of the atmosphere.

Although their concentrations are small, water vapor, carbon dioxide, ozone, and other trace gases have profound effects on weather and climate.

Ozone (O_3) is a toxic, highly reactive pollutant which is produced in the lower atmosphere by the action of the sun on oxides of nitrogen and by electrical discharges, such as lightning. The greatest concentration of ozone is found between 50,000 and 100,000 feet above the earth's surface. The upper ozone layer is beneficial for the most part because ozone absorbs harmful ultraviolet radiation from the sun. This filtering process at high levels protects plants and animals on the earth's surface. However, direct exposure of aircrews and passengers to the toxic properties of O_3 can be a problem during high altitude flights. These will be discussed in Part III, Aviation Weather Hazards.

Liquid or solid particles that are small enough to remain suspended in the air are known as particulates or aerosols. Some of these are large enough to be seen, but most are not. The most obvious particulates in the atmosphere are water droplets and ice crystals associated with fog and clouds. Other sources of particulates include volcanoes, forest fires, dust storms, industrial processes, automobile and aircraft engines, and the oceans, to name a few. Particulates are important because they intercept solar and terrestrial radiation, provide surfaces for condensation of water vapor, reduce visibility, and, in the worst cases, can foul engines.

Section B

ATMOSPHERIC PROPERTIES

Since the atmospheric "system" is mainly a mixture of gases, its description is commonly given in terms of the state of the gases that make up that mixture. The three fundamental variables used to describe this state are temperature, density, and pressure.

TEMPERATURE

Temperature is defined in a number of ways; for example, as a measure of the direction heat will flow, or as simply a measure of "hotness" or "coldness." Another useful interpretation of temperature is as a measure of the motion of the molecules. Kinetic energy is energy that exists by virtue of motion. A molecule possesses kinetic energy proportional to the square of its speed of movement; temperature is defined as the average of the kinetic energy of the many molecules that make up a substance. The greater the average kinetic energy, the greater the temperature. (Figure 1-2)

A temperature of absolute zero is the point where all molecular motion ceases. The corresponding temperature scale is known as the absolute or Kelvin scale. You will be introduced to the details of the more familiar Fahrenheit and Celsius temperature scales in the next chapter. For the moment, the Kelvin scale will serve our purposes.

On the absolute or Kelvin (°K) temperature scale, the temperature where all molecular motion ceases is 0°K. The melting point of ice is 273°K (0°C) and the boiling point of water is 373°K (100°C).

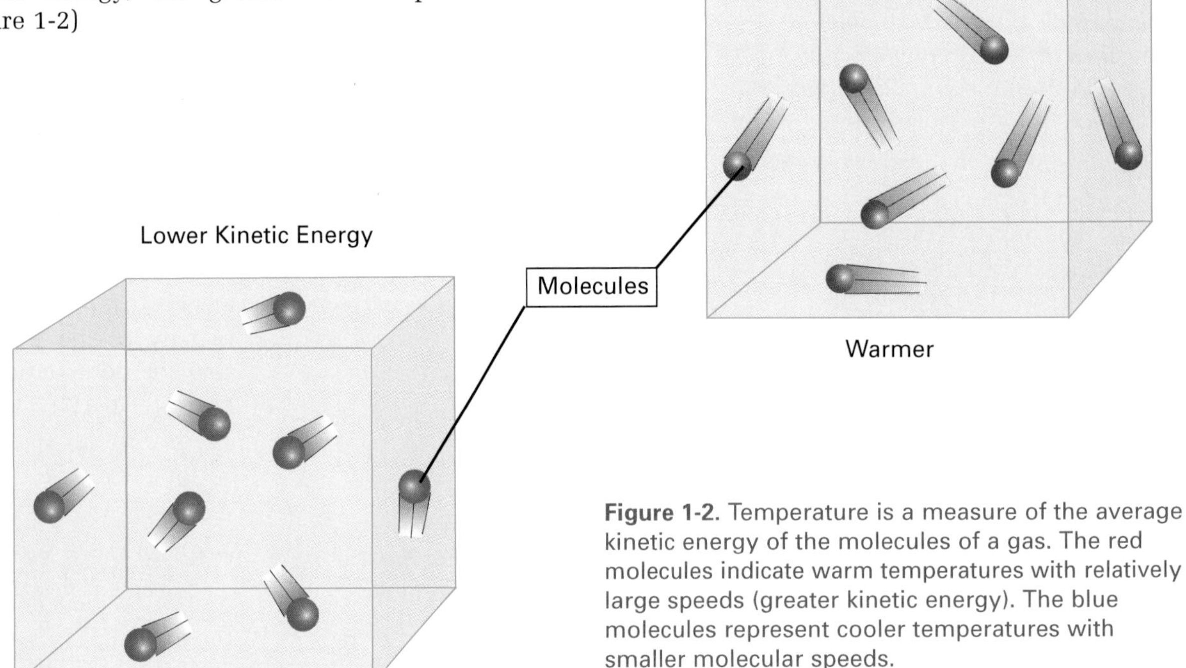

Figure 1-2. Temperature is a measure of the average kinetic energy of the molecules of a gas. The red molecules indicate warm temperatures with relatively large speeds (greater kinetic energy). The blue molecules represent cooler temperatures with smaller molecular speeds.

DENSITY

Density of a gas is the mass of the molecules in a given volume. If the total mass of molecules in that volume decreases, the density decreases. If the mass remains the same but the volume increases, the density also decreases. The units of density are expressed in terms of mass per unit volume. (Figure 1-3)

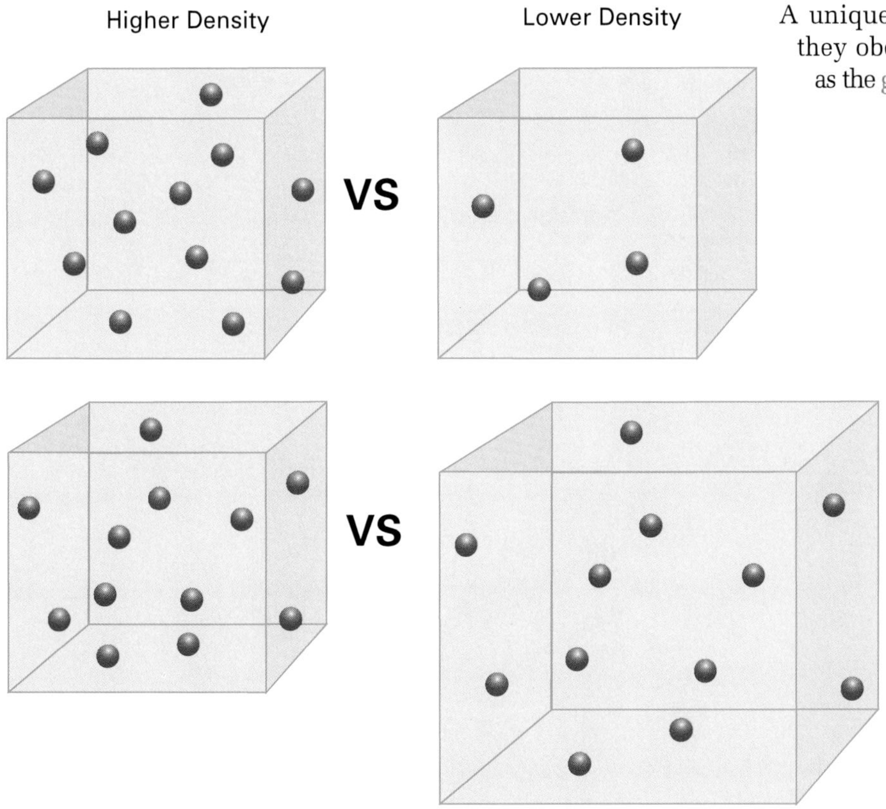

Higher Density

Lower Density

VS

VS

Figure 1-3. In this figure, the mass within each volume is represented by a number of molecules, each with the same mass. The density is proportional to the number of molecules within the box divided by the total volume. The figure shows that density is decreased when gas molecules are removed or when the volume is increased.

PRESSURE

Pressure is the force exerted by the moving molecules of the gas on a given area (say a square inch or square foot). Pressure at a point acts equally in all directions. A typical value of atmospheric pressure at sea level is 14.7 pounds per square inch (1013.25 millibars, 29.92 inches of mercury).

THE GAS LAW

A unique characteristic of gases is that they obey a physical principle known as the gas law, which can be written as:

$$\frac{P}{DT} = R$$

In this equation P is pressure, D is density, T is the absolute temperature, and R is a constant number which is known from experiment and theory. The equation above simply states that the ratio of pressure to the product of density and temperature is always the same. For example, if the pressure changes, then either the density or the temperature, or both, must also change in order for the ratio to remain constant. Figure 1-4 illustrates the application of the gas law and three simple ways to lower the pressure in the vessels by varying the temperature or the density.

The gas law makes the measurement of the gaseous state of the atmosphere much simpler. If we know any two of the three variables that describe the gas, we can always calculate the third. In practice, we usually measure pressure and temperature and deduce the density from the gas law.

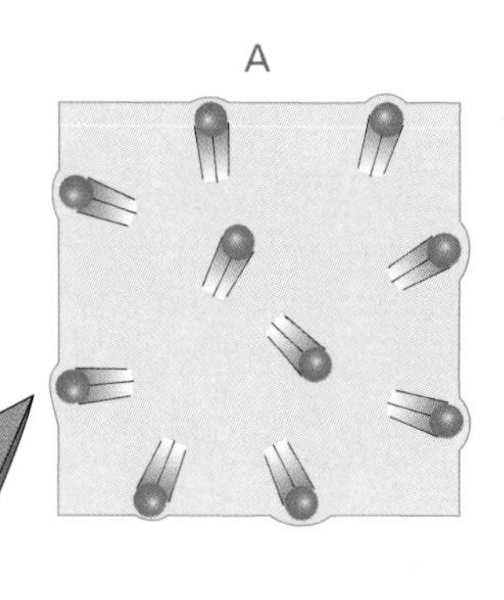

Figure 1-4. Pressure is force per unit area. Pressure is exerted by the collective force of the molecules colliding with the sides of the vessels. When the density is kept constant (A), the only way to lower the pressure is to reduce the temperature. The molecules become less energetic and exert less force on the vessel. When the temperature and volume of the vessel remain constant (B), the pressure can only be reduced by removing gas. Although the molecules remain energetic, there are fewer of them, so the force they exert on the sides of the vessel is reduced. When the temperature and mass of the molecules in the vessel remain the same (C), the pressure can only be lowered by increasing the volume of the vessel. The molecules then exert their collective force over a larger area.

Section C

ATMOSPHERIC STRUCTURE

The brief introduction to the atmospheric composition and the gas law has provided you with some useful vocabulary and other tools to describe the atmosphere. In this section, you will apply those tools to examine the structure of the atmosphere.

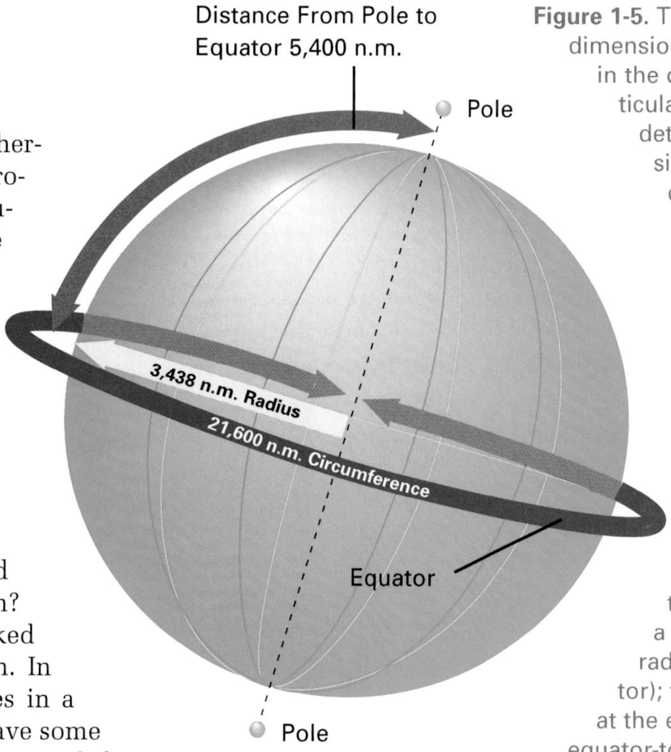

DIMENSIONS

In much of the material in this and later chapters, we will be concerned with the size of the atmosphere and its phenomena. "How big? How high? How far?" are common questions asked in regard to atmospheric description. In order to keep distances and altitudes in a meaningful context, it is helpful to have some "measuring sticks" for reference. Some of the most useful are the dimensions of the earth. (Figure 1-5)

Figure 1-5. The earth and its dimensions. The numbers in the diagram are particularly useful for the determination of the sizes of atmospheric circulation systems such as the large cyclones that move across the earth's surface. The figure shows the most frequently referenced dimensions: the average radius (the earth is really a flattened sphere with a slightly greater radius at the equator); the circumference at the equator and the equator-to-pole distance. The latter estimates are based on one degree of arc of a great circle being approximately equal to 60 n.m.

The units used in this text will be those commonly used in aviation meteorology in the United States. These are given below with some useful conversions.

LENGTH

1 inch (in) = 2.54 centimeters (cm)
1 foot (ft) = 12 in = 30.5 cm = 0.305 meter (m)
1 statute mile (s.m.) = 5,280 ft = 0.87 n.m. = 1.61 kilometers (km)
1 nautical mile (n.m.) = 6,080 ft = 1.15 s.m. = 1.85 km = 1 minute of arc of a great circle
1 degree of arc of a great circle = 60 n.m. = 111 km

AREA

1 square foot (ft^2) = 144 (in^2) = 0.09(m^2)

VOLUME

1 cubic inch (in^3) = 16.39(cm^3)

SPEED

1 Knot (kt) = 1 n.m. per hour = 1.15 s.m. per hour = 1.85 km per hour = 0.51 m per second

PRESSURE

1 inch of mercury (in. Hg.) = 33.865 millibars (mb)
1 mb = 1 hectoPascal (hPa) = 100 Pascals (Pa) = 0.0295 (in. Hg.)
1 atmosphere = 29.92 in. Hg. = 1013.25 mb =14.7 pounds per square inch (lbs/in^2) = 101,325 Pascals (Pa)

You will soon see that the atmosphere is much thinner than it is wide. The atmospheric properties such as temperature, density, and pressure, vary much more rapidly in the vertical direction than they do in the horizontal direction. For this reason, we begin our examination of atmospheric structure with a close look at the "typical" vertical structure.

Figure 1-7. A view of the atmosphere from space. The blue layer shows where air molecules are of sufficient density to scatter blue light from the sun. Note the relatively small vertical dimension of the blue layer compared to its great horizontal dimension.(Photo courtesy of Lunar and Planetary Institute, NASA photo.)

ATMOSPHERIC LAYERS

An important specification of the atmosphere is its thickness; that is, the distance between the surface of the earth and the "top" of the atmosphere. Although it is technically considered a fluid, it doesn't have a well-defined upper surface as does water. The atmosphere is highly compressible and it just "fades away" with increasing altitude.

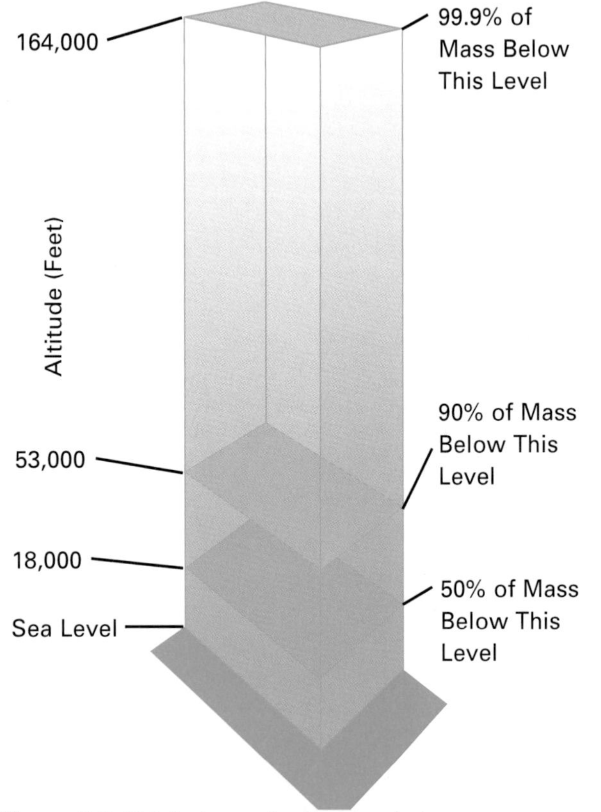

Figure 1-6. Total atmospheric mass below specific altitudes.

We can, however, consider an approximate "top" based on the density of the atmosphere. The density at an altitude of 164,000 feet (about 27 n.m.) is only about one-thousandth of the sea level density. In terms of the total mass of the atmosphere, 99.9% of the mass lies below that altitude. (Figure 1-6) There is no question that this altitude is close to the "top." The important conclusion here is that, in comparison to the average radius of the earth (3,438 n.m.), this is a relatively small distance. The atmosphere is a very thin layer compared to its horizontal extent, analogous to the cover on a baseball. Figure 1-7 gives a view from space that dramatically illustrates this characteristic.

TEMPERATURE LAYERS

To further analyze the atmosphere, we can build a reference model by dividing the envelope of gases into layers with similar properties. By far, the most common model of atmospheric structure is one which divides the atmosphere into layers according to the way that temperature changes with altitude. (Figure 1-8)

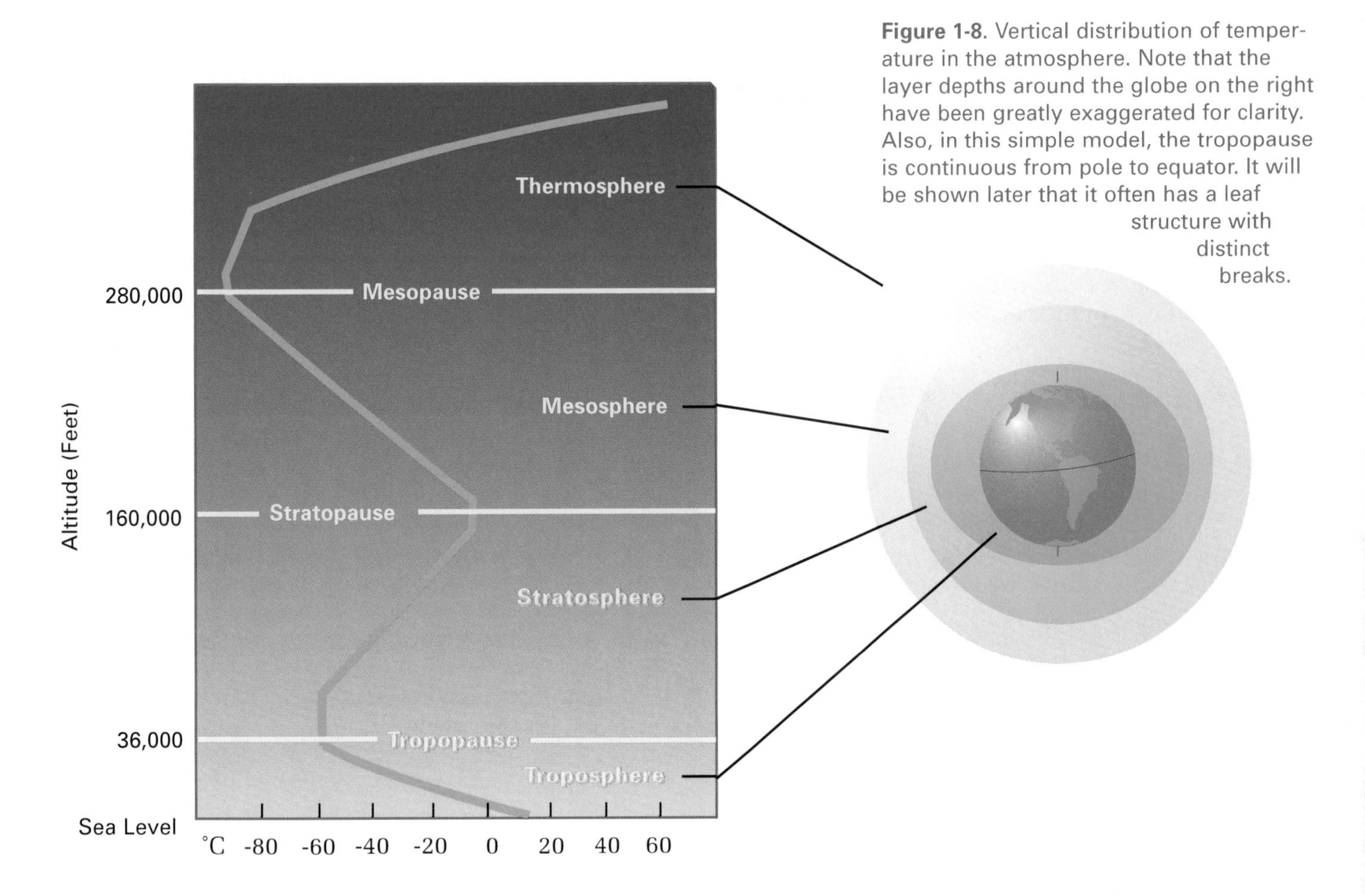

Figure 1-8. Vertical distribution of temperature in the atmosphere. Note that the layer depths around the globe on the right have been greatly exaggerated for clarity. Also, in this simple model, the tropopause is continuous from pole to equator. It will be shown later that it often has a leaf structure with distinct breaks.

In the lowest layer of the atmosphere, or troposphere, the average temperature decreases with altitude. The great majority of the clouds and weather occurs in the troposphere.

The top of the troposphere is about 36,000 feet above mean sea level (MSL) in middle latitudes. This upper boundary (a level, not a layer) is known as the tropopause. The temperature usually decreases to a minimum value at this altitude. The tropopause is a very important atmospheric feature for pilots because of its connection to a variety of weather phenomena such as jet streams, clear air turbulence, and thunderstorms. The altitude of the tropopause varies with the season and location over the globe. The tropopause is lower near the poles and in the winter; it is higher near the equator and in the summer.

As we move upward into the stratosphere, temperature tends to change slowly at first and then increases with altitude. As evidenced by the name of the layer, air in the stratosphere is confined to move more or less horizontally in strata or layers. In contrast, in the troposphere (from the word, trope, meaning "turn" or "change"), there are often strong vertical air motions. We will see in a later chapter that the "stability" of the stratosphere and "instability" of the troposphere are related directly to the variation of temperature with altitude in those layers.

At the top of the stratosphere is the stratopause. It occurs at an altitude of about 160,000 feet MSL. The temperature reaches a maximum value at this height. Immediately above is the mesosphere, a layer where the temperature again decreases with height. The mesosphere extends to a height of slightly more than 280,000 feet MSL, where the mesopause and the coldest temperatures in the diagram are located.

The highest layer in our model atmosphere is the thermosphere. Temperatures generally increase with altitude in this layer. However, the meaning of air temperature is not so clear. The number of air molecules is so small at these very high levels that an "average kinetic energy" of the air molecules doesn't have much meaning. Objects in space at such heights have temperatures that are more closely related to radiation gain on the sun-facing side of the object and radiation loss on the opposite side.

OTHER LAYERS

Figure 1-9 shows a variety of other atmospheric layer designations that are also commonly used to describe the vertical structure of the atmosphere.

The ozone layer is found in the lower stratosphere. It is characterized by a relatively high concentration of O_3 with maximum concentrations near 80,000 feet MSL. Ozone forms when oxygen atoms and oxygen molecules combine. The increase in temperature with altitude in the stratosphere is due to the absorption of solar radiation by the ozone in that layer.

The ozone hole is a region of the ozone layer with lower-than-normal O_3 concentration. It is especially noticeable over the South Pole in spring months (September-December). The ozone hole is created when pollutants, in particular, man-made chlorofluorocarbons (CFCs), reach stratospheric levels. Solar radiation is intense enough to break the CFCs down so that the chlorine is free to destroy ozone molecules.

The ionosphere is a deep layer of charged particles (ions and free electrons) that extends from the lower mesosphere upward through the thermosphere. The production of charged particles occurs at those altitudes because incoming solar radiation has sufficient energy to displace electrons from atoms and molecules. AM radio waves are reflected and/or absorbed by different sublayers of the ionosphere. Radio communications may be greatly influenced by variations in the lower part of the ionosphere at sunrise and sunset and during periods of greater solar activity.

Figure 1-9 also shows a curve representing the variation of atmospheric pressure with altitude. An important characteristic of pressure is that it always decreases with altitude. This property is

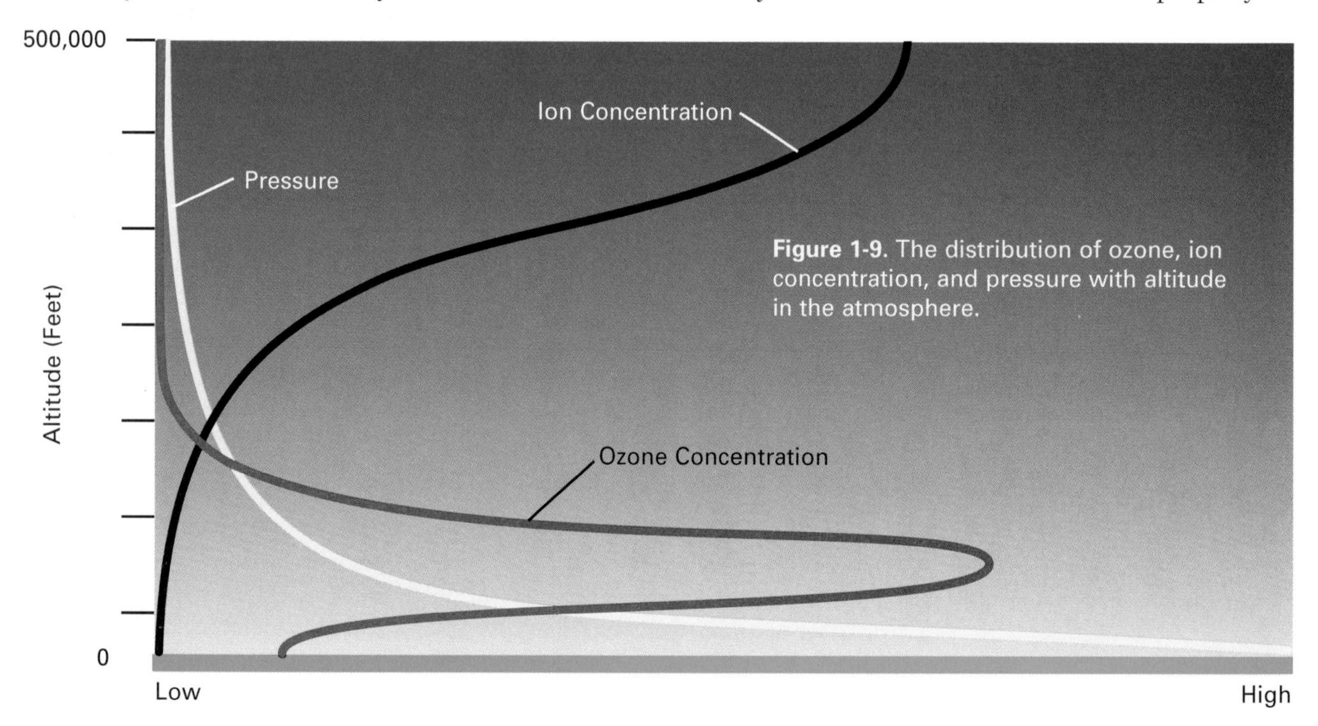

Figure 1-9. The distribution of ozone, ion concentration, and pressure with altitude in the atmosphere.

used to determine aircraft altitude from atmospheric pressure, an important topic of Chapter 3.

Another notable feature of the pressure curve is the rapid decrease in pressure just above the ground and the very gradual decrease at the higher levels. This further demonstrates the compressibility of the atmosphere and the lack of a well-defined upper surface discussed earlier.

In the lower troposphere, pressure decreases about one inch of mercury (about 34 mb) for each thousand feet of altitude gain.

There are several important physiological effects related to flight at high levels because of the decrease in pressure with altitude. The concentration of oxygen in the mixture of atmospheric gases is proportional to total atmospheric pressure. Oxygen concentration therefore decreases with height in the same manner as total pressure. These decreases in pressure and oxygen concentration determine the following critical altitudes for pilots and crew:

1. Supplemental breathing oxygen is recommended above 10,000 feet MSL during the day and 5,000 feet MSL at night.
2. Above 12,500 feet MSL, supplemental oxygen is required for the crew on flights of more than 30 minutes in duration.
3. Above 14,000 feet MSL, the flight crew is required to use supplemental oxygen
4. In unpressurized environments above approximately 40,000 feet, supplemental oxygen must be supplied under pressure.
5. At approximately 63,000 feet, the pressure exerted by gases escaping from body fluids exceeds the atmospheric pressure. In an unpressurized environment, the fluids will vaporize (the blood will "boil"). The pilot of an unpressurized aircraft must wear a full pressure suit above 50,000 feet MSL.

Similar to pressure, the vertical distribution of atmospheric density doesn't really lend itself to precise layer classifications. In general, density decreases with height, reflecting our earlier observations that most of the mass of the atmosphere is concentrated in the lowest layers. Density is, none-the-less, important in aviation applications. Aircraft performance is directly dependent on the mass of the atmosphere and performance degrades when the density is low. This is clearly the case at high levels in the atmosphere. But you don't have to fly at stratospheric levels to experience problems due to lower-than-normal density. There are situations when density is critically low near the ground because of very high surface temperatures. An expanded discussion of the effects of these conditions and the concept and use of density altitude is presented in a later chapter.

STANDARD ATMOSPHERE

> **In the ISA troposphere, the temperature decreases 2C° for each 1,000-foot increase in altitude.**

The standard atmosphere, also called the international standard atmosphere (ISA), is an idealized atmosphere with specific vertical distributions of pressure, temperature, and density prescribed by international agreement. The standard atmosphere is used for several aerospace applications, not the least of which is determining altitude from pressure altimeters (Chapter 3). The ISA for the lower stratosphere and troposphere is shown graphically in figure 1-10. In the remaining text, we will focus most of our attention on these lowest layers of the atmosphere where the majority of aircraft operations take place.

It is helpful to keep in mind that, although the ISA is a useful tool for aviation, there are large variations from standard conditions in the real atmosphere. The standard atmosphere is most representative of average mid-latitude conditions, at least in the troposphere and lower stratosphere. As shown in figure 1-10, the troposphere is actually colder and the tropopause is lower than ISA over the poles and, respectively, warmer and higher than ISA over the equator. As we describe the atmosphere and its variations in subsequent chapters, the standard atmosphere will serve as a helpful reference. Detailed altitude, temperature, and pressure tabulations for the standard atmosphere are given in the Appendix.

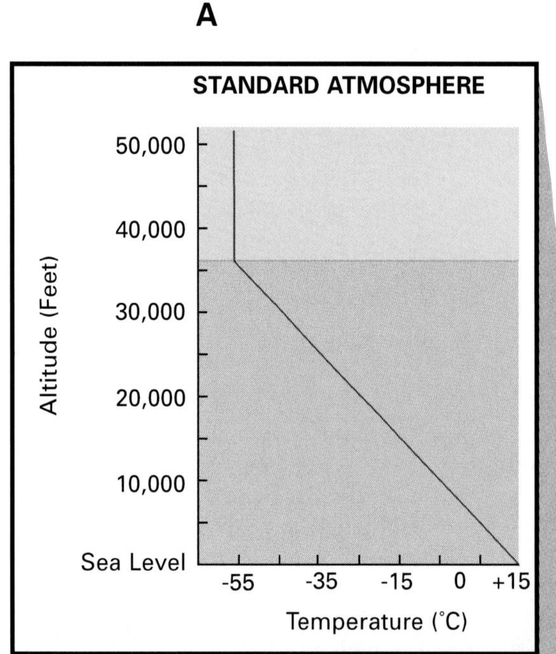

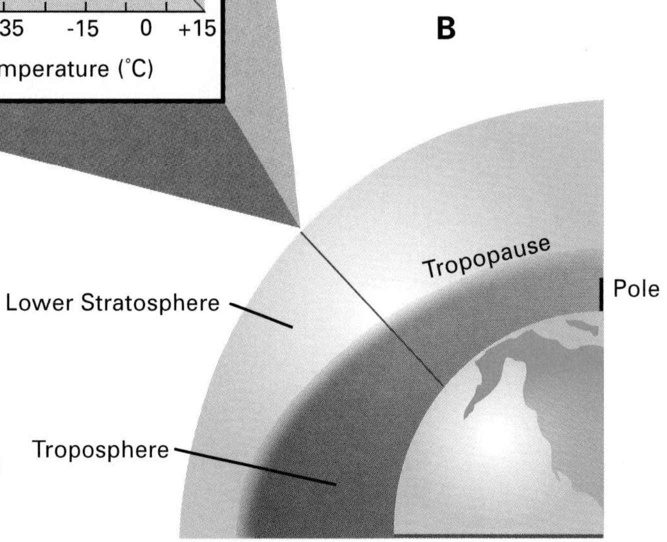

Figure 1-10. The standard atmosphere temperatures in the troposphere and lower stratosphere are plotted versus altitude in diagram A. Diagram B shows the variation of tropopause height between equator and pole. The line through the atmosphere in B indicates the standard atmosphere shown in diagram A is most representative of observed mid-latitude conditions. As in similar diagrams, the vertical dimension of the atmosphere is greatly exaggerated.

SUMMARY

In this chapter, you have started to build the background of basic concepts and vocabulary necessary for the study of aviation weather. You should now be aware of the average composition, structure, and dimensions of the atmosphere. What may have seemed at first to be a rather complicated picture has been simplified by constructing a "model" of the atmospheric structure based mainly on the variation of temperature with altitude. It will soon be clear that the model not only aids in learning and remembering the basic atmospheric structure, but it also will help you understand the causes of many atmospheric phenomena.

KEY TERMS

Absolute Zero
Aerosols
Atmosphere
Carbon Dioxide
Density
Gas Law
International Standard Atmosphere
Ionosphere
Kinetic Energy
Mesopause
Mesosphere
Nitrogen
Oxides of Nitrogen
Oxygen
Ozone
Ozone Hole
Ozone Layer
Particulates
Pressure
Standard Atmosphere
Stratopause
Stratosphere
Temperature
Thermosphere
Tropopause
Troposphere
Water Vapor

CHAPTER QUESTIONS

1. Examine the ISA in the Appendix and determine the approximate altitudes where atmospheric pressure decreases to one-half and one-quarter of the sea level value.

2. A dry gas is in a closed vessel.
    1. What happens to the pressure of the gas if the density remains the same and the temperature goes up?
    2. How do you keep the pressure inside a vessel constant when you increase the temperature?
    3. How do you decrease the pressure and keep the density constant?

3. You place an empty one gallon aluminum can in the unheated cargo compartment of your aircraft. Just before closing the compartment, you place an air tight seal on the can. After takeoff, you climb from sea level to 10,000 feet MSL for the cruise portion of your flight. Your aircraft is unpressurized and the atmosphere (pressure and temperature) is the same as standard atmospheric conditions. What happens to the can? Why?

4. Find a book or manual that deals with the physiology of flight and look up the definitions of "anoxia" and "hypoxia."

    1. What are typical symptoms of hypoxia?
    2. How long can one typically operate without supplementary oxygen at 15,000 feet MSL? 20,000 feet MSL? 30,000 feet MSL?

5. You are constructing a world globe with terrain in relief. You want to make it to scale.

    1. If the globe is two feet in diameter, how high above "sea level" would the top of Mt. Everest (29,028 feet MSL) protrude in inches?
    2. Would you describe the surface of the earth as "smooth"? From what perspective?
    3. What is the approximate pressure at the top of Mt. Everest?

6. You have just taken off from an airport locat ed at sea level. Conditions are exactly the same as prescribed by the International Standard Atmosphere. What will be your outside air temperature (OAT) at the following altitudes?
    1. 1,000 feet?
    2. 1,500 feet?
    3. 3,300 feet?
    4. 7,400 feet?
    5. 32,000 feet?

7. What is the potential impact of supersonic transports (SSTs) on the ozone layer?

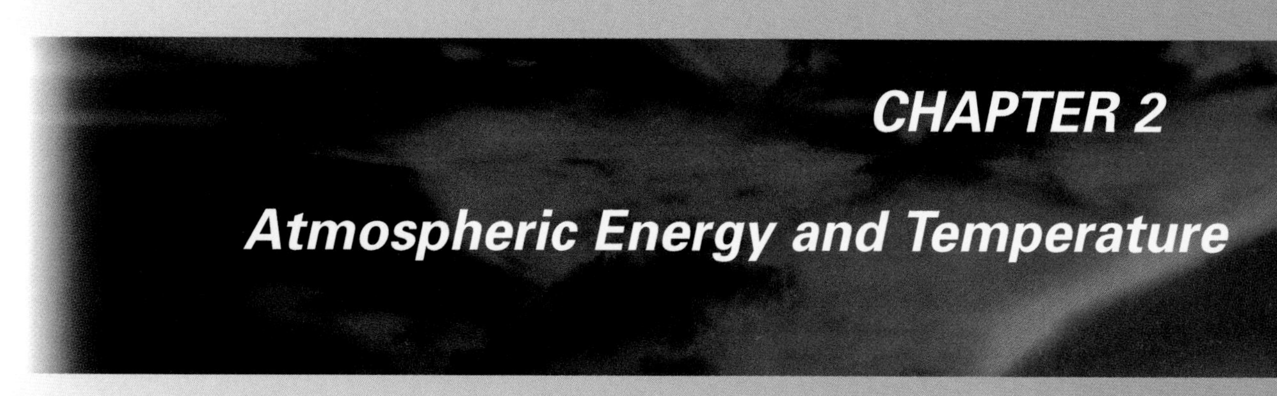

CHAPTER 2

Atmospheric Energy and Temperature

Introduction

In this chapter, we continue to build a basic reference model of the atmosphere. Now we turn our attention to the energy that drives the atmosphere. Of particular interest is the source of atmospheric energy (the sun). We are interested in the details of energy exchange and resulting atmospheric temperatures. These pieces of information are important parts of the foundation of your understanding of winds and weather.

When you complete this chapter, you will understand important sun-earth relationships and their seasonal and daily variations, modes of energy transfer between the sun and the earth, and between the earth and the atmosphere. You will also learn some practical aspects of measuring temperature and details of global temperature patterns.

Section A

ENERGY TRANSFER

The atmospheric "system" that was briefly described in Chapter 1 operates much like a heat engine. Solar energy enters the system and undergoes a series of energy conversions, finally producing winds, clouds, and precipitation. In order for these processes to be sustained, heat energy must not only be transferred from the sun, but must also be redistributed within the atmosphere. That supply and redistribution is accomplished by three energy transfer processes:

Radiation is the transfer of energy by electromagnetic waves.

Conduction is the transfer of energy through molecular motion.

Convection is the transfer of energy through the movement of mass.

In meteorology, we commonly reserve the term "convection" for vertical movements of the atmosphere and "advection" for horizontal movements. There are other processes that also account for the transfer of heat energy that are explained in later chapters. These include the absorption and release of heat

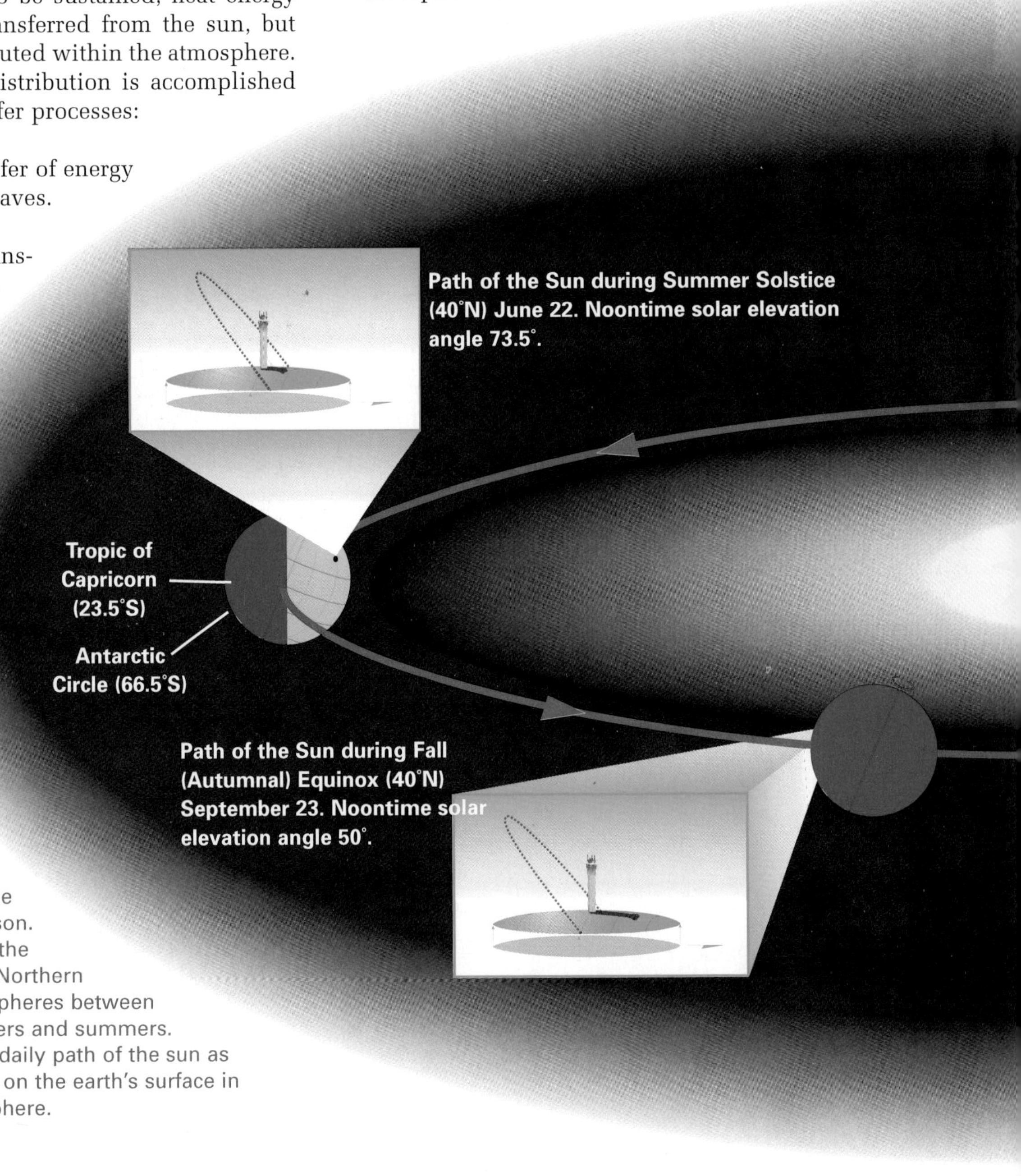

Figure 2-1. Positions of the earth in its orbit around the sun on the first day of each season. Note the changes in the shaded areas of the Northern and Southern Hemispheres between their respective winters and summers. The insets show the daily path of the sun as seen from a location on the earth's surface in the Northern Hemisphere.

associated with changes of state of H_2O (such as through evaporation and condensation) and the storage and movement of heat by ocean currents.

RADIATION

Conduction and convection/advection require mass for energy transfer; therefore, the transfer of energy from the sun across nearly empty space must be accomplished primarily by radiation. Solar radiation received at the earth's surface varies with latitude and season for several reasons: the earth is a sphere (slightly flattened at the poles); it rotates on its axis and orbits the sun; and the incoming energy from the sun must pass through the earth's atmosphere (figure 2 - 1).

SUN-EARTH GEOMETRY

The intensity of solar radiation received at any one point on the earth's surface depends on the location of the sun relative to that point. That location, in turn, depends on the following:

1. Time of day. The earth rotates on its axis once every 24 hours.

2. Time of year. The earth orbits around the sun once every 365.25 days.

3. Latitude. The axis of the earth is tilted 23.5°to the plane of its orbit about the sun.

The geometry of these controls is summarized in figure 2-1. The basic day-night variation in radiation is due to the rotation of the earth each day. But the lengths of the days and nights are not usually equal. The orbit of the earth causes each pole to be tilted toward the sun during half the year and away from the sun during the other half. This causes a variation of the length of the day at each earth latitude. At noon on the first days of spring and fall (the equinoxes), the sun's rays are perpendicular to the earth's surface at the equator. On these dates, the length of daylight is the same (12 hours) everywhere on earth.

On the first day of summer and the first day of winter (the solstices), the noonday sun reaches its highest and lowest latitudes, respectively. The longest day of the year is at summer solstice and the shortest day is at winter solstice. North of 66.5° north latitude (Arctic Circle) and south of 66.5° south latitude (Antarctic Circle) there is at least one day when the sun does not rise and one day when it does not set. This effect reaches a maximum at the poles where there are six months of darkness and six months of light. The low sun angles produce unique visibility hazards at high latitudes.

The influence of the changing position of the sun relative to the earth is illustrated in terms of the solar elevation angle (angle of the sun above the horizon) in figure 2-2. If that angle is small, solar energy is spread over a broad surface area, minimizing heating. This condition is typical near sunrise and sunset, and at high latitudes, especially in winter.

When the solar elevation angle is large, solar energy is concentrated in a smaller area, maximizing heating. These conditions are typical at noon, in the summer, and at low latitudes. In fact, the noon elevation angle will reach 90° (the sun is directly overhead) twice during the year between latitudes 23.5° north (Tropic of Cancer) and 23.5° south (Tropic of Capricorn).

Figure 2-2. The influence of solar elevation angle on the concentration of energy received at the surface. Light beams indicate parallel rays of energy from the sun near the equinox. While the same amounts of solar energy strike the earth at the equator and near the pole, that energy is spread over a much larger surface area near the pole (small solar elevation angle) than at the equator (large solar elevation angle).

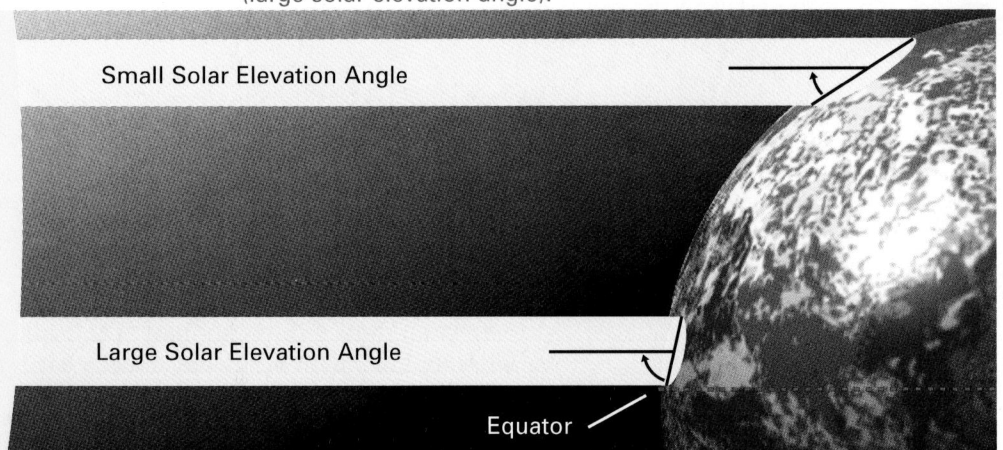

RADIATION PROCESSES

You were introduced to the concept of absolute zero in Chapter 1. Absolute zero is the temperature where all molecular motion ceases. It is also an important reference point for the understanding of electromagnetic radiation. Electromagnetic energy radiates from any object that has a temperature above absolute zero. The higher the temperature, the greater the radiation. In fact, a basic law of physics is that the total radiation emitted from an object is proportional to T^4, where T is the object's temperature in C° above absolute zero. This means that an object twice as warm as another object would radiate 16 times (2^4) as much radiation. Two good examples of the application of this law are the earth and the sun. The sun has an effective radiating temperature of about 6000°C above absolute zero while the earth's radiating temperature is only about 288°C above absolute zero. Therefore, a unit area of the sun radiates 188,379 times the energy of the same area on the cooler earth ($6000^4/288^4$ = 188,379) and, of course, the sun's total surface area is much larger than the earth's. Radiated energy travels at a speed of 186,000 statute miles per second (about 300,000 kilometers per second) in a vacuum.

The noontime solar elevation angle (e) for a station at latitude L is given by the following equation:

$$e = 90 - (L - L_p),$$

Where L_p is the solar declination; that is, the latitude where the noon sun is directly overhead. For example, on the first day of Northern Hemisphere summer, the noon sun is overhead at 23.5°north. The noontime solar elevation angle at Denver (40°north) is

$$e = 90 - (40 - 23.5) = 73.5°$$

On the same day, a station in the Southern Hemisphere at 40°south (-40°)would have a noontime solar elevation angle of

$$e = 90 - (-40 - 23.5) = 26.5°.$$

The speed of any simple wave (c) is related to frequency and wavelength as

$$c = f \times L.$$

With electromagnetic energy, we are fortunate because c is the speed of light. Since it is a constant, we can describe the characteristics of an electromagnetic wave in terms of either wavelength or frequency; that is, given one, the other can always be determined.

This speed is often referred to as the speed of light even though light is just one of many types of electromagnetic radiation.

In many respects, electromagnetic radiation behaves as collection of waves, each with different characteristics. You may have observed similar combinations of waves in other situations. When observing the surface of the ocean, for example, it is easy to visualize the motion of a particular patch of water as being influenced by a number of distinct waves, all of which are present at the same time. There may be very long swells combined with shorter waves caused by the wind, the passage of a ship, or by the presence of a pier.

With regard to electromagnetic radiation, if you have seen the separation of a beam of white light into its respective colors (red through blue), then you have seen the individual wave components of white light. Each color that makes up the white light may be uniquely described in terms of a wave. Terminology that we use to describe waves is reviewed in figure 2-3.

Applying these ideas to visible radiation, red light has relatively long wavelengths and low frequencies. Frequencies lower than red are called infrared (IR). Blue light has relatively short wavelengths and high frequencies. Frequencies higher than blue are called ultraviolet (UV).

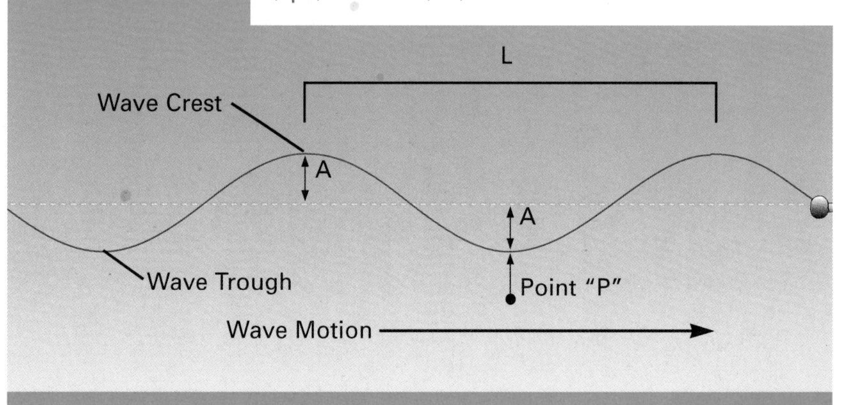

Figure 2-3. This diagram represents a train of waves on a rope tied to a fixed object. A person standing at the other end of the rope has put wave energy into the rope by moving it vertically. The waves are moving from left to right. Wavelength (L) is the distance between two successive, identical wave features, such as two wave crests. Wave amplitude (A) is half the distance between the lowest and highest points of the wave. The wave frequency (f) is the number of waves that pass some fixed point (for example, point "P")in a given time interval. Units of frequency are cycles per second (cps) or Hertz (Hz).

The amplitude (A) of a wave is related to the energy of the wave. To be precise, wave energy is proportional to the square of the amplitude (A^2) of the wave. A useful diagram to illustrate the energy of electromagnetic radiation is a spectrum; that is, a graph of electromagnetic wave energy (A^2) for all electromagnetic waves versus their wavelength. Two examples of energy spectra are shown in figure 2-4.

As we saw earlier, the total energy radiated by any object is proportional to T^4. This property is shown in figure 2-4. The area under each curve in the diagram is proportional to the total energy in all wavelengths. The radiation from the sun is hundreds of thousands of times more energetic than that from the earth.

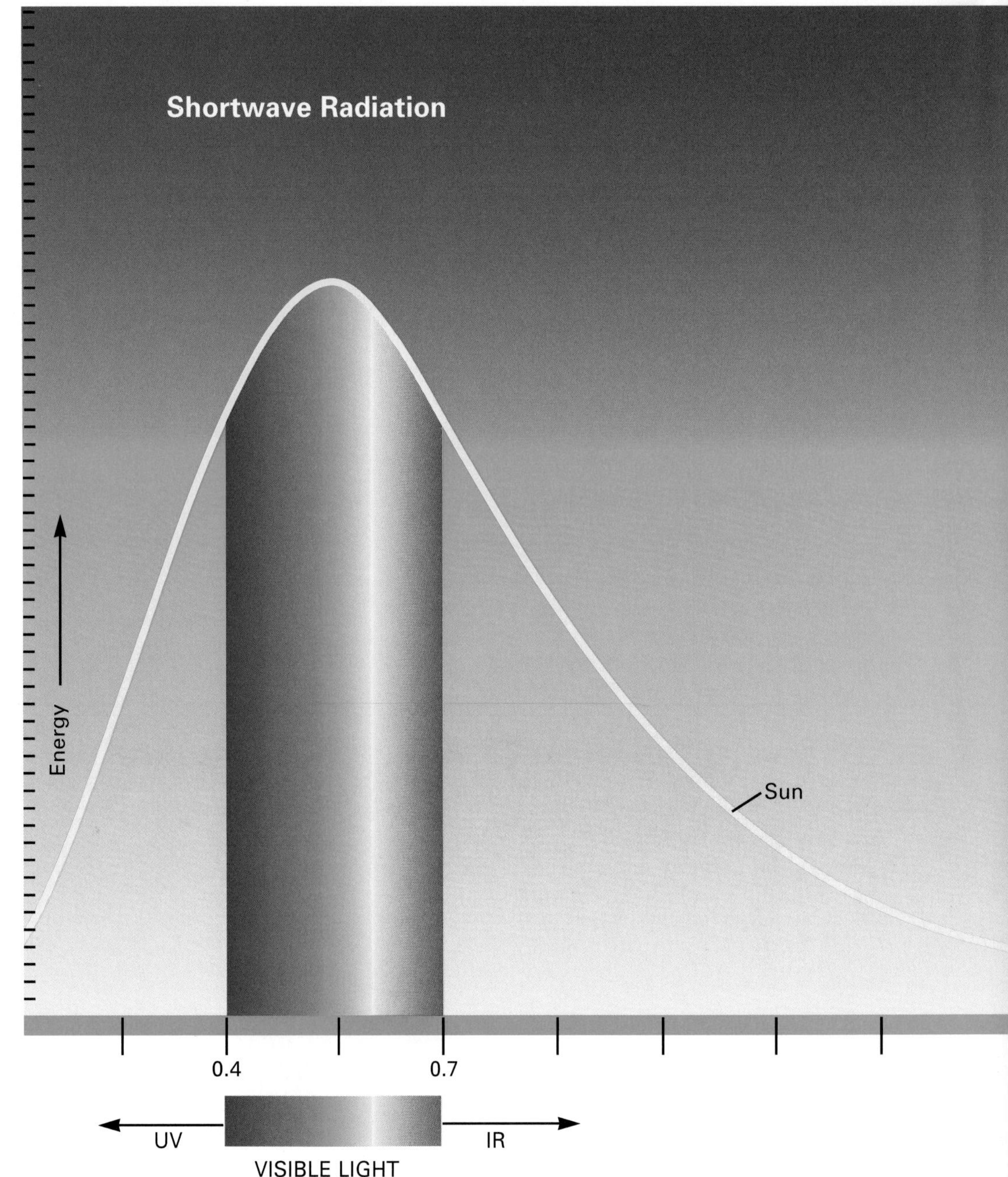

Figure 2-4. Spectra of radiation for the sun and the earth. Note that the values on the energy scale for the earth's spectrum are much less than that of the sun. Also note that the earth's spectrum has its maximum energy at much longer wavelengths than that of the sun.

A perfect radiating body emits energy in all possible wavelengths, but the wave energies are not emitted equally in all wavelengths; a spectrum will show a distinct maximum in energy at a particular wavelength depending on the temperature of the radiating body. As the temperature increases, the maximum radiation occurs at shorter and shorter wavelengths. For example, as shown in figure 2-4, the maximum energy radiated in the solar spectrum is at significantly shorter (visible) wavelengths with a large contribution in the UV region. The maximum in the terrestrial spectrum is at longer wavelengths, well into the IR region.

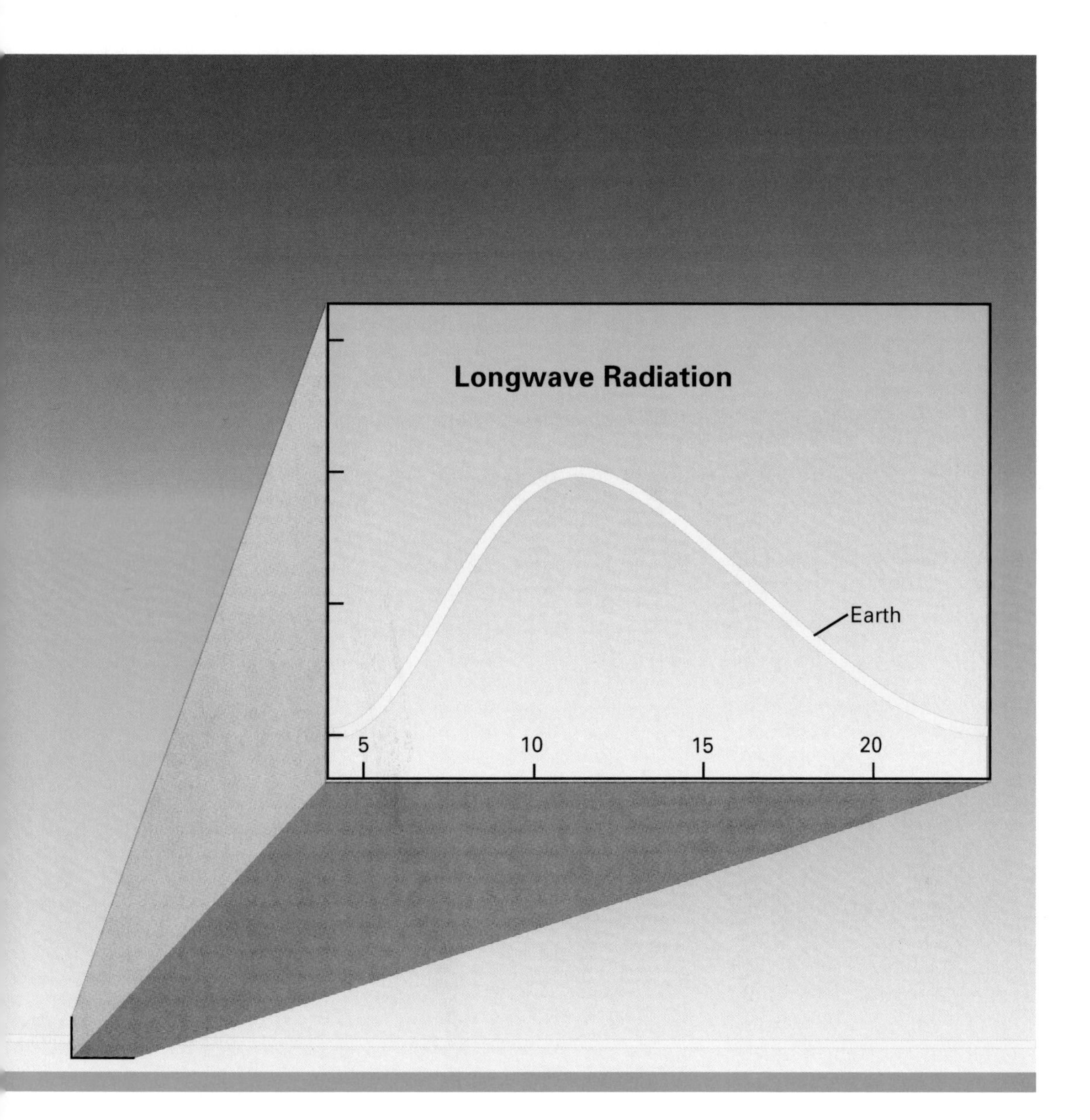

Wavelength (millionths of a meter) ⟶

These properties of solar and terrestrial radiation are important in explaining many temperature variations that occur in the atmosphere. As solar energy enters the earth's atmosphere, nearly 20% is absorbed by the atmospheric gases and clouds. Additionally, there is a loss of approximately 30% of the incoming solar radiation due to reflection and scattering by clouds and other particulates and reflection from the earth's surface. This loss is the albedo of the earth and its atmosphere. The remaining solar radiation is absorbed by the earth. (Figure 2-5)

The earth's surface becomes the primary energy source for the atmosphere. Energy is transferred to the atmosphere by the processes terrestrial radiation, conduction, convection/advection, and by evaporation. As we discussed earlier in this chapter, the transfer of heat through evaporation will be covered in a later chapter.

TERRESTRIAL RADIATION

Because the earth has a temperature well above absolute zero, it is continually losing infrared radiation. During the day, the loss of terrestrial radiation is offset by the receipt of solar radiation, so the temperature increases. But at night, there is no solar input and the earth continues to radiate, cooling significantly. This daily or diurnal variation in the temperature of the surface of the earth is critical in producing important day-to-night changes in wind, ceiling, and visibility. In later chapters on weather hazards, we will also see how radiation processes near the earth's surface can contribute to the production of frost on aircraft wings and to the development of strong low-level wind shear.

An important control of surface temperature and, therefore, of terrestrial radiation, is the heat capacity. This refers to the amount of heat energy that is necessary to raise the temperature of a substance by a certain amount. The surface of the earth is made up of a variety of substances with different heat capacities which cause substantial temperature differences across the earth's surface. A good example is water which has about four times the heat capacity of a typical dry soil. The ability of water to absorb large amounts of solar energy is increased because radiation can penetrate to a greater depth in water than in soil, and because water can mix easily, spreading the energy around. This means that if equal amounts of solar radiation fall on equal masses of water and soil, for example along a coastline, the water temperature increases much more slowly than the nearby land temperature. At night, the water, with its great reservoir of heat, cools more slowly than the land. The resulting land-sea temperature differences in both of these cases are crucial in understanding diurnal wind patterns such as sea and land breezes.

The temperature near the earth's surface also depends on other properties of the surface. For example, snow reflects a large fraction of incoming solar radiation and gives up infrared radiation easily; these influences help keep the temperatures low over snow surfaces under clear skies.

Terrestrial radiation behaves differently than solar radiation because it is emitted in the infrared portion of the spectrum. Whereas the atmosphere is highly transparent to much of solar radiation, certain atmospheric gases easily absorb the infrared radiation from the earth. When these gases are present, they absorb then reemit the energy, part upward and part downward. The IR energy that returns to the earth reduces the loss of energy from the surface.

One of the most important of these IR-absorbing gases is water vapor. An example of the influence of water vapor on nighttime cooling is seen in the differences between summertime overnight lows in the humid Southeastern U.S. and the drier West.

Figure 2-5. The source of energy for the atmosphere is the sun. Approximately 51% of the energy striking the top of the atmosphere is actually absorbed at the earth's surface. The solar radiation scattered and reflected into space (30%) is the earth's albedo.

Although daytime highs may be the same in both locations, nighttime minimum temperatures are often 20°F or more higher in the Southeast because of the large amounts of water vapor in the air. The presence of clouds at night increases the capture of infrared radiation, further restricting nighttime cooling.

The capture of terrestrial radiation by certain atmospheric gases is called the greenhouse effect, and the gases are called greenhouse gases. Like a greenhouse, once the energy is in the atmosphere, its escape is hindered. The concern over global warming is based upon measured increases of greenhouse gases due to natural and man-made pollutants such as carbon dioxide, methane, and chlorofluorocarbons, or CFCs.

CONDUCTION

You probably have experienced the effects of conduction when you have left a spoon in a bowl of hot soup. The spoon heats up as conduction transfers the heat energy along the handle.

Since air is a poor conductor, the most significant energy transfer by conduction in the atmosphere occurs at the earth's surface. At night, the ground cools because of radiation; the cold ground then conducts heat away from the air immediately in contact with the ground. During the day, solar radiation heats the ground which heats the air next to it by conduction. These processes are very important in the production of a variety of weather phenomena, including wind, fog, low clouds, and convection.

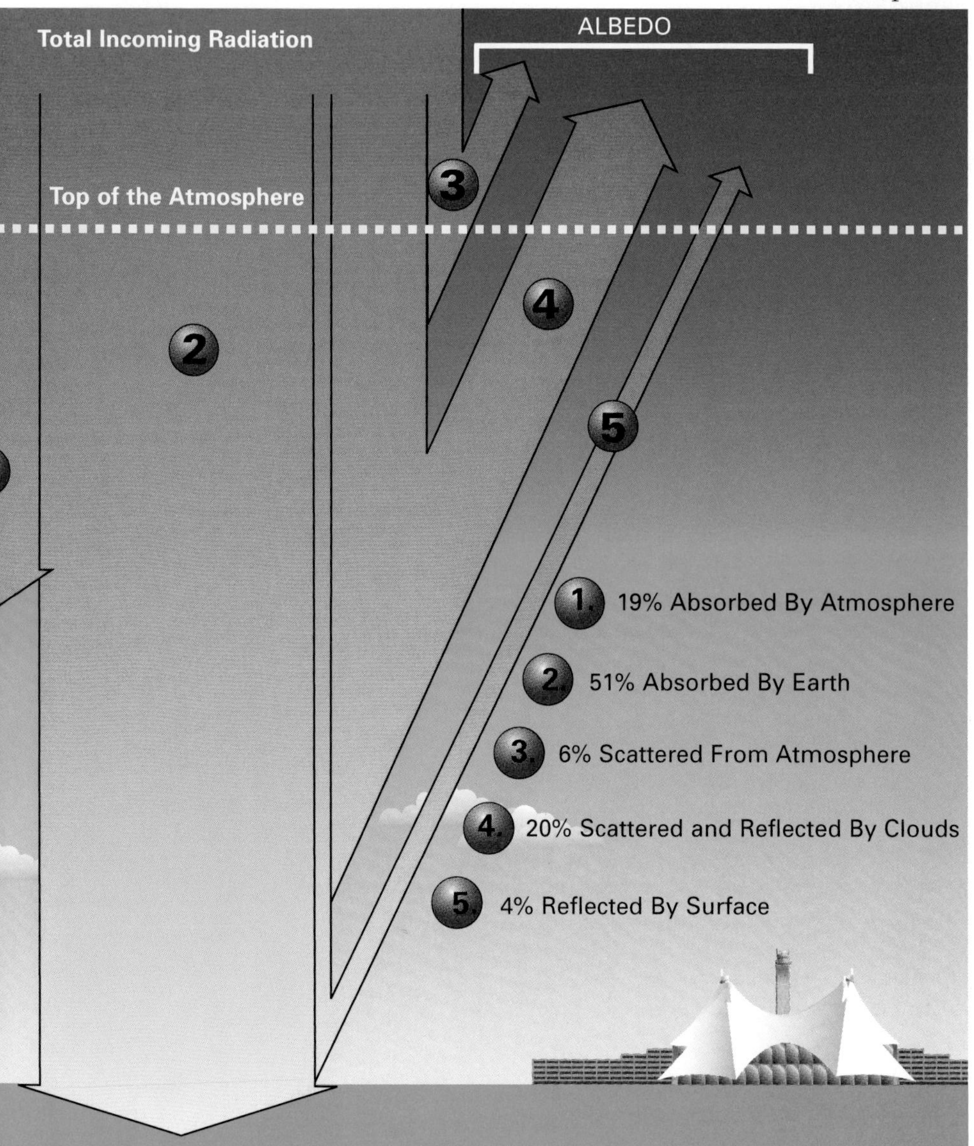

ADVECTION AND CONVECTION

If we were to depend on solar and terrestrial radiation alone for all energy transfer, the earth and atmosphere would become extremely cold in the polar regions and extremely hot near the equator. Fortunately, advection and convection (including the transport of water vapor and its latent heat), and the transfer of heat by ocean currents compensate for the unequal distribution of radiant energy. Advection includes the north-south movements of large warm and cold air masses. Convection includes the large scale ascent and descent of air masses and the smaller scale development of thunderstorms. These topics are examined closely in later chapters.

The most obvious effects of the sun's energy are seen in the distribution of temperatures within the atmosphere.

Section B

TEMPERATURE

In this section, after a brief introduction to common temperature scales, measurements, and terminology, we examine the global patterns of tropospheric temperatures.

TEMPERATURE SCALES

Temperature scales common to aviation are Fahrenheit (°F) and Celsius (°C). Figure 2-6 shows the relation of these scales to each other.

Conversions from Celsius to Fahrenheit are simple if you remember that there are 100 Celsius degrees and 180 Fahrenheit degrees between the melting and boiling points of water at sea level. Knowledge of that ratio, 100/180 or 5/9, and the one point on the scale where temperatures are the same (-40°C = -40°F) allows simple conversions.

({°C + 40} x 9/5) - 40 = °F

Example: T = 20°C = ?°F

({20°C +40} x 9/5) - 40 = 68°F

Conversion from Fahrenheit to Celsius:

({°F + 40} x 5/9) - 40 = °C

Example: T = 23°F = ?°C

({23°F +40} x 5/9) - 40 = -5°C

Note that both conversions are the same except the factor 9/5 (1.80) is used to convert from °C to °F and 5/9 (0.56) from °F to °C.

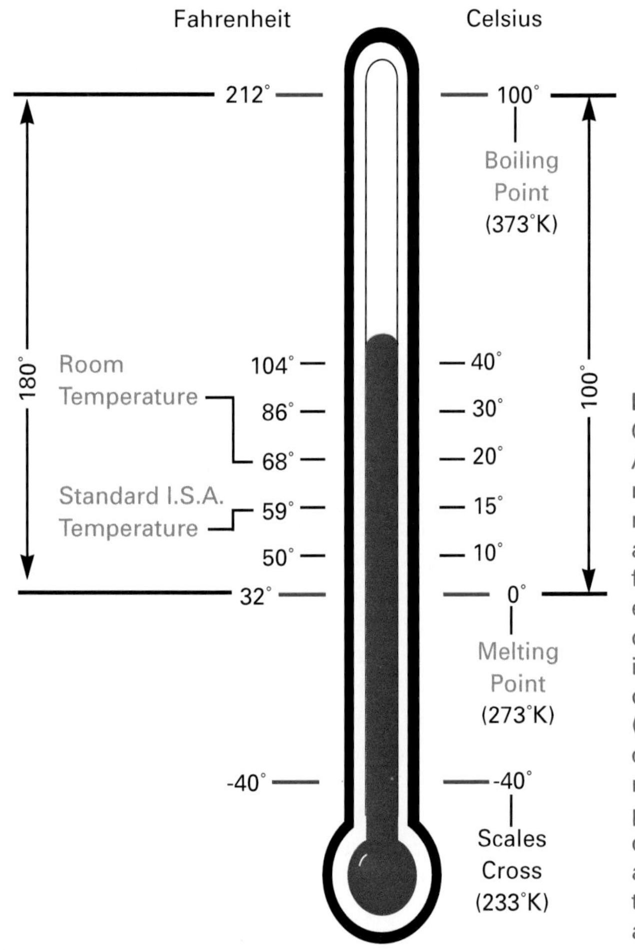

Figure 2-6. Fahrenheit and Celsius temperature scales. Although the conversion formula are the most direct connections between scales, there are several memory devices for quick estimates. Note for every change of 10 Celsius degrees, there is a corrsponding change of 18 Fahrenheit degrees: (0°C + 10) = 10°C = (32°F + 18) = 50° F. Also note common reference points: melting and boiling points of pure water at sea level, standard sea level temperature, and room temperature. Kelvin temperatures are provided as a reference.

TEMPERATURE MEASUREMENTS

A temperature frequently referred to in meteorological applications is the surface air temperature (often called "surface temperature"). This is the temperature of the air measured at 1.5 meters (about 5 feet) above the ground. It is usually measured in a standard instrument shelter ("in the shade") to protect the thermometer from direct solar radiation but allow the free ventilation of outside air. (Figure 2-7)

Other common temperatures used in aviation meteorology are those measured in the "free atmosphere"; that is, above the earth's surface. Such temperatures are usually identified as temperatures aloft, as upper air temperatures, or with reference to the height or pressure level where they are measured, such as the 300 mb temperature at Miami.

Another aviation related air temperature measurement is indicated air temperature (IAT) which is the temperature of the air as measured by the temperature probe on the outside of the aircraft. The air impacting on the probe is heated by compression so that the probe senses an IAT that is higher than the temperature of the undisturbed atmosphere. Also, the rush of air over the outside air temperature probe creates friction, which causes further heating and a still higher reading. Compression and friction heating are more apparent in high speed aircraft (over 180 knots) than in low speed aircraft. The outside air temperature (OAT) (also called the true air temperature) is determined by correcting the measured or indicated air temperature for compression and friction heating.

Figure 2-7. Surface air temperatures are regularly measured in an instrument shelter such as the one shown on the right. Upper air temperatures are measured with instruments carried aloft by sounding balloons as shown on the left.

Upper air temperatures are usually measured directly with freely rising, instrumented balloons (a sounding) or by aircraft. Soundings are also made remotely with instruments called "radiometers" that are placed aboard satellites to measure the radiant energy emitted from the earth, clouds, and various atmospheric gases. The temperature is determined through the relationship between radiation and temperature that was discussed previously.

Later, in the chapter on atmospheric moisture, you will be introduced to three other temperatures (wet bulb, dewpoint, and frostpoint) that will be useful in the discussion of fog, cloud, and precipitation formation.

GLOBAL TEMPERATURE DISTRIBUTION

In Chapter 1, the general features of temperature variations with height in the atmosphere were introduced (for example, troposphere, stratosphere, mesosphere, thermosphere). Then we looked more closely at the standard atmosphere, a detailed model of the vertical temperature distribution. Now we expand our temperature model by considering another dimension, the horizontal distribution of average temperatures.

SURFACE TEMPERATURES

Figure 2-8 shows surface air temperatures around the world. Two notable features are the large changes in temperature from summer and winter, and the large temperature decrease from equator to pole. These patterns are due largely to variations in day length which, in turn, are due to changes in solar elevation angle with latitude and season.

Ocean currents, land-sea differences, and the presence of mountains tend to modify the large scale temperature patterns over some areas of the globe. For example, the surface air temperatures over continents are colder than nearby oceans in wintertime and warmer in summer. The average temperature at Seattle, Washington, in January is 38°F while the average at Chicago, Illinois (which is actually farther south) is 26°F.

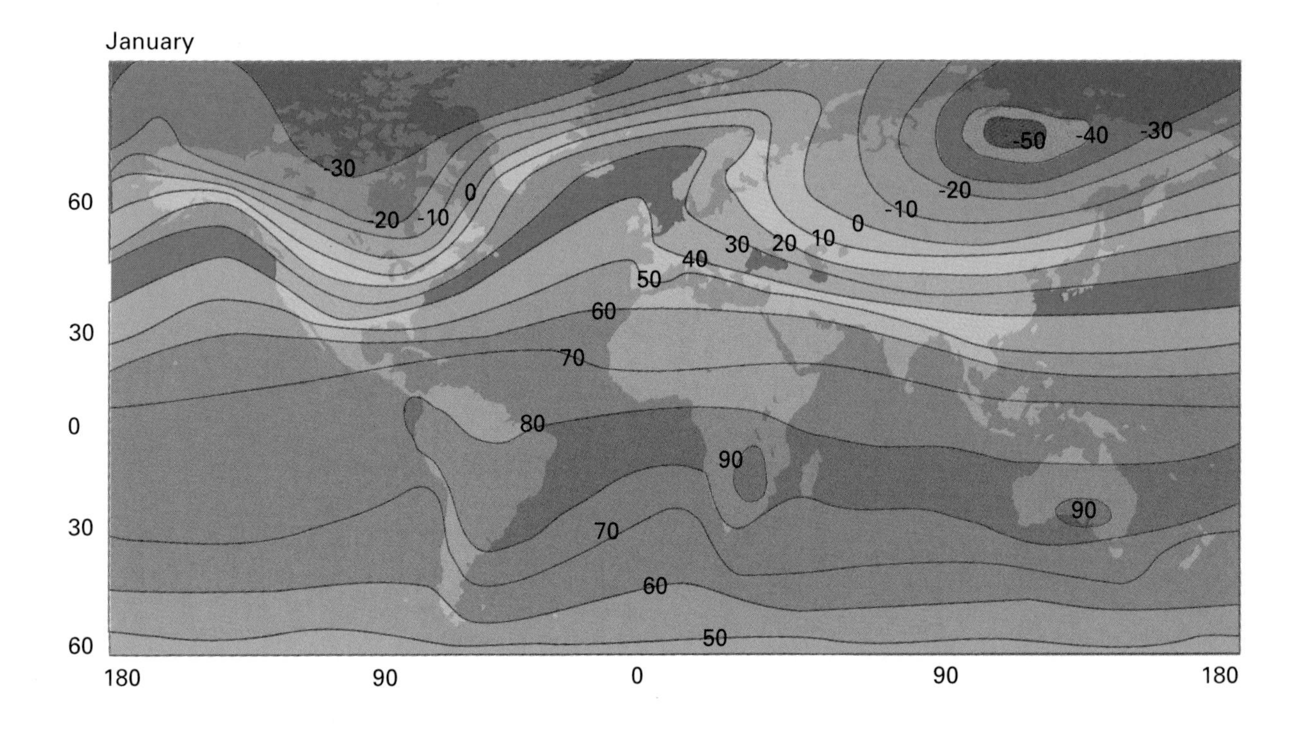

Figure 2-8. Average surface temperatures (°F) for the world, January (Left) and July (Right). The lines of equal temperature are called isotherms.

Other significant patterns in figure 2-8 are regions of weak and strong temperature gradients. A temperature gradient is defined as the change of temperature divided by the distance over which the change occurs. Where the isotherms are close together, the gradients are relatively strong (large temperature change over a small distance). Notice that surface temperatures are not evenly distributed between the cold poles and the warm equator. The large pole-to-equator temperature gradients occur in midlatitudes. Furthermore, these gradients are stronger in winter when the polar regions are in darkness. It will be seen that these abrupt transition zones between warm and cold air are favorite locations of the development of large storms.

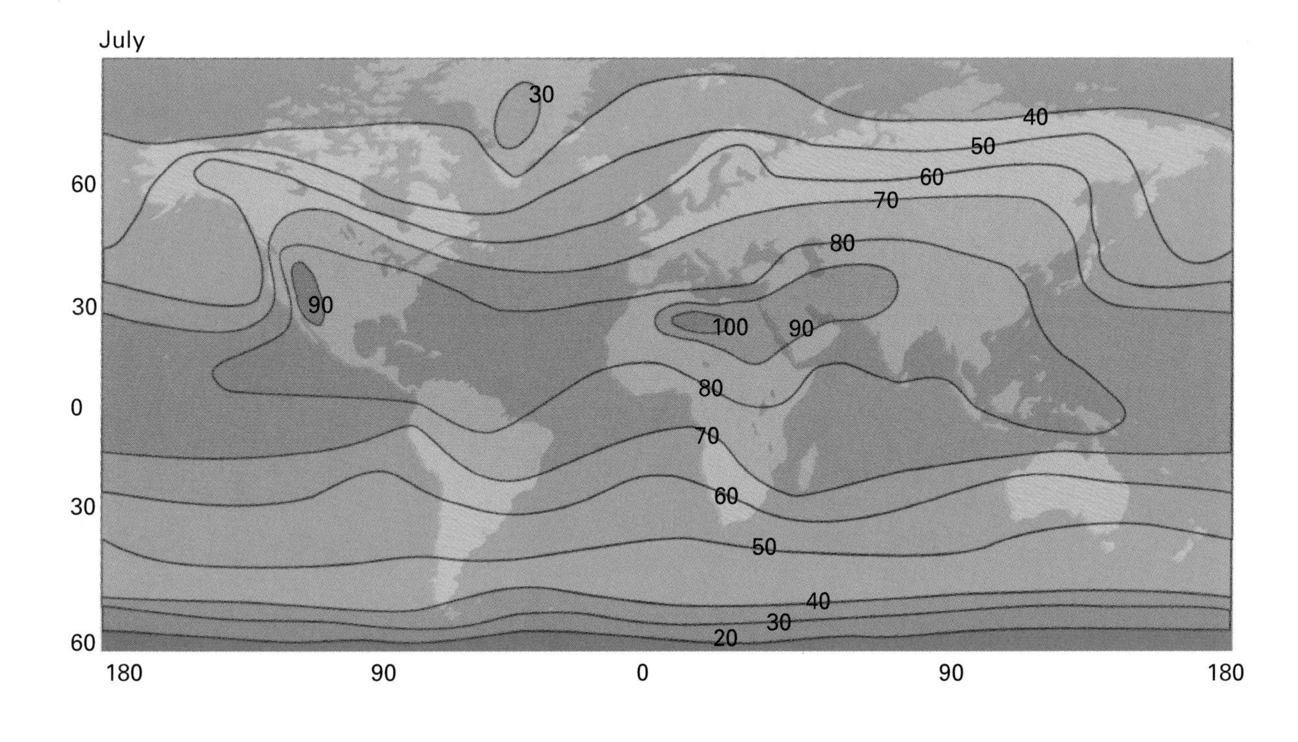

UPPER AIR TEMPERATURES

In the free atmosphere, the impact of the sun-earth geometry on the temperature distribution is large. However, as we move away from the earth, the direct influence of the heating and cooling of the surface becomes less obvious, especially above the tropopause. Advection, convection, and the absorption of radiation by ozone becomes more important in the determination of the temperature distribution. The result is that we generally see cold poles and a warm equator below the tropopause level. However, above the tropopause level the picture changes. Figure 2-9 illustrates the average horizontal distribution of temperatures near 18,000 feet MSL (500 mb) in the mid-troposphere and near 53,000 feet MSL (100 mb) in the lower stratosphere for January and July.

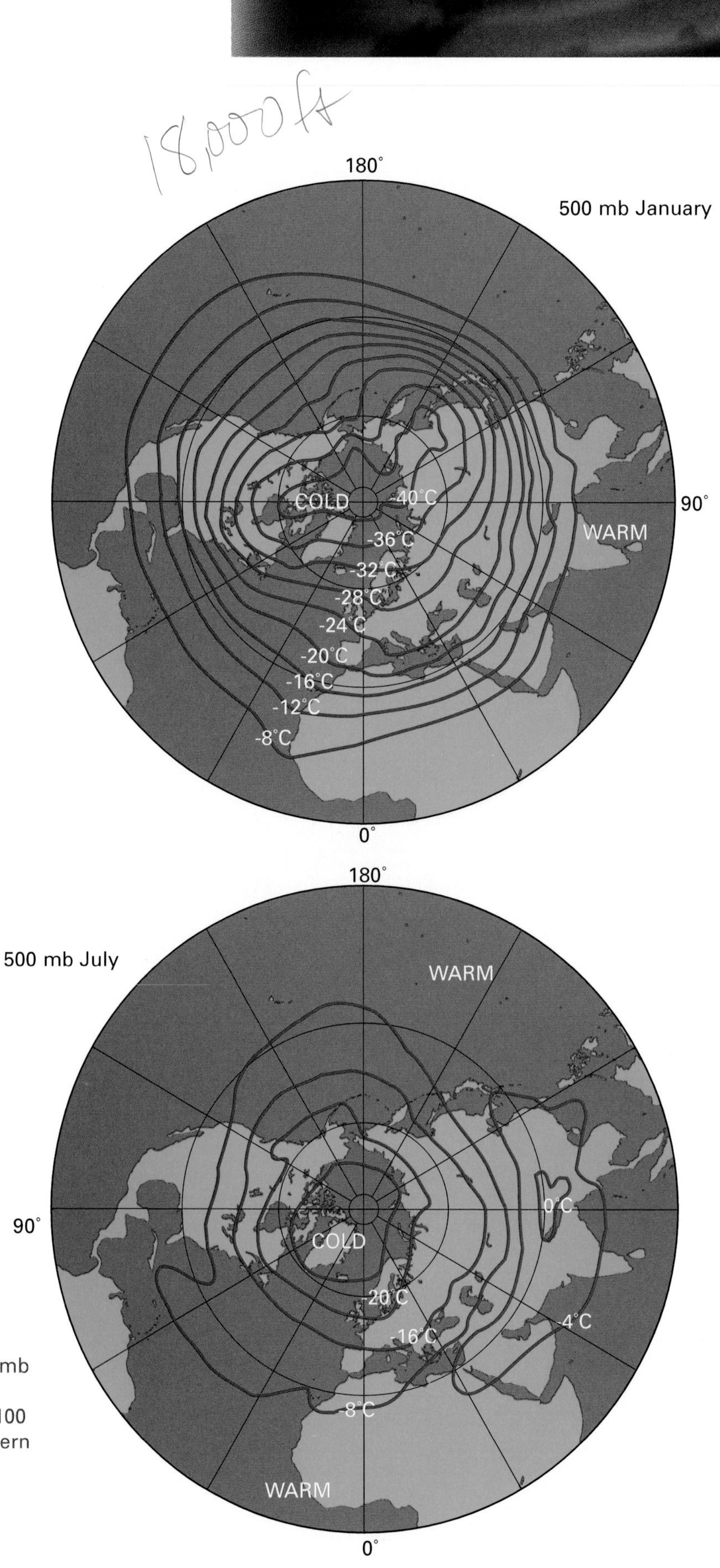

Figure 2-9. Left: Temperatures (°C) at 500 mb (about 18,000 feet MSL) for the Northern Hemisphere. Right: Temperatures (°C) at 100 mb (about 53,000 feet MSL) for the Northern Hemisphere. Relatively cold and warm regions are labeled.

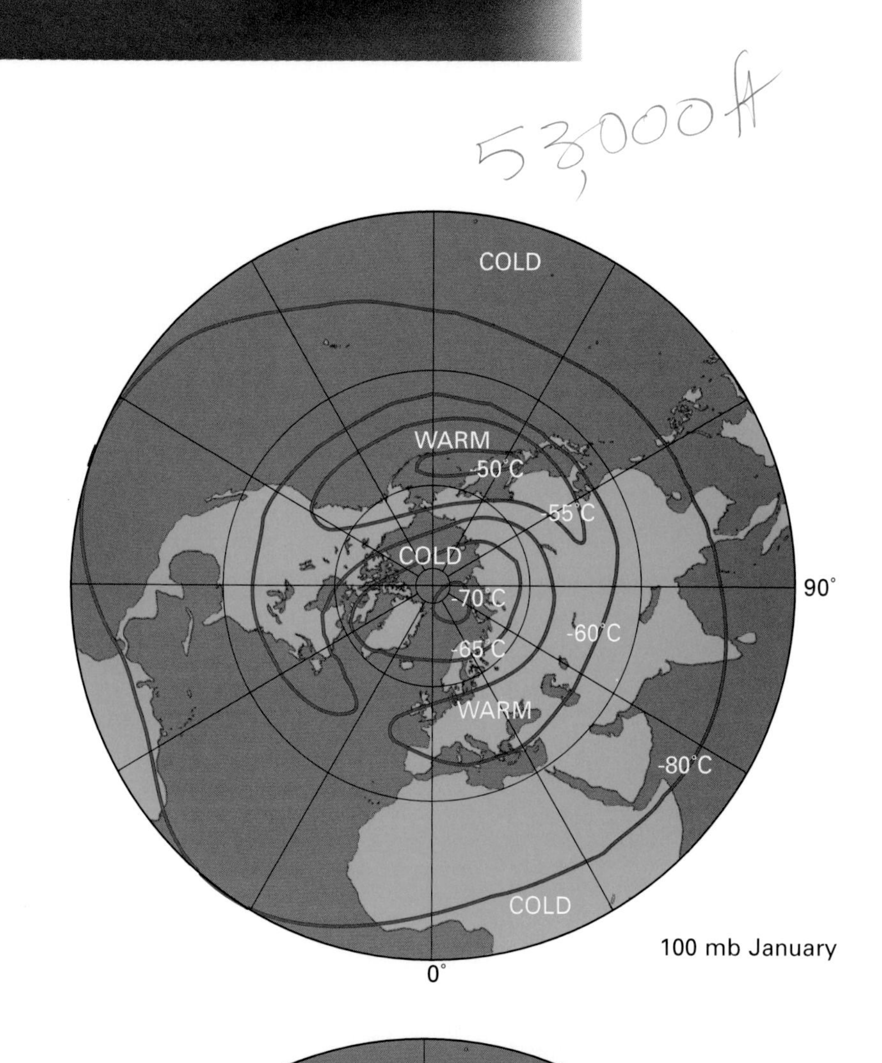

100 mb January

The temperature charts at 18,000 feet have features similar to figure 2-8, except of course, the temperatures are colder. Similarly, temperatures near 53,000 feet are much lower than surface air temperatures. However, there is a more significant difference in the stratospheric charts. In summer, the stratospheric equatorial temperatures are much colder than polar temperatures. In winter, the pattern is a little more complicated because the winter pole is in darkness, so the absorption of solar radiation by ozone is nil. The result is that both equatorial and polar regions are cold at 100 mb and there is a mid-latitude band of relatively warm temperatures. These temperature differences are important in explaining the characteristics of bands of strong winds (jet streams) found near the tropopause and in the stratosphere.

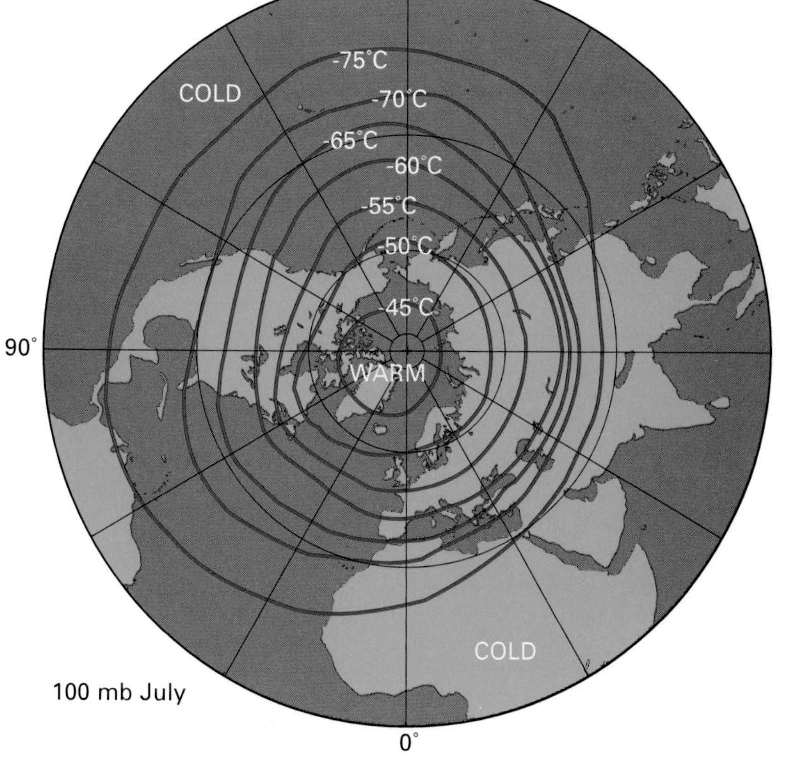

100 mb July

SUMMARY

The observed temperatures of the earth and the atmosphere are produced primarily by solar energy that is transferred across space via radiation. Solar energy is redistributed in the earth-atmosphere system by terrestrial radiation, conduction, and convection/advection. Solar and terrestrial radiation have distinctly different characteristics. A variety of global temperature patterns result when these differences are combined with changes in sun-earth geometry, variations in the nature of the earth's surface, the presence certain variable gases, and atmospheric motions. In subsequent chapters we will see that these temperature patterns and their changes are the root causes of atmospheric winds and weather.

KEY TERMS

Absolute Zero
Advection
Albedo
Amplitude
Boiling Point
Celsius
Conduction
Convection
Diurnal Variation
Equinox
Fahrenheit
Frequency
Greenhouse Effect
Heat Capacity
Indicated Air Temperature (IAT)
Infrared (IR)
Melting Point
Outside Air Temperature (OAT)
Radiation
Room Temperature
Solar Declination
Solar Elevation Angle
Solstice
Sounding
Spectrum
Speed of Light
Standard Sea Level Temperature
Surface Air Temperature
Temperature Gradient
Terrestrial Radiation
True Air Temperature (TAT)
Ultraviolet (UV)
Upper Air Temperature
Wavelength

CHAPTER QUESTIONS

1. Convert the following temperatures from °C to°F.
   1. -60
   2. -40
   3. -15
   4. 5
   5. 35
2. Convert the following from °F to °C.
   1. -453
   2. -100
   3. 0
   4. 25
   5. 113

3. Compute solar elevation on the first day of Northern Hemisphere winter for the following locations.
   1. Brownsville, Texas.
   2. Seattle, Washington.
   3. Barrow, Alaska.
   4. Mexico City, Mexico.
   5. Panama City, Panama.
   6. Melbourne, Australia.
   7. the South Pole.

4. Repeat question number 3, but compute the solar elevation for the first day of the Northern Hemisphere Equinox.

5. About how long does it take solar radiation to reach the earth?

6. In the next chapter, you will find that differences between ISA temperatures and actual temperatures cause errors in pressure altimeter readings. In order to get an idea of how different ISA can be from real conditions, examine figures 2-8 and 2-9 and determine the maximum positive and negative differences between ISA temperatures and the average temperatures across the globe for the following:
   1. the surface.
   2. 18,000 feet MSL.
   3. 53,000 feet MSL.

7. Use an ordinary thermometer to measure the air temperature at heights of 2 inches, 4 inches, 20 inches, and 5 feet above the ground on a hot afternoon (be sure the sun doesn't shine directly on the thermometer) and on a clear, calm night. Plot your results on a piece of graph paper.
   Contrast and explain the results.

8. On a clear, calm morning, just before sunrise, measure the temperature of the air about an inch above the top surface of the wing of a small aircraft. Note the height of the point of measurement above the ground. Move away from the airplane and measure the air temperature at the same level in the open. Explain the results.

9. The free atmosphere is continually losing more energy than it gains by radiation, both day and night. Despite this, average air temperatures stay about the same. Why?

10. The high temperatures at Boston, Massachusetts and at Boise, Idaho were both 85°F on a day where the weather was clear in both locations. The next morning, the low temperature at Boston was 78°F while the low at Boise was 53°F. These were again followed by identical highs of 85°F. There were no major weather changes during the period. Give a reasonable explanation for the temperature differences.

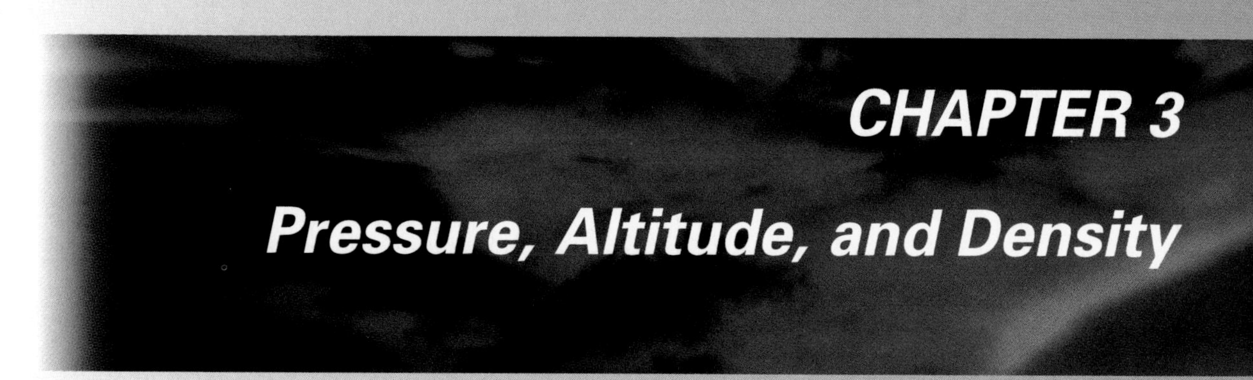

CHAPTER 3

Pressure, Altitude, and Density

Introduction

Pressure and its variations have important applications for aviation, ranging from measurements of altitude and airspeed to the prediction of winds and weather. This chapter focuses on several of these applications. When you complete the chapter, you will have a good physical understanding of atmospheric pressure, altimetry, and density altitude. Furthermore, you will develop important background knowledge about the global patterns of atmospheric pressure. This information will prove useful in the next chapter when we examine the causes and characteristics of atmospheric winds.

Section A

ATMOSPHERIC PRESSURE

Pressure was defined generally in Chapter 1 as the force exerted by the vibrating molecules of the gas on a given area. This force arises because the molecules are moving about randomly at speeds proportional to their temperature above absolute zero. The pressure exerted by atmospheric gases has the same general meaning; however, because of a special circumstance in the atmosphere, there is an additional, more useful definition. Atmospheric pressure may also be defined as the weight of a column of the atmosphere with a given cross-sectional area. (Figure 3-1)

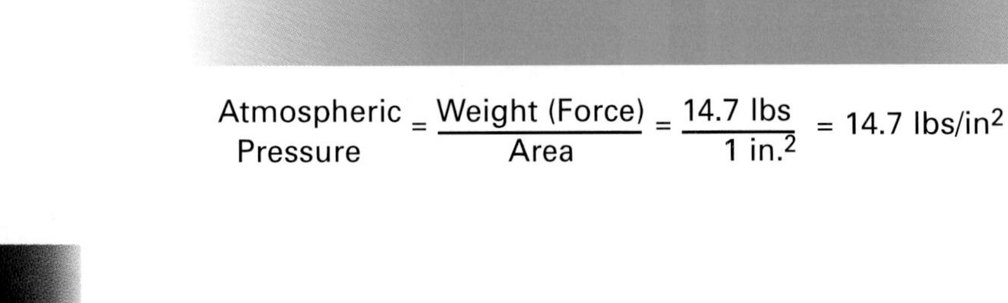

$$\text{Atmospheric Pressure} = \frac{\text{Weight (Force)}}{\text{Area}} = \frac{14.7 \text{ lbs}}{1 \text{ in.}^2} = 14.7 \text{ lbs/in}^2$$

Figure 3-1. Atmospheric pressure as the weight of a single column of air. In the standard atmosphere, a one square inch column of air the height of the atmosphere weighs 14.7 pounds at sea level. The pressure is 14.7 pounds per square inch (lbs/in^2). The column extends from the point of measurement at sea level to the top of the atmosphere.

The special circumstance that permits this definition for the atmosphere is the balance between the downward-directed gravitational force and an upward-directed force caused by the decrease of atmospheric pressure with altitude. This is called hydrostatic balance and is illustrated in figure 3-2.

PRESSURE MEASUREMENTS

The definition of atmospheric pressure as the weight of a column of air per unit area is demonstrated nicely in the construction of one of the

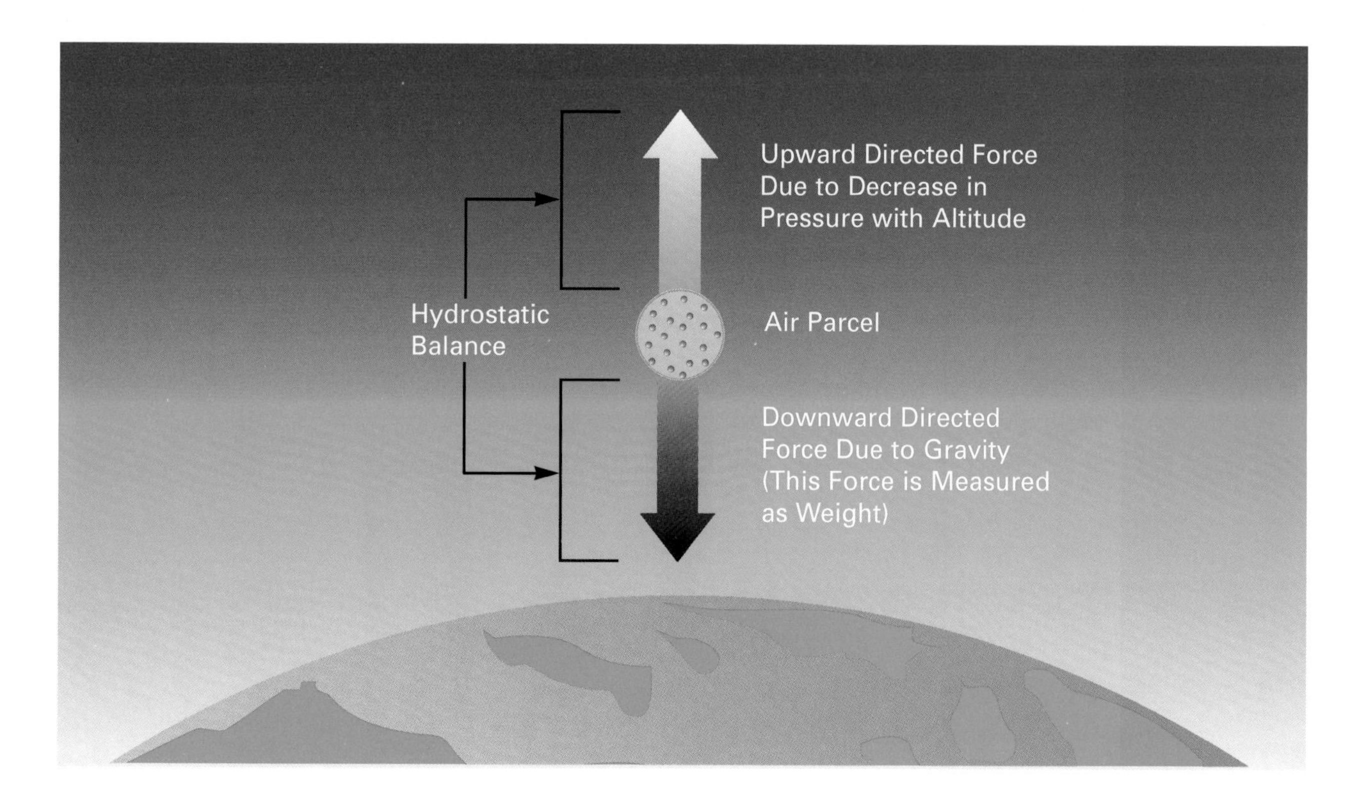

Figure 3-2. Hydrostatic balance. The air parcel resists any change in vertical movement because the forces acting on it tend to be equal and opposite. However, the parcel can still be accelerated horizontally under these circumstances.

most basic devices for the measurement of pressure: the mercurial barometer. We measure atmospheric pressure similar to the way we determine the weight of an object on a familiar balance scale. The pressure (weight) of the atmosphere is balanced against the weight of the mercury. (Figure 3-3)

As the weight of the atmosphere changes, the height of the mercury column also changes. Appropriately, the word barometer is derived from the Greek word *baros* which means weight.

Our examination of the structure of the mercurial barometer helps to explain why atmospheric pressure is commonly expressed in units of length (inches of mercury) as well as the units of force per area (pounds per square inch). As seen in figure 3-3, "length" refers to the height of the top of the column of mercury above the surface of the mercury reservoir of the barometer. A sea level pressure of 14.7 lbs/in^2 will force the mercury to a height of 29.92 inches above the reference. Recall from Chapter 1 that pressure decreases about one inch of mercury per 1,000 feet. This means that the mercury column would be about 28.92 inches high at an altitude of 1,000 feet MSL.

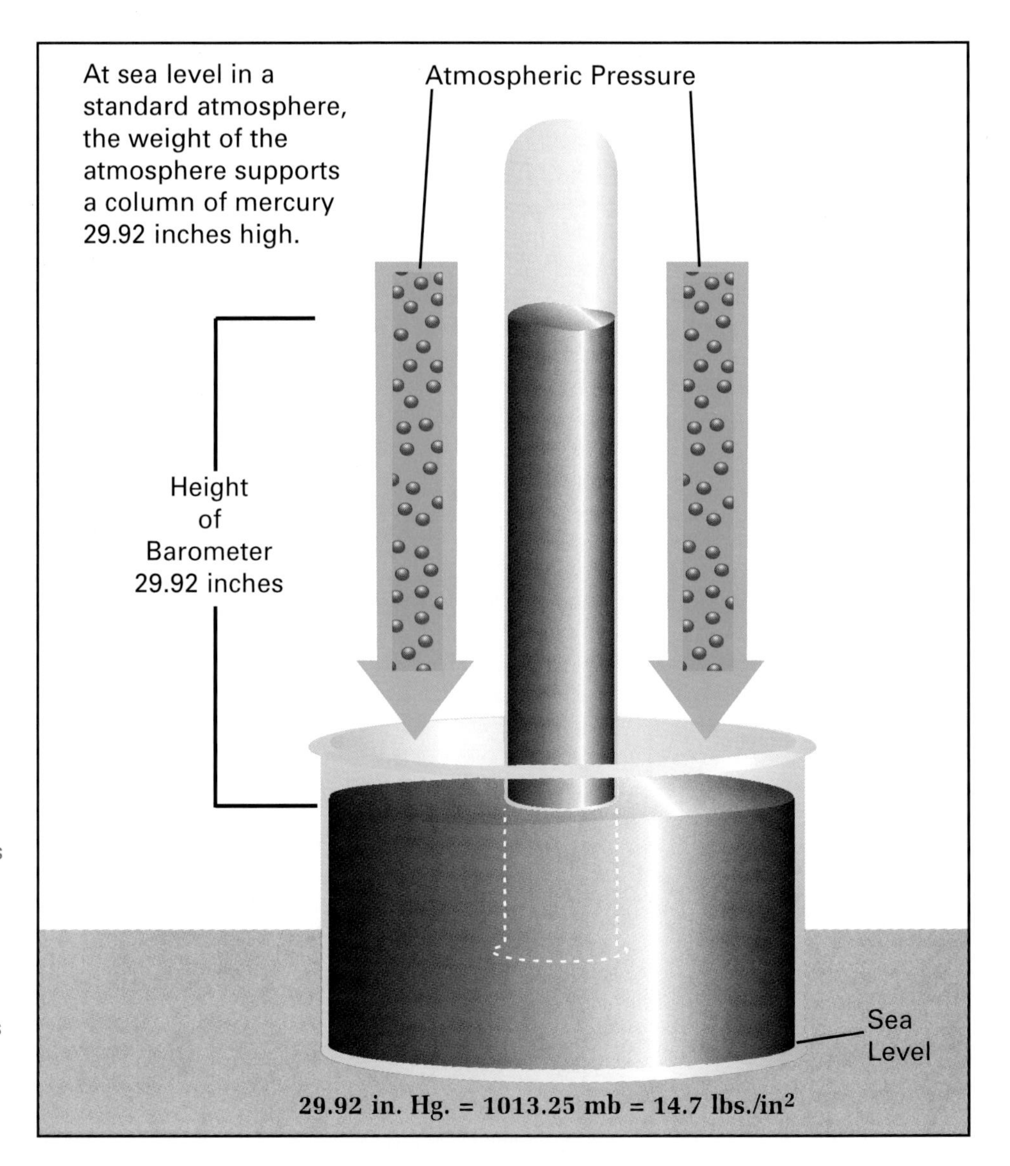

Figure 3-3. A mercurial barometer is constructed by pouring mercury into a tube closed on one end. The tube is then inverted into a reservoir of mercury open to the atmosphere. The mercury flows back out of the tube until the weight of the remaining mercury column is balanced by the pressure (weight) of the atmosphere over the mercury reservoir.

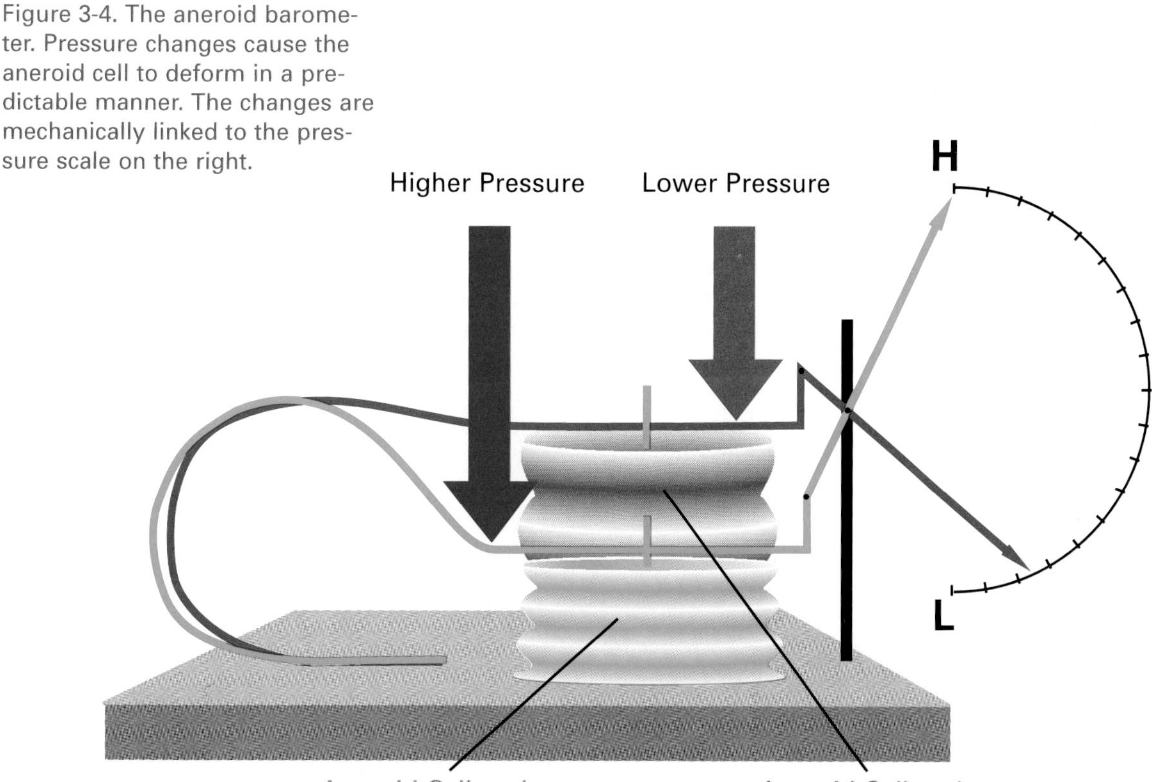

Figure 3-4. The aneroid barometer. Pressure changes cause the aneroid cell to deform in a predictable manner. The changes are mechanically linked to the pressure scale on the right.

Liquids other than mercury can be used to construct barometers; however, mercury has the advantage of being very dense, which keeps the size of the barometer manageable. For example, a water barometer would have to be nearly 34 feet high to register standard sea level pressure.

Although it provides accurate measurements, the mercurial barometer is not very useful outside the observatory or laboratory. The instrument is fragile, it must be kept upright, and if the reservoir is ruptured, one must be concerned about the toxicity of mercury. For these reasons, another pressure instrument, known as the aneroid barometer, is more frequently used outside the laboratory. In contrast to the mercurial barometer, the aneroid barometer has no liquid. Rather, it operates on differences in air pressure between the atmosphere and a closed vessel (an aneroid cell). Again, the root of the word helps us remember the principle of operation. Literally, *aneroid* means "not wet." As shown in figure 3-4, the aneroid barometer is a partial vacuum container that is strong enough not to collapse under pressure, but flexible enough so that its shape will change a specified amount as atmospheric pressure increases or decreases. The change in shape is linked mechanically to an indicator that shows the pressure value. Although not as accurate as the mercurial barometer, the aneroid barometer has several advantages. It is small and rugged; that is, it can withstand strong g-forces due to atmospheric turbulence and maneuvering.

Section B

CHARTING ATMOSPHERIC PRESSURE

Pilots and meteorologists pay careful attention to the horizontal distribution of atmospheric pressure because horizontal differences in pressure are related to the development of wind. Storms and fair weather areas also have distinctive pressure patterns which are important aids for weather diagnosis and prediction. Such pressure patterns are normally identified by inspecting charts which show the horizontal distribution of atmospheric pressure.

STATION AND SEA LEVEL PRESSURE

Surface pressure measurements are most useful if they can be compared with nearby measurements at the same altitude. Over land areas, the direct comparison of station pressures are usually difficult because weather stations are often at different altitudes. (Figure 3-5)

Even slight differences in altitude are important because the change of pressure over a given vertical distance is always much greater than the change of pressure over the same horizontal distance. For example, near sea level, a station

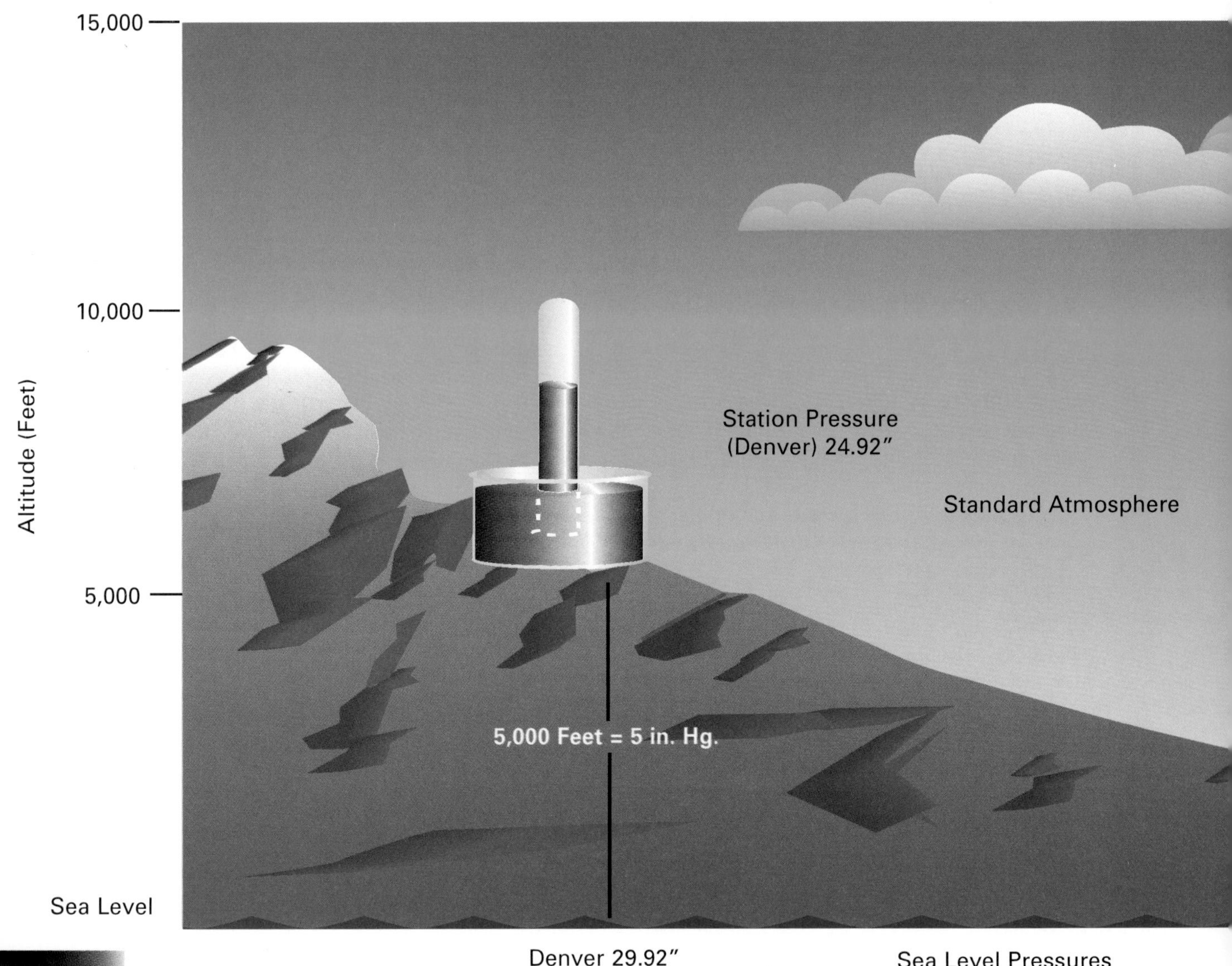

elevation difference of only 100 feet will cause a difference in station pressure of about 1/10 inch of mercury. If that vertical difference was erroneously reported as a horizontal pressure difference, it would imply an unrealistically strong horizontal wind. In order to correct for such altitude differences, station pressures are extrapolated to sea level pressure. Sea level pressure is the atmospheric pressure measured or estimated at an elevation equal to mean sea level. This extrapolated sea level pressure (station pressure corrected for elevation) is used by pilots to determine altitude. It also is used in aviation reports to depict the atmospheric pressure of a reporting location.

Figure 3-5. Station pressure and sea level pressure. New Orleans, Louisiana, is located near sea level. In the standard atmosphere, the station pressure at New Orleans is equal to the sea level pressure, or about 29.92 in. Hg. In comparison, the altitude of Denver, Colorado, is about 5,000 feet MSL. Since atmospheric pressure decreases one inch of mercury per 1,000 feet, the station pressure at Denver is about 24.92 in. Hg. in the standard atmosphere. The sea level pressure at Denver is 29.92 in. Hg.

Station Pressure
(New Orleans) 29.92"

New Orleans 29.92"

Sea level pressure can be approximated if you know station pressure and elevation. Simply add one inch of mercury to the station pressure for every 1000 feet of station elevation. For example, if the station pressure is 27.50 inches of mercury and the station elevation is 2,500 feet, the sea level pressure is approximately 30.00 inches of mercury. At National Weather Service stations, more precise estimates of sea level pressure are made by also using the station temperature. This more precise computation modifies the sea level pressure estimates to account for density differences between the standard atmosphere and actual conditions. Sea level pressures are commonly reported in the U.S. and Canada, but elsewhere, station pressure or some related measurement may be reported.

SEA LEVEL PRESSURE PATTERNS

Figure 3-6 shows the global patterns of average sea level pressure for January and July. A chart which shows pressure as well as other meteorological conditions at the surface of the earth is referred to generally as a surface analysis chart.

Another important property illustrated in figure 3-6 is the pressure gradient. A pressure gradient is a difference in pressure over a given distance. A pressure difference of 4.0 mb per 100 n.m. is an example of a moderate horizontal pressure gradient in mid-latitudes. If the pressure gradient is strong, isobars will be close together; if the gradient is weak, the isobars will be spaced far apart. A pressure gradient on a sea level pressure chart is

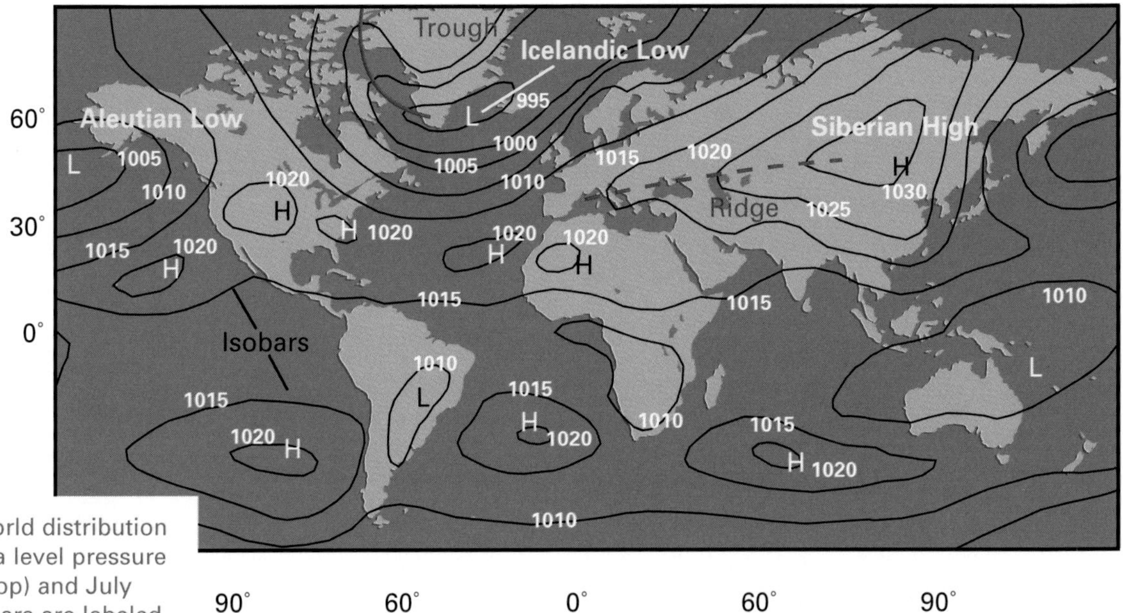

Figure 3-6. World distribution of average sea level pressure for January (top) and July (bottom). Isobars are labeled in millibars. Examples of a trough, a ridge, are shown.

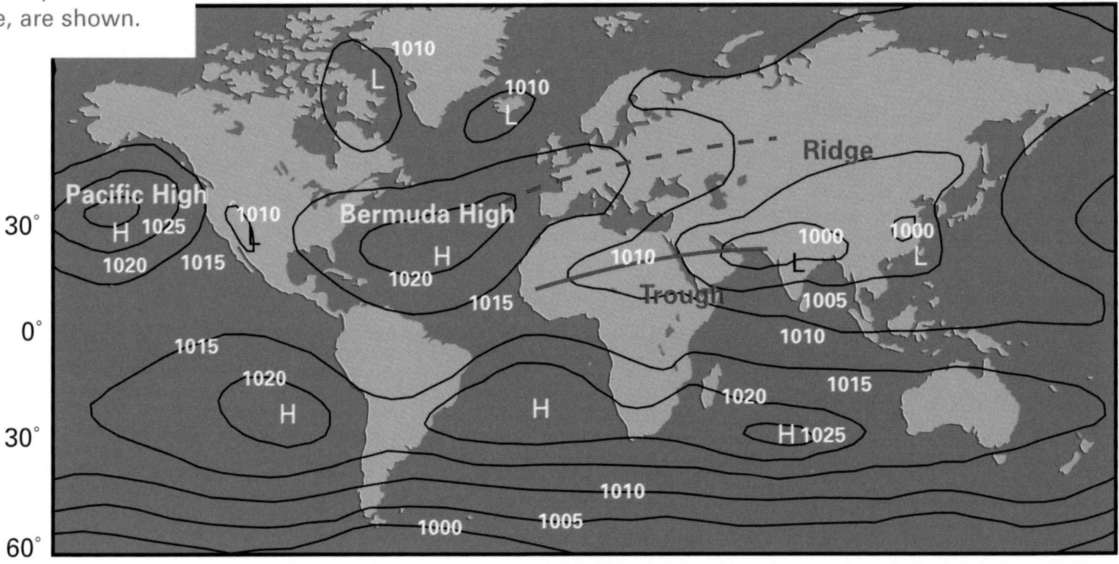

Some useful terms are shown in figure 3-6. The lines on the charts are isobars, or lines of constant pressure. A high pressure center or a high (H) on a weather chart is a location where the sea level pressure is high compared to its surroundings. A ridge is an elongated region of relatively high pressure. Similarly, a low pressure center or low (L) is a roughly circular area with a lower sea level pressure in the center as compared to the surrounding region; a trough is an elongated region of relatively low pressure. These features are to a surface analysis chart what mountains and valleys are to a topographical chart.

comparable to the height gradient on a topographical chart; the stronger the height gradient, the steeper the slope. In the next chapter, you will learn how to relate the pressure gradient to wind speed and direction.

On a surface analysis chart, the solid lines that depict sea-level pressure patterns are called isobars.

Also shown on figure 3-6 are some of the larger semi-permanent pressure systems and their common geographical names. These include the Bermuda High, the Aleutian Low, the Siberian High, the Icelandic Low, and the Pacific High. Later in the text, we will examine the unique wind and weather associated with these systems. Notice the range of sea level pressures in figure 3-6. Typical values vary from less than 995 mb in the Icelandic Low in January to more than 1030 mb in the Siberian High.

On a surface analysis chart, close spacing of the isobars indicates strong pressure gradient.

CONSTANT PRESSURE CHARTS

Highs, lows, troughs, and ridges in the atmospheric pressure field frequently extend above the earth's surface, often well into the stratosphere. For flight planning purposes, it is useful to examine weather charts at various altitudes to determine the general wind, temperature, and weather conditions.

The weather charts commonly used to show the distribution of pressure above the earth's surface are slightly different than sea level pressure charts. This is because, above the earth's surface, meteorologists find it easier to deal with heights on pressure surfaces rather than pressure on height surfaces. The difference between the two types of "surfaces" is easy to interpret. A constant height surface is simply a horizontal plane where the altitude (MSL) is the same at all points.

Extreme low sea level pressure values are found in hurricanes (lowest reported: 870 mb or 25.69 in. Hg.); extreme high values occur in very cold wintertime high pressure areas (highest reported: 1083.8 mb or 32.00 in. Hg.).

In contrast, a constant pressure surface is one where the pressure is the same at all points. Like the ocean's surface, a constant pressure surface is not necessarily level. Many upper air weather charts that you use are called constant pressure analysis charts or, simply, constant pressure charts. The relationship between isobars on a constant height surface and heights of a constant pressure surface is illustrated in figure 3-7.

The interpretation of a constant pressure chart is identical with the sea level pressure chart as far as highs, lows, troughs, ridges, and gradients are concerned. The main difference is one of terms used to describe the elements. On constant pressure surfaces, lines of constant height are called contours rather than isobars. Gradients are height gradients rather than pressure gradients. An example of a constant pressure chart is given in figure 3-8.

Figure 3-7. From our previous discussion of sea level pressure you know that the pressure at points A and B are determined by the weight of the column of atmosphere above those points. At 10,000 feet MSL the pressure at points C and D are also determined by the weight of their respective columns above the 10,000-foot level (dashed line). Notice the solid line representing the 700 mb pressure surface. At point C, the pressure at 10,000 feet is lower than 700 mb because the height of the 700 mb surface is lower than 10,000 feet. Similarly, the pressure at 10,000 feet at point D is higher than 700 mb because the 700 mb surface is higher than 10,000 feet. This demonstrates the relationship between pressure on a constant altitude surface and heights on a constant pressure surface.

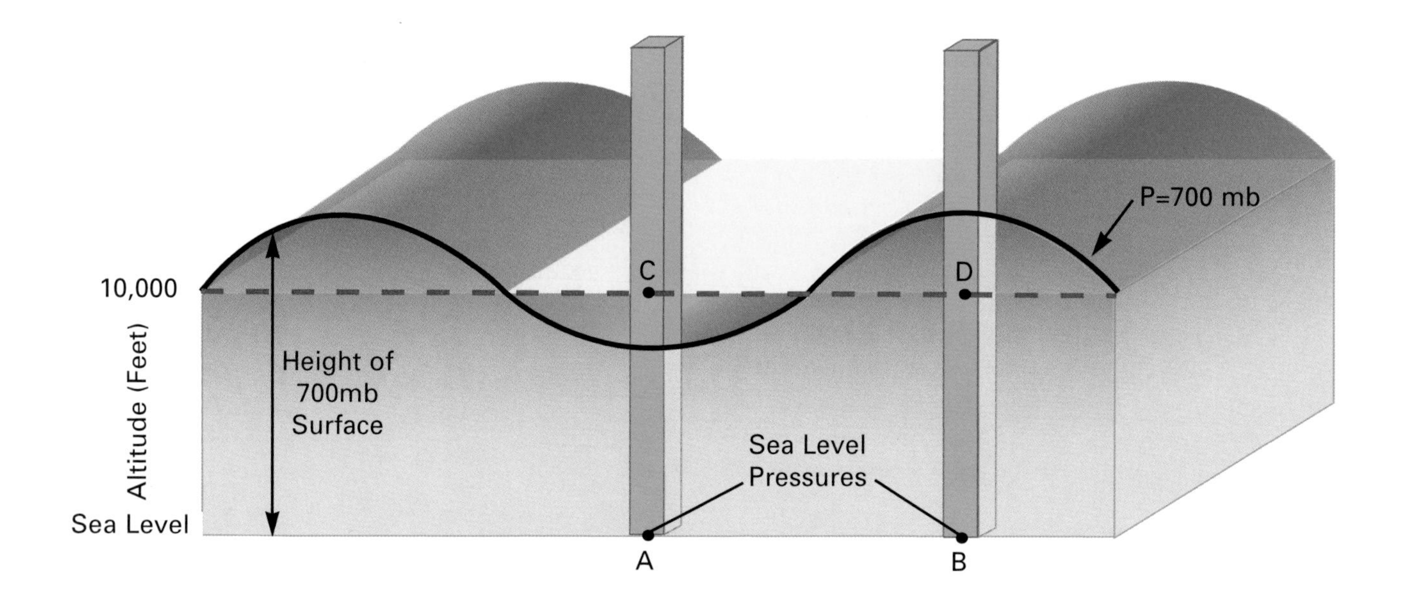

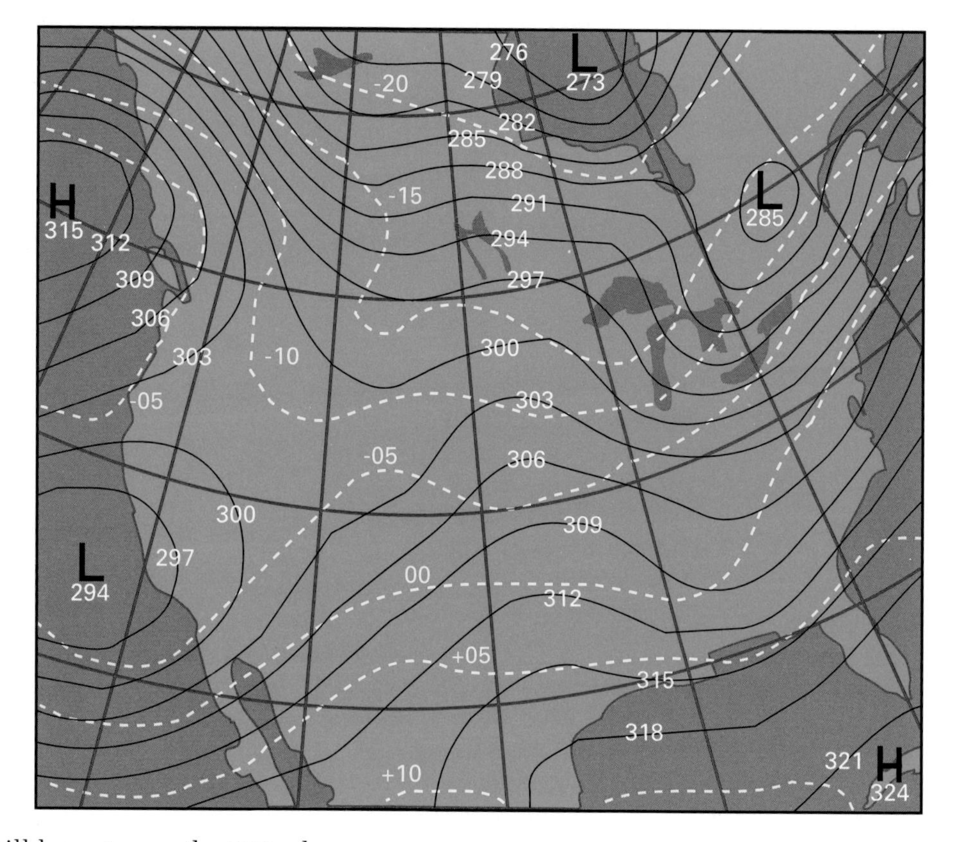

Figure 3-8. 700 mb constant pressure chart. Solid lines are contours labeled in decameters (300 = 3000 m). Dashed lines are isotherms labeled in °C.

Later in the text, you will learn to use the 700 mb chart and other constant pressure charts for flight planning. There are only a few pressures for which constant pressure charts are regularly constructed and it would be helpful to learn these and their pressure altitudes for later reference. Pressure altitude is the altitude of a given pressure surface in the standard atmosphere. (Figure 3-9.) Pilots can determine pressure altitude by setting the standard sea level pressure, 29.92 in. Hg., in the aircraft altimeter (more on this later in this chapter).

Pressure	**Pressure Altitude**	
Millibars	Feet	Meters
850	4,781	1,457
700	9,882	3,012
500	18,289	5,574
300	30,065	9,164
250	33,999	10,363
200	38,662	11,784

Figure 3-9. Left: Constant pressure levels for which analysis charts are usually available to pilots. Right: Pressure altitudes indicated in both feet and meters. These pressure altitudes are exact. For quick memorization, it is convenient to round the heights to the nearest 1,000 feet (for example, the 700 mb pressure altitude becomes 10,000 feet).

Pressure altitude is the altitude indicated when the altimeter setting is 29.92.

Section C

THE PRESSURE ALTIMETER

Perhaps the most important aviation application of the concept of atmospheric pressure is the pressure altimeter. The altimeter is essentially an aneroid barometer that reads in units of altitude rather than pressure. This is possible by using the standard atmosphere to make the conversion from pressure to altitude. (See Appendix) A schematic diagram of a pressure altimeter is shown in figure 3-10.

If the actual state of the atmosphere is the same as the standard atmosphere, then the pressure altitude is equivalent to the actual altitude. However, this is usually not the case. Altimeter indications may be inexact if the actual atmospheric conditions are nonstandard. This is true even if the altimeter is in perfect working condition and accurately calibrated. Therefore, you must always be aware of the difference between the altitude measured by your altimeter (indicated altitude) and the actual altitude of your aircraft above mean sea level (true altitude) or above the ground (absolute altitude).

There are three specific altimeter errors caused by nonstandard atmospheric conditions.

1. Sea level pressure different from 29.92 inches of mercury.
2. Temperature warmer or colder than standard temperature.
3. Strong vertical gusts.

The first altimeter error arises because the standard atmosphere is based on a fixed sea level pressure of 29.92 inches of mercury (1013.25 mb)

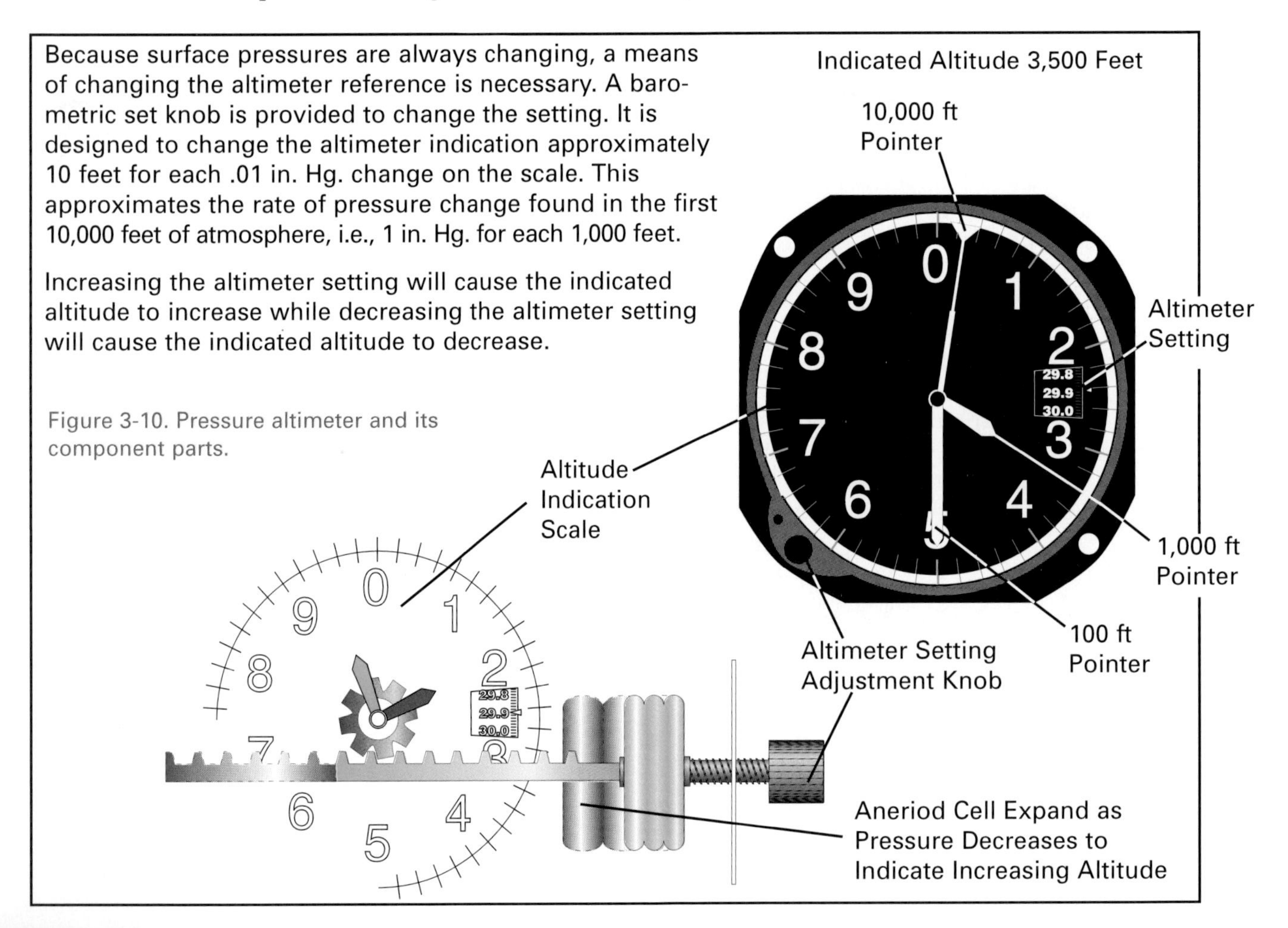

Figure 3-10. Pressure altimeter and its component parts.

and the actual sea level pressure may vary widely from that value. Figure 3-11 shows that the true altitude of an aircraft will be higher than the indicated altitude when sea level pressure is higher than standard. True altitude will be lower than indicated when sea level pressure is lower than standard. This problem is responsible for the well-known rule of thumb:

High to Low, Look out Below

air. If the altimeter is set to station pressure, it will read zero on the ground and indicate absolute altitude in the air. Since sea level pressure changes from place to place, the pilot must be alert to changes in altimeter setting enroute. With flight levels at and above 18,000 feet in U.S. airspace, altimeters are set to 29.92 inches of mercury.

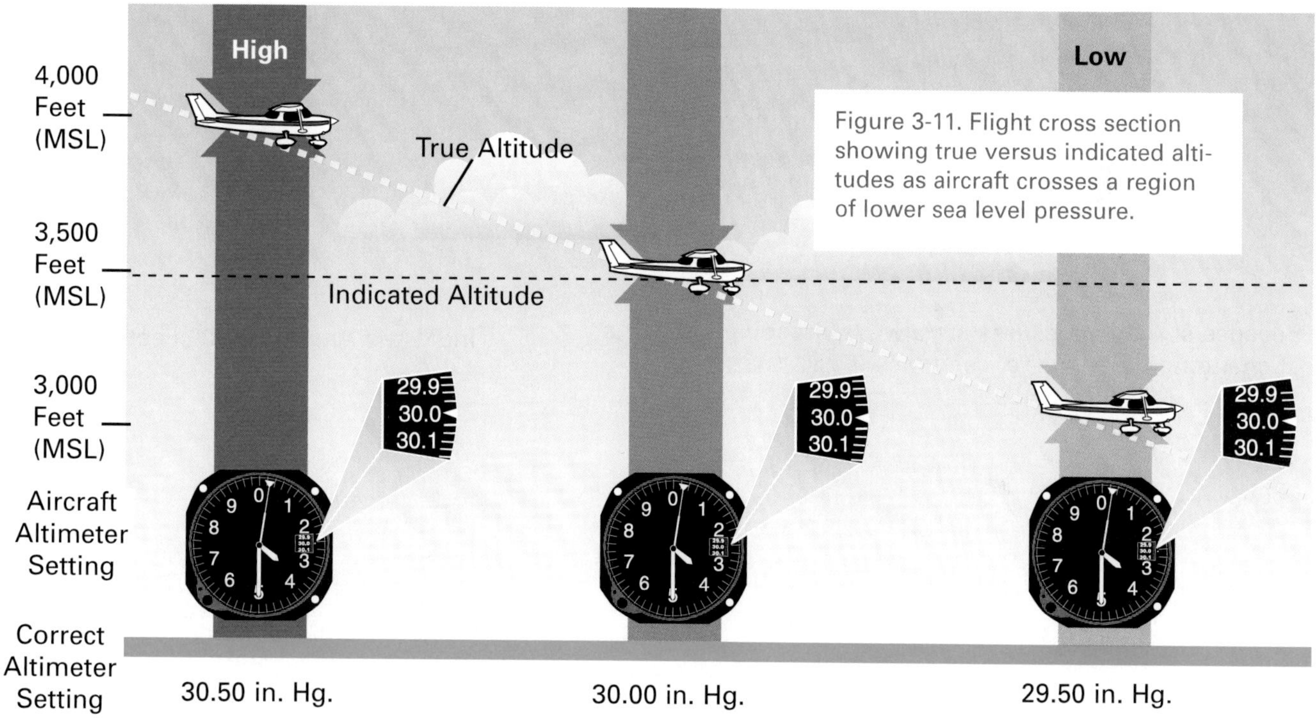

Figure 3-11. Flight cross section showing true versus indicated altitudes as aircraft crosses a region of lower sea level pressure.

This variable sea level pressure is usually taken care of by adjusting the altimeter to the proper altimeter setting. This is the sea level pressure determined from the station pressure and the standard atmosphere. For altitudes below 18,000 feet in the U.S., this adjustment ensures that the altimeter will read the field elevation when the aircraft is on the ground and indicate the approximate altitude above mean sea level when the aircraft is in the

Altimeter setting is the value to which the barometric pressure scale on the altimeter is set so the altitude indicates true altitude at field elevation.

The second pressure altimeter error occurs when atmospheric temperatures are warmer or colder than the standard atmosphere. A problem arises in these cases because atmospheric pressure decreases with altitude more rapidly in cold air than in warm air.

This means that a correct altimeter setting only ensures a correct altitude on the ground. Once you are in the air, the indicated and true altitude will be equal only if the atmospheric temperatures are standard. If the atmosphere is colder than standard, your indicated altitude will be higher than true altitude. If the atmosphere is warmer than standard, the indicated altitude will be lower than the true altitude. (Figure 3-12) Temperature errors are generally smaller than those associated with variations in sea level pressure. For example, if the actual temperature was 10C° warmer than standard, the true altitude would be about 4% higher. This is only 40 feet at 1,000 feet MSL. But the error increases with height. At 12,000 feet MSL, it is about 500 feet. Flight over high mountains in bad weather, requires close attention to possible temperature errors.

Remember, the pressure altimeter will not automatically show exact altitude in flight. It is pilot's responsibility to ensure terrain avoidance.(USAF, 1990)

The third pressure altimeter error that arises because of nonstandard atmospheric conditions is caused by rapid changes in vertical movements of the air. These changes upset the balance of forces that allows atmospheric pressure to be related directly to altitude. Some errors may be expected in the extreme conditions of thunderstorms and in strong mountain waves. More on this problem is presented in Part III on aviation weather hazards.

The altimeter errors discussed above are all related to atmospheric conditions. Other errors may arise due to instrument problems. These include improper calibration, friction, lag, improper instrument location, and temperature changes of the instrument. These are beyond the scope of

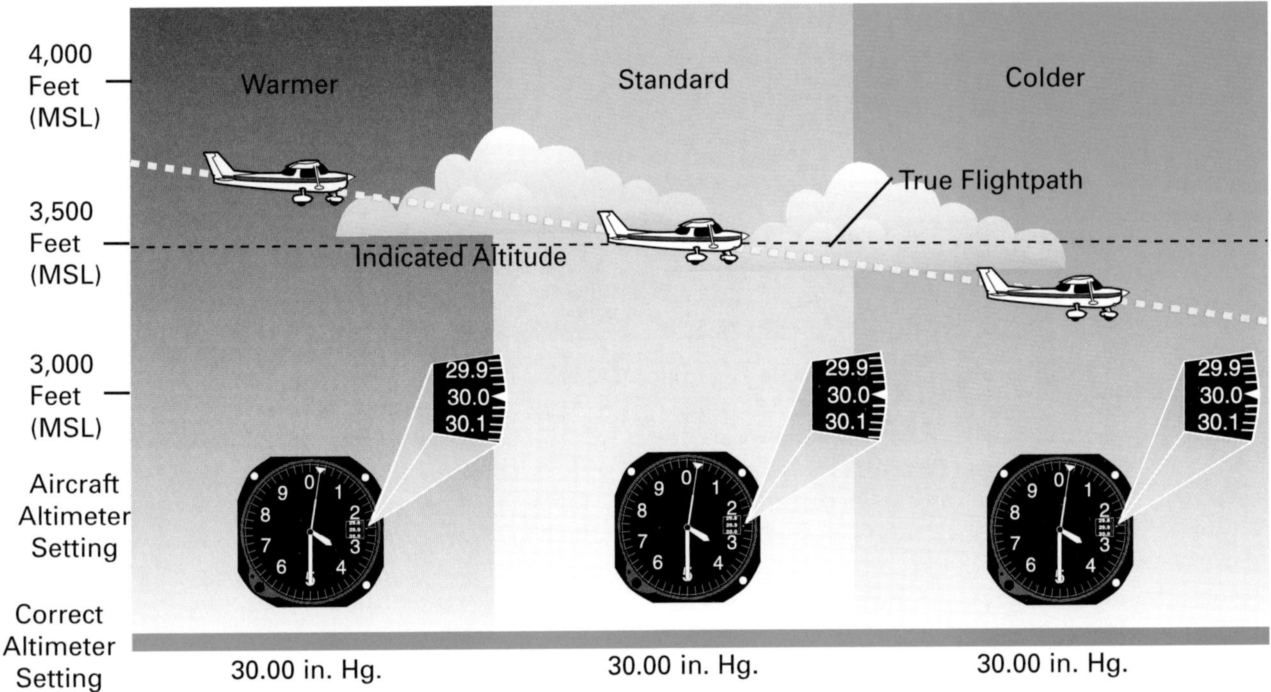

Figure 3-12. Flight cross section from a warmer to a colder airmass. Notice that standard atmospheric conditions only occur near the center of the diagram. The aircraft true altitude is higher than indicated in the warmer air and lower than indicated in the colder air.

On warm days pressure surfaces are raised and the indicated altitude is lower than true altitude.

this text and the reader is referred to other sources for details. (Jeppesen, 1995)

Sea level pressures and altimeter settings at airports and weather stations are made available with other weather information in standard coded formats. Some of these are shown in figure 3-13.

Figure 3-13. Reporting codes for aviation weather information. Report type, date, time, pressure, temperature, and altimeter settings are explained here. Other items such as wind, sky cover, weather, visibility, and dewpoint will be explained in subsequent chapters. Complete code breakdowns are given in the Appendix.

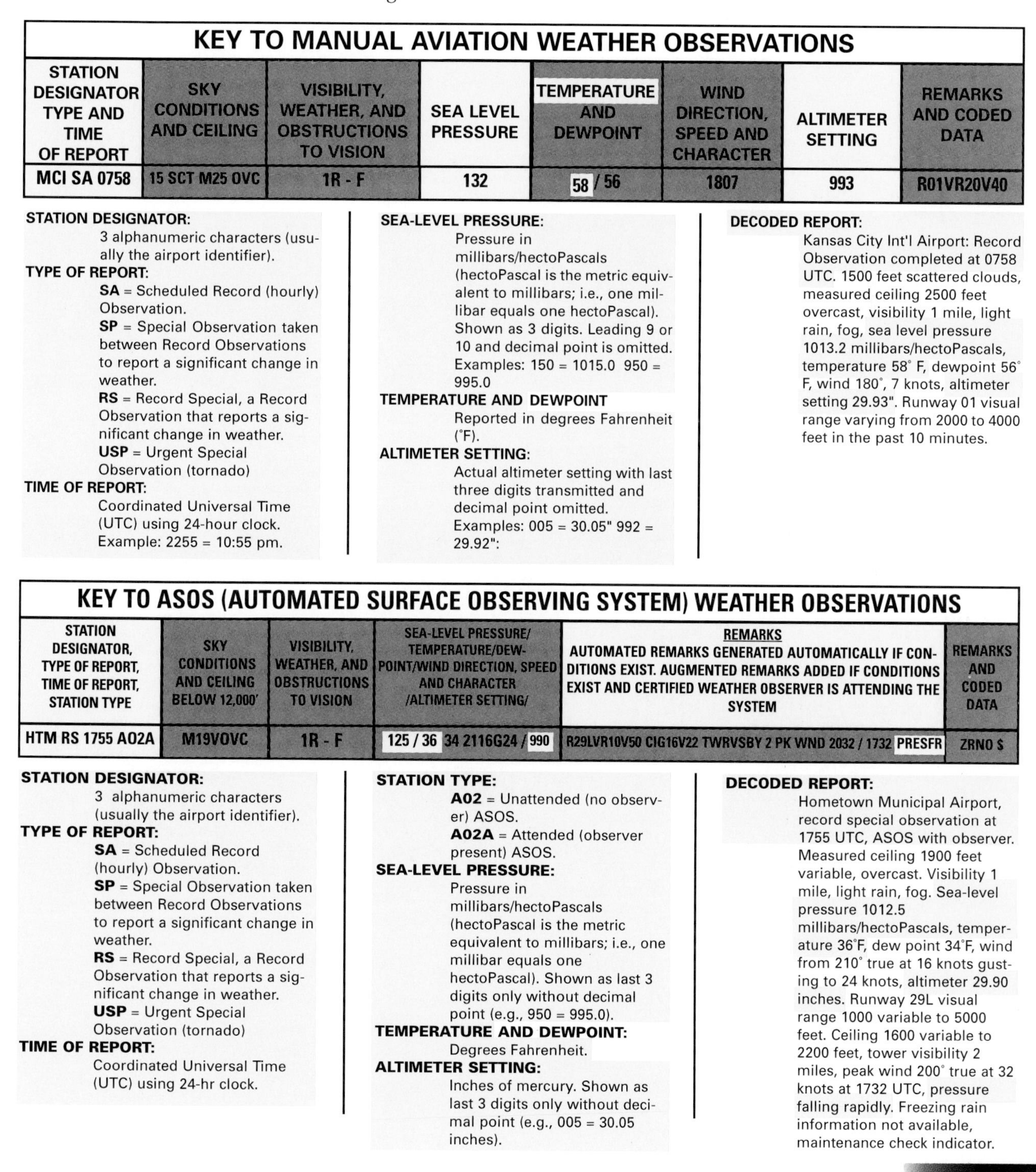

KEY TO MANUAL AVIATION WEATHER OBSERVATIONS

STATION DESIGNATOR TYPE AND TIME OF REPORT	SKY CONDITIONS AND CEILING	VISIBILITY, WEATHER, AND OBSTRUCTIONS TO VISION	SEA LEVEL PRESSURE	TEMPERATURE AND DEWPOINT	WIND DIRECTION, SPEED AND CHARACTER	ALTIMETER SETTING	REMARKS AND CODED DATA
MCI SA 0758	15 SCT M25 OVC	1R - F	132	58 / 56	1807	993	R01VR20V40

STATION DESIGNATOR:
3 alphanumeric characters (usually the airport identifier).

TYPE OF REPORT:
SA = Scheduled Record (hourly) Observation.
SP = Special Observation taken between Record Observations to report a significant change in weather.
RS = Record Special, a Record Observation that reports a significant change in weather.
USP = Urgent Special Observation (tornado)

TIME OF REPORT:
Coordinated Universal Time (UTC) using 24-hour clock.
Example: 2255 = 10:55 pm.

SEA-LEVEL PRESSURE:
Pressure in millibars/hectoPascals (hectoPascal is the metric equivalent to millibars; i.e., one millibar equals one hectoPascal). Shown as 3 digits. Leading 9 or 10 and decimal point is omitted.
Examples: 150 = 1015.0 950 = 995.0

TEMPERATURE AND DEWPOINT
Reported in degrees Fahrenheit (°F).

ALTIMETER SETTING:
Actual altimeter setting with last three digits transmitted and decimal point omitted.
Examples: 005 = 30.05" 992 = 29.92":

DECODED REPORT:
Kansas City Int'l Airport: Record Observation completed at 0758 UTC. 1500 feet scattered clouds, measured ceiling 2500 feet overcast, visibility 1 mile, light rain, fog, sea level pressure 1013.2 millibars/hectoPascals, temperature 58° F, dewpoint 56° F, wind 180°, 7 knots, altimeter setting 29.93". Runway 01 visual range varying from 2000 to 4000 feet in the past 10 minutes.

KEY TO ASOS (AUTOMATED SURFACE OBSERVING SYSTEM) WEATHER OBSERVATIONS

STATION DESIGNATOR, TYPE OF REPORT, TIME OF REPORT, STATION TYPE	SKY CONDITIONS AND CEILING BELOW 12,000'	VISIBILITY, WEATHER, AND OBSTRUCTIONS TO VISION	SEA-LEVEL PRESSURE/ TEMPERATURE/DEW-POINT/WIND DIRECTION, SPEED AND CHARACTER /ALTIMETER SETTING/	REMARKS AUTOMATED REMARKS GENERATED AUTOMATICALLY IF CONDITIONS EXIST. AUGMENTED REMARKS ADDED IF CONDITIONS EXIST AND CERTIFIED WEATHER OBSERVER IS ATTENDING THE SYSTEM	REMARKS AND CODED DATA
HTM RS 1755 A02A	M19VOVC	1R - F	125 / 36 34 2116G24 / 990	R29LVR10V50 CIG16V22 TWRVSBY 2 PK WND 2032 / 1732 PRESFR	ZRNO $

STATION DESIGNATOR:
3 alphanumeric characters (usually the airport identifier).

TYPE OF REPORT:
SA = Scheduled Record (hourly) Observation.
SP = Special Observation taken between Record Observations to report a significant change in weather.
RS = Record Special, a Record Observation that reports a significant change in weather.
USP = Urgent Special Observation (tornado)

TIME OF REPORT:
Coordinated Universal Time (UTC) using 24-hr clock.

STATION TYPE:
A02 = Unattended (no observer) ASOS.
A02A = Attended (observer present) ASOS.

SEA-LEVEL PRESSURE:
Pressure in millibars/hectoPascals (hectoPascal is the metric equivalent to millibars; i.e., one millibar equals one hectoPascal). Shown as last 3 digits only without decimal point (e.g., 950 = 995.0).

TEMPERATURE AND DEWPOINT:
Degrees Fahrenheit.

ALTIMETER SETTING:
Inches of mercury. Shown as last 3 digits only without decimal point (e.g., 005 = 30.05 inches).

DECODED REPORT:
Hometown Municipal Airport, record special observation at 1755 UTC, ASOS with observer. Measured ceiling 1900 feet variable, overcast. Visibility 1 mile, light rain, fog. Sea-level pressure 1012.5 millibars/hectoPascals, temperature 36°F, dew point 34°F, wind from 210° true at 16 knots gusting to 24 knots, altimeter 29.90 inches. Runway 29L visual range 1000 variable to 5000 feet. Ceiling 1600 variable to 2200 feet, tower visibility 2 miles, peak wind 200° true at 32 knots at 1732 UTC, pressure falling rapidly. Freezing rain information not available, maintenance check indicator.

KEY TO AWOS (AUTOMATED WEATHER OBSERVING SYSTEM) OBSERVATIONS

STATION DESIGNATOR, TYPE OF REPORT, TIME OF REPORT, STATION TYPE	SKY CONDITIONS AND CEILING BELOW 12,000'	VISIBILITY	TEMPERATURE / DEWPOINT/WIND DIRECTION, SPEED AND CHARACTER/ ALTIMETER SETTING /	REMARKS AUTOMATED REMARKS GENERATED AUTOMATICALLY IF CONDITIONS EXIST. AUGMENTED REMARKS ADDED IF CONDITIONS EXIST AND CERTIFIED WEATHER OBSERVER IS ATTENDING THE SYSTEM
HTM RS 1755 AWOS	M20 OVC	1V	36 / 34 / 2015G25 / 990	P010 / VSBY 1/2V2 WND 17V23 / WEA: RF

LOCATION IDENTIFIER:
3 alphanumeric characters (usually the airport identifier).

TYPE OF REPORT:
SA = Scheduled record (routine) observation. All observations identified as SA. Most are transmitted at 20-minute intervals (approximately 15, 35, and 55 minutes past each hour).

TIME OF REPORT:
Coordinated Universal Time (UTC) using 24-hour clock.

STATION TYPE:
AWOS = Automated Weather Observing System site. **Note:** In the future, some systems will use "AO" designators.

TEMPERATURE AND DEWPOINT:
Reported in degrees Fahrenheit.

ALTIMETER SETTING:
Inches of mercury. Shown as last 3 digits only without decimal point (e.g., 30.05 inches = 005).

DECODED REPORT:
Hometown Municipal Airport, observation at 1755 UTC, AWOS report. Measured ceiling 2000 feet overcast. Visibility 1 mile variable. Temperature 36 degrees (F), dewpoint 34 degrees (F), wind from 200 degrees true at 15 knots gusting to 25 knots, altimeter setting 29.90 inches. Precipitation accumulation during past hour 0.10 inch. Visibility variable between 1/2 and 2 miles. Wind direction variable from 170 degrees to 230 degrees true. Observer reports light rain (R-) and fog (F).

KEY TO METAR (NEW AVIATION ROUTINE WEATHER REPORT) OBSERVATIONS

TYPE OF REPORT	STATION DESIGNATOR, TIME OF REPORT	WIND	VISIBILITY,	WEATHER AND OBSTRUCTIONS TO VISIBILITY	SKY CONDITIONS	TEMPERATURE / DEWPOINT	ALTIMETER SETTING	REMARKS
METAR	KSEA 1250Z	08032G45KT	1/2SM R32L/ 1200 FT	TSRA	SCT 008 OVC012CB	15 / 08	A2995	RMK

TYPE OF REPORT:
There are two types of report - the METAR which is a routine observation report and SPECI which is a Special METAR weather observation. The type of report, METAR or SPECI, will always appear in the report header or lead element of the report.

STATION DESIGNATOR:
The METAR code uses ICAO 4-letter station identifiers. In the contiguous 48 states, the 3-letter domestic station identifier is prefixed with a "K"; i.e., the domestic identifier for Seattle is SEA while the ICAO identifier is KSEA. Elsewhere, the first two letters of the ICAO identifier indicate what region of the world and country (or state) the station is in. For Alaska, all station identifiers start with "PA"; for Hawaii, all station identifiers start with "PH."

TIME:
The time the observation is taken is transmitted as a four digit time group appended with a Z to denote Coordinated Universal Time (UTC). Example: 1250Z.

TEMPERATURE/DEWPOINT:
Temperature and dewpoint are reported in a two-digit form in degrees Celsius. Temperatures below zero are prefixed with an "M."
Examples:
15/08 - temperature 15 degrees, dewpoint 8 degrees
00/M02 - temperature zero degrees, dewpoint minus 2 degrees

ALTIMETER:
Altimeter settings are reported in a four-digit format in inches of mercury prefixed with an "A" to denote the units of pressure.
Example:
A2995 - twenty-nine point nine-five inches of mercury

DECODED REPORT:
Routine observation report for Seattle, Washington at 1250 UTC. Wind from 080 degrees at 32 knots with gusts to 45 knots. Visibility 1/2 statute mile, runway 32 left visual range 1,200 feet, thunderstorm with moderate rain. There are scattered clouds at 800 feet and overcast cumulonimbus clouds at 1,200 feet. Temperature 15 degrees Celsius, dewpoint 8 degrees Celsius, altimeter setting 29.95 inches of mercury. Remarks: recent weather event, thunderstorm began 24 minutes past the hour, rain began 24 minutes past the hour.

Section D

DENSITY

As you know from your studies of the physics of flight, aircraft performance depends critically on atmospheric density. An aircraft operating at 20,000 feet MSL in the standard atmosphere has about one half of the density as at sea level. At 40,000 feet MSL, this decreases to approximately one quarter of the sea level value, and to about one tenth at 60,000 feet MSL. At high atmospheric densities, aircraft performance is enhanced, while at low atmospheric densities, performance deteriorates.

These problems can be handled to some extent with special aircraft and powerplant designs and by attention to aircraft operation. One example is the ER-2, the stratospheric reconnaissance aircraft. With its glider-like design, it handles much better in a high altitude, low density environment than does a conventionally designed aircraft.

Pressure altitude and density altitude have the same value at standard temperature.

Density altitude increases about 120 feet (above pressure altitude) for every one C° increase in temperature above standard. It can be calculated on your flight computer or from a density altitude chart.

DENSITY ALTITUDE

Difficulties with flight in low density conditions are not restricted to extreme altitudes. There is a significant deterioration of performance for aircraft operating in high surface temperature conditions, especially at airports with elevations well above sea level. Specifically, longer-than-usual takeoff rolls are required and climbouts are slower than at sea level. These conditions are usually described in terms of the density altitude of the atmosphere in the vicinity of the airport.

Density altitude is the altitude above mean sea level at which a given atmospheric density occurs in the standard atmosphere. It can also be interpreted as pressure altitude corrected for nonstandard temperature differences.

In warmer-than-standard surface conditions, you would say that the density altitude is "high"; that is, operation of your aircraft in a high density altitude condition is equivalent to taking off from a higher airport during standard conditions. In a high density altitude situation, the actual density at the surface is found well above the airport altitude in the standard atmosphere. (Figure 3-14)

The precise effects of a given density altitude on takeoff distance and climb rate are presented in most aircraft flight manuals. For example, under the conditions given here, the takeoff distance for a light aircraft (not supercharged) would be increased by about 60% and the climb rate would be decreased by 40% of that required at sea level. Obviously, a short runway on a hot day in a high, narrow mountain valley offers large problems.

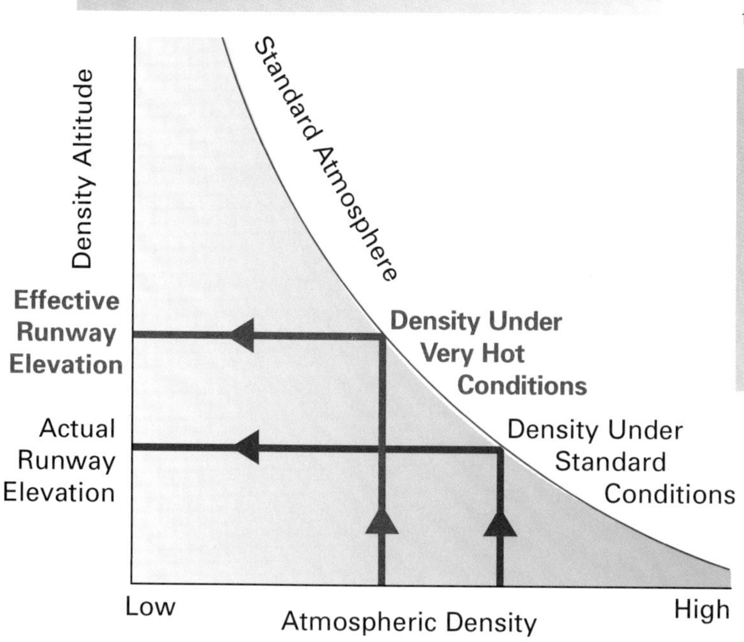

Figure 3-14. The heavy, solid curve shows how density decreases with altitude in the standard atmosphere. Under standard conditions, the surface density would correspond with a density altitude equal to the elevation of the airport. Under very warm conditions, the density is lower and it corresponds with a higher density altitude.

SUMMARY

Atmospheric pressure is an essential component of aviation weather basics. An understanding of pressure is the foundation for understanding such diverse and important topics as altimetry, winds, and storms. In this chapter, you have learned about the useful relationship between atmospheric pressure and the weight of the atmosphere and how that relationship allows us to measure pressure and altitude. Details about the distribution of average sea level pressure around the globe, as well as the terminology and methods for the interpretation of atmospheric pressure charts at the surface and aloft should now be part of your growing knowledge of aviation weather.

You have gained valuable insight into the effects of atmospheric variations in pressure and temperature on the accuracy of pressure altimeter measurements. You have been introduced to standard weather reports available to pilots. In particular, you have learned where to find locations of the reporting stations, times, pressures, temperatures, and altimeter settings in those reports. Finally, you have become familiar with the concept of density altitude and its impact on aircraft performance.

KEY TERMS

Absolute Altitude
Altimeter Errors
Altimeter Setting
Aneroid Barometer
Aneroid Cell
Atmospheric Pressure
Constant Pressure Charts
Contours
Density Altitude
Height Gradient
High
Hydrostatic Balance
Indicated Altitude
Isobars
Low
Mercurial Barometer
Pressure Altimeter
Pressure Altitude
Pressure Gradient
Ridge
Sea Level Pressure
Station Pressure
Surface Analysis Chart
Trough
True Altitude

CHAPTER QUESTIONS

1. Under certain conditions, your ears act as sensitive aneroid barometers and can cause discomfort. Document those conditions and explain the procedures for minimizing the problem.

2. What is the pressure exerted by an atmospheric column over one square foot at sea level in the standard atmosphere?

3. Even if a mercurial barometer could be designed so that it was not so bulky and fragile, it wouldn't work well in an airplane. Discuss.

4. You are flying at an indicated altitude of 3,000 feet MSL over a region where there is a strong high pressure area at the surface. Sea level pressure is 1046 mb. Your altimeter is set at 29.92. For simplicity, assume that the atmosphere is at standard temperature and there are no other errors in measurement. What is your true altitude?

5. Realistically, the situation in question 4 is commonly associated with a shallow, very cold airmass in winter. Discuss this added effect on altitude measurements with a pressure altimeter.

6. Find the range (high and low) of average sea level pressure over the earth's surface from figure 3-6. Convert the pressure to inches of mercury. Now assume that weather disturbances move across an airport (actual field elevation 1,000 feet MSL), causing the pressure to vary between the highest and lowest values of sea level pressure. Except for pressure changes, assume the atmosphere is standard. What errors in field elevation would arise if your altimeter remained at 29.92 inches?

7. Calculate the density altitude for an airport located at 2,000 feet MSL with an altimeter setting of 29.92 and a temperature of 95°F. Use a flight computer or density altitude chart.
    1. Do the computation for the same airport, but for a temperature of 104°F and an altimeter setting of 29.80 in. Hg.
    2. Research and determine the increase in takeoff distance and the decrease in climb rate for an aircraft specified by your instructor.

8. Why can't a water pump raise water higher than about 34 feet under standard atmospheric conditions?

CHAPTER 4

Wind

Introduction

The motion of air is important in many weather-producing processes. Moving air carries heat, moisture, and pollutants from one location to another — at times in a gentle breeze, occasionally in a pure hurricane. Air movements create favorable conditions for the formation and dissipation of clouds and precipitation; in some cases, those motions cause the visibility to decrease to zero; in others, they sweep the skies crystal clear. Winds move atmospheric mass and therefore affect changes in atmospheric pressure. As you will see, these pressure changes modify the winds. All of these factors create reasons for the changeable nature of not only the wind, but also weather.

In flight, winds can have significant effects on navigation. Chaotic air motions cause turbulence which, at least, is uncomfortable and, at worst, can be catastrophic. Should atmospheric winds change suddenly over a short distance, flight may not be sustainable. Without question, as a pilot, you must understand air motions for efficient and safe flight.

In this chapter, we consider the causes and characteristics of horizontal motions of the atmosphere. The chapter material provides you with a practical understanding of important relationhips between the wind, atmospheric pressure, and the earth's rotation. You will also gain some insight into the important influences of friction between the moving air and the earth's surface. When you complete the chapter, you will not only have an understanding of the fundamental causes of the wind, but you will also know how wind is measured and you will be able to interpret general wind conditions from isobars and contours on weather charts.

Section A

WIND TERMINOLOGY, CODES, AND MEASUREMENTS

To a pilot, the concept of motion in three dimensions comes much easier than to ground-bound people. For example, after takeoff and during climbout, you are aware of your movement across the ground as well as your climbing attitude. Similarly, when air moves from one location to another, it can simultaneously move both horizontally and vertically.

As the pilot finds it convenient to describe and measure aircraft position changes and altitude changes separately, so does the meteorologist find it helpful to separate descriptions of horizontal air movements and vertical air motions. A practical reason for this separation is that horizontal motions are much stronger than vertical motions with the exceptions of a few turbulent phenomena, such as thunderstorms and mountain lee waves. Also, horizontal motions are easier to measure than vertical motions. We separate them here, reserving the term wind for horizontal air motions. Vertical motions will be discussed in the next chapter. (Figure 4-1)

It is common to refer to the "wind velocity" when describing the wind. This term is often erroneously interpreted as "wind speed." This is not the case. Wind velocity is a vector quantity. A vector quantity has a magnitude and a direction, as opposed to scalar quantities, such as temperature and pressure, which only have magnitude. The magnitude of the wind velocity is the wind speed, usually expressed in nautical miles per hour (knots), statute miles per hour (m.p.h.), kilometers per hour, or meters per second. The wind direction is the direction from which the wind is blowing, measured in degrees, or to eight or sixteen points of the compass, clockwise from true

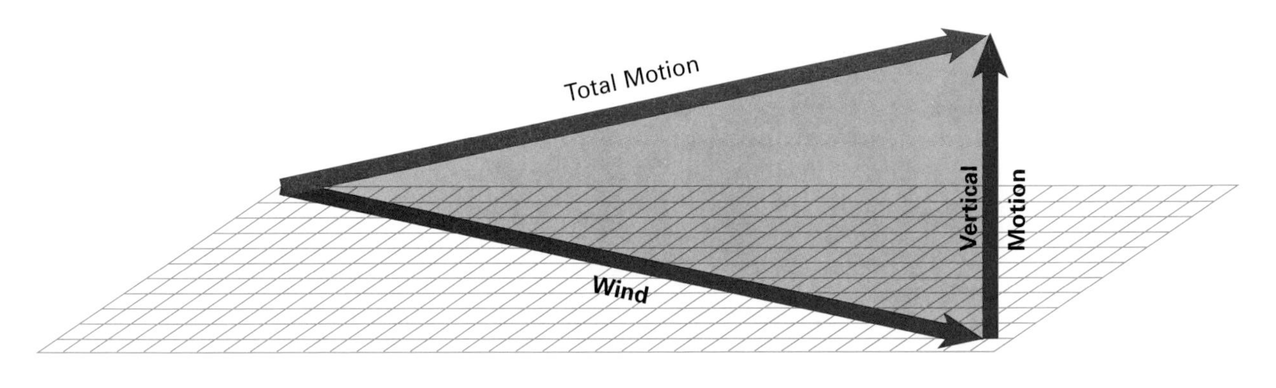

Figure 4-1 The total atmospheric motions are composed of horizontal motions (wind) and vertical motions.

north (360°). For example, a "westerly" wind blows from the west and has a direction of 270°. Note that meteorologists *always* state their wind directions relative to true north ("true") rather than magnetic north ("magnetic"). Air traffic controllers, on the other hand, always state wind direction in terms of magnetic north, unless specifically stated otherwise.

Wind velocity is measured at the surface by several different methods. The most common techniques use anemometers and wind vanes. (Figure 4-2) For winds aloft, measurement techniques include the tracking of free balloons, doppler radar, aircraft navigation systems, and satellite.

Figure 4-2. Two surface wind measurement systems. On the left, wind speed is determined from the rotation speed of a cup anemometer while the direction is measured with a wind vane. Note, the vane points *into* the wind. On the right, wind speed and direction are determined from a single airplane-shaped device known as an aerovane. The speed of rotation of the propeller determines the wind speed while the "airplane" also flies *into* the wind.

In weather reports of surface weather conditions, several different wind reports may be given. (Figure 4-3) All of these relate to winds measured by a standard instrument which is usually located 30 feet above the ground and away from any obstructions. Reported wind speeds and directions are usually one- or two-minute averages. This average wind speed is also referred to as the sustained speed. A gust is reported when there is at least a ten-knot variation between instantaneous peaks and lulls during the last ten minutes. A squall is reported when there is a sudden increase of wind speed by at least 15 knots to a sustained speed of 20 knots or more for a one-minute period. The peak wind speed is the maximum wind speed since the last hourly observation.

KEY TO MANUAL AVIATION WEATHER OBSERVATIONS (SA)

STATION DESIGNATOR TYPE AND TIME OF REPORT	SKY CONDITIONS AND CEILING	VISIBILITY, WEATHER, AND OBSTRUCTIONS TO VISION	SEA-LEVEL PRESSURE	TEMPERATURE AND DEW POINT	WIND DIRECTION, SPEED AND CHARACTER	ALTIMETER SETTING	REMARKS AND CODED DATA
MCI SA 0758	15 SCT M15 OVC	1R - F	132	/58/56	/1807	/993/	R01VR20V40

WIND DIRECTION, SPEED & CHARACTER:
Direction in tens of degrees from true north, speed in knots. 0000 = calm. G = gusts. Q = squalls. Peak speed of gusts in the past ten minutes follows G or Q. WSHFT in Remarks = windshift occurred at time indi cated. Example: 3627G40 = 360° at 27 peak gusts 40 knots.

DECODED REPORT:
Kansas City Int'l Airport: Record Observation completed at 0758 UTC. 1500 feet scattered clouds, measured ceiling 2500 feet over cast, visibility 1 mile, light rain, fog, sea level pressure 1013.2 millibars/hectoPascals, tempera ture 58°F, dewpoint 56°F, wind 180°, 7 knots, altimeter setting 29.93". Runway 01 visual range varying from 2000 to 4000 feet in the past 10 minutes.

KEY TO ASOS (AUTOMATED SURFACE OBSERVING SYSTEM) WEATHER OBSERVATIONS

STATION DESIGNATOR, TYPE OF REPORT, TIME OF REPORT, STATION TYPE	SKY CONDITION AND CEILING BELOW 12,000	VISIBILITY, WEATHER AND OBSERVATIONS TO VISION	SEA-LEVEL PRESSURE/ TEMPERATURE / DEWPOINT / WIND DIRECTION, SPEED AND CHARACTER ALTIMETER SETTING	REMARKS: AUTOMATED REMARKS GENERATED AUTOMATICALLYIF CONDITIONS EXIST. AUGMENTED REMARKS ADDED IF CONDITIONS EXIST AND CERTIFIED WEATHER OBSERVER IS ATTENDING THE SYSTEM	STATUS REMARKS SYSTEM GENERATED
HTM RS 1755 AO2A	M19V OVC	1R - F	125/36/34/2116G24/990/	R29LVR10V50 CIG 16V22 TWR VSBY 2 PK WIND 2032 1732 PRESFR	ZRNO $

WIND DIRECTION, SPEED AND CHARACTER:
Direction in tens of degrees from **true** north. Voice broadcast in degrees from **magnetic**. Speed in knots. **0000** = calm. **E** = estimated. **G** = gusts. **Q** = squalls. Variable wind, peak wind, wind shift: see Remarks.

DECODED REPORT:
Hometown Municipal Airport, record special observation at 1755 UTC, ASOS with observer. Measured ceiling 1900 feet variable, overcast. Visibility 1 mile, light rain, fog. Sea-level pressure 1012.5 millibars/hectoPascals, temperature 36°F, dewpoint 34°F, wind from 210° true at 16 knots gusting to 24 knots, altimeter 29.90 inches. Runway 29L visual range 1000 variable to 5000 feet. Ceiling 1600 variable to 2200 feet, tower visibility 2 miles, peak wind 200° true at 32 knots at 1732 UTC. pressure falling rapidly. Freezing rain information not available, maintenance check indicator.

Figure 4-3. Reporting codes for aviation weather information. Wind reports are explained here. Other items are explained in previous and subsequent chapters. The complete code for these reports is located in the appendix.

KEY TO AWOS (AUTOMATED WEATHER OBSERVING SYSTEM) OBSERVATIONS

STATION DESIGNATOR, TYPE OF REPORT, TIME OF REPORT, STATION TYPE	SKY CONDITION AND CEILING BELOW 12,000′	VISIBILITY	TEMPERATURE / DEWPOINT / WIND DIRECTION, SPEED AND CHARACTER ALTIMETER SETTING	REMARKS: AUTOMATED REMARKS GENERATED AUTOMATICALLY IF CONDITIONS EXIST. AUGMENTED REMARKS ADDED IF CONDITIONS EXIST AND CERTIFIED WEATHER OBSERVER IS ATTENDING THE SYSTEM
HTM SA 1755 AWOS	M20 OVC	1V	36/34/2015G25/990	P010/VSBY 1/2 V2 WND 17V23/WEA: R-F

WIND DIRECTION, SPEED & CHARACTER:

Direction in tens of degrees from true north, except voice broadcast is in degrees magnetic. Speed in knots. **0000** = calm. **G** = gusts. See Automated Remarks for variable direction.

AUTOMATED REMARKS:

WND V = variable wind direction.

DECODED REPORT:

Hometown Municipal Airport, observation at 1755 UTC, AWOS report. Measured ceiling 2000 feet overcast. Visibility 1 mile variable. Temperature 36 degrees (F), dewpoint 34 degrees (F), wind from 200 degrees true at 15 knots gusting to 25 knots, altimeter setting 29.90 inches. Precipitation accumulation during past hour 0.10 inch. Visibility variable between 1/2 and 2 miles. Wind direction variable from 170 degrees to 230 degrees true. Observer reports light rain (R–) and fog (F).

KEY TO METAR (NEW AVIATION ROUTINE WEATHER REPORT) OBSERVATIONS

TYPE OF REPORT	STATION DESIGNATOR, TIME OF REPORT	WIND	VISIBILITY,	WEATHER, AND OBSTRUCTIONS TO VISIBILITY	SKY CONDITIONS	TEMPERATURE / DEWPOINT	ALTIMETER SETTINGS	REMARKS
METAR	KSEA 125ØZ	Ø8Ø32G45KT	1/2SM R32L/12ØØFT	TSRA	SCTØØ8 OVCØ12CB	15 / Ø8	A2995	RMK RE TSB24 RAB24

WIND:

The wind is reported as a five digit group (six digits if speed is over 99 knots). The first three digits is the direction the wind is <u>blowing from</u> in ten's of degrees, or "VRB" if the direction is variable. The next two digits is the speed in knots, or if over 99 knots, the next three digits. If the wind is gusty, it is reported as a "G" after the speed followed by the highest gust reported.

Examples:

13008KT - wind from 130 degrees at 8 knots

08032G45KT - wind from 080 degrees at 32 knots with gusts to 45 knots.

VRB04KT - wind variable in direction at 4 knots

00000KT - wind calm

210103G130KT - wind from 210 degrees at 103 knots with gusts to 130 knots.

If the wind direction is variable by 60 degrees or more and the speed is greater than 6 knots, a variable group consisting of the extremes of the wind direction separated by a "V" will follow the prevailing wind group.

Example:

32012G22KT 280V350

DECODED REPORT:

Routine observation report for Seattle, Washington at 1250 UTC. Wind from 080 degrees at 32 knots with gusts to 45 knots. Visibility 1/2 statute mile, runway 32 left visual range 1,200 feet, thunderstorm with moderate rain. There are scattered clouds at 800 feet and overcast cumulonimbus clouds at 1,200 feet. Temperature 15 degrees Celsius, dewpoint 8 degrees Celsius, altimeter setting 29.95 inches of mercury. Remarks: recent weather event, thunderstorm began 24 minutes past the hour, rain began 24 minutes past the hour.

Section B

CAUSES OF WIND

What makes the wind blow? The concise answer is found in the basic physical principle that governs all motions; that is, the conservation of momentum. Newton stated this principle quite simply:

If an object of mass, M, is subjected to an unbalanced force (F_{total}), it will undergo an acceleration, A, that is:

$$F_{total} = MA$$

We already applied this principle when we defined atmospheric pressure as the weight of the atmosphere. To do this, we took advantage of the fact that the atmosphere is in hydrostatic balance. That condition is a special case of the above statement of Newton's principle. A mass of air is not accelerated either upward or downward if the total forces acting on it are balanced.

With regard to horizontal motions of air, imbalances are common and horizontal accelerations occur. In this case, the acceleration is the change of the speed and/or direction of a mass of air as it moves along its path. F_{total} is the sum of all of the horizontally directed forces which act on a particular mass of air. Figure 4-4 illustrates how a parcel is affected if it is under the influence of balanced and unbalanced forces.

With this information, our question, "What makes the wind blow?" can be stated better as two questions: What are the forces that affect the air parcels? What are the causes of the forces?

The most important forces that affect air motions are

1. Pressure gradient force
2. Coriolis force
3. Frictional force

Because a "mass" of air is not an easy thing to visualize, meteorologists have found it useful to introduce the concept of an air parcel. An air parcel is a volume of the atmosphere that is small enough so that its mass can be treated as if it were concentrated at a single point. Practically speaking, you can think of an air parcel as being "bigger than a bread box, but smaller than a boxcar."

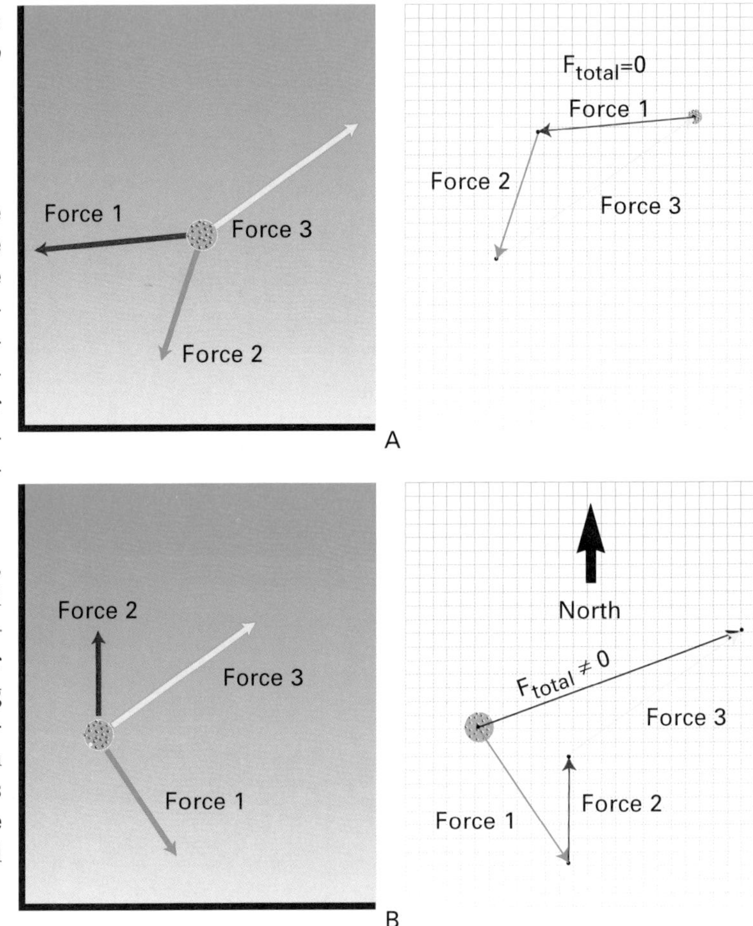

Figure 4-4. On the left are shown two parcels of air, each under the influence of three individual forces. Each force is represented as a vector with a magnitude and a direction. The force diagram constructed on the right shows the sum of the forces determined by simply adding the tail of one vector onto the head of the next. In diagram A, the head of the last vector ends up at the tail of the first. Therefore, the sum of the forces is zero. The parcel will not be accelerated. This is not true in diagram B. There is a gap between the first and last vector. The sum of the forces is not zero; therefore, the parcel is accelerated to the northeast, as indicated by the purple vector.

Section C

PRESSURE GRADIENT FORCE

The concept of a "gradient" was previously introduced in connection with discussions of temperature and pressure. Recall that a "pressure gradient" is simply the difference in pressure between two points divided by the distance between the points.

The fact that a pressure gradient has an influence on air movement is obvious when you deal with gases under pressure. For example, if you inflate a tire, you establish a pressure gradient across the thickness of the tire. If you puncture the tire, the air accelerates from the inside to the outside; that is, toward lower pressure. The larger the pressure difference, the greater the acceleration through the opening. The force involved here is known as the pressure gradient force. An example of this force and how it can be created in another fluid (water) is shown in figure 4-5.

In a similar way, the atmosphere causes air parcels to be accelerated across the surface of the earth toward low pressure; that is, when a horizontal pressure gradient force exists. Notice that, since we are dealing specifically with the wind, we only need to consider horizontal pressure differences. When you study vertical air motions later in the text, the vertical pressure gradient force will be considered.

CAUSES OF PRESSURE GRADIENTS

The horizontal pressure gradient force is a root cause of wind. While both Coriolis force and the frictional force require motion before they become effective, pressure gradient force does not. Since pressure gradients are so essential to air motion, it is helpful to know how they develop.

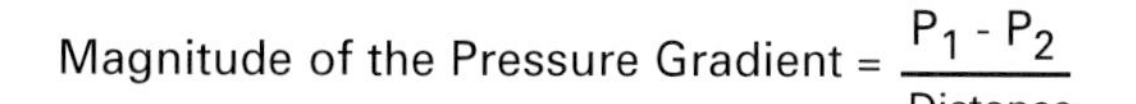

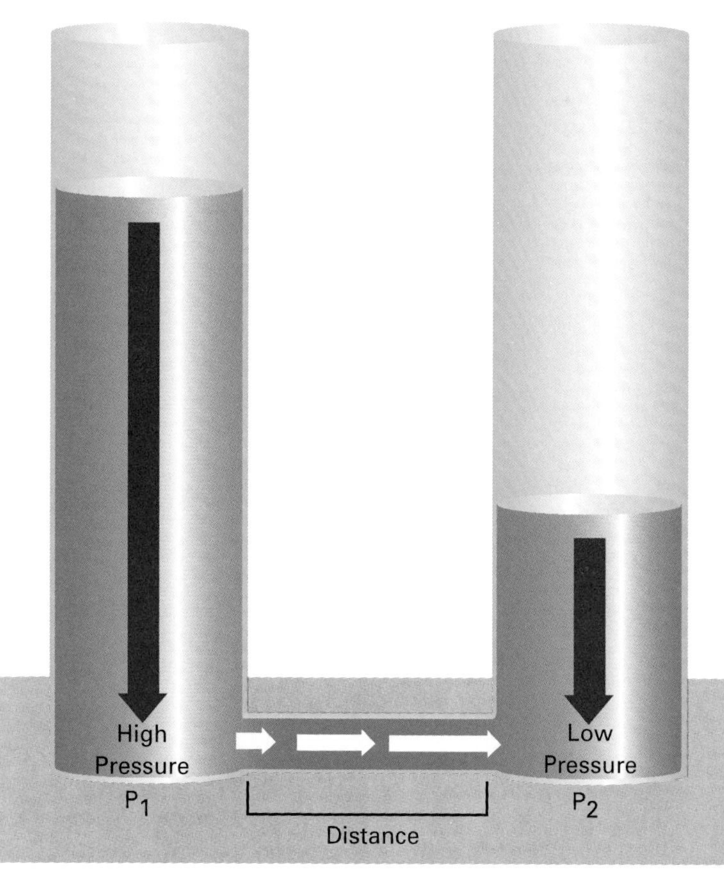

Figure 4-5. A pressure gradient develops in the pipe connecting the two reservoirs. This is due to the difference in water pressure generated by the difference in the depth (weight) of each water column, divided by the horizontal distance between the columns. The resulting pressure gradient force accelerates water through the pipe from the deep reservoir (high pressure) to the shallow reservoir (low pressure).

The upper diagram in figure 4-6 shows a coastline sometime in mid-morning in the summer when the temperature of the land and sea are equal. In this ideal situation, there is no horizontal pressure gradient and no air movement across the coastline. As the sun continues to heat the earth's surface later in the morning (lower diagram in figure 4-6) the land temperature exceeds the water temperature. This happens because of the high heat capacity of water compared to land. In Chapter 2, you saw that it took much more energy to raise the temperature of water than that of dry soil. This is an example of the creation of a horizontal temperature gradient by differential heating.

The warmer land surface heats the overlying air by conduction and convection. The result is that the column of air over the land swells. That is, the mass of the heated air expands into a deeper layer than an equivalent amount of mass in the cooler air column over water. This is expected since we know from Chapter 3 that pressure decreases more rapidly with height in cold air than in warm air. As shown in figure 4-6, this process causes a horizontal pressure gradient aloft; warm air at that level starts to move toward the lower pressure over the sea.

There is more. In figure 4-6, as soon as the mass leaves the upper part of the heated column, the weight of that column (measured at the surface) decreases and the surface pressure goes down over the land. This creates a second horizontal pressure gradient between the columns of air, except that this gradient is at the bottom of the columns where the lower pressure is over the land. Air at this level starts to move across the coastline from the sea toward the land.

This is an interesting and useful result. By simply creating a temperature difference between two locations where the air was originally at rest, the air has been caused to move in one direction aloft and in the opposite direction at the surface. In general, the movement of air which results from differential heating is called a thermal circulation. Thermal circulations have two horizontal branches; an upper branch which is called the return flow, and a lower branch. The example shown in figure 4-6 is a sea breeze. Note the name indicates the source of the lower branch of the circulation; that is, the sea breeze blows from the sea. Some other thermal circulations are the land breeze, mountain breeze, and valley breeze. These will be described in a later chapter.

Thermal circulations of the type described above occur over distances of ten miles to a hundred miles or so. On larger scales, circulations are found which also have their roots in differential solar heating; however, the warm and cold air masses created by this process are often carried far from their sources. Additionally, in these cases, the rotation of the earth is important. As you would expect, the result is a bit more complex; but, with the help of Newton and a few simple diagrams, understandable.

The pressure gradient force is always directed perpendicular to the isobars toward lower pressure.

In a thermal circulation, the stronger the temperature gradient, the stronger the pressure gradient, and the stronger the wind.

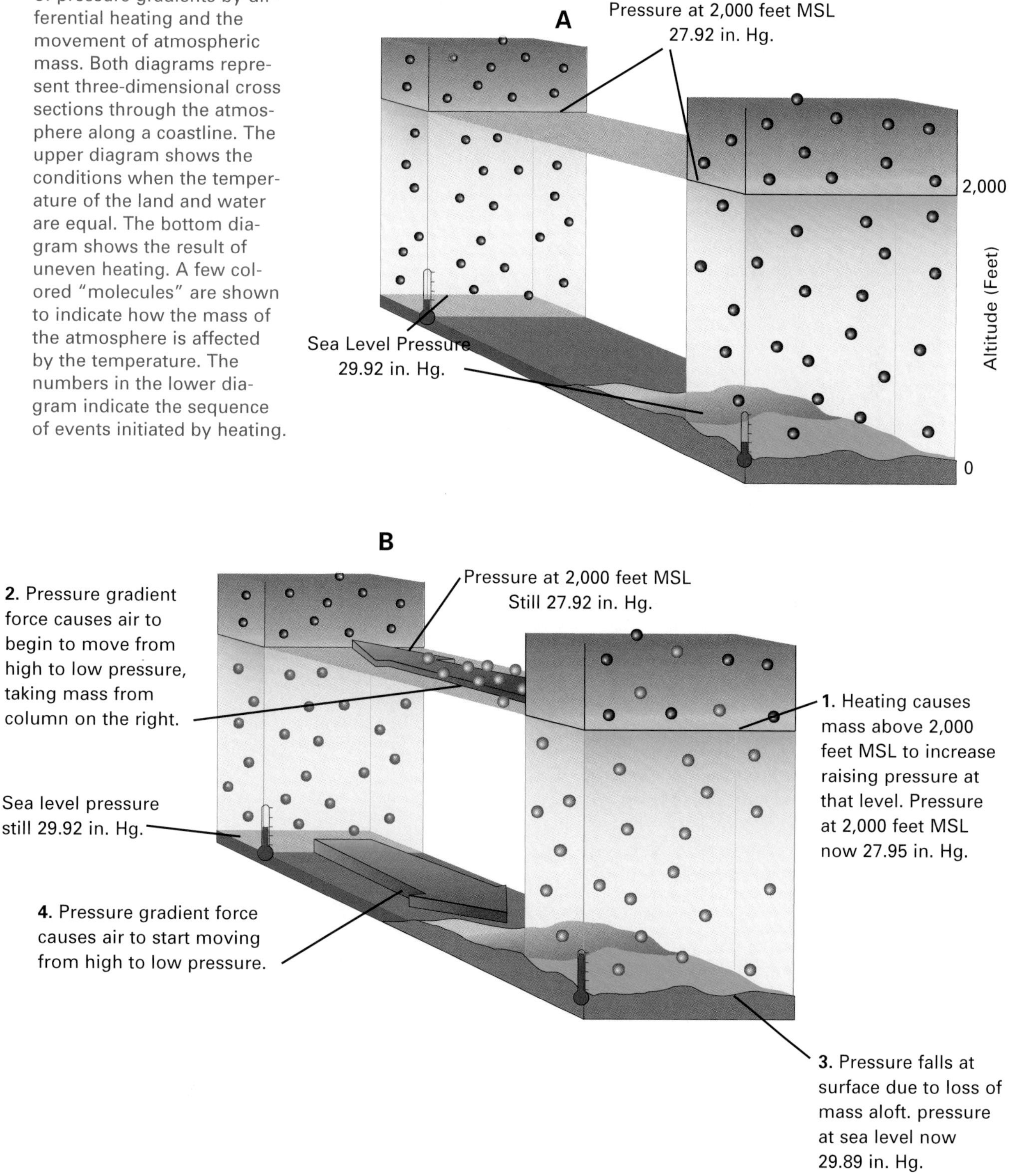

Figure 4-6. The development of pressure gradients by differential heating and the movement of atmospheric mass. Both diagrams represent three-dimensional cross sections through the atmosphere along a coastline. The upper diagram shows the conditions when the temperature of the land and water are equal. The bottom diagram shows the result of uneven heating. A few colored "molecules" are shown to indicate how the mass of the atmosphere is affected by the temperature. The numbers in the lower diagram indicate the sequence of events initiated by heating.

Section D

CORIOLIS FORCE

We live and fly in a rotating frame of reference. To us, the earth is fixed and the sun and stars move across the sky. Of course, you know that the movement of these celestial bodies is due to the rotation of the earth. Since we normally observe all motions from this rotating frame of reference, the effect of that rotation must be taken into account when we explain the observed motions. This is usually done by introducing the concept of Coriolis force, which is named for one of the first scientists to make an in-depth study of this effect.

Coriolis force affects all objects moving across the face of the earth. It influences ocean currents and the paths of airplanes. Most importantly, from our meteorological perspective, as soon as air begins to move, it is influenced by Coriolis force. Although a rigorous treatment of Coriolis force is beyond the scope of this book, some important properties can be demonstrated with a simple experiment.

In figure 4-7, A man is shown standing on a merry-go-round which is rotating counterclockwise (ccw). The direction of rotation is determined by looking at the merry-go-round from the top. Imagine you are standing on the opposite side of the merry-go-round throwing a ball to the person in figure 4-7. As shown in the picture, you observe that the ball misses the target to the right. To you, a "force" acts on the ball, causing it to accelerate; that is, to change from the intended direction by curving to the right.

A person standing off to the side of the merry-go-round observes that the ball flies in a straight line after it leaves your hand. That person sees immediately that the ball is not being deflected, rather it is the position of you and the catcher that changes during the time it takes the ball to travel across the merry-go-round.

If the rotation rate is increased, the position changes are greater and the deflection of the ball is greater, as observed by those on the merry-go-round. But to the observer

Figure 4-7. Coriolis force explains to the thrower why the ball appears to curve to the right of the intended target.

off to the side, the path is still straight. Note that if the direction of rotation is clockwise (cw), the deflection of the ball is to the left.

When we attempt to describe the motion of the atmosphere (or anything else) relative to the rotating earth, we must also consider Coriolis force. However, things become a little more involved because the earth is a rotating sphere, rather than a rotating disk. In the case of the merry-go-round, it did not matter where the thrower and the catcher were located on the rotating platform. For a fixed rotation rate and a constant speed of the ball, Coriolis force was the same everywhere on the platform. This is because the axis of rotation of the merry-go-round was vertical; that is, perpendicular to the platform across which the ball was moving. This is not the case with the earth. For a fixed rotation rate and speed of the ball, Coriolis force is different at different latitudes. The variation is illustrated in figure 4-8.

If, as shown in figure 4-8, our merry-go-round is attached to the earth at the North Pole, it rotates counterclockwise. Note that the axis of the earth and the axis of rotation of the merry-go-round are parallel at the pole. If we let the earth's rotation rate and the speed of the ball be the same as in the previous example, then (considering only Coriolis force) we would see the same effect on the ball. At the equator, the situation is different. In that location, the axis of the earth is perpendicular to the axis of the merry-go-round, so the merry-go-round does not rotate about its

Figure 4-8. For a given speed of the ball and a constant rotation rate, Coriolis force is a maximum at the poles and zero at the equator.

vertical axis. Coriolis force is zero at the equator and the ball moves in a straight line. Between equator and pole, Coriolis force decreases from a maximum at the poles to zero at the equator.

Fortunately, the earth rotates much more slowly than our merry-go-round (one rotation per 24 hours); therefore, Coriolis force is much weaker. This is why we do not see all baseballs curving to the right in the Northern Hemisphere and to the left in the Southern Hemisphere. The effect is there, but it is only significant when an object, such as a parcel of air, moves over large distances (several hundred miles or more), allowing the weak force time to act. In the next section, we will see the impact of Coriolis force on very large atmospheric circulations. For smaller distances and times (for example, in a sea breeze), other forces such as pressure gradient, are much stronger. In those cases, the deflective effect of Coriolis force is not very noticeable, if at all.

Another aspect of Coriolis force is that it is opposite in the Southern Hemisphere. Motion there is deflected to the left. This difference between hemispheres is understood when our view of the earth is taken from the South Pole; the earth has clockwise rotation.

The most important characteristics of the Coriolis force may be summarized as follows:

1. Coriolis force always acts 90° to the right of the wind in the Northern Hemisphere and 90° to the left in the Southern Hemisphere. Therefore, Coriolis force affects only wind direction, not wind speed.

2. Although Coriolis force does not affect the wind speed, it depends on the wind speed. It requires the air to be moving. If the wind speed is zero, the Coriolis force is zero. The greater the wind speed, the greater the Coriolis force.

3. Coriolis force depends on the latitude. For a given wind speed, Coriolis force varies from zero at the equator to a maximum at the poles.

4. Although Coriolis force affects air motion on all scales, in comparison to other forces, its effect is minimal for small scale circulations and very important for large scale wind systems.

Navigation across large distances requires corrections for the influence of Coriolis force. Aircraft tracks must be corrected to the left in the Northern Hemisphere and to the right in the Southern Hemisphere to counteract the deflection due to the earth's rotation. These curved paths, in turn, affect some instruments used for celestial navigation, requiring further corrections for position computations. (USAF, 1985).

Section E

GEOSTROPHIC BALANCE

A very useful characteristic of the atmosphere is that the pressure gradient force and the Coriolis force tend to balance each other when the scales of atmospheric circulations are large enough. This means that when air travels over distances of hundreds of miles or more (the farther the better), Coriolis and pressure gradient forces tend to be equal in magnitude, but opposite in direction. This condition is known as geostrophic balance. The related wind is the geostrophic wind. It is quite helpful in understanding the characteristics of wind and it provides a good approximation to the actual wind.

"Geostrophic" is a useful memory device because the root of the word literally means "earth-turning," an obvious reference to the Coriolis effect of the earth's rotation. Since Coriolis force depends on the wind speed, geostrophic balance can only happen when the wind is already blowing. Because Coriolis force always acts 90° to the right of the wind, this balance can only occur when the pressure gradient force acts 90° to the left of the wind. This condition is shown in figure 4-9. Some of the properties of the geostrophic wind are summarized as follows:

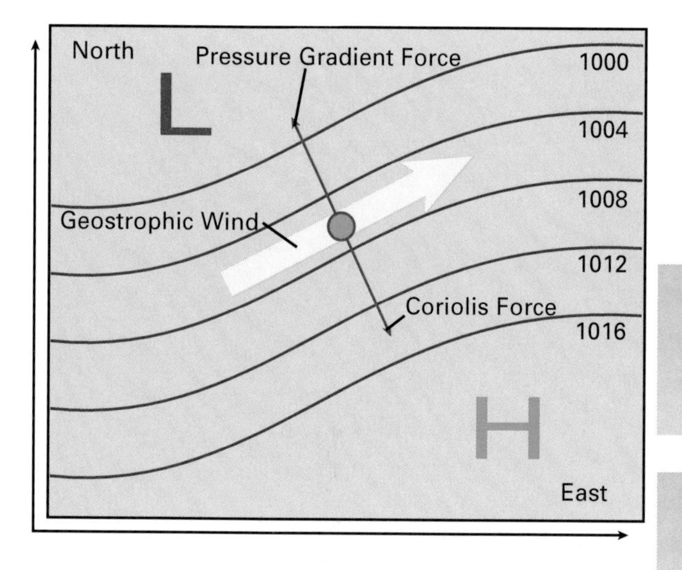

Figure 4-9. Solid purple lines are isobars (mb). Solid red arrows indicate forces acting on a parcel of air (small circle). The velocity of the parcel is indicated by the broad, yellow arrow. Geostrophic balance occurs when the pressure gradient and Coriolis forces are equal in magnitude and opposite in direction and are the only forces acting on the parcel of air.

1. The wind that blows under such a balance is parallel to the isobars (or contours on a constant pressure surface) with the lowest pressure on the left, looking downwind. This condition is easily remembered with a well-known rule of thumb (Buys-Ballot's Law): *With your back to the wind, the low pressure is on your left in the Northern Hemisphere.*

2. The stronger the pressure gradient, the stronger the wind speed.

3. In the Northern Hemisphere, winds tend to blow counterclockwise around low pressure centers (cyclones) and clockwise around high pressure centers (anticyclones). Therefore, counterclockwise motion is often described as cyclonic flow and clockwise motion as anticyclonic flow. The wind flow directions around lows and highs are reversed in the Southern Hemisphere.

4. Geostrophic balance does not occur in small scale circulations such as sea breezes, thunderstorms, tornadoes, and dust devils because the pressure gradient force is much greater than the Coriolis force.

When the isobars on the surface analysis chart are close together, the pressure gradient force is large and wind speeds are strong.

In the Northern Hemisphere, Coriolis force deflects the air to the right until it is parallel to the isobars.

Winds do not blow directly from large scale high pressure areas to low-pressure areas because of Coriolis force.

ESTIMATING WINDS FROM ISOBARS AND CONTOURS

The geostrophic wind is a practical tool for the interpretation of large scale weather charts. It allows you to estimate the winds from the pressure field. This is very convenient because, typically, there are more pressure and altimeter setting reports than there are direct wind measurements. Examples of the approximate agreement of observed winds with the pressure field (as you would expect with geostrophic balance) are shown in figure 4-10.

> **The 500 mb constant pressure chart is suitable for flight planning at FL 180. Observed temperature and wind information give approximate conditions along the proposed route.**

Several characteristics of geostrophic winds are apparent in figure 4-10: weak winds in areas with weak pressure gradients; strong winds in regions with strong pressure gradients; counterclockwise circulations of air around lows; and clockwise circulations around highs.

Evidence that the observed winds are approximately geostrophic over large scales is even more noticeable on constant pressure charts because friction effects are smaller. (Figure 4-11)

Figure 4-10. An abbreviated surface analysis chart. Three pieces of information are plotted on the chart at each station location indicated by a small circle: temperature, observed pressure, and wind. Also, isobars are drawn every four millibars on the basis of the plotted pressure reports.

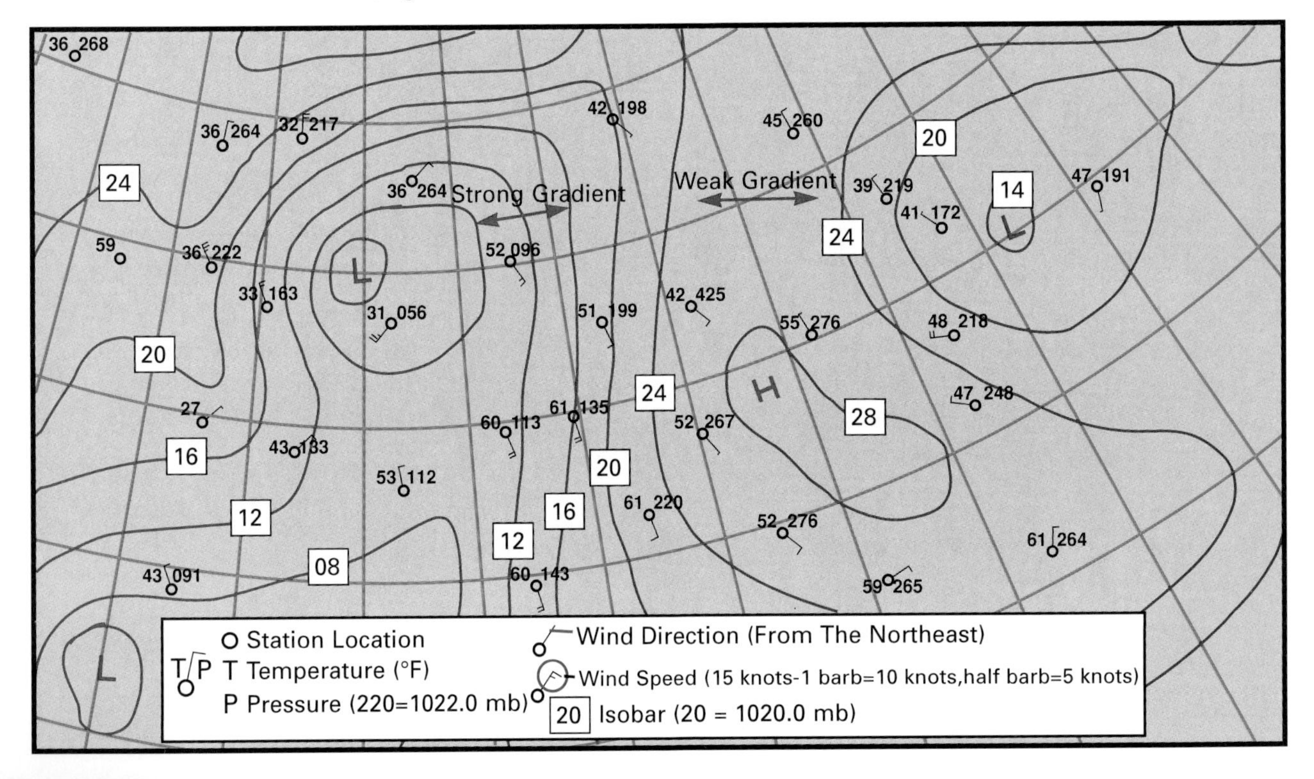

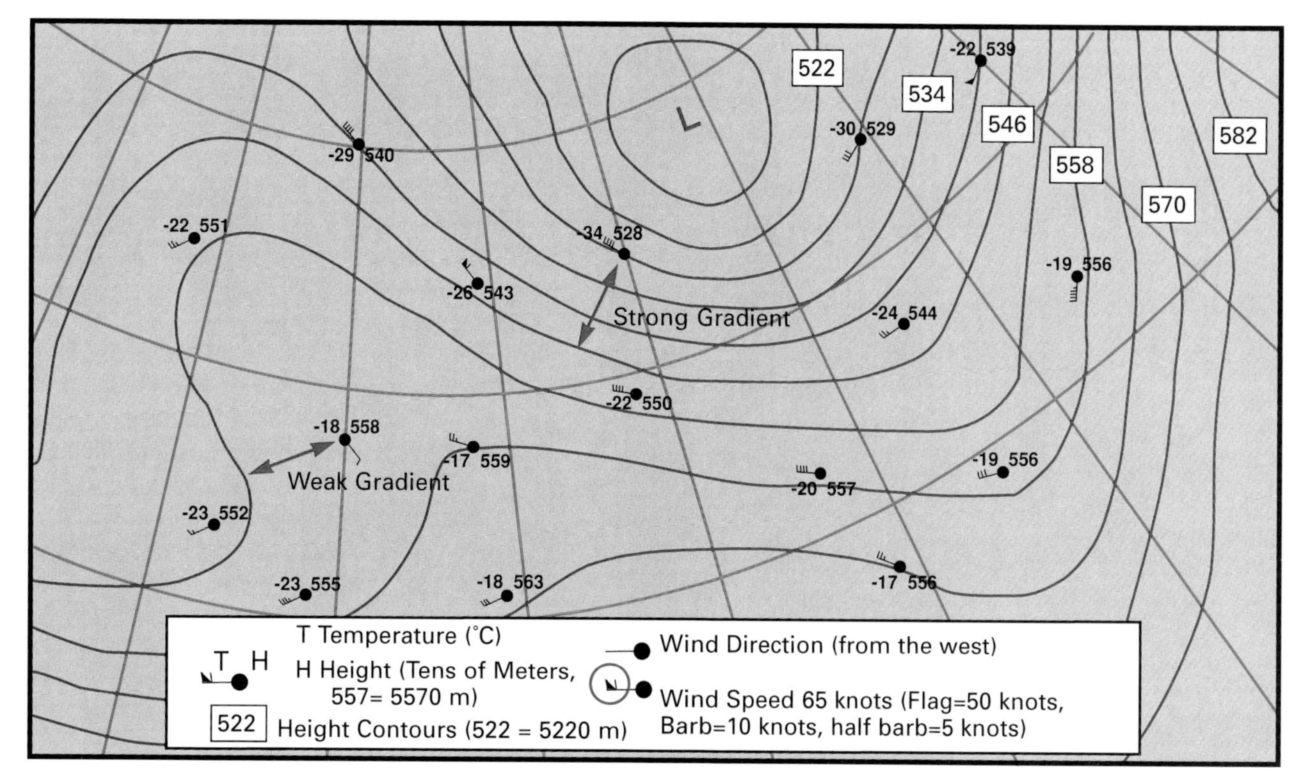

Figure 4-11. 500 mb constant pressure chart. The information plotted around the station locations includes observed 500 mb heights, temperatures, and winds. The lines on the charts are 500 mb height contours analyzed every 60 meters on the basis of the observed data. Notice that the winds are strongest where the height gradient is strongest and that the winds are nearly parallel to the contours. These features are evidence of airflow that is nearly in geostrophic balance.

D-VALUES

The geostrophic wind is also a useful navigational tool, especially over water and when there are not many other navigational aids available. By determining the pressure gradient along your track, you can determine the crosstrack component of the geostrophic wind and estimate your drift. Important measurements for this calculation are your pressure altitude (PA) measured with your pressure altimeter set at 29.92 inches, and your true altitude (TA) measured, for example, with a radar altimeter. The difference between the two (TA—PA) is known as the D-value. The crosstrack geostrophic wind is proportional to the gradient in D-values along your flight track. Greater details about D-value measurements and procedures for drift calculations are given in navigation texts and manuals (USAF, 1985).

Section F

FRICTION

The root of the word friction is another useful memory device. It comes from the Latin word meaning "rub." Friction is the force that resists the relative motion of two bodies in contact. Friction also occurs within fluids, such as the atmosphere, and at the interface between fluids and solids (skin friction). In your studies of aeronautics, you have been introduced to drag as one of the primary forces affecting aircraft in flight. Drag is the resistance of the atmosphere to the relative motion of the aircraft. Drag includes skin friction as well as form drag which is caused by turbulence induced by the shape of the aircraft.

Meteorologists use the term surface friction to describe the resistive force that arises from a combination of skin friction and turbulence near the earth's surface. The primary effects of surface friction are experienced through the lowest 2,000 feet of the atmosphere. This is called the atmospheric boundary layer. It is a transition zone between large surface frictional effects near the ground and negligible effects above the boundary layer. (Figure 4-12)

To understand the influence of friction, consider the following hypothetical situation. The wind at anemometer level (about 30 feet AGL) is initially in geostrophic balance (only pressure gradient

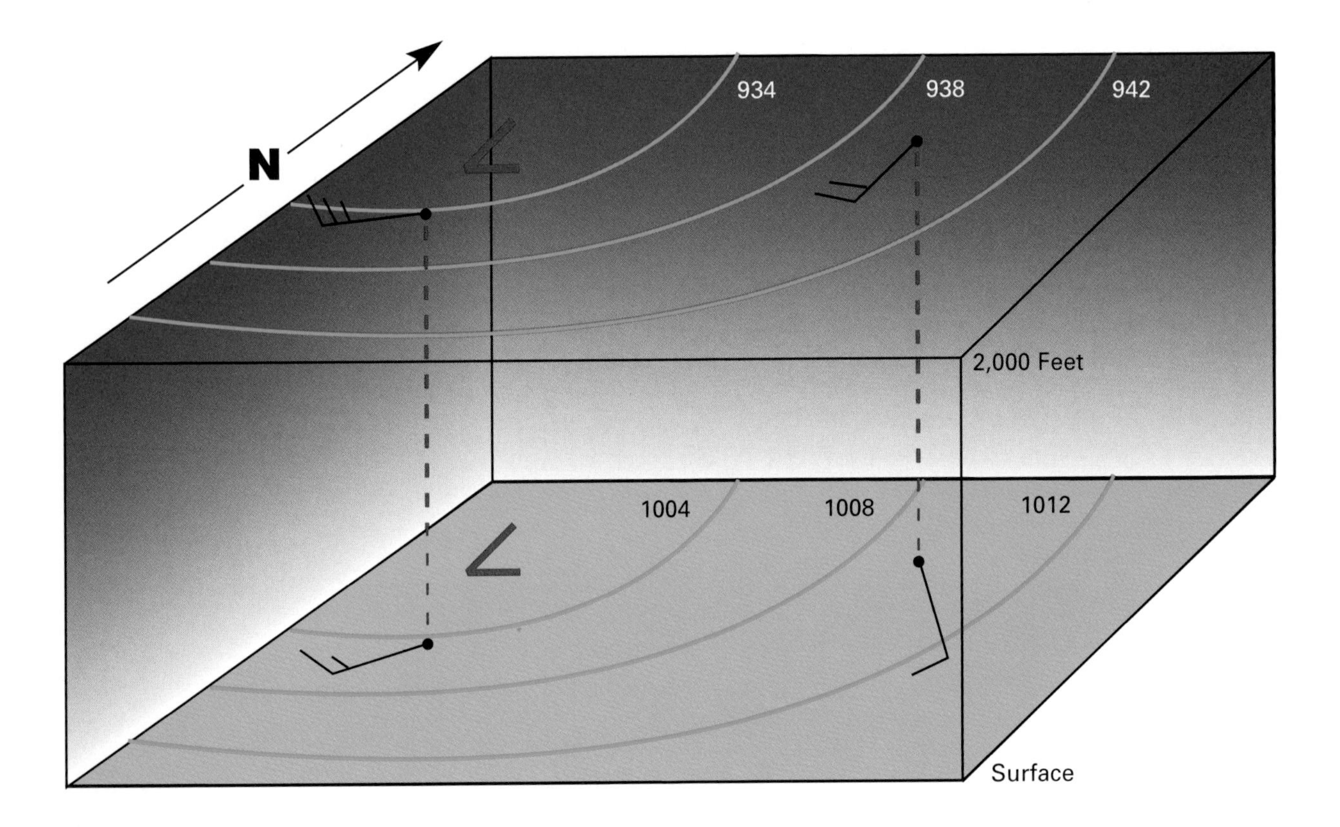

Figure 4-12. The influence of surface friction is greatest in the atmospheric boundary layer, which is typically the lowest 2,000 feet of the troposphere. Friction effects are at a maximum at the surface and decrease to a minimum at the top of the boundary layer. Effects illustrated here are changes in wind speed and wind direction with increasing altitude. Isobars are labeled in millibars.

and Coriolis forces exist). We then "turn on" the friction caused by the earth's surface and let all forces come into balance (pressure gradient, Coriolis, and friction). In the final balance, the wind speed is less than its original geostrophic value. Furthermore, the wind is no longer parallel to the isobars. It blows slightly across the isobars toward lower pressure. The angle between the wind and the isobars varies from about 10° over water to about 45° over land, depending on the roughness of the surface.

Keep in mind that this behavior is only approximate. It assumes, ideally, that a balance of forces is maintained. It doesn't include the effects of mountainous terrain. Also, in actual conditions, the slowing of the wind near the ground causes the air to form turbulent eddies that cause fluctuations in surface wind speed and direction. Details of these effects are described in a later chapter. The following is a summary of the effects of friction in the boundary layer.

Because of the decrease of surface frictional forces with height, the winds at 2,000 feet AGL tend to parallel the isobars. At the surface, winds cross the isobars at an angle toward lower pressure and are weaker.

1. Winds increase with altitude in the atmospheric boundary layer, with the greatest increases just above the surface.

2. The wind changes direction clockwise (veers) with increasing altitude.

3. When winds near the surface are strong, the boundary layer is turbulent and winds are gusty. As you descend through the boundary layer to land on a windy day, the air becomes rougher as you get closer to the ground.

4. The boundary layer is deeper during the day and in the warmer months of the year. It is shallower at night and during the colder months

5. The boundary layer is deeper over rough terrain.

6. Winds near the ground tend to spiral counterclockwise into a cyclone and spiral clockwise out of anticyclones.

7. Some of the effects listed here may be masked in stormy conditions.

Wind is caused by pressure differences and modified by the earth's rotation and surface friction.

Section G

OTHER EFFECTS

In most situations, pressure gradient force, Coriolis force, and friction explain the dominant, large scale characteristics of the winds. However, you should be aware of a few other influences that can modify that picture, sometimes signifcantly.

WIND PRODUCTION BY VERTICAL MOTIONS

In general, when an air parcel moves vertically for any reason, it carries its horizontal winds (actually horizontal momentum) to a different altitude, where it is mixed with the surroundings. This process changes the winds at the new altitude, causing the pressure gradient to change. One of the most frequent ways this occurs is when mechanical or thermal turbulence causes the boundary layer to become well-mixed. Stronger winds are brought from the top of the boundary layer to the ground, producing gustiness.

In some other atmospheric circulations, the effects of these processes can be quick and large, producing very strong horizontal winds at the surface. Examples occur in airflow over mountains and in thunderstorms. (Figure 4-13) Details about these phenomena, and flight hazards associated with them, are discussed later in the text.

ACCELERATED AIRFLOW

When air moves along a curved path, even if it is travelling at a constant speed, it is subjected to an acceleration; that is, the direction of motion is constantly changing along the path. This is known as centripetal acceleration. It is due to an *imbalance* in forces. When discussing this effect, some find it more convenient to refer to a "force" that produces the centripetal acceleration — the centrifugal force. In either case, large scale wind speeds are slightly modified from what we would expect according to geostrophic balance.

As the radius of the curved circulation becomes smaller, the effects of centrifugal force become larger. Where the scale is so small that the pressure gradient force is much larger than the

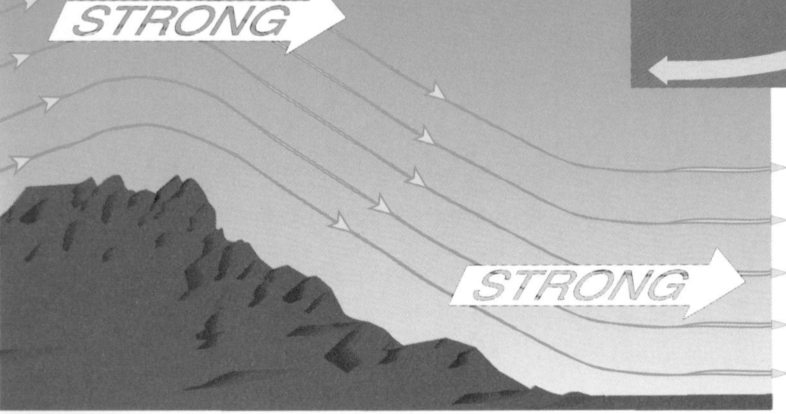

Figure 4-13. Two examples of the modification of winds at the surface by the vertical movement of air. Strong horizontal winds are carried down to the ground on the downwind side of a mountain. A strong vertical current is generated by a rain shower and creates strong horizontal winds near the ground.

Coriolis force, the pressure gradient and centrifugal forces may be in cyclostrophic balance and produce cyclostrophic winds. The most dramatic examples of these are dust devils and tornadoes. (Figure 4-14) More details of these particular phenomena are given in later chapters.

The geostrophic and other balances of forces that act on air parcels are idealizations. They are simple models that help us explain the causes and many of the characteristics of the wind and anticipate its behavior. The air is often accelerated by small imbalances in the sum of the forces. These are the sources of large changes in the wind and, as you will see, the production of weather in short time periods.

Figure 4-14. Tornado (left) and dust devil (right). These circulations have winds that tend toward cyclostrophic balance. Because Coriolis force is much smaller than the pressure gradient force, winds can circulate either clockwise or counterclockwise.Tornado photograph courtesy of NOAA. Dust Devil photograph courtesy of Dan Tyler.

SUMMARY

The basic properties of horizontal motions of the atmosphere have been examined in this chapter. You should now understand that air responds to pressure gradients by being accelerated toward lower pressure. Furthermore, pressure gradients are caused by temperature gradients and the movement of atmospheric mass by the winds. Once the air is in motion, Coriolis force becomes important, especially in large scale atmospheric circulations.

The wind that results when Coriolis force is exactly in balance with the pressure gradient force is the geostrophic wind. Because the near balance of these two forces is common, the geostrophic wind has proven to be a very useful estimate of actual wind in a variety of applications ranging from the interpretation of isobars and contours on weather charts, to navigation. Friction modifies the geostrophic balance, especially in the atmospheric boundary layer where its effect is apparent in cross-isobar airflow, turbulence, and gusty winds.

Your knowledge of the basic causes and characteristics of wind will be of great value as you examine vertical motions, clouds, and weather in the next two chapters and, subsequently, specific atmospheric circulations.

KEY TERMS

Acceleration
Anticyclone
Anticyclonic Flow
Boundary Layer
Centrifugal Force
Coriolis Force
Cyclone
Cyclonic Flow
Cyclostrophic Balance
Cyclostrophic Winds
D-Value
Differential Heating
Drag
Form Drag
Friction
Geostrophic Balance
Geostrophic Wind
Gust
Horizontal Pressure Gradient Force
Parcel
Peak Wind
Pressure Gradient Force
Return Flow
Scalar
Sea Breeze
Skin Friction
Squall
Surface Friction
Sustained Speed
Thermal Circulation
True North
Vector
Veer
Wind
Wind Direction
Wind Speed
Wind Velocity

CHAPTER QUESTIONS

1. A series of weather reports are listed at the bottom of the page. Decode all wind information.

2. With the guidance of your instructor, obtain a surface analysis chart with well-defined high and low pressure systems. Select five or ten widely separated weather reporting stations for which wind data are available. Construct a table to record the answers of each of the questions below for each station.
   1. What are the observed wind speeds and directions?
   2. What is the most likely wind direction at 2000 feet AGL?
   3. Which of the locations has the strongest surface geostrophic winds?

3. With the guidance of your instructor, select a 700 mb constant pressure chart for, preferably, a winter day. Select five observing stations for which wind data are plotted and in regions where the contours have different directions and gradients. Answer each of the questions below for each station.
   1. What are the observed wind speeds and directions?
   2. What are the geostrophic wind directions?
   3. Which of the locations has the strongest geostrophic winds?
   4. What is the approximate flight level of the chart?

4. Do some research on aircraft navigation and document the effects of Coriolis force in detail.

5. If the weather chart in question number 2 was in the Southern Hemisphere and the pressure pattern remained exactly the same, how would the winds be different? Redraw the map to illustrate your answer.

6. In the text, the development of a sea breeze was explained. Make a similar explanation of a land breeze, starting with (ideally) calm conditions after sunset. Draw appropriate diagrams. Be sure you explain how the pressure gradients are created both near the surface and aloft.

7. What is the bearing of the center of a nearby low pressure area (assume it is circular) from your location if your measured wind direction at 2,000 feet AGL is
   1. northwest?
   2. 240°?
   3. south?
   4. 090°?

8. Would your answers to question number 7 be any different if the wind directions were measured at the airport weather station instead of 2,000 feet?

9. A Foucault pendulum is often used to demonstrate the Coriolis effect. Do some research and describe what a Foucault pendulum is and how it works.

SFO SA 1555 M13 OVC 8 132/58/54/2815G25/992
DEN SP 1020 W5X 1/4F 180/68/64/1706/006
IAD SA 0953 AO2 CLR BLO 120 M 101/42/41/0000/991
LBL SA 0354 AWOS E80 OVC 3F 59/58/1302/017
METAR KBNA 1310Z 34024G40KTS 3/4SM +TSRA OVC020CB 28/23 A300
GLS SA 1455 AMOS E100OVC 7 80/73/0706/996 PK WND 13 028

CHAPTER 5

Vertical Motion and Stability

Introduction

In the previous chapter, we concentrated on the causes and characteristics of the wind; that is, the horizontal part of three-dimensional atmospheric motions. In this chapter, we examine vertical atmospheric motions. Although vertical motions are often so small that they are not felt by the pilot, they are still important in aviation weather. Very slow upward motions play a key role in the production of clouds and precipitation, and therefore, in the creation of flight hazards, such as poor visibilities, low ceilings, and icing. Gentle downward motions dissipate clouds and contribute to fair weather. Also, the atmosphere is not limited to weak vertical movements. Occasionally, turbulent upward and downward motions are large enough to cause injury, damage, and loss of aircraft control. Clearly, understanding the nature of vertical motions is a useful addition to your aviation weather knowledge. When you complete this chapter, you will understand not only how vertical motions are produced, but also what the important effects of atmospheric stability are on those motions.

Section A

VERTICAL MOTIONS

As we saw in the previous chapter, when an air parcel moves from one location to another, it typically has a horizontal component (wind) and a vertical component which is called vertical motion. Because of the hydrostatic balance of the atmosphere, vertical motions are usually much smaller than horizontal motions. However, there are some important exceptions. Small imbalances between the gravitational force and the vertical pressure gradient force arise in circulations such as thunderstorms and cause large vertical accelerations and vertical motions. (Figure 5-1)

CAUSES

Air may move upward or downward due to a number of causes. The most frequent causes are convergence and divergence, orography, fronts, and convection.

CONVERGENCE/DIVERGENCE

Convergence corresponds to a net inflow of air into a given area. It may occur when wind speed slows down in the direction of flow and/or when opposing airstreams meet. Divergence is the net outflow from a given area. Winds may diverge when the wind speed increases in the direction of the flow and/or when an airstream spreads out in the downstream direction. If either one of these processes occurs at some point in the atmosphere, air moves upward or downward. This interaction is common to all fluids; that is, motion in one region usually causes motions in a nearby region. You have observed this property many times in liquids; for example, when you dip a bucket of water from a lake, you don't leave a hole in the surface. The surrounding water rushes in to replace it.

The effect of convergence and divergence of the wind on vertical motions is easy to visualize near the earth's surface. (Figure 5-2) When surface winds converge, the inflowing air is removed by rising motions aloft.

Vertical Velocities	Feet Per Minute
Average for the entire Atmosphere	0
Nimbostratus Cloud	10
Weak Thermal	200
Growing Cumulus Cloud	500+
Thunderstorm	2,000+
Strong Thunderstorm	5,000+
Extreme Mountain Wave	5,000+
Extreme Clear Air Turbulence	5,000+

Figure 5-1. Examples of typical and extreme vertical motions. **Notes:** 100 f.p.m. is approximately equal to one knot. Extreme cases of 10,000 f.p.m. have been reported. **Caution:** What appears to be nimbostratus may contain regions of strong convection.

Embedded thunderstorms are obscured by massive cloud layers and cannot be seen.

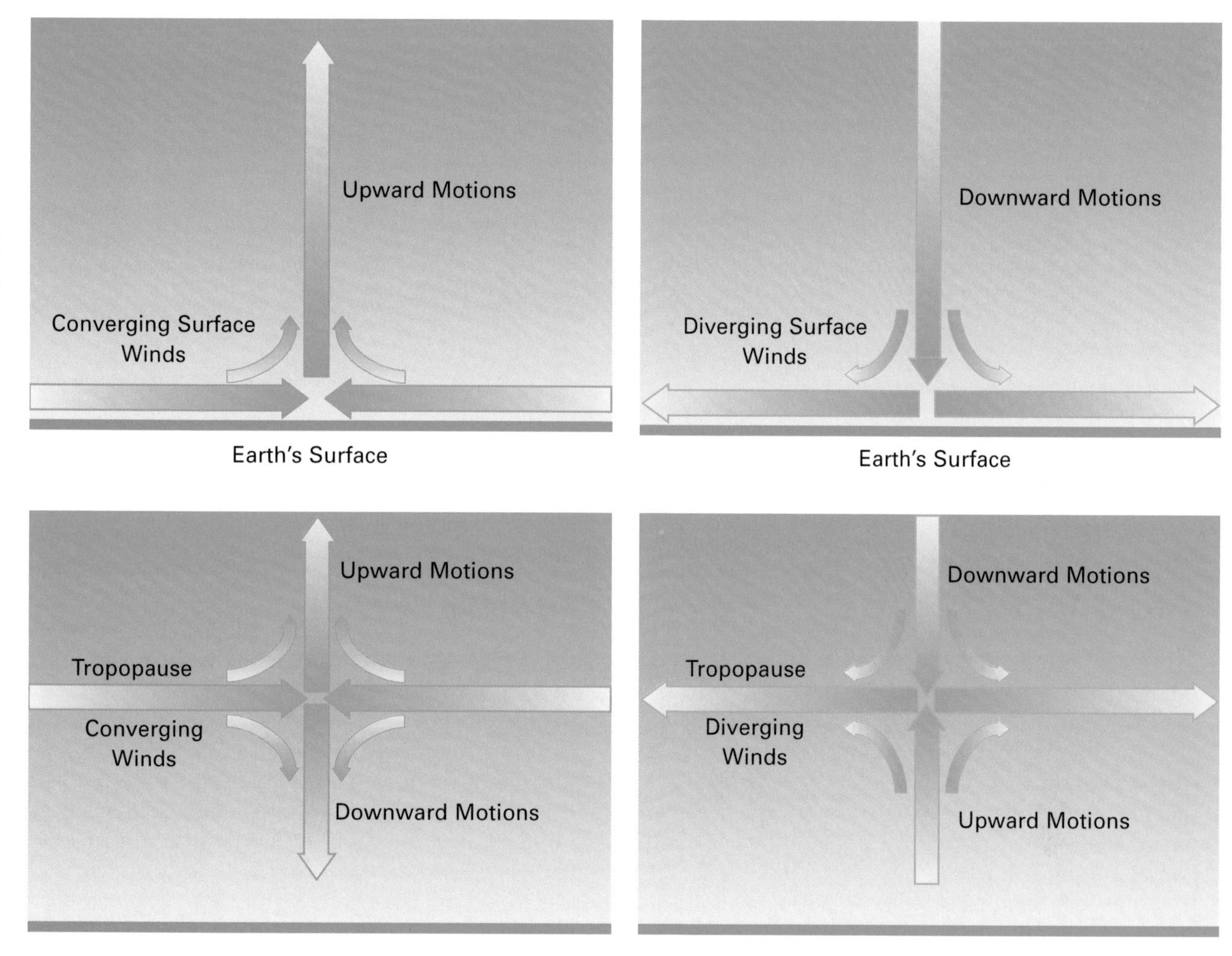

Figure 5-2. Patterns of vertical motions caused by convergence and divergence near the ground (top) and near the tropopause (bottom).

Conversely, divergence of surface winds causes air to sink from aloft to replace the air being removed at lower levels. Figure 5-2 also shows those vertical motions which may develop when the convergence or divergence occurs near the tropopause.

From the previous chapter on wind, recall that for large scale flow (nearly in geostrophic balance), the effect of friction causes the surface winds to blow across the isobars at a slight angle toward lower pressure. This means that around large low pressure areas, surface winds spiral into the centers (convergence). For high pressure areas, they spiral outward from the centers (divergence). Therefore the air rises in low pressure areas and sinks in high pressure areas. (Figure 5-3)

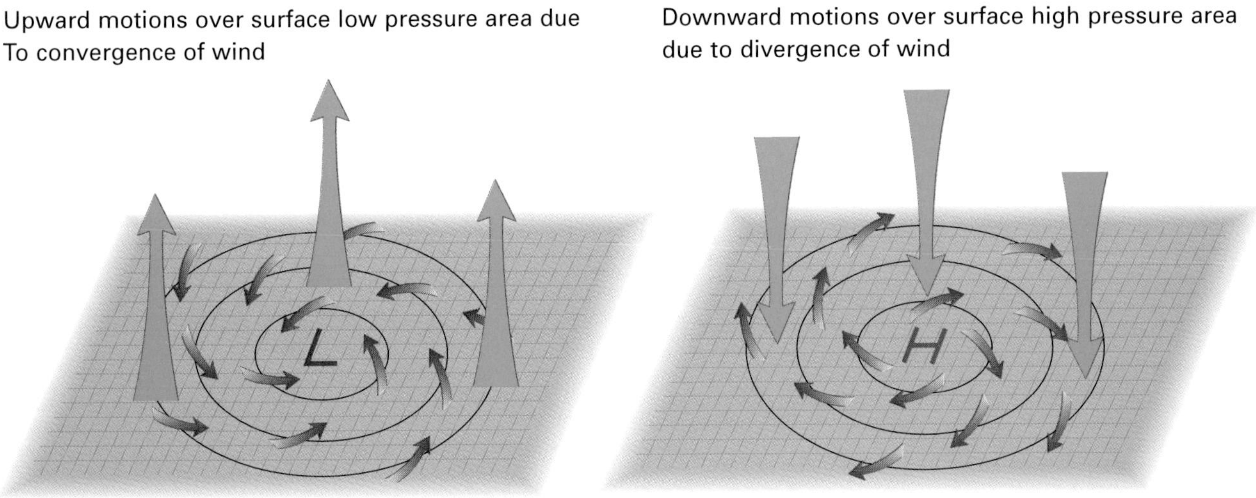

Figure 5-3. Left: Convergence of surface wind in the low pressure area causes upward motion. Right: Divergence of surface winds in a high pressure area causes downward motions.

OROGRAPHY

Air can be forced upward or downward by a barrier in its path. A simple example is orographic lifting. When wind encounters a mountain or hill, it is simply pushed upward. On the downwind or lee side of the mountain, air moves downward. The strength of the vertical velocities depends on the speed of the wind perpendicular to the mountain side and the steepness of the terrain. (Figure 5-4)

FRONTS

When the atmosphere itself creates an obstacle to the wind, a similar barrier effect can be produced. For instance, when a cold airmass is next to a warm airmass, a narrow, sloping boundary is created between the two. This is called a front. If either airmass moves toward the other, the warm air moves upward in a process called frontal lifting or, in some special cases, overrunning. (Figure 5-5) Air can also descend over fronts. Fronts and their influence on aviation weather are discussed in greater detail in Part II of this text.

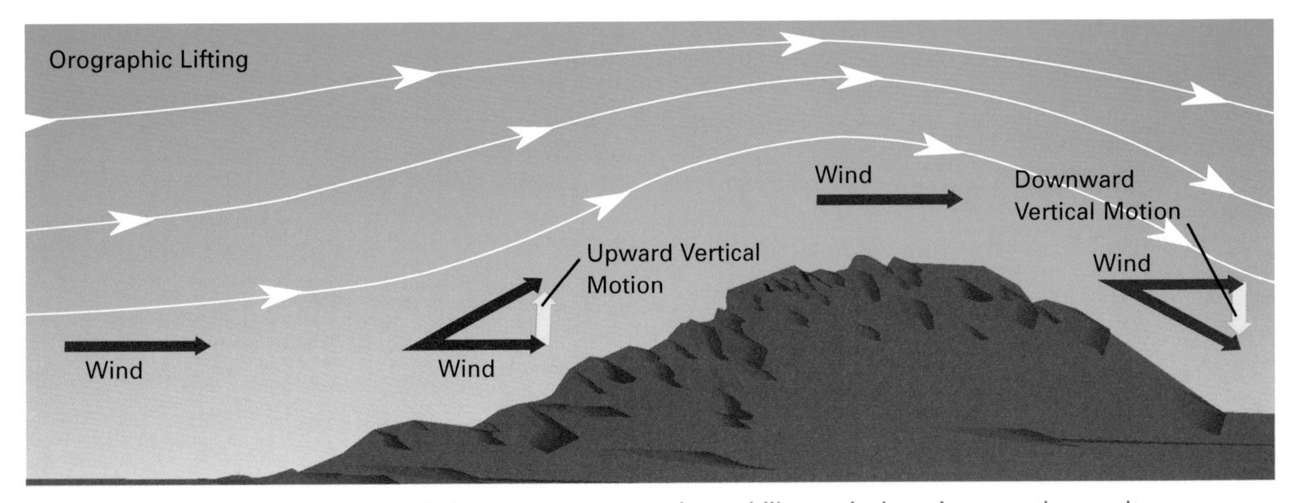

Figure 5-4. When the wind encounters mountains or hills, vertical motions are the result.

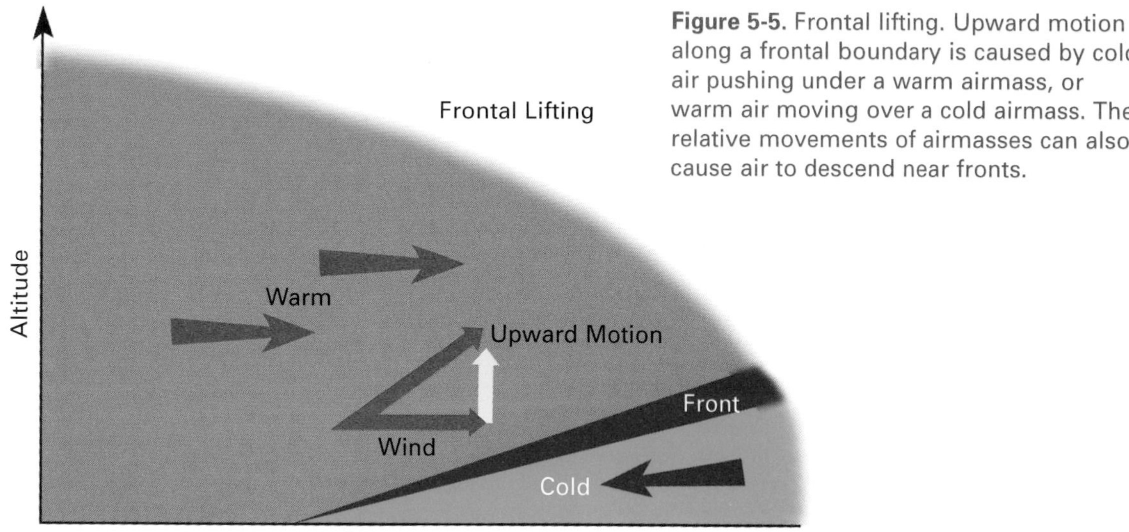

Figure 5-5. Frontal lifting. Upward motion along a frontal boundary is caused by cold air pushing under a warm airmass, or warm air moving over a cold airmass. The relative movements of airmasses can also cause air to descend near fronts.

CONVECTION

If air at a particular level in the atmosphere is warmer than its surroundings, it will rise. As you know from Chapter 2, this is a form of convection. Although this word is used to describe a wide variety of processes involving vertical motions, it is most often used in reference to rising warm air and/or the clouds and weather associated with that process. We will use the latter meaning for convection throughout the remainder of the text.

As warm air rises in the convective lifting process, the surrounding air sinks. (Figure 5-6) Convection, as described here, occurs under unstable atmospheric conditions. Stability and instability are discussed at length in the next section.

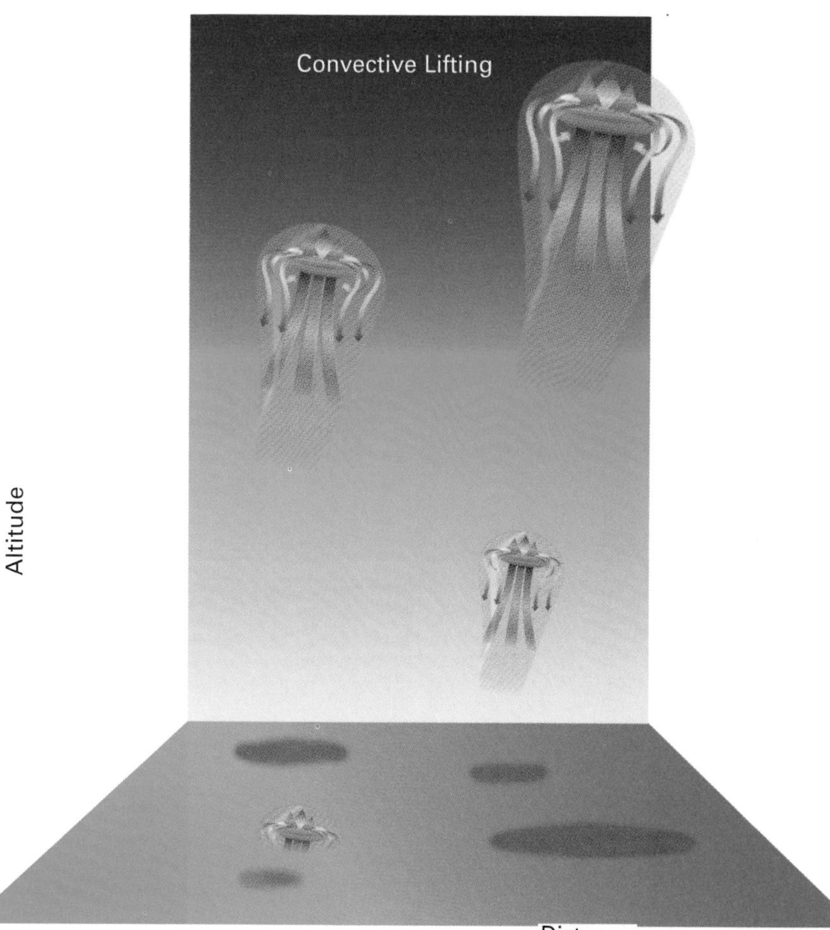

Figure 5-6. Convective lifting. Air rises in discrete bubbles when the atmosphere is unstable.

MECHANICAL TURBULENCE

When wind blows over the ground, friction causes the air to slow down in the lowest layers. The greater speeds at higher altitudes causes the air to roll up into irregular circulations about horizontal axes. (Figure 5-7) These chaotic eddies are swept along with the wind, producing downward motions on their downwind side and upward motions on their upwind side. This is known generally as mechanical turbulence. Rough air experienced when landing on windy days is caused by these small scale circulations. More details about low-level turbulence are given in Part III.

WAVE MOTIONS

Under certain circumstances, air may be disturbed by small scale wave motions; that is, parcels of air may be caused to oscillate vertically. (Figure 5-8) Such oscillations that move away from the source of the disturbance are called atmospheric gravity waves because the earth's gravity plays an important role in producing them. A mountain lee wave is one type of gravity wave. More details are given later in the text.

Whereas convection occurs under unstable atmospheric conditions, gravity waves occur under stable conditions. Meteorologists use the concept of atmospheric stability to deal with gravity effects on vertical motions. We must examine the meaning and application of stability before we consider more of the details of gravity waves or convection.

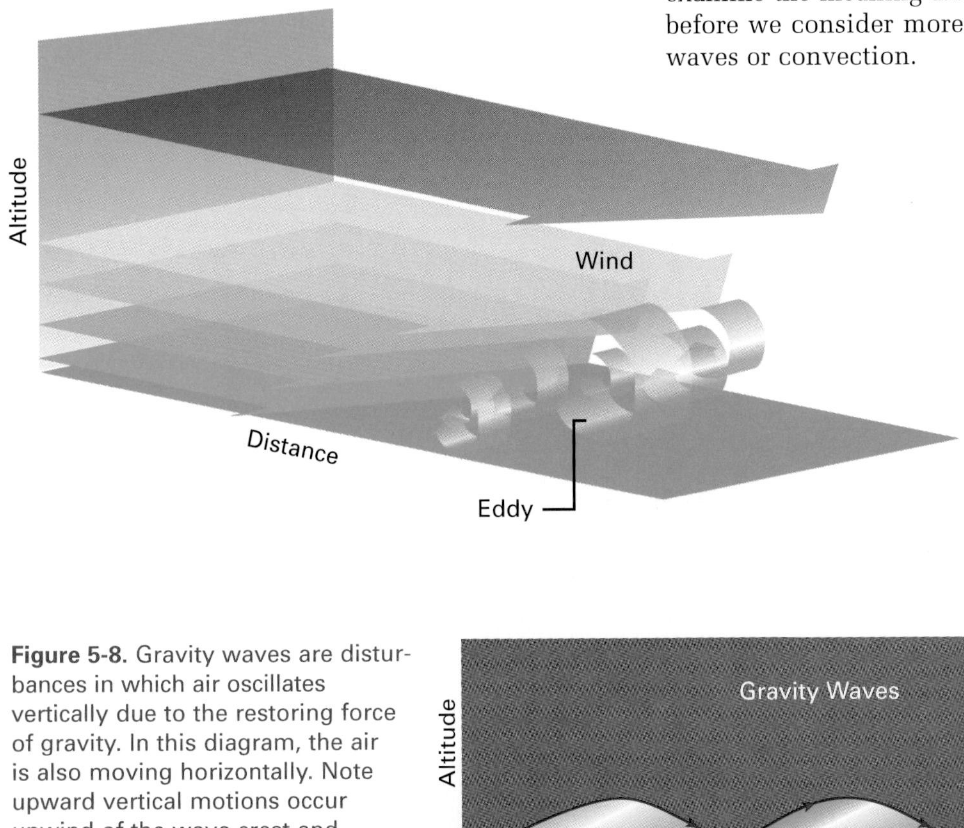

Figure 5-7. The rapid increase of wind velocity near the earth's surface causes mechanical turbulence. Turbulent eddies produce fluctuating vertical motions as they are swept along by the winds. The eddies have a three dimensional structure that is constantly being stretched and deformed by the wind.

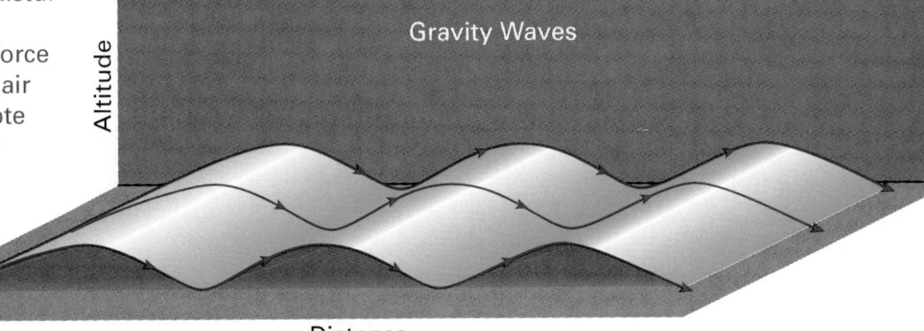

Figure 5-8. Gravity waves are disturbances in which air oscillates vertically due to the restoring force of gravity. In this diagram, the air is also moving horizontally. Note upward vertical motions occur upwind of the wave crest and downward vertical motions occur downwind.

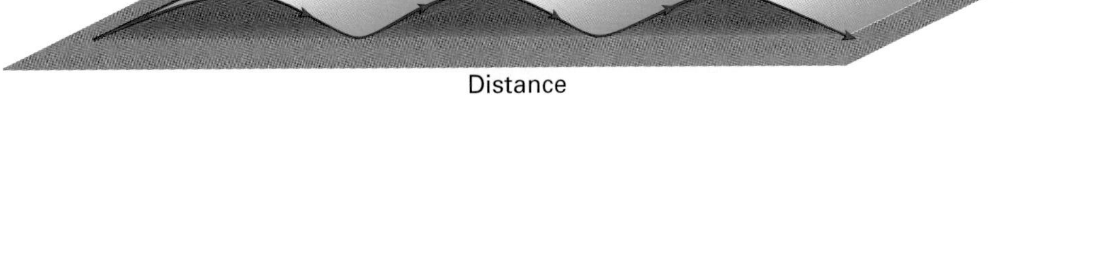

Section B

STABILITY

Stability is a general concept applied to explain the behavior of mechanical systems. When discussing such systems we may describe them as either stable or unstable. A stable system may be defined as one that, if displaced or distorted, tends to return to its original location and/or configuration. On the other hand, an unstable system is one that tends to move away from its original position, once it has been displaced. A system with neutral stability remains in its new position after it is displaced; that is, it neither returns to, nor is it accelerated away from, its original position. Some simple examples of stable, unstable and neutral systems are shown in figure 5-9.

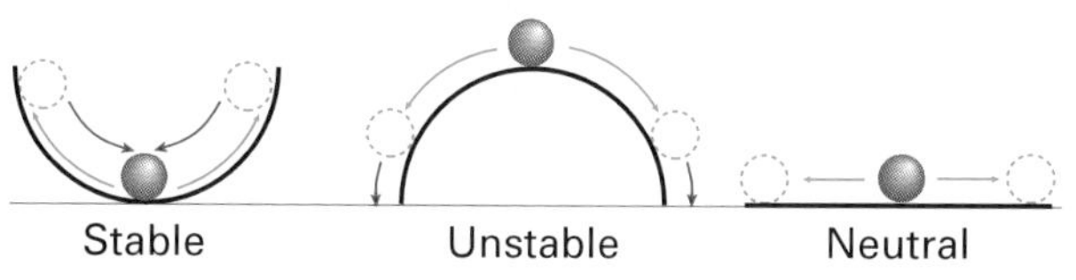

Figure 5-9. Examples of stable, unstable, and neutral systems. Blue arrows indicate the initial, small displacement of a marble. Red arrows indicate the subsequent motion. In the neutral case there is no subsequent motion.

In the remainder of this section, the concept of stability and instability are applied to the atmosphere to understand and anticipate the influence of gravity on the development of vertical motions in the atmosphere.

ATMOSPHERIC STABILITY

As applied here, atmospheric *stability* is a condition that makes it difficult for air parcels to move upward or downward. In contrast, atmospheric *instability* is a condition that promotes vertical motions. Similar to the mechanical systems described above, atmospheric stability is determined by considering the behavior of a parcel of air after it receives a small vertical displacement. When an air parcel is displaced (upward or downward) and forces develop that cause it to return to its initial position, the parcel is said to be stable. On the other hand, if the forces develop that cause the parcel to accelerate away from its original position, the parcel is unstable. (Figure 5-10)

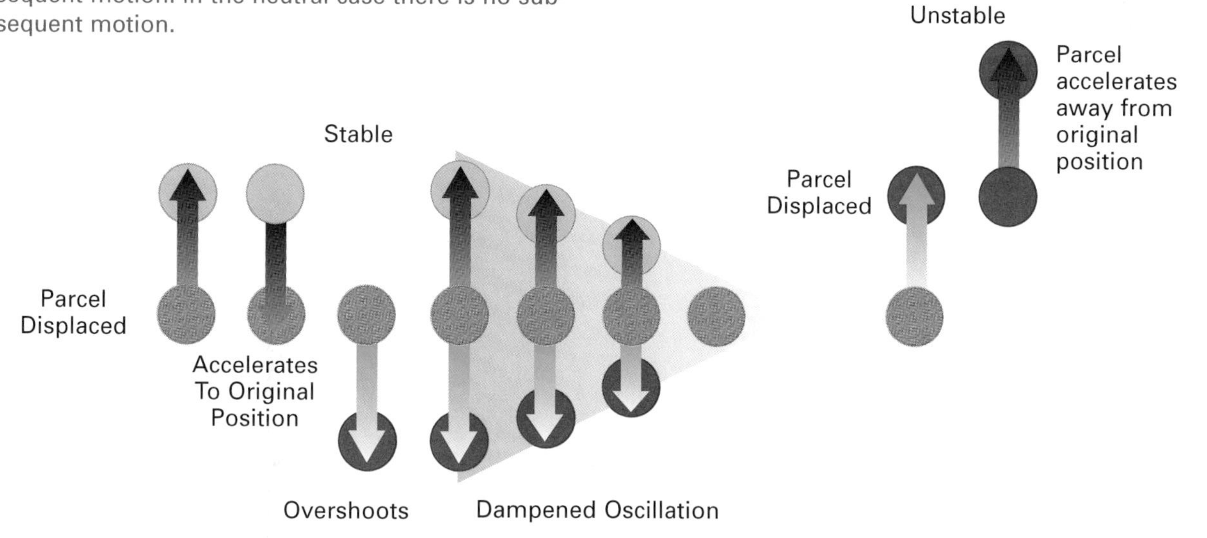

Figure 5-10 Air parcel stability. In the lefthand diagram, the displaced parcel is stable. After it is displaced, it oscillates around its original position, not unlike a pendulum. In the righthand diagram, the displaced parcel is unstable; once it is given a push, it accelerates away from its original position.

BUOYANCY

In order to completely understand stability and instability of air parcels, we must understand the forces that arise when they are displaced. There actually is one primary force that must be considered: buoyancy force. Buoyancy is the property of an object that allows it to float on the surface of a liquid, or ascend through and remain freely suspended in a compressible fluid such as the atmosphere (Huschke, 1959).

Archimedes' Principle applies here. It simply states that when an object is placed in a fluid (liquid or gas), it will be subjected to a positive or negative force depending on whether the object weighs more or less than the fluid it displaces. This is where the gravitational force enters the picture. Weight is a force, defined as the product of mass and gravity. Since gravity varies little across the surface of the earth, differences in weight depend mainly on differences in mass or, in meteorological terms, density. Archimedes' Principle can be thought of as the bowling ball/balsa wood-in-the-bucket-of-water concept. (Figure 5-11)

The density of a displaced parcel of air can be easily related to its temperature because the pressure of a displaced parcel adjusts to the pressure of its surroundings. Recall from the gas law that if two volumes of air are at the same pressure, the one with the lower density is warmer. This allows us to state Archimedes' Principle for air parcels in terms of temperature and combine it with the concept of stability.

1. If a parcel of air is displaced upward and becomes warmer than its surroundings, it is positively buoyant. It will accelerate upward (away from its original position); it is unstable.

2. If a parcel of air is displaced upward and is colder than its surroundings, it is negatively buoyant. It will be accelerated downward (back to its original position); it is stable.

DETERMINING ATMOSPHERIC STABILITY

Although mentally moving parcels of air around in the atmosphere helps us understand the concept of stability, in practice, it is not very convenient. Instead, meteorologists evaluate the stability of the atmosphere by taking atmospheric soundings and analyzing the soundings to determine stability conditions. Although you won't be doing this very often yourself, you will be using information and terminology that comes from these analyses. Therefore, it is helpful to become familiar with the analysis process.

Figure 5-11. The bowling ball in a bucket of water is much more dense than the water it displaces. It is said to have negative buoyancy. A downward-directed force accelerates the ball to the bottom of the bucket. If you push a piece of balsa wood down into the bucket of water, it is much less dense than the water it displaces; there is an upward-directed force. It has positive buoyancy. Once you release the block of balsa wood, it accelerates to the surface of the water.

DRY ADIABATIC PROCESS

Whenever a parcel of air changes its altitude, its temperature changes. The reason this happens is that it must change its pressure to match the pressure of its surroundings; so when an air parcel rises, it lowers its pressure by expanding. When the air parcel descends, it compresses.

In order to expand against its surroundings as it rises to a higher altitude (a lower pressure), an air parcel must use energy. The major source of parcel energy is in the motion of its molecules. As you know, the energy of this motion is measured by the temperature. Therefore, as the air parcel moves upward, its temperature decreases because it uses some of this energy to expand. Similarly, if an air parcel descends, its temperature goes up. This temperature change process, *cooling by expansion* and *warming by compression*, is called the dry adiabatic process. (Figure 5-12) It is called a "dry" process because it does not consider the influences of evaporation and condensation. This will be a topic of the next chapter.

Since pressure always decreases with height,

1. adiabatic cooling will always accompany upward motion.
2. adiabatic heating will always accompany downward motion.

Unsaturated (cloud-free) air flowing upslope will cool at the rate of approximately 3C° per 1,000 feet.

The rate of temperature change associated with a dry adiabatic process is a constant: 3C° per 1,000 feet (5.4F° per 1,000 feet).

SOUNDINGS AND LAPSE RATES

In its most common form, an atmospheric sounding is simply the measurement of atmospheric conditions along a vertical line through an atmospheric layer at a given time and location. You have already seen graphs of average atmospheric soundings of temperature, pressure, and density in earlier chapters to describe the structure of the atmosphere. Those averages are based on twice-a-day soundings made 365 days per year at hundreds of locations throughout the world. Instruments are carried to about 100,000 feet MSL by free, unmanned balloons called radiosondes or rawinsondes.

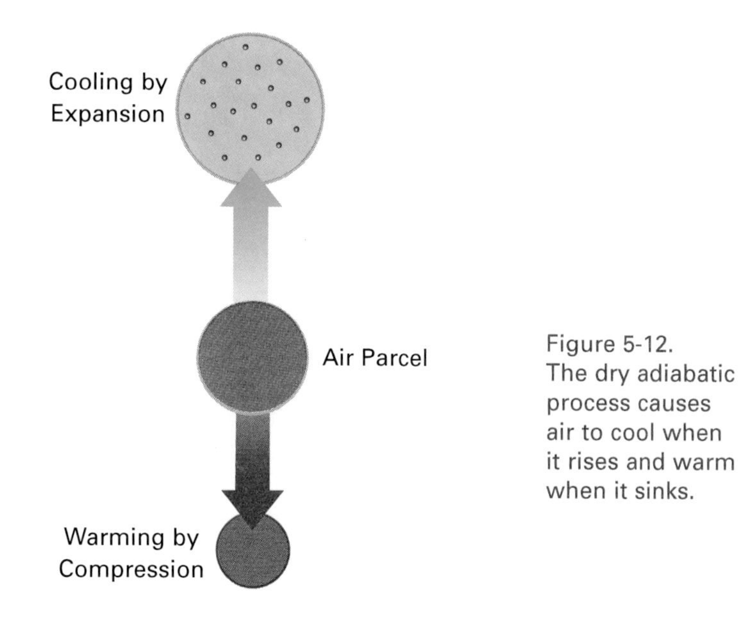

Figure 5-12.
The dry adiabatic process causes air to cool when it rises and warm when it sinks.

Figure 5-13. Radiosonde sounding system. Winds are determined by tracking the balloon from a ground station or satellite. Pressure, temperature, and moisture measurements are made by the small instrument package which also carries a transmitter to send the information to a tracking station. Typical radiosondes rise at about 1,000 feet per minute.

(Figure 5-13) The instruments also provide profiles of atmospheric moisture and are tracked to provide winds at different altitudes.

An important stability measurement that can be determined from a sounding is the change of temperature with altitude. When defined as follows, it is known as the lapse rate (LR).

$$LR = \frac{T(bottom) - T(top)}{DELZ}$$

Where T(bottom) is the temperature at the bottom of a layer, T(top) is the temperature at the top of the layer, and DELZ is the thickness of the layer. Notice that LR can be positive or negative. If temperature decreases with increasing altitude, LR is positive. From earlier discussions, you already know that the tropospheric lapse rate in the standard atmosphere is 2C° per 1,000 feet.

Newer atmospheric sounding systems include microwave radars called wind profilers, vertically pointing, ground-based radiometers, lasers (Lidars), space-based (satellite) radiometers, and devices based on sound measurements (Sodars and Radio Acoustic Sounders).

The rate at which the temperature of a dry parcel of air decreases as it ascends is also a useful reference in stability determinations. This is known as the dry adiabatic lapse rate (DALR). As noted in the last section, it is equal to 3C° per 1,000 feet. Figure 5-14 gives examples of these and other lapse rates that you may encounter in a particular sounding.

In describing any layer in the atmosphere, the altitudes of the top of the layer and the bottom (base) are useful pieces of information. For example, in figure 5-14, the base of the inversion is relatively cold and the top is relatively warm.

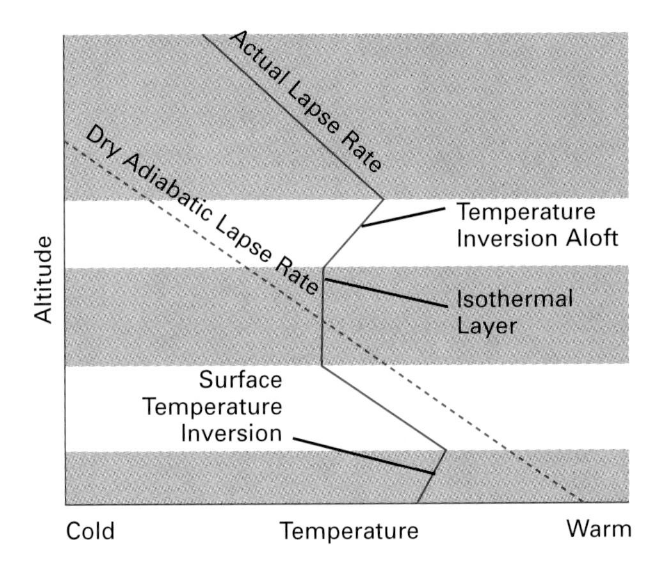

Figure 5-14. Temperature sounding showing layers with different lapse rates (solid lines) and the dry adiabatic lapse rate (dashed line). Of particular interest is the isothermal layer (no change in temperature with height), and the inversion layer (temperature increases with height).

As you will see, cloud layers are often found near inversion bases. It is also helpful to know if inversion layers are next to the ground. These "ground-based" or surface-based inversions often form at night and may be the source of serious wind shear problems.

STABILITY

The determination of the stability of an atmospheric layer is a straightforward procedure, given a sounding and your knowledge of DALR and LR. The steps are given below and the stability criteria are shown graphically in figure 5-15. An example is given in figure 5-16.

1. Select the layer in the sounding in which you are interested.
2. Within the layer, compare LR and DALR.
3. Determine which of the following stability criteria given below are satisfied.

If	Then
LR > DALR	Absolutely Unstable
LR = DALR	Neutral
LR < DALR	Stable

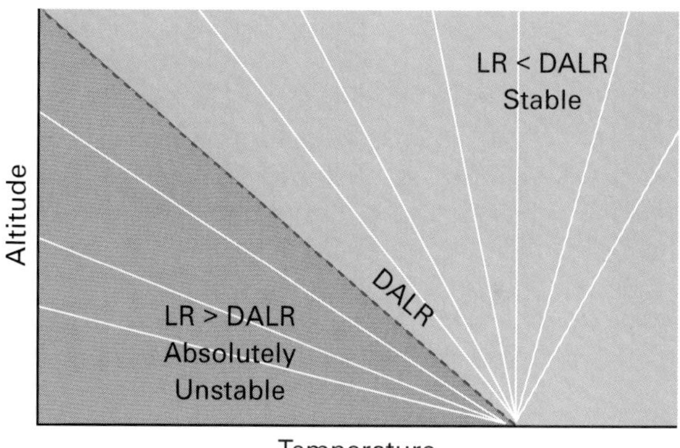

Figure 5-15. Stability criteria and lapse rates. In a dry atmosphere, all sounding curves that are steeper than the curve labeled DALR (dashed) are absolutely unstable. All of those for which DALR is steeper, are stable. If a sounding curve and the curve labeled DALR are parallel, conditions are neutral. Inversions (negative LRs) are very stable. Conditions where LR > DALR are sometimes called "superadiabatic."

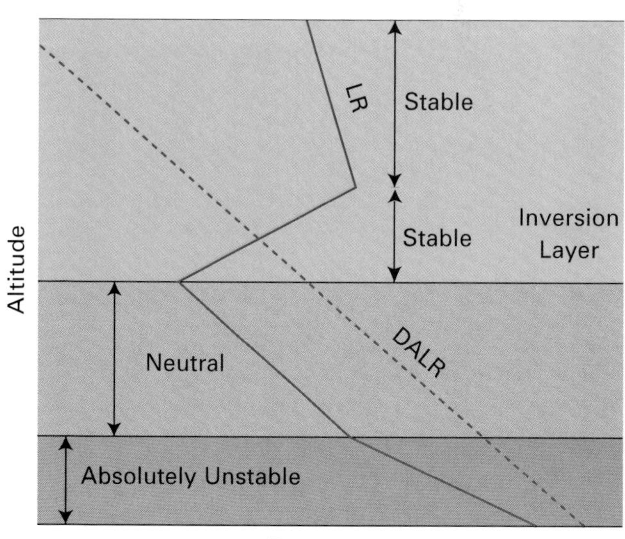

Figure 5-16. A sounding with layers of different stabilities is shown. Although both of the top two layers are stable, the inversion layer is more stable.

We have discussed two different ways of determining whether an air parcel is stable or unstable. The first involved determining whether the temperature of an air parcel would be warmer, colder or the same as its surroundings after it was pushed upward. The second simply required the comparison of the actual (measured) lapse rate with the dry adiabatic lapse rate. Figure 5-17 gives an example that demonstrates that the two methods to determine stability are equivalent.

In figure 5-17, the diagram on the left shows an absolutely unstable situation by both measures; LR is greater than DALR and when the parcel reaches 2,000 feet, it is one degree warmer than its surroundings. On the right, the situation is stable; LR is less than DALR and when the parcel reaches 2,000 feet, it is one degree colder than its surroundings.

If a cloud forms or dissipates during the upward or downward displacement of an air parcel, the stability of the air parcel will be modified. Moisture influences on stability are examined in the following chapter.

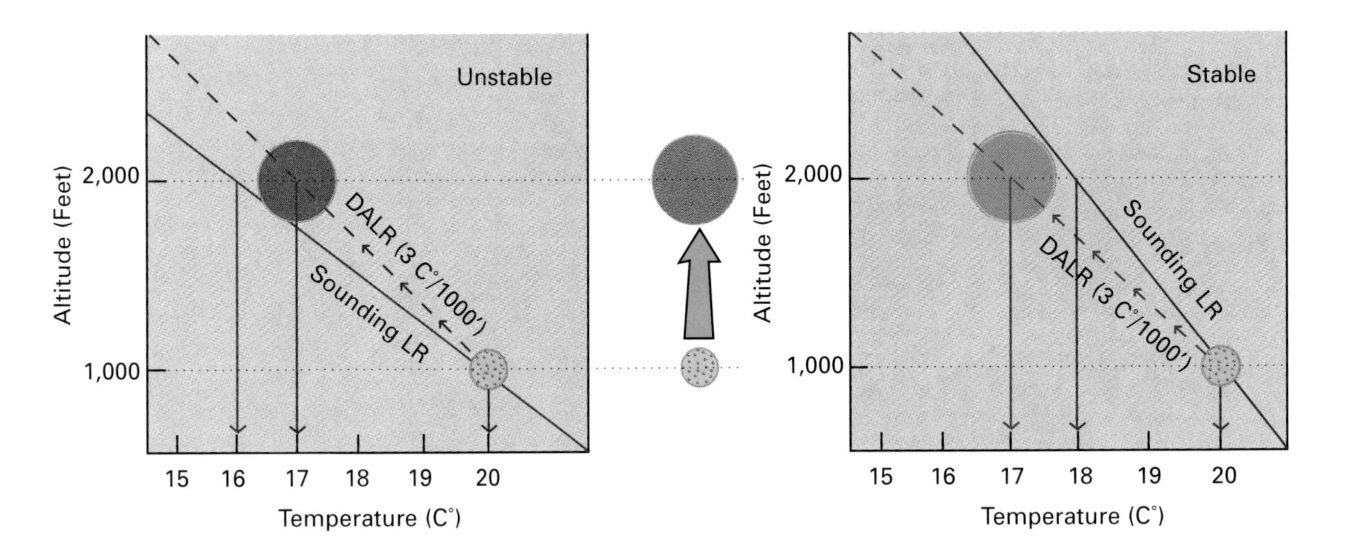

Figure 5-17. The effect of the difference of LR and DALR on stability. The center diagram indicates the displacement of a parcel of air from an altitude of 1,000 feet to an altitude of 2,000 feet. Two temperature-altitude graphs are shown on either side. The sloping solid lines (LR) indicate the actual sounding and the sloping dashed lines (DALR) indicate the temperatures a parcel would have as it is lifted adiabatically from 1,000 feet.

Section C

STABILITY AND VERTICAL MOTION

Sections A and B have, respectively, given information about the development of vertical motions and the meaning and evaluation of stability. In this section, we look at how stability (and instability) influences vertical movements of the atmosphere.

A stable atmosphere does not necessarily prevent air from moving vertically, but it does make that movement more difficult. In a stable atmosphere, air parcels must be given "outside" help if they are to continue their ascent or descent. This "help" could come from convergence/ divergence, orographic lifting, or frontal lifting. The air must be pushed or pulled. The stable environment is constantly working against vertical movements, so in most (but not all!) cases, vertical motions are very small and the airflow is smooth.

A stable airmass is more likely to have smoother air than an unstable airmass.

In an unstable atmosphere, convection is the rule. Air rises because it is positively buoyant. Aside from an initial "kick" from surface heating, or from any of the processes mentioned above, no outside help is needed. The air rises because it is warmer than its surroundings.

In comparison with vertical motions in a stable environment, unstable vertical movements are large and the airflow is turbulent. The differences in unstable and stable environments are best seen in differences in cloud forms. (Figure 5-18)

The formation of either predominantly stratiform or predominantly cumuliform clouds depends upon the stability of the air being lifted.

Other visual indications of stable conditions are the presence of fog, smoke, and haze. These conditions are often present near the ground in the nighttime hours and in the winter when a surface-based inversion is present.

Figure 5-18. Differences in vertical motions between stable and unstable conditions are seen in the smooth, layered stratiform cloud (stable), above, and the very turbulent-looking cumuliform cloud of great vertical development (unstable), left. A thorough discussion of cloud causes and types is given in the next chapter.

Conditions favorable for the formation of a surface based temperature inversion are clear, cool nights with calm or light winds.

Unstable air is associated with good visibilities and rough low-level flying conditions in the lower atmosphere in the afternoon and especially in the summer. At these times, ground temperatures tend to be much warmer than air temperatures; conditions in the lower troposphere near the ground are often absolutely unstable.

The stability of an airmass is decreased by heating it from below.

In many situations, it happens that an unstable layer of air is capped by a strong stable layer, perhaps an inversion. In this case, freely rising warm air in the unstable layer is arrested when it reaches the stable layer. This effect is often made visible by vertically developing clouds, the tops of which flatten out at the base of a stable layer. (Figure 5-19)

Stability and vertical motions are intimately connected in another way. Not only does stability affect vertical motions as described above, but vertical motions affect stability. Sinking motions tend to make the atmosphere more stable and rising motions tend to make it less stable. Therefore, in high pressure areas where air is generally descending, the atmosphere is more often stable. Low pressure areas are more often associated with upward motions and unstable air. We will see the influences of these interactions on clouds and weather in the following chapters.

Figure 5-19. The low-level source of air feeding the cumulus clouds is an unstable layer. The tops of the clouds flatten out as they reach an overlying stable layer.

SUMMARY

Vertical motions in the atmosphere are critical for aviation because of their role in the production of turbulence, clouds, and associated phenomena. You have learned that upward and downward motions are forced by fronts, mountains, warm surfaces, and where airstreams converge and diverge. Additionally, the resulting vertical motions are magnified or suppressed, depending on the atmospheric stability. The understanding of stability has required you to study and understand the concepts of buoyancy and the adiabatic process. With these tools, you have learned how atmospheric stability is evaluated by examining atmospheric temperature soundings. The information in this chapter is basic to later discussions of a wide variety of topics ranging from clouds and weather of large-scale cyclones, to thunderstorms, to small-scale clear air turbulence.

KEY TERMS

Adiabatic Cooling
Adiabatic Heating
Archimedes' Principle
Atmospheric Instability
Atmospheric Stability
Buoyancy
Convective Lifting
Convergence
Divergence
Dry Adiabatic Lapse Rate
Dry Adiabatic Process
Front
Frontal Lifting
Gravity Waves
Inversion Layer
Isothermal Layer
Lapse Rate
Mechanical Turbulence
Negative Buoyancy
Neutral Stability
Orographic Lifting
Positive Buoyancy
Radiosonde
Rawinsonde
Sounding
Stable
Surface-Based Inversion
Unstable
Vertical Motion

CHAPTER QUESTIONS

1. A wind of 25 knots blows against a mountain side that slopes upward from sea level to 2,000 feet AGL over a distance of 5 miles.
    1. What will the magnitude of the upward vertical velocity be along the side of the mountain in f.p.m.? (Hint: In this case, vertical velocity = slope x wind speed. Be sure to keep your units consistent).
    2. If the temperature at the base of the hill is 50°F, what is the temperature of air parcels as they reach the top of the hill in °C (consider only adiabatic processes)?
2. What happens to air parcels displaced downward in a stable atmosphere? Why?
3. You are flying over a flat plain when you experience a significant upward gust (turbulence). List the possible causes of the upward vertical motion.

4. A simple temperature sounding is shown below. Match the layers with the appropriate letters from the sounding. A layer may have more than one letter.

   1. Stable ___
   2. Unstable ___
   3. Neutral ___
   4. Inversion ___
   5. Isothermal ___

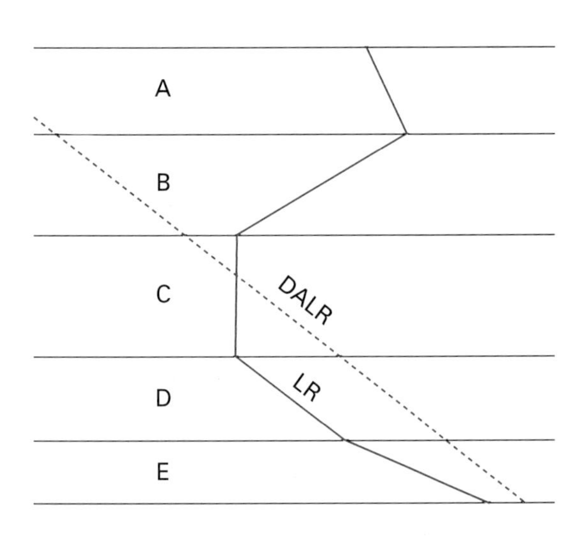

5. You are flying at 4,000 feet AGL over a city during the day. There are no clouds present. As you descend to land at a nearby airport, you notice that, although visibility was very good at 4,000 feet, there is a marked decrease just below 2,500 feet AGL. As you make your final approach, you observe a distinct increase in visibility near the ground. Give a reasonable explanation of the meteorological situation on the basis of what you know about stability.

6. Explain how air can rise when it is stable.

7. The air temperature at a particular station located at sea level is 77°F. The current sounding shows that the lapse rate is 2.5C° per 1,000 feet in the lowest 2,000 feet of the atmosphere.

   1. What is the temperature at 1,500 feet AGL?

   2. Say a parcel of air is lifted from the surface to 1,500 feet AGL. What is the parcel temperature at that level? Is it stable or unstable? Explain why, from the perspective of parcel temperature and lapse rate.

8. The surface air temperature at your airport is 86°F. There is no sounding, but an aircraft reports 80°F at 1,300 feet AGL on climbout. Do you expect turbulence on takeoff? Explain.

CHAPTER 6

Atmospheric Moisture

Introduction

In Chapter 1, we learned that water vapor is a variable gas, occupying only a small percentage of the volume of the gases in the atmosphere. Although water vapor is around us in small and fickle quantities, it has major consequences, not the least of which include icing, thunderstorms, freezing rain, downbursts, whiteouts, frost, and lightning.

In this chapter, we look at the basics of atmospheric moisture, a term which is used here to imply the presence of H_2O in any one or all of its states: water vapor, water, or ice. We examine the transformation between states and the importance of air temperature in the transformation process. When you complete this chapter, you will understand the causes and effects of state changes, how clouds form and dissipate, and how precipitation is produced. You will also know how clouds and precipitation are classified and observed.

Section A

MOISTURE CHARACTERISTICS

Even over tropical oceans, where the greatest concentrations of water vapor are found, it still amounts to only a few percent of the total volume of the atmosphere. In other areas, especially over continents and at high latitudes, day-to-day variations in water vapor may be very large, while the average concentration is very small. Despite this limited and widely varying availability, water vapor is the *stuff* that weather is made of – without it, the pilot's problems with the atmosphere would be greatly reduced and the aviation meteorologist would probably be out of business. But this is not the case. Many of the flight hazards encountered in aviation operations owe their existence to the presence of water vapor, water, and ice in the atmosphere.

In order to discuss moisture in the atmosphere, you must first become aware of certain definitions and basic physical processes. That essential information is developed in this section and applied in succeeding sections to explain the presence of clouds and precipitation.

STATE CHANGES

The three states that H_2O can take in the normal range of atmospheric temperatures and pressures are shown graphically in figure 6-1. Water vapor is a colorless, odorless, tasteless gas in which the molecules are free to move about, as in any gas. In the liquid state (water), molecules are restricted in their movements in comparison to water vapor at the same temperature. As a solid (ice), the molecular structure is even more rigid, and the freedom of movement is greatly restricted.

A change of state refers to the transition from one form of H_2O to another. These transitions have specific names as indicated in figure 6-1. The state changes indicated on the top of the diagram all involve the transition of the H_2O molecules to a higher energy state. They

The processes by which water vapor is added to unsaturated air are evaporation and sublimation.

are melting (ice to water), evaporation (water to vapor), and sublimation (ice directly to vapor without water as an intermediate state).

On the bottom of figure 6-1, all of the transitions take the H_2O molecules to a lower energy state. They are condensation (vapor to water), freezing (water to ice), and deposition (vapor directly to ice without water as an intermediate state).

As with most other substances, when H_2O is in one particular state, a change in molecular motion always corresponds to a temperature change. However, things are a little different when a state change occurs. At that point, there is a large change in molecular motion that does not correspond with a measurable temperature change. For example, if water evaporates, energy is used by the molecules to jump to the higher

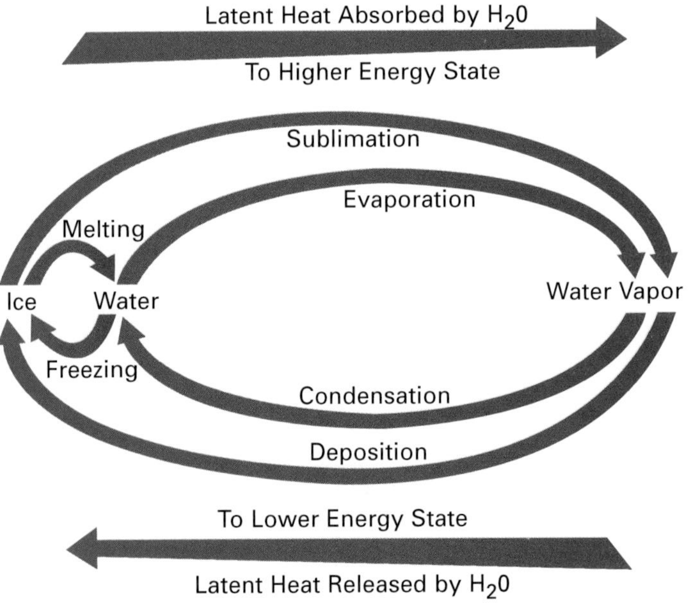

Figure 6-1. Changes of State of H_2O. The latent heat energy involved in the ice-to-water state change is only about 17% of the latent heat required in the water-to-vapor change.

energy associated with the water vapor state. (Figure 6-1) If condensation occurs, energy is lost by H_2O molecules as they return to water, a lower energy state.

The amount of heat energy that is absorbed or released when H_2O changes from one state to another is called latent heat. It is absorbed and "hidden" (not measurable as a temperature change) in H_2O molecules in the higher energy states and released as sensible heat (heat that can be felt and measured) when the molecules pass to lower energy states. (Figure 6-1)

You have probably had experiences related to these processes. For example, when you step out of a swimming pool on a hot day, you experience cooler temperatures as heat is taken from your skin to supply the water with the energy to evaporate. So-called swamp coolers are very popular in dry climates; they operate on the principle of cooling by evaporation. The evaporative cooling effect is not limited to water. For example, alcohol evaporates readily. If you place rubbing alcohol or some other alcohol-based substance on your skin, you will experience cooling as sensible heat is removed so the liquid can make the jump to the vapor state in the evaporation process.

The evaporation (also called vaporization) of fuel in the carburetor of your aircraft engine results in cooling. When that evaporative cooling is combined with cooling caused by expansion of air in the carburetor, induction icing may result even when the OAT is above freezing. This problem is covered in greater detail later in the text.

The release of latent heat when condensation occurs is a major energy source for many meteorological circulations. For example, heat released in the condensation process during cloud formation is an important factor in the production of the greater instability and stronger vertical motions of thunderstorms. Similarly, the heat taken from tropical oceans by evaporation becomes the primary energy source for hurricanes when it is released in condensation.

VAPOR PRESSURE

In the mixture of atmospheric gases, each individual gas exerts a partial pressure. When all of the partial pressures are added together, they equal the total atmospheric pressure (29.92 inches of mercury at sea level in the standard atmosphere). The partial pressure exerted by water vapor (H_2O in gaseous form) is called vapor pressure (VP). It is the force per unit area exerted by the molecules of water vapor in the atmosphere. It is proportional to the amount of water vapor in the atmosphere. For example, if water vapor is added to the atmosphere, vapor pressure increases.

An important condition with respect to the presence of water vapor in the atmosphere is known as saturation. It occurs when the same amount of molecules are leaving a water surface as are returning. The vapor pressure exerted by the molecules of water vapor in this equilibrium condition is known as saturation vapor pressure (SVP). (Figure 6-2)

Figure 6-2. Cross section of the interface between water and water vapor. The red dots indicate H_2O molecules leaving the water surface and the blue dots indicate H_2O molecules returning to the water surface. When the same number of molecules are going in both directions, the condition is said to be saturated.

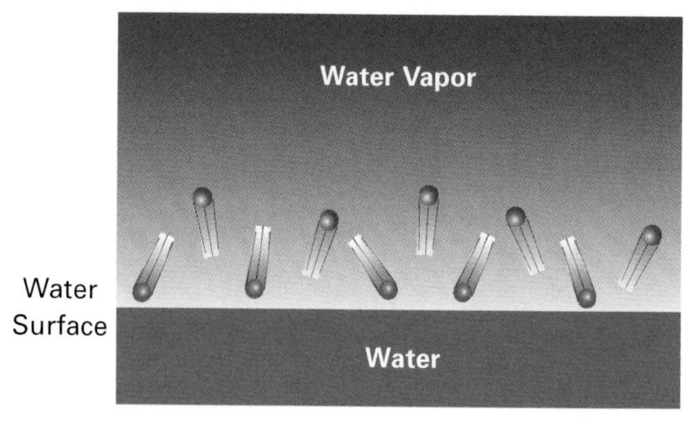

If the atmosphere is saturated at sea level under standard conditions, the pressure (SVP) exerted by the water vapor molecules in the atmosphere is only about 0.5 in.Hg. versus 29.92 in.Hg. for the total pressure.

The significance of the concept of saturation is that it serves as a practical upper limit for the amount of water vapor the atmosphere can hold at a given temperature. SVP is a measure of that upper limit and depends primarily on the temperature. Saturation has many practical applications and is very useful in explaining several atmospheric processes involving changes of state. From theory, the relationship between SVP and temperature is well-known, so we can estimate SVP by simply measuring the temperature. (Figure 6-3)

TEMPERATURE		SATURATION VAPOR PRESSURE (inches of mercury)	
(°F)	(°C)	Ice	Water
-40	-40	.004	.006
-22	-30	.011	.015
-4	-20	.030	.037
14	-10	.077	.085
32	0	.180	.180
50	10		.362
68	20		.690
86	30		1.253
104	40		2.179

Figure 6-3. The dependence of saturation vapor pressure (SVP) on temperature.

The amount of water vapor needed to saturate the air largely depends on air temperature.

Figure 6-3 shows two very important characteristics of SVP. First, saturation vapor pressure increases rapidly as the temperature increases. Consider, for example, air that is saturated; a cloud is present. There is no evaporation because the air is at full capacity for water vapor. If we heat the air, we increase that capacity and the cloud evaporates. More details on cloud formation/dissipation processes are given in later sections.

The second important characteristic of figure 6-3 is that at temperatures below 32°F (0°C), there are two possible saturation vapor pressures for each temperature. This is because water does not necessarily freeze at those temperatures (it is supercooled). The value of SVP in this temperature range depends on whether the surface is ice or water. SVP is lower over ice than over water at the same temperature.

The concept of saturation vapor pressure helps explain why water boils. Boiling occurs when SVP equals the total air pressure. This occurs at standard sea level pressure when the water temperature is raised to 100°C. At that point, bubbles of water vapor form throughout the water and rise to the surface.

Recall from an earlier chapter that, without a pressure suit, your bodily fluids will vaporize at very high altitudes. Similarly, water boils at lower temperatures at high altitudes because the atmospheric pressure is lower at those levels. In other words, the temperature does not have to be raised as much as it does at sea level to make the saturation vapor pressure equal the total air pressure.

RELATIVE HUMIDITY

It is often useful to determine how close the atmosphere is to saturation. This information can help you anticipate the formation of clouds or fog. This is done by measuring the amount of water vapor in the atmosphere in terms of actual VP and then estimating SVP from a temperature measurement. (Figure 6-3) The degree of saturation is then computed by taking the ratio of VP and SVP and multiplying it by 100. The result is called relative humidity (RH). It expresses the amount of water vapor actually in the air as a percentage of the amount required for saturation.

$$RH(\%) = (VP/SVP) \times 100$$

It is important to remember that RH is *relative* to SVP, and that SVP depends on the temperature. RH tells us nothing about the actual amount of water vapor in the air. For example, saturated air at -4°F in Alaska only has about one twentieth of the water vapor in saturated air at 68°F in Florida, although RH = 100% in both cases. This has some important ramifications. For example, suppose the air outside your cockpit is saturated (RH = 100%) at a temperature of -10°C. If that air is brought into the cockpit for ventilation, and heated to 10°C (50°F) along the way, the cockpit humidity will be less than 25 percent.

DEWPOINT TEMPERATURE

Dewpoint is the temperature at which condensation first occurs when air is cooled at a constant pressure without adding or removing water vapor. Dewpoint temperature is always less than the air temperature, with one exception. When the air is saturated (RH = 100%), the temperature and dewpoint are equal.

Dewpoint refers to the temperature to which air must be cooled to become saturated.

For aviation meteorology, dewpoint is reported rather than relative humidity. Dewpoint is extremely useful in predicting precipitation amounts, thunderstorms, and icing. However, changes in relative humidity are helpful for anticipating clouds, fog, and low visibilities. A very useful quantity that relates RH and dewpoint is the temperature-dewpoint spread (also called the dewpoint depression). It is the difference between the air temperature and dewpoint. When the temperature-dewpoint spread is small, the RH is high. When the spread is very large, the RH is low. (Figure 6-4)

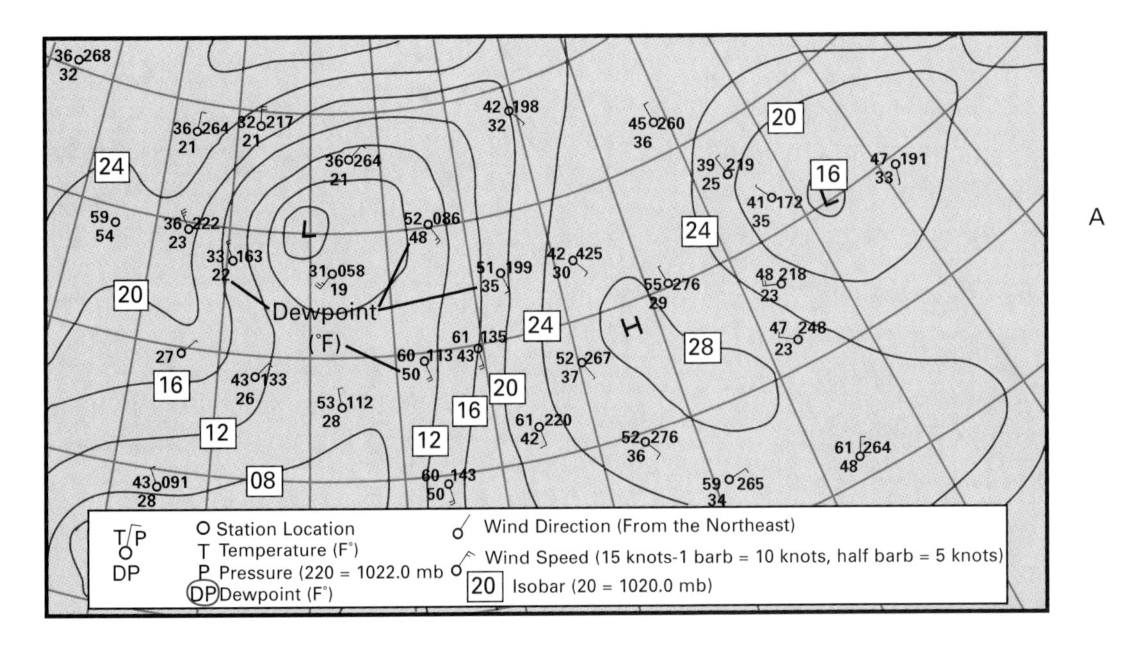

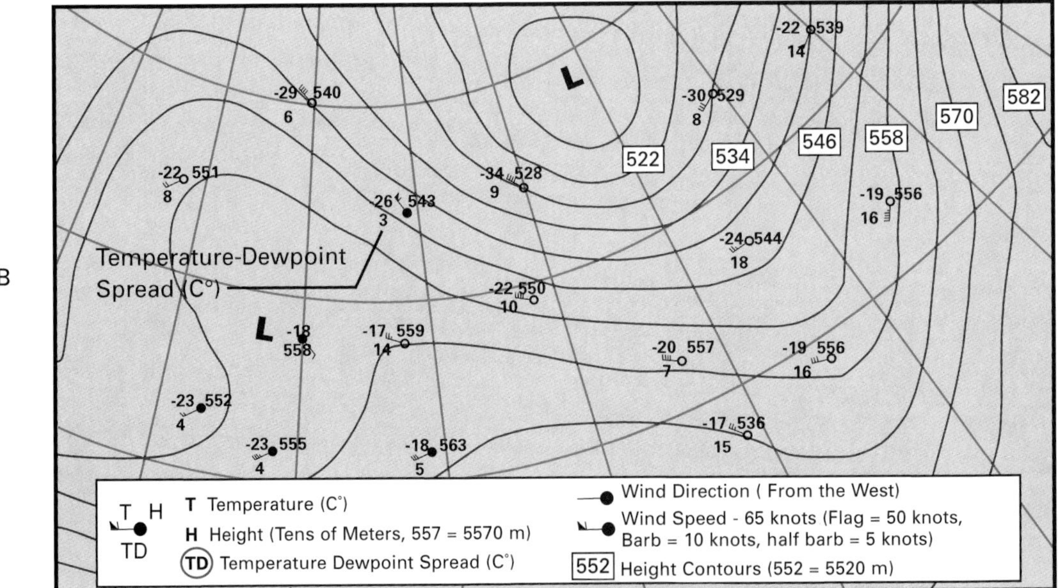

Figure 6-4. Surface analysis chart (A) and 500 mb constant pressure chart (B) showing plotted dewpoint (on the surface analysis chart) and temperature-dewpoint spread (on the 500 mb constant pressure chart). Wind, temperature, pressure (A) and heights (B) as described in the previous chapter are also shown.

Dew is a condensation product that forms when the ground or other object (such as the wings of a parked airplane) loses energy through nighttime (nocturnal) radiation. Cooling reduces the temperature of the thin layer of air next to the object. When the temperature reaches its dewpoint, dew condenses on the colder surface. If the temperature falls below 32°F (0°C) after dew is present, it will freeze (white dew).

In contrast with white dew, white frost is a deposition product. It forms under the same conditions favorable for dew, except that the dewpoint temperature is below 32°F (0°C). In this case, the critical temperature is the frostpoint, rather than the dewpoint. In operational usage, the term dewpoint is always used, regardless of the temperature.

Direct measurements of dewpoint are difficult. However, there are some practical indirect methods. One of the most common ways to determine dewpoint is from measurements with an instrument called a psychrometer. This instrument consists of two thermometers. One measures the air temperature (dry bulb thermometer). The second thermometer is covered with a wick saturated with water (wet bulb thermometer). When the wick is ventilated, evaporation occurs and the temperature decreases. The wet bulb temperature is the lowest temperature that can be reached by evaporative cooling. The difference between dry bulb and wet bulb temperatures (wet bulb depression) will be large if the atmosphere is dry and zero if the atmosphere is saturated. The relationship between dry bulb temperature, wet bulb depression, and dewpoint (and relative humidity) are well known. Reference tables are provided in the appendix.

Frost forms when the temperature of the collecting surface is at or below the dewpoint of the adjacent air and the dewpoint is below freezing. Frost is considered hazardous to flight because it spoils the smooth flow of air over the wings, thereby decreasing lifting capability.

Section B

CLOUDS

A cloud is a suspension of water droplets and/or ice crystals in the atmosphere. In and around clouds, condensation, sublimation, evaporation, latent heat release, snow, rain, hail, and several other important processes and hazards occur that are unique to the saturated atmosphere. In this section, we examine the causes and types of clouds.

CLOUD FORMATION

The three requirements for cloud formation are:

1. Water vapor

2. Condensation nuclei

3. Cooling

WATER VAPOR

Clouds don't form in dry air. Their development requires adequate water vapor so that a reasonable amount of cooling will lead to a change of state from vapor to water droplets or ice crystals. Clouds are always more likely to form in air with high RH (small temperature-dewpoint spread) than in drier air.

CONDENSATION NUCLEI

Condensation nuclei are microscopic particles, such as dust and salt, that provide surfaces on which water vapor can condense to water or sublimate to ice. These are also called "hygroscopic" nuclei because they have an affinity for water. Without these particles, it would be more difficult for the state changes to occur. Fortunately, an adequate number of condensation nuclei are almost always present in the atmosphere.

Fog is usually more prevalent in industrial areas because of an abundance of condensation nuclei from combustion products.

COOLING

If air is not already saturated, either more water vapor must be added to bring it to saturation or it must be cooled in order to form a cloud. In either case, the RH must be raised to 100 percent. You know from our previous discussions that cooling is an effective way to reach saturation because the capacity of air to hold water vapor decreases as the temperature goes down.

Cooling for cloud formation is usually the result of one of two processes:

1. Contact of the air with a cold surface
2. Adiabatic expansion

Cooling by contact with the earth's surface is primarily responsible for fog, dew, and frost. This contact cooling is the process by which heat is conducted away from the warmer air to the colder earth. You should be aware that, regardless of whether clouds form, contact cooling always causes the stability of the air to increase.

Contact cooling occurs when warm air is advected over a relatively cool surface. If there is adequate moisture, so-called advection fog forms. A classic example of the formation of advection fog occurs when air from the warm Gulf Stream waters of the Atlantic moves over the colder waters of the Labrador current. Another example is found along the California coast where northwesterly winds bring warm, moist Pacific air across the colder California current and upwelling coastal waters. Both of these regions are notorious for low clouds and fog. (Figure 6-5)

Advection fog is most likely to occur in coastal areas.

Figure 6-5. A visible image from a Geostationary Operational Environmental Satellite (GOES), shows the tops of fog and low clouds caused by advective cooling over the cold water along the west coast of the United States.

Contact cooling can also occur at night after the ground cools due to terrestrial radiation. Contact cooling effects are greater under conditions of clear skies and light winds. When the temperature is reduced to the dewpoint by such cooling, so-called radiation fog will form. If the fog is very shallow (less than 20 feet deep), it is called ground fog. Radiation fog is a common phenomenon in wintertime when nights are long. It is often found in river valleys (valley fog) where cool air pools and moisture is abundant. Good examples are found in the valleys of the Appalachian Mountains in the eastern U.S. and the San Joaquin Valley in California. (Figure 6-6) Specific flight hazards associated with fog are presented in Part III.

The great majority of clouds that occur away from the earth's surface (in the free atmosphere) form in air that is cooled by adiabatic expansion; that is, in air that is moving upward. Whenever you see a cloud in the free atmosphere, most likely the air is either moving upward or has recently been moving upward. If a cloud is dissipating, it is often moving downward. It also may be mixing with its dry surroundings. In either case, the cloud is evaporating.

Conditions favorable for the formation of radiation fog over a land surface are clear skies, little or no wind, and a small temperature-dewpoint spread.

Figure 6-6. Valley fog is caused when nighttime radiational cooling along the mountainsides produces cool, dense air which subsequently drains into the valley. River valleys with ample supplies of moisture are favorite locations for valley fog during the cold months.

Adiabatic cooling can also contribute to cloud formation in the vicinity of sloping terrain. This is why there is increased cloudiness on the upwind (windward) side of mountains and decreased cloudiness on the downwind (leeward) side of the mountains. The drier, downwind side of the mountain is often described as a rainshadow. (Figure 6-7)

Contact cooling always produces stable air; whereas, adiabatic cooling can be associated with either stable or unstable air. The air that rises rapidly in thunderstorms is unstable. In contrast, wintertime fogs over the western plains of the U.S. often develop when moist air flows northward from the Gulf of Mexico. The air is cooled adiabatically, as it moves upslope, and cooled by contact with the cold ground, especially after sunset. This is sometimes called upslope fog. (Figure 6-8)

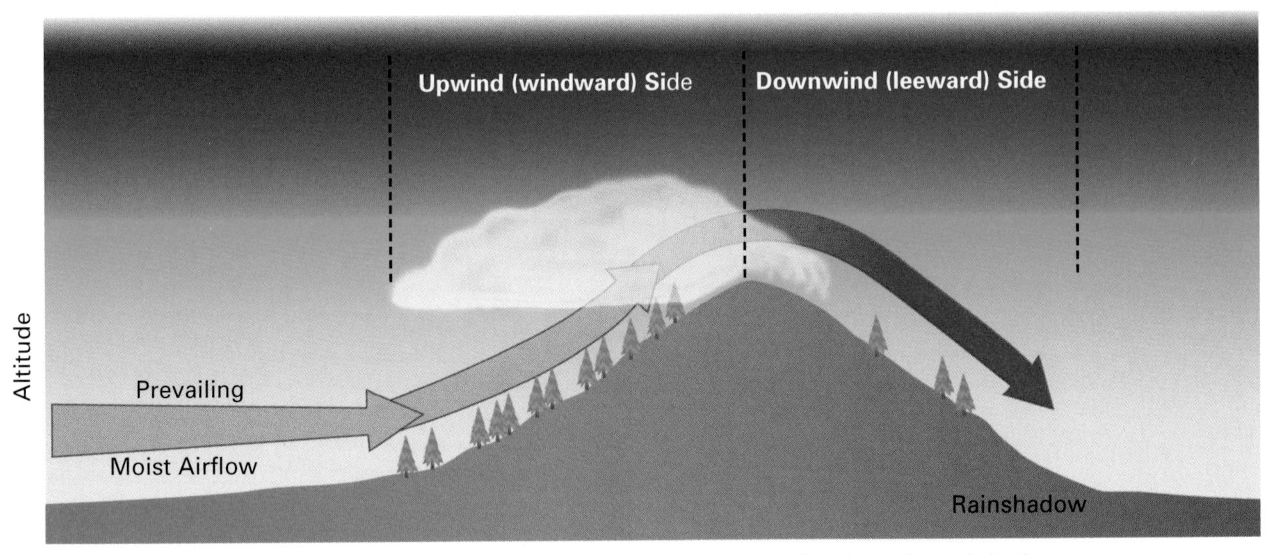

Figure 6-7. When moist air flows over a mountain range, clouds and precipitation normally form on the upwind (windward) side.

The types of fog that depend upon wind in order to exist are advection fog and upslope fog.

Figure 6-8. In winter, warm, moist air from the Gulf of Mexico is carried northward. The air cools adiabatically as it moves over rising terrain, resulting in upslope fog. Nocturnal radiation and contact cooling stabilizes the air and further reduces the air temperature to the dewpoint.

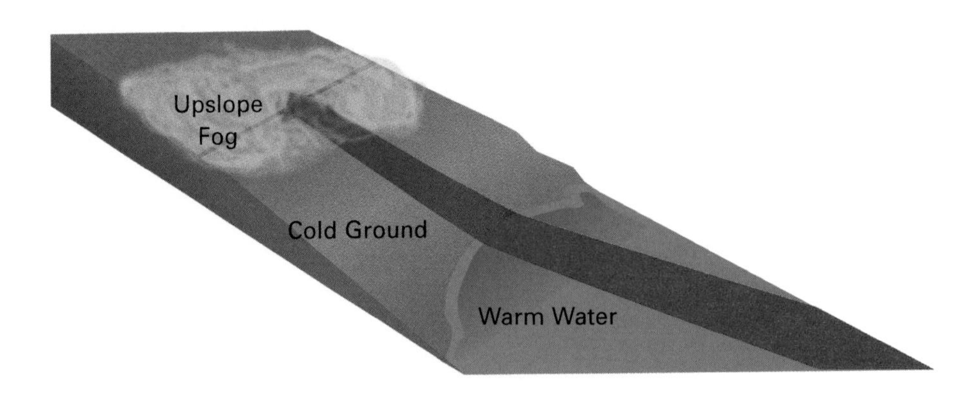

Clouds can also form when warm, moist air is mixed with cold air. A good example of this process is steam fog. When very cold air moves over warm water, air very close to the surface picks up water vapor from the strong evaporation. Because the moist air in contact with the surface is warmer than the overlying cold air, convection develops, causing the moist air to mix with the cold air aloft. With the mixing, the air temperature is reduced below the dewpoint of the moist air. Condensation occurs in a shallow layer of wispy, plume-like columns. This process explains why steam fog is also called evaporation fog or sea smoke. (Figure 6-9)

Figure 6-9. Steam fog is common over unfrozen water bodies in the cold months of the year.

LATENT HEAT AND STABILITY

When air becomes saturated, further cooling results in a change of state (cloud formation) and a release of latent heat. This is particularly important when an air parcel is rising and cooling adiabatically. The additional heat can cause important changes in stability.

When condensation occurs in a parcel of rising air, adiabatic cooling is partially offset by warming due to the release of latent heat. Keep in mind that latent heat never completely offsets adiabatic cooling. A saturated parcel continues to cool as it rises, but at a slower rate than if it were dry. This is called a **saturated adiabatic process** and the rate of cooling of a rising, saturated parcel is called the moist or **saturated adiabatic lapse rate** (SALR).

The rate of cooling of an air parcel during ascent in saturated conditions (in a cloud) is always less than in dry conditions.

Although DALR is a constant 3C° per 1,000 feet, SALR is variable. It is about the same as DALR at extremely cold temperatures (-40°C), but is only a third of DALR value at very hot temperatures (100°F). This is because saturated air holds much more water vapor at high temperatures, so there is much more latent heat to release when condensation occurs.

The saturated adiabatic lapse rate varies between about 3C° per 1,000 feet for very cold temperatures and 1C°per 1,000 feet for very hot temperatures.

The condensation level is the level at which a cloud forms in rising air. Below the condensation level the rising parcel cools at the DALR and above that level it cools at the SALR. Since parcel stability depends on the difference between the temperature of the parcel and the temperature of its surroundings, a saturated, rising parcel will be less stable than a dry parcel, all other conditions being the same. Note that *less stable* does not necessarily mean *unstable*. The difference is apparent in the cloud forms described in the next section. (Figure 6-10)

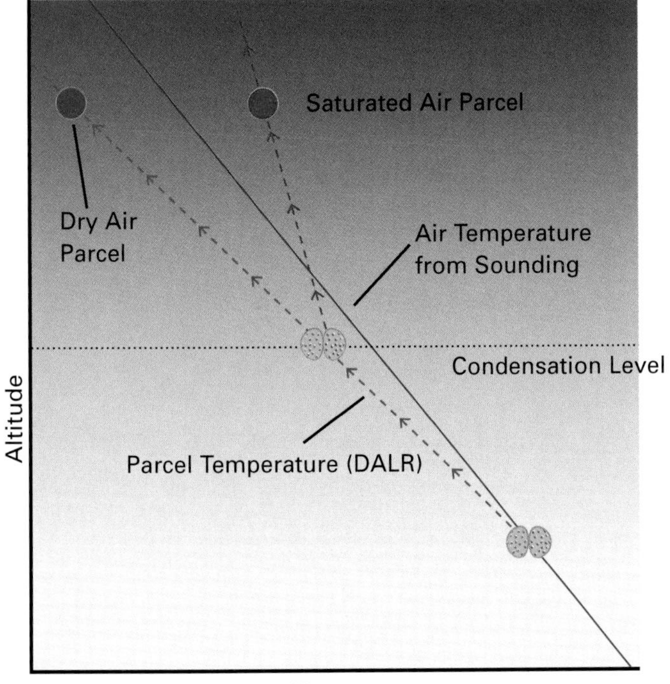

Figure 6-10. If two parcels of air are lifted the same distance, starting from the same level and the same initial temperature, the parcel of air which becomes saturated will be warmer than the unsaturated parcel. Note, in this case, that the dry parcel remains colder than its surroundings (stable), but above the condensation level, the saturated parcel becomes warmer (unstable).

Figure 6-10 illustrates a fourth stability classification (in addition to absolutely unstable, neutral, and stable). In the example shown in the figure, if the parcel stays dry, it is stable; but, if it reaches saturation it becomes unstable. The outcome depends on whether latent heat is released. This is a case of conditional instability. The "condition" is that the air parcel must become saturated in order to be unstable.

CLOUD OBSERVATIONS

Clouds are important indicators of the state of the atmosphere. They give visual clues about imminent weather changes, including the advance or retreat of large scale weather systems, winds, turbulence, and stability. Learning to observe clouds properly is essential for the safety of flight.

STANDARD OBSERVATIONS

Clouds are regularly observed and reported at weather stations. As a pilot, you will be using these observations for flight planning and to make in-flight decisions. It is important that you develop the ability to interpret an observation as if you were the observer who reported it. Cloud observations are made from the ground, where the view of the sky is different than what you see from the cockpit. In order to derive the most useful information from a cloud report, you must learn to "stand in the shoes of the observer."

The technical description of all of the clouds present in the sky at a particular location is called the sky condition. A complete observation of sky condition includes:

1. Cloud height(s)

2. Amount of sky covered by clouds

3. The type(s) of clouds

CLOUD HEIGHT

A cloud layer refers to clouds with bases at approximately the same level. A cloud layer may be a continuous sheet of clouds, or it may be made up of many individual clouds. It is sometimes called a cloud deck. There may be one or more cloud layers reported in a given observation.

Cloud height refers to the height of the base of a cloud layer above *ground* level (AGL). Heights are reported in hundreds of feet in the United States. Clouds with bases 50 feet AGL or less are reported as fog. When cloud bases are very low and ragged and/or when fog is present, the reported cloud height corresponds to the **vertical visibility**; that is, the distance that an observer or some remote sensing device can "see" into the cloud. (Figure 6-11)

The heights of cloud bases can be estimated or measured. The measurements are determined from aircraft reports, by means of a ceilometer or with a ceiling height balloon. A ceilometer is a device with a vertical pointing light beam that locates and measures the point where the beam is brightest. The ceiling height balloon is a small balloon with a known rate of ascent. The height of the cloud base is determined by timing the balloon from its release to the point where it disappears in the cloud.

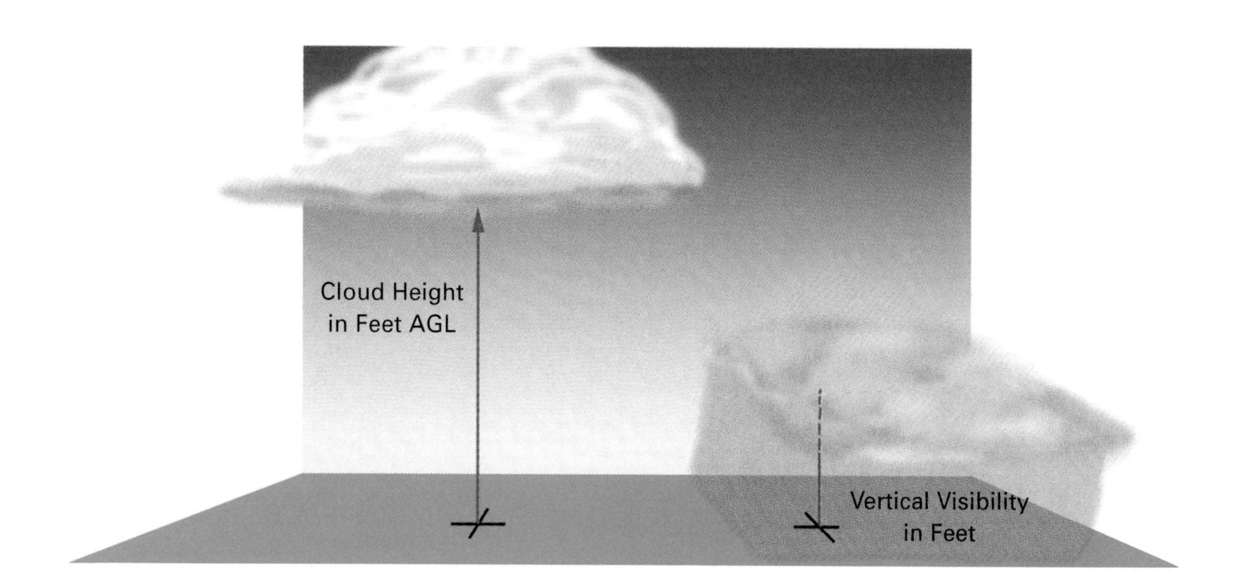

Figure 6-11. Cloud height is the height of the base of the cloud above the ground. When the cloud base is not clearly defined or when the sky is completely obscured, the reported cloud height is actually the vertical visibility.

CLOUD AMOUNT

Cloud amount refers to the amount of sky covered by each cloud layer. It is usually observed in tenths or eighths of the **celestial dome**, which is the hemisphere of sky observed from a point on the ground. Cloud amount is reported as the total cloud cover at and below the layer in question. "At and below" implies a cumulative amount. Keep in mind that a surface observer can only see the lowest cloud deck. (Figure 6-12)

When no clouds are present (less than 1/10), the sky condition is reported as **clear (CLR)**. If the cloud amount for a particular layer is 1/10 to 5/10, the coverage is designated **scattered (SCT)**. If the cloud amount is 6/10 to 9/10, the cloud layer is **broken (BKN)**. If the coverage is 10/10, the layer is **overcast (OVC)**. (Figure 6-12)

VISIBILITY

Horizontal visibilities play an important role in the classification of certain sky conditions. Furthermore, sky conditions and visibility are critical for the specification of certain flight restrictions. For these reasons, it is important to pause here and define some useful terminology.

Visibilities reported in standard weather reports are horizontal surface visibilities; that is, they are measured by an instrument or a person standing on the ground or some convenient point of

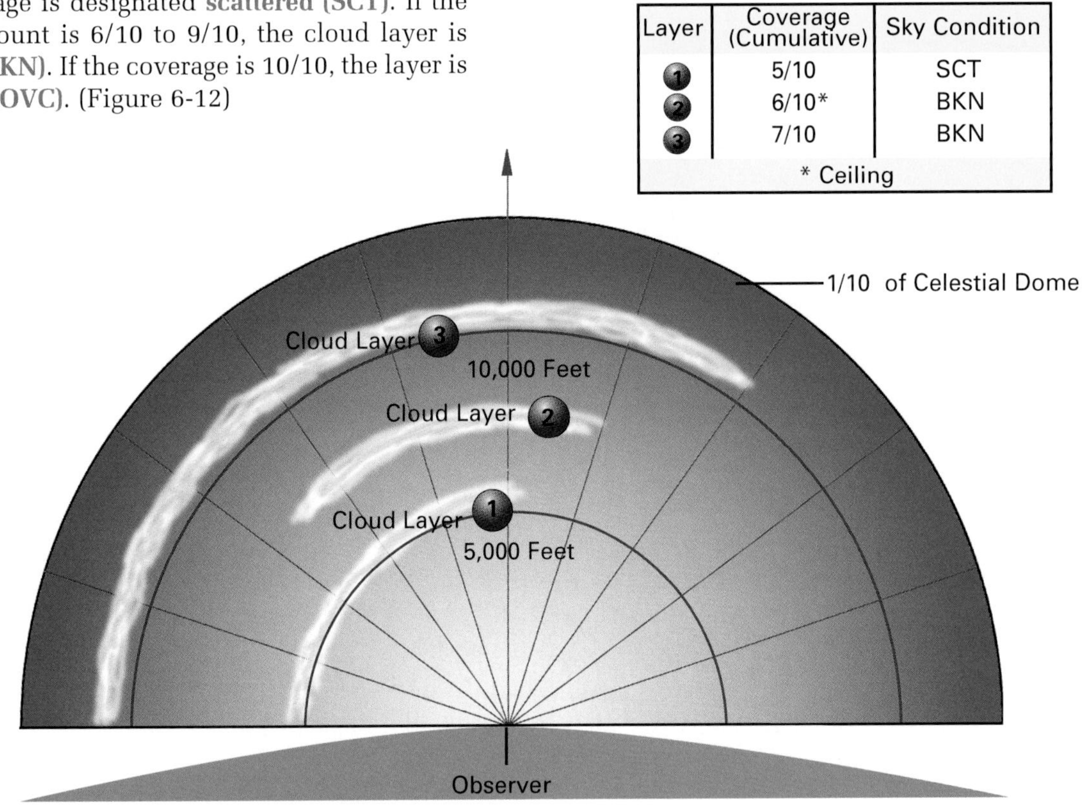

Layer	Coverage (Cumulative)	Sky Condition
1	5/10	SCT
2	6/10*	BKN
3	7/10	BKN

* Ceiling

Figure 6-12. Cross section showing a ground observer's view of the celestial dome. The thick white curves represent cloud layers at different altitudes. Each "slice" of the celestial dome which intersects a cloud layer visible to the observer represents a tenth of total cloud amount. Cloud amounts are reported for each cloud layer in the order of lowest to highest. The observed coverage of each succeeding layer is added to the cumulative coverage of the lowest layers.

measurement, such as the roof of a building or a tower. The point of measurement is often near the local weather station. Tower visibility is the horizontal visibility determined from the control tower. In reports, it is identified as such only when the official surface visibility is determined from a different location.

Prevailing visibility is the greatest horizontal distance over which objects or bright lights can be seen and identified over at least half of the horizon circle. Prevailing visibility is taken as the representative visibility at a particular location. Be aware that visibilities may be lower in other sectors of the horizon circle. (Figure 6-13) These are usually reported. Standard reporting formats for visibility, as well as sky condition, weather, and precipitation are presented at the end of this chapter.

In the U.S., prevailing visibility is reported in statute miles and fractions. Runway visibility (RVV) is the visibility from a particular location along an identified runway. It also is reported in statute miles and fractions. Runway visual range (RVR) is the maximum horizontal distance down a specified instrument runway that a pilot can see and identify standard high intensity lights. It is reported in hundreds of feet.

At this point in our discussion, it is helpful to look at some of the criteria and terminology used to define visual and instrument weather conditions. Weather conditions are divided into four classifications: visual flight rules (VFR), marginal visual flight rules (MVFR), instrument flight rules (IFR), and low instrument flight rules (LIFR). Meteorologists define these classifications in terms of ceiling and visibility. (Figure 6-14)

VFR: ceiling > 3,000 feet AGL and visibility > 5 s.m.

MVFR: ceiling 1,000 to 3,000 feet AGL and /or visibility 3 to 5 s.m.

IFR: ceiling < 1,000 feet AGL and /or visibility < 3 s.m.

LIFR: ceiling < 500 feet AGL and /or visibility < 1 s.m.

Figure 6-14. Terminology and criteria used to define visual and instrument weather conditions.

More information about weather conditions is contained in the chapter on instrument meteorological conditions.

Figure 6-13. In this example, an observer can see and identify objects (day) or bright lights (night) at 5 s.m. except to the east where fog restricts visibility to 2 s.m. The prevailing visibility in this case is 5 s.m.

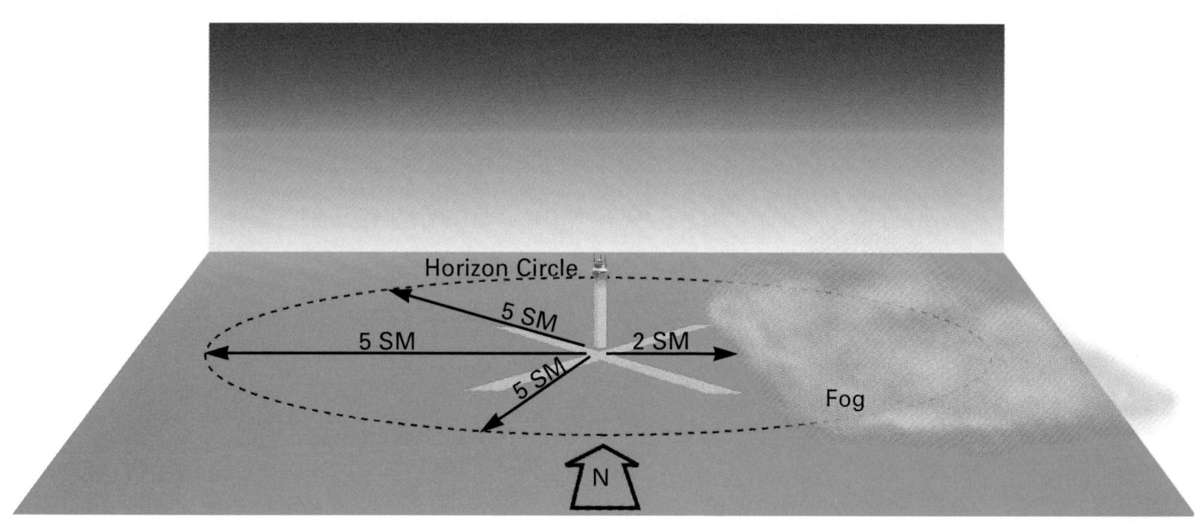

OBSCURATIONS AND CEILINGS

If the sky is totally hidden by a surface-based obscuring phenomena, such as fog, smoke, dust, or precipitation that reduces the prevailing visibility to less than seven miles, the sky is reported as obscured (X). If the sky is only partly hidden (9/10 or less) by a surface-based obscuring phenomenon, it is called partially obscured (-X). (Figure 6-15)

Cloud layers may be reported as thin. This means that half or more of the coverage is transparent; that is, blue sky, higher cloud layers, or stars can be seen through the layer. For example, a thin, scattered layer at 10,000 feet AGL is reported as 100-SCT (read as: ten thousand thin scattered).

The lowest cloud layer or obscuration which has 6/10 or greater coverage, and is not reported as thin or partially obscured, is designated as a

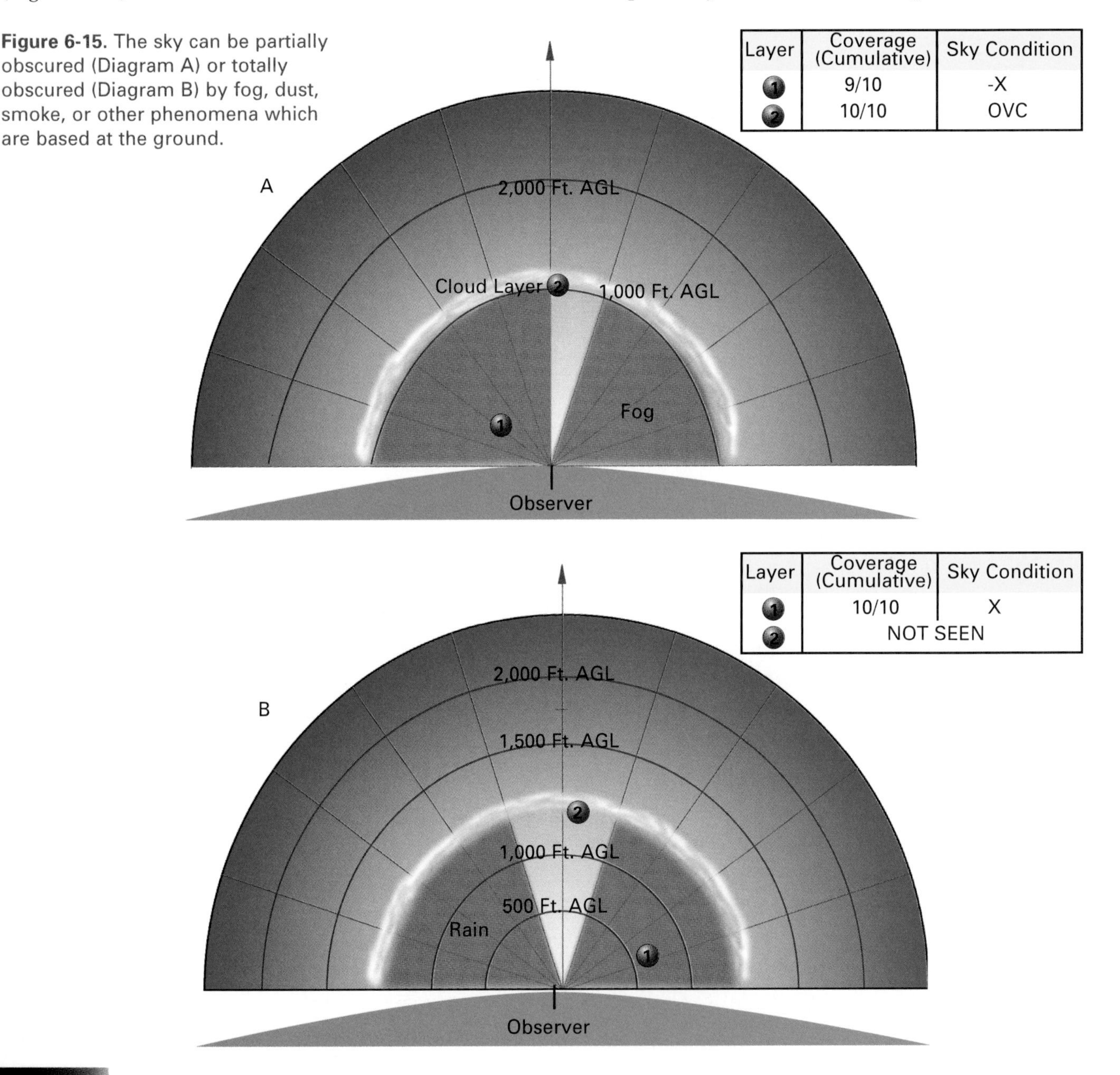

Layer	Coverage (Cumulative)	Sky Condition
1	9/10	-X
2	10/10	OVC

Layer	Coverage (Cumulative)	Sky Condition
1	10/10	X
2	NOT SEEN	

Figure 6-15. The sky can be partially obscured (Diagram A) or totally obscured (Diagram B) by fog, dust, smoke, or other phenomena which are based at the ground.

ceiling. This is important to pilots because it has implications regarding VFR, IFR, and the associated flight rules. When a cloud layer is designated as a ceiling, a letter precedes the ceiling height indicating that it is measured (M), estimated (E), or indefinite (W). (Figure 6-16)

A ceiling is defined as the height above the earth's surface of the lowest layer of clouds or obscuring phenomena reported as broken or overcast, and not classified as thin or partial.

When clouds are observed in flight, a pilot weather report (PIREP) should be made of the heights of bases and tops of individual cloud layers whenever possible. Remember, weather observing stations are few and far between and surface observers do not have the in flight perspective. Therefore, PIREPs are invaluable sources of cloud information. All cloud heights observed in flight are reported in feet MSL not AGL. A PIREP of a broken layer of clouds with a base at 2,500 feet MSL and a top at 3,300 MSL would be coded: 025 BKN 033.

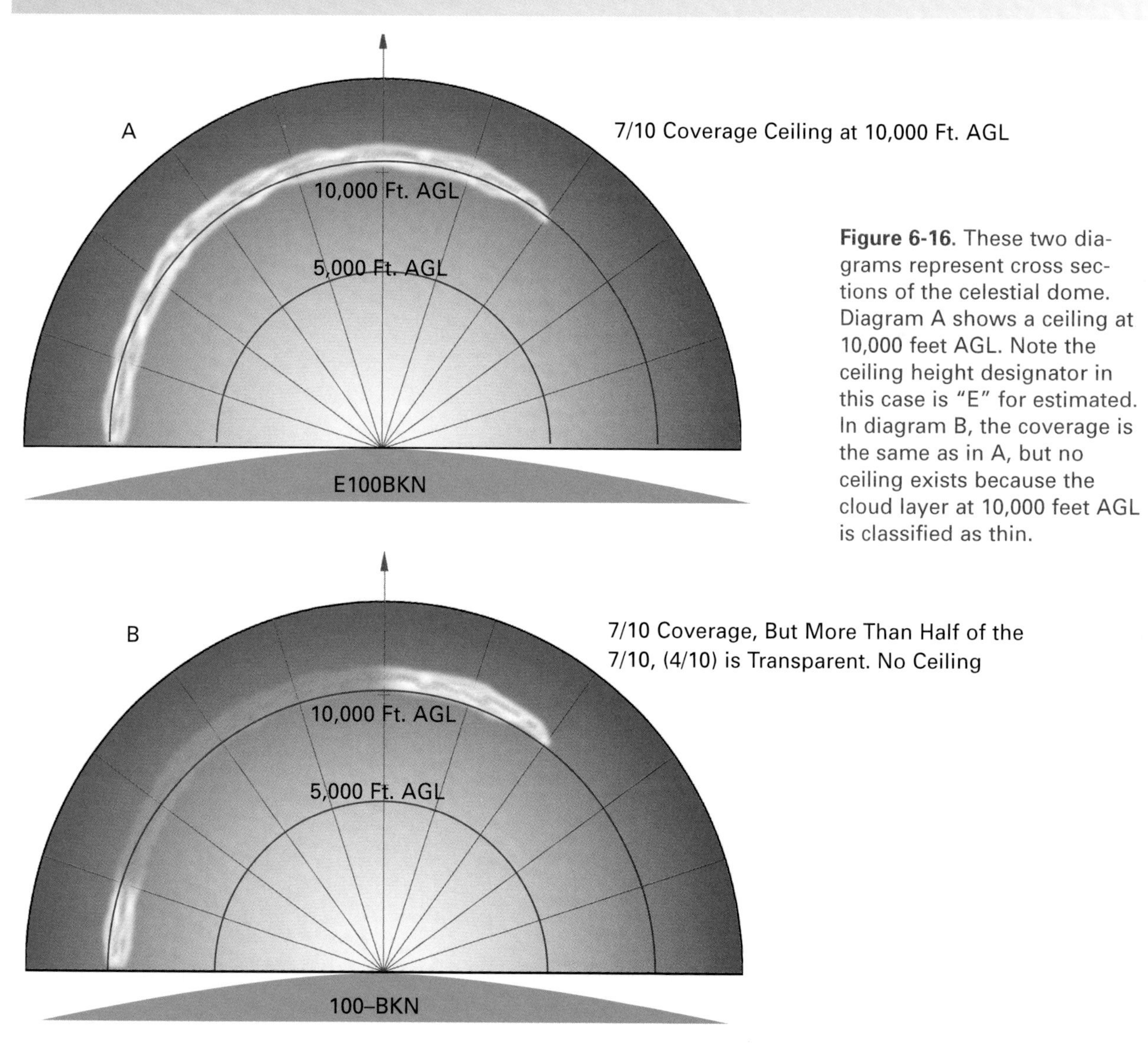

Figure 6-16. These two diagrams represent cross sections of the celestial dome. Diagram A shows a ceiling at 10,000 feet AGL. Note the ceiling height designator in this case is "E" for estimated. In diagram B, the coverage is the same as in A, but no ceiling exists because the cloud layer at 10,000 feet AGL is classified as thin.

CLOUD TYPE

Cloud type is determined on the basis of what a cloud looks like: its height, shape, and behavior. Clouds are classified as low, middle, or high clouds according to the height of their bases above the ground, and as clouds with vertical development. The ten basic clouds in these categories are listed in figure 6-17 and illustrated in figure 6-18.

LOW CLOUDS	MIDDLE CLOUDS	HIGH CLOUDS
< 6,500 feet AGL	6,500 to 20,000 feet AGL	> 20,000 feet AGL
stratus (ST)	altocumulus (AC)	cirrocumulus (CC)
stratocumulus (SC)	altostratus (AS)	cirrostratus (CS)
nimbostratus (NS)		cirrus (CI)
cumulonimbus* (CB)		
cumulus* (CU)		

* = Clouds with Vertical Development

Figure 6-17. International cloud classifications. Altitude ranges of cloud bases are given for each cloud category. Variations occur depending on season and geographical location.

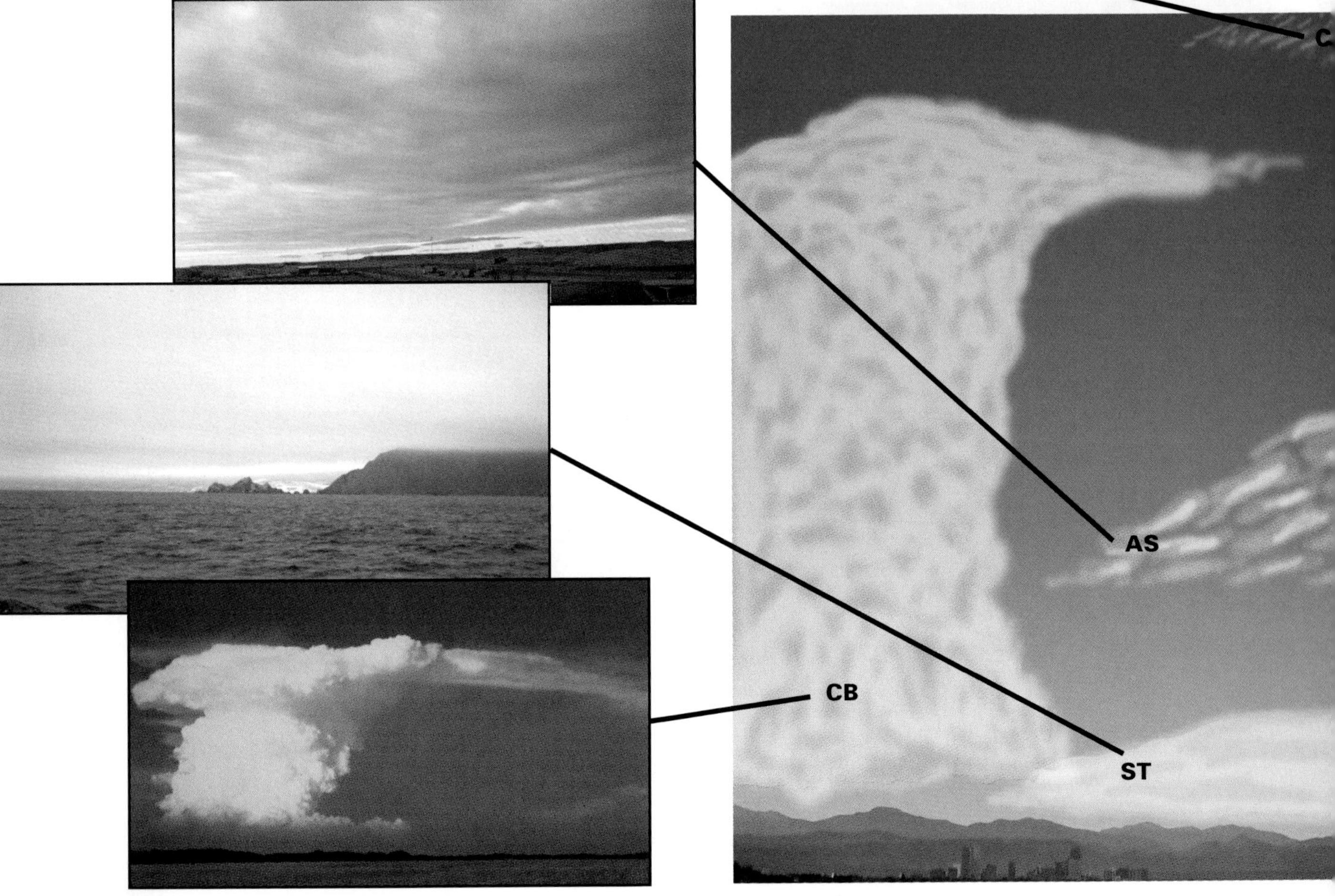

Figure 6-18. Ten basic cloud types. International Cloud abbreviations are shown. The CB is a cloud of great vertical development. Notice in the center diagram that the CB base is in the low cloud range while its top is in the high cloud range. Also note that the CB is not drawn to scale. Typically, a single CB is about as wide as it is tall. (See CB Photograph)

Cloud names are based on the following terms: cumulus (heap), stratus (layer), nimbus (rain), and cirrus (ringlet). In addition, the prefix "alto-" designates middle clouds; and the word "cirrus" and the prefix "cirro-" indicate high clouds.

The four families of clouds are high, middle, low, and those with extensive vertical development.

The distinctive feathery appearance of high clouds is due to the fact that those clouds are composed primarily of ice crystals. Between temperatures of 0°C and -40°C, middle and low clouds are composed of water droplets and ice crystals. The proportion of droplets is much greater at higher temperatures. At temperatures above 0°C clouds are composed entirely of water droplets.

A high cloud is composed mostly of ice crystals.

Keep in mind that cloud height categories are approximate. For example, at high latitudes and in the winter, the tropopause and cirriform clouds tend to be lower, occasionally dipping into the middle-cloud range. Also, the difference between AS and NS is not always precise. Altostratus bases often lower with time; if precipitation begins, the cloud may be identified as NS even though the cloud base may not yet be in the low-cloud range.

Cumuliform clouds in all categories are indicative of some instability. However, well-developed CU, and especially CB, indicate great instability. Although their bases are usually in the low-cloud height range, well-developed CU and CB tops commonly extend well into middle- and high-cloud ranges.

The stability of the air before lifting occurs determines the structure or type of clouds which form as a result of air being forced to ascend.

Because of the extreme dryness in some areas, such as the intermountain western U.S., air must rise 10,000 feet AGL or more to reach the condensation level. High-based CU or CB are common in that area in the summer.

Since cumulus clouds usually develop in air rising from the ground, the heights (H) of cumulus cloud bases (not altocumulus or cirrocumulus) can be accurately estimated from the measured surface temperature (T) and dewpoint (DP) with the formula

$$H = (T-DP)/4.4,$$

where H is in thousands of feet and T and DP are in °F. For example, if cumulus clouds are present with a surface temperature of 72°F and a dewpoint temperature of 50°F, the cloud bases will be at approximately 5,000 feet AGL.

There are many other cloud forms besides the ten basic types discussed here. Some, such as lenticular clouds, are important visual indicators of possible turbulence and other flight hazards. These important cloud variations will be presented later in the text.

OTHER USEFUL CLOUD OBSERVATIONS

The development of your ability to estimate cloud heights and amounts, and to identify the basic cloud types is essential in connecting classroom aviation meteorology to cockpit aviation meteorology. Clouds are caused by specific temperature, wind, vertical motion, and moisture conditions.

Their visual identification will help you anticipate, identify, and avoid many potential aviation hazards. You want to be able to "read the sky" from your aircraft as easily as you read this book.

Cloud watching is a good exercise to develop your skills, especially when you compare your observations with the official reports from a nearby airport. Another benefit of such an activity is that there is much more useful information in cloud observations than simply height, amount, and type. For example, by noting the time it takes clouds to grow, dissipate, or simply move across the sky, you will begin to understand time scales and life cycles. Furthermore, wind directions and relative wind speeds aloft can be estimated by watching the movement of clouds in various layers. Movement and the size of clouds (especially CU) can also be judged by watching cloud shadows. In later chapters, we will expand this list of meaningful cloud features.

Satellites provide us with almost continuous cloud observations from geostationary orbits at 22,000 miles above the equator and from polar orbits a few hundred miles above the surface. Onboard radiometers measure visible light reflected from the clouds and the earth and infrared radiation emitted by the earth, clouds, and certain gases in the atmosphere. Some examples are shown in figure 6-19.

In figure 6-19, the image in A is a visible picture, which depends on reflected light. Except for the lack of color, the cloud and ground features seen in A are about what you would see by eye from the same viewpoint. Visible images are usually only made in daylight.

Diagram B is an image determined from measurements of IR radiation. This is essentially a picture of the pattern of the temperature of the earth's surface when skies are clear or the temperature of the tops of the clouds when the sky is cloudy. Because the highest clouds are generally the coldest, they are easy to identify. In this image, the coldest temperatures correspond with the brightest white and the warmest temperatures correspond with black. Infrared images can be gathered 24 hours per day.

Figure 6-19. Examples of satellite images from the Geostationary Orbiting Environmental Satellite (GOES). Diagram A is a visible image. Diagram B is an infrared image.

A

B

Section C
PRECIPITATION

For surface observations, precipitation is formally defined as any or all forms of particles, whether liquid or solid, that fall from the atmosphere and reach the ground. The point of view of this definition is important. Weather reports and forecasts of surface conditions don't necessarily include the falling particles that an aircraft in flight may encounter. These often evaporate before they reach the ground.

Precipitation contributes to many aviation weather problems. It can reduce ceiling and visibility, affect engine performance, increase braking distance, and cause severe wind shear. Under the right temperature conditions, precipitation can freeze on contact, affecting flight performance and aircraft ground handling. Your knowledge of the characteristics and causes of precipitation provides the necessary background to understand and deal with these and related flight hazards.

PRECIPITATION CAUSES

It is a common error for newcomers to meteorology to make the broad assumption that, since 100% RH means that clouds are present and the presence of precipitation is associated with clouds, then 100% RH must mean precipitation! This is *NOT* necessarily true. Yes, a cloud usually forms when the atmosphere is saturated, but as you know from your own experience,

MOST CLOUDS DON'T PRECIPITATE.

This statement is based on three important facts about clouds and precipitation:

1. Precipitation particles (water and ice) must be much larger than cloud particles so they can fall out of the cloud and exist long enough to reach the ground.
2. Most of the time, processes that produce small cloud particles are not very effective in producing large precipitation particles.
3. Efficient precipitation-producing processes mainly occur in certain cloud types (NS, CB).

In order to understand these limitations, we must examine the processes by which water droplets and ice crystals grow. There are three ways by which precipitation-size particles can be produced.

Condensation/deposition refers to the processes by which cloud particles are initially formed. For precipitation to occur, water droplets or ice crystals simply continue to grow through the addition of more molecules of water vapor by the same processes until the particles are large enough to fall out of the cloud.

The second growth process is called coalescence. In this process, two or more droplets collide and merge into a larger droplet. This happens because the initial sizes of the cloud water droplets are different. The larger drops fall faster, growing as they collide and capture the smaller ones. (Figure 6-20)

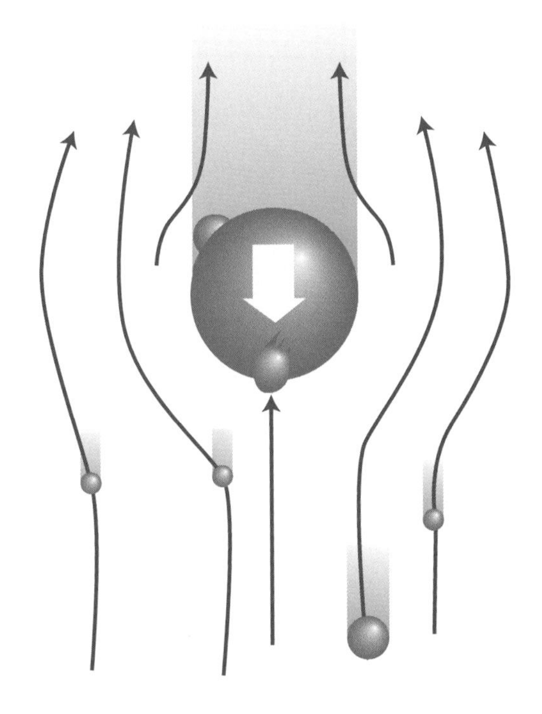

Figure 6-20. Large droplets fall faster than smaller particles, capturing them as they descend.

Although important, neither of the cloud growth processes discussed above can account for all precipitation. These processes are too slow to allow much precipitation to fall within the normal lifetime of a rain cloud.

There are two notable exceptions. Very low stratus and fog are known to produce precipitation that is very light and composed of very small droplets. In many of these cases, the coalescence process is efficient enough to generate the precipitation. The cloud is so close to the ground, that the droplets do not evaporate before they reach the surface. Another exception is found in the tropical regions where large condensation nuclei (salt) from the oceans result in some large cloud droplets. These are large enough and numerous enough for the coalescence process to work efficiently. But outside the tropics, conditions are different.

In middle and high latitudes, especially over continents, condensation nuclei are much smaller, water droplets are smaller and more numerous, and clouds are colder. Coalescence alone cannot produce significant precipitation under these conditions. However, they do favor another growth process.

The third way that cloud particles grow to precipitation size is called the ice-crystal process. It can only operate in regions where water droplets and ice crystals coexist. For ice crystals to be present, the temperature must be below 0°C.

Supercooled water droplets can exist in this subzero environment. They are common in great quantities at temperatures between 0°C and -10°C and droplets have been observed at temperatures near -40°C. The existence of supercooled water droplets illustrates why 0°C is technically referred to as the "melting" point rather than the "freezing" point.

Supercooled water droplets are a primary cause of aircraft icing.

The reason the ice-crystal process works is that it takes less water vapor molecules to reach saturation over ice than over water at a given subzero temperature. Therefore, when water and ice coexist at the same temperature, air in contact with the ice crystals can be saturated (RH = 100%), while the same air in contact with the water droplets is unsaturated (RH<100%). The result of this mixed environment is that ice crystals grow from the water vapor given up by the evaporating water droplets. This process is very efficient, allowing the crystals to grow rapidly to precipitation-size particles in relatively short time periods. (Figure 6-21)

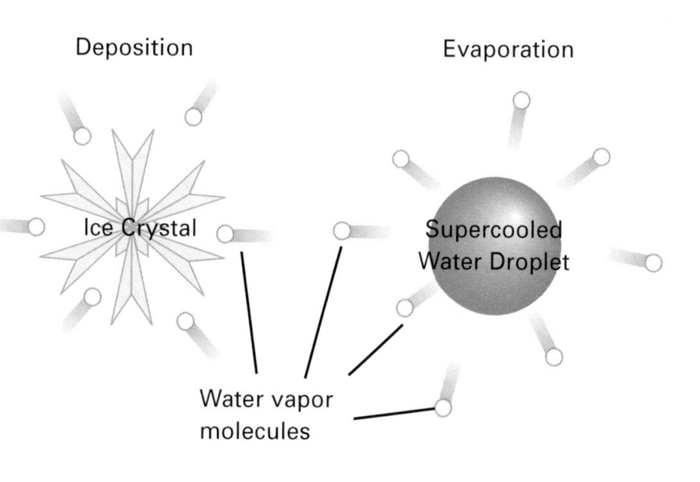

Figure 6-21. Ice crystals grow at the expense of supercooled water droplets in the ice crystal process.

The result of all of this is that, in middle and high latitudes, most precipitation begins as snow. It now becomes clear why most clouds don't precipitate. The clouds may be too warm; that is, they have temperatures above 0°C so ice crystals cannot exist. The clouds (such as cirrus) also may be too cold so there are no water droplets present. In either case, the ice crystal process will not work.

PRECIPITATION CHARACTERISTICS

A complete precipitation report includes type, intensity, and amount. As a pilot, you should know how these observations are made so you can better interpret the weather you experience and the weather reported from other areas.

TYPES

The most common precipitation types include drizzle, rain, rain showers, freezing drizzle, freezing rain, snow, snow showers, snow grains, ice pellets, and hail.

Drizzle is distinguished by very small droplets (diameters less than 0.02 inches). It is commonly associated with fog or low stratus clouds. Rain has larger droplets which fall faster. Rain falls at a relatively steady rate; that is, it starts, changes intensity, and stops gradually. Rain showers refer to liquid precipitation that starts, changes intensity, and stops suddenly. The largest liquid precipitation droplets (about 0.2 inches) and greatest short-term precipitation amounts typically occur with rain showers associated with cumulus clouds and thunderstorms.

Freezing drizzle and freezing rain fulfill the definitions given above except they freeze upon contact with the ground or other objects, such as trees, power lines, and aircraft. Freezing rain produces black ice. This refers to difficult-to-distinguish clear ice on black pavement — a serious hazard for aircraft on the ground. Conditions under which freezing rain or drizzle form are illustrated in figure 6-22.

Ice pellets (or sleet) are transparent, globular, solid grains of ice that are formed from the freezing of raindrops before they reach the ground; they may also be hailstones smaller than 0.2 inches in diameter.

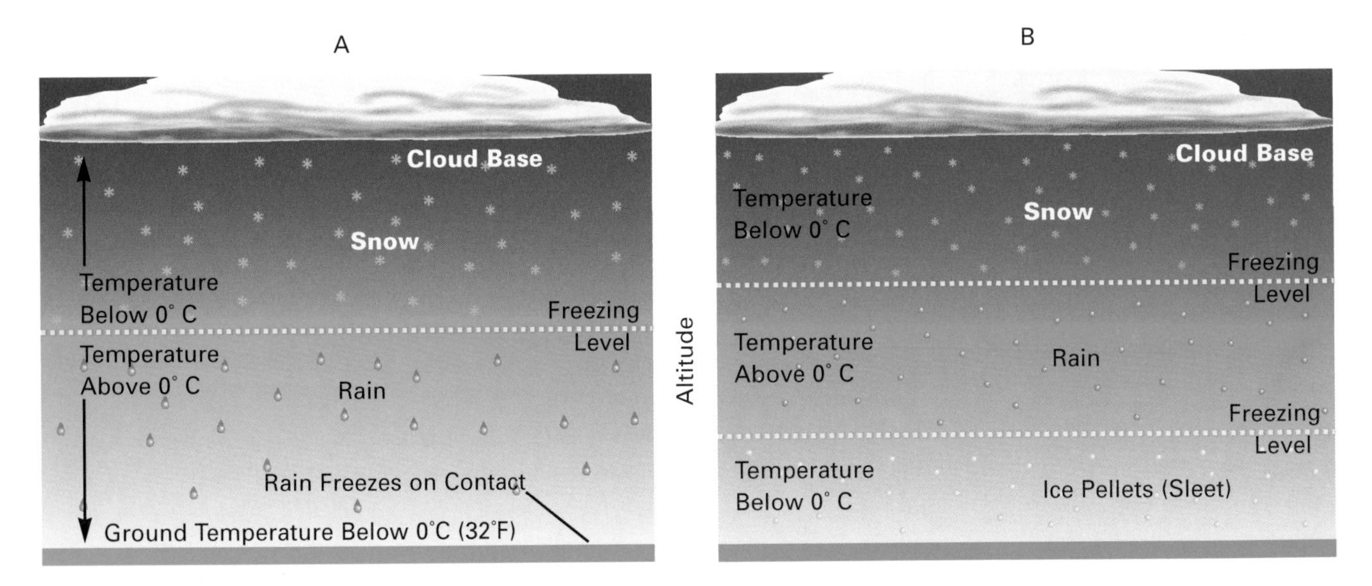

Figure 6-22. Diagram A illustrates the formation of freezing rain (or drizzle). The freezing level is well above the ground. Snow melts as it falls through that level and falls as rain to the surface. The rain freezes on contact with the ground where the temperatures are colder than 0°C. In diagram B, conditions are nearly the same, except that the rain freezes before it reaches the ground as it falls into a shallow layer of cold air. The result is ice pellets (sleet).

The presence of ice pellets at the surface is evidence that there is a temperature inversion with freezing rain at a higher altitude.

Snow is precipitation composed of ice crystals. Snow and snow showers are distinguished in the same manner as are rain and rain showers. Snow grains are the solid equivalent of drizzle. They are very small, white, opaque particles of ice. As distinguished from ice pellets (sleet, small hail) they are flatter and they neither shatter nor bounce when they strike the surface.

Hail is precipitation in the form of balls or irregular lumps of ice (0.2 inches or more in diameter) that are produced in the strong updrafts of cumuliform clouds. These are almost always cumulonimbus clouds. Hailstone sizes range to more than five inches in diameter with weights of more than one and one half pounds. The formation of hail and its flight hazards are examined in greater detail in the chapter on thunderstorms.

There are two other precipitation phenomena that you should be able to recognize. One is virga, which is precipitation that evaporates between the base of the cloud and the ground. It is often associated with cumuliform clouds. In the upper troposphere, cirrus fallstreaks are ice crystals that descend from cirrus clouds. (Figure 6-23) They are also called mare's tails or cirrus uncinus.

Figure 6-23. Virga appears as a curtain of precipitation descending from the cloud base.

INTENSITY AND AMOUNT

When precipitation is reported at the surface, the observer estimates the intensity of the precipitation as light, moderate, or heavy. These intensities correspond with the fall rate at the time of the observation. Moderate rain is more than 0.10 to 0.30 inches per hour (.01 to .03 inches per six minutes). Light is less than this range and heavy is greater. With drizzle or snow, intensities are roughly related to visibility. As the intensity of drizzle or snowfall increases, visibility decreases. (Figure 6-24)

Intensity	Visibility (Statute Miles)
Light	> or = 5/8
Moderate	5/16 to < 5/8
Heavy	< 5/16

Figure 6-24. Drizzle or snowfall intensity and associated visibility.

The amount of precipitation is usually expressed as the depth of water that would have accumulated over a given period of time if the water hadn't run off, soaked into the ground, or evaporated. If precipitation has occurred, but the amount is too small to be measured (the smallest reportable amount is .01 inches), then a trace is reported. On the other end of the scale, very large rainfall amounts can occur in certain situations. Figure 6-25 shows some record rainfall amounts for periods varying from a minute to a year.

Observation Period	Rainfall Total (inches)
1 Minute	1.2
42 Minutes	12.0
12 Hour	53.0
1 Month	366.0
1 Year	1042.0

Figure 6-25. Record rainfall rates observed throughout the world.

Precipitation intensity can also be determined by the strength of weather radar echoes. The video integrator processor (VIP) level is a commonly available radar reflectivity intensity scale used to judge the strength of radar echoes and the associated rate of precipitation. (Figure 6-26)

If snow has occurred, snow depth, the depth of the snow actually on the ground is reported. In the U.S., snow depth is measured and reported in inches. The snowfall amount is also converted to a water equivalent for the precipitation report. This is necessary because equal snowfalls can have substantially different water contents. A typical value is 10 inches of snow to 1 inch of water (10:1), but ratios of 2:1 and 20:1 are not unusual, depending on the air temperature.

A heavy snow warning indicates a snowfall of 4 inches or more in 12 hours or 6 inches in 24 hours. A blizzard denotes sustained winds of 35 m.p.h. (30.4 kts) and considerable falling or blowing snow, frequently reducing visibility to less than 1/4 s.m.

REPORTS

Standard aviation weather observations document precipitation types, intensities, and amounts. This information is reported together

Figure 6-26. VIP levels and precipitation rates.

VIP Level	Precipitation Intensity	Rainfall Rate (inches per hour)	
		Stratiform	Convective
1	Light	< 0.1	< 0.2
2	Moderate	0.1 to 0.5	0.2 to 1.1
3	Heavy	0.5 to 1.0	1.1 to 2.2
4	Very Heavy	1.0 to 2.0	2.2 to 4.5
5	Intense	2.0 to 5.0	4.5 to 7.1
6	Extreme	> 5.0	> 7.1

with a number of other measurements related to atmospheric moisture. These include sky condition, weather, visibility, dewpoint, and related remarks. Examples of the typical coded reports of these variables are given in figure 6-27.

Figure 6-27. Reporting codes for aviation weather information. Sections corresponding to topics discussed in this chapter are highlighted.

KEY TO MANUAL AVIATION WEATHER OBSERVATIONS

STATION DESIGNATOR TYPE AND TIME OF REPORT	SKY CONDITIONS AND CEILING	VISIBILITY, WEATHER, AND OBSTRUCTIONS TO VISION	SEA LEVEL PRESSURE	TEMPERATURE AND DEWPOINT	WIND DIRECTION, SPEED AND CHARACTER	ALTIMETER SETTING	REMARKS AND CODED DATA
MCI SA 0758	15 SCT M25 OVC	1R - F	132	58 / 56	1807	993	R01VR20V40

SKY CONDITION AND CEILING:

Sky condition contractions are for each layer in ascending order. Numbers preceding con tractions are base height in hun dreds of feet above ground level (AGL). Sky condition contrac tions are

(- = Thin):

CLR = Clear: Less than 0.1 sky cover.

SCT = Scattered layer aloft: 0.1 through 0.5 sky cover.

BKN = Broken layer aloft: 0.6 through 0.9 sky cover (consti tutes a ceiling layer).

OVC = Overcast layer aloft: More than 0.9 sky cover, or 1.0 sky cover (constitutes a ceiling layer).

X = Surface-based obscuration (all of sky is hidden by surface-based phenomena; constitutes a ceiling layer).

-X = Surface-based partial obscuration (0.1 or more, but not all, of sky is hidden by sur face- based phenomena).

Ceiling designator: A letter pre ceding height of a layer; identi fies a ceiling and indicates how ceiling was obtained.

M = Measured. **E** = Estimated.

W = Indefinite. Vertical visibility into a surface-based obscura tion.

V following height = variable ceiling.

VISIBILITY:

Reported in statute miles and fractions. **V** = Variable.

WEATHER & OBSTRUCTIONS TO VISION:

- = Light. (no sign) = Moderate. + = Heavy.

Weather:

T+	Severe thunderstorm
T	Thunderstorm
R	Rain
RW	Rain shower
L	Drizzle
ZR	Freezing rain
ZL	Freezing drizzle
A	Hail
IP	Ice pellets
IPW	Ice pellet shower
S	Snow
SW	Snow shower
SP	Snow pellets
SG	Snow grains
IC	Ice crystals

Obstructions to Vision:

BD	Blowing dust
BN	Blowing sand
BS	Blowing snow
BY	Blowing spray
D	Dust
F	Fog
GF	Ground fog
H	Haze
IF	Ice fog
K	Smoke

TEMPERATURE AND DEWPOINT:

Reported in degrees Fahrenheit (°F).

REMARKS:

Runway Visibility (RVV) or Runway Visual Range (RVR)

Runway visibility (RVV) is the visibility from a particular loca tion along an identified runway and is reported in miles and fractions of miles. Runway visu al range (RVR) is the maximum horizontal distance down a specified instrument runway at which a pilot can see to identify standard high intensity runway lights, reported in hundreds of feet. The VV and VR reports are for a 10 minute period preced ing observation time. Runway number precedes the reports. V = Variable.

DECODED REPORT:

Kansas City Int'l Airport: Record Observation completed at 0758 UTC. 1500 feet scattered clouds, measured ceiling 2500 feet overcast, visibility 1 mile, light rain, fog, sea level pressure 1013.2 millibars/hectoPascals, temperature 58° F, dewpoint 56° F, wind 180°, 7 knots, altimeter setting 29.93". Runway 01 visual range varying from 2000 to 4000 feet in the past 10 minutes.

KEY TO ASOS (AUTOMATED SURFACE OBSERVING SYSTEM) WEATHER OBSERVATIONS					
STATION DESIGNATOR, TYPE OF REPORT, TIME OF REPORT, STATION TYPE	SKY CONDITIONS AND CEILING BELOW 12,000'	VISIBILITY, WEATHER, AND OBSTRUCTIONS TO VISION	SEA-LEVEL PRESSURE/ TEMPERATURE / DEWPOINT / WIND DIRECTION, SPEED AND CHARACTER /ALTIMETER SETTING/	REMARKS AUTOMATED REMARKS GENERATED AUTOMATICALLY IF CONDITIONS EXIST. AUGMENTED REMARKS ADDED IF CONDITIONS EXIST AND CERTIFIED WEATHER OBSERVER IS ATTENDING THE SYSTEM	REMARKS AND CODED DATA
HTM RS 1755 AO2A	M19VOVC	1R - F	125 / 36 / 34 2116G24 / 990	R29LVR10V50 CIG16V22 TWRVSBY 2 PK WND 2032 / 1732 PRESFR	ZRNO $

SKY CONDITION AND CEILING BELOW 12,000' AGL:

Sky condition contractions are for each layer in ascending order. Numbers preceding contractions are base height in hundreds of feet above ground level (AGL).

CLR BLO 120 = Less than 0.1 sky cover below 12,000'

SCT = Scattered: 0.1 to 0.5 sky cover.

BKN = Broken: 0.6 to 0.9 sky cover.

OVC = Overcast: More than 0.9 sky cover. A letter preceding the height of a base identifies a ceiling layer and indicates how ceiling height was determined.

M = Measured

W = Indefinite

E = Estimated

X = Obscured sky

The letter **V** is added immediately following the height of a base to indicate a variable ceiling: see Remarks.

VISIBILITY:

Reported in statute miles and fractions from **<1/4** through **10+**.

V = variable: see Remarks.

PRESENT WEATHER:

TORNADO (when augmented).

T = Thunder (when augmented): see Status Remarks.

R = Liquid precipitation that does not freeze (e.g., rain).

P – = Light precipitation in unknown form.

ZR = Liquid precipitation that freezes on impact (e.g., freezing rain): see Status Remarks.

A = Hail (when augmented).

S = Frozen precipitation other than hail (e.g., snow).

+ = Heavy. No sign = Moderate. **–** = Light.

OBSTRUCTIONS TO VISION:

Reported only when visibility is less than 7 statute miles.

F = Fog **H** = Haze

VOLCANIC ASH (when augmented).

TEMPERATURE AND DEWPOINT:

Degrees Fahrenheit.

REMARKS:

Can Include:

RVR (Runway Visual Range), **VOLCANIC ASH, VIRGA, TWR VSBY** (Tower visibility), **SFC VSBY** (Surface visibility), **VSBY V** (Variable visibility), **CIG V** (Variable ceiling), **WSHFT** (Windshift), **PK WND** (Peak wind), **WND V** (Variable wind direction), **PCPN** (Precipitation amount), **PRESRR** (Pressure rising rapidly), **PRESFR** (Pressure falling rapidly), **PRJMP** (Pressure jump), **B** (Time weather began), **E** (Time weather ended).

DECODED REPORT:

Hometown Municipal Airport, record special observation at 1755 UTC, ASOS with observer. Measured ceiling 1900 feet variable, overcast. Visibility 1 mile, light rain, fog. Sea-level pressure 1012.5 millibars/hectoPascals, temperature 36°F, dew point 34°F, wind from 210° true at 16 knots gusting to 24 knots, altimeter 29.90 inches. Runway 29L visual range 1000 variable to 5000 feet. Ceiling 1600 variable to 2200 feet, tower visibility 2 miles, peak wind 200° true at 32 knots at 1732 UTC, pressure falling rapidly. Freezing rain information not available, maintenance check indicator.

KEY TO AWOS (AUTOMATED WEATHER OBSERVING SYSTEM) OBSERVATIONS				
STATION DESIGNATOR, TYPE OF REPORT, TIME OF REPORT, STATION TYPE	SKY CONDITIONS AND CEILING BELOW 12,000'	VISIBILITY	TEMPERATURE / DEWPOINT /WIND DIRECTION, SPEED AND CHARACTER/ ALTIMETER SETTING/	REMARKS AUTOMATED REMARKS GENERATED AUTOMATICALLY IF CONDITIONS EXIST. AUGMENTED REMARKS ADDED IF CONDITIONS EXIST AND CERTIFIED WEATHER OBSERVER IS ATTENDING THE SYSTEM
HTM RS 1755 AWOS	M20 OVC	1V	36 / 34 / 2015G25 / 990	P010 / VSBY 1/2V2 WND 17V23 / WEA:RF

SKY CONDITION AND CEILING:
Sky condition contractions are for each layer in ascending order. Numbers preceding contractions are base heights in hundreds of feet above ground level (AGL).
CLR BLO 120 = No clouds below 12,000 ft.
SCT = Scattered: 0.1 to 0.5 sky cover.
BKN = Broken: 0.6 to 0.9 sky cover.
OVC = Overcast: More than 0.9 sky cover.
X = Obscured sky **-X** = Partially obscured
A letter preceding the height of a base identifies a ceiling layer and indicates how ceiling height was determined.
M = Measured **W** = Indefinite

VISIBILITY:
Reported in statute miles and fractions. Visibility greater than 10 not reported. **V** = variable: see Automated Remarks

TEMPERATURE AND DEW POINT:
Reported in degrees Fahrenheit.

PRESENT WEATHER/OBSTRUCTIONS TO VISION: Reported only when observer is available. See Augmented Remarks. In the future, some systems will report precipitation, fog, and haze in the body of the observation.

AUTOMATED REMARKS:
Precipitation accumulation reported in hundredths of inches (e.g., P110 = 1.10 inches; P010 = 0.10 inch).

AUGMENTED REMARKS:
"WEA:" indicates manual observer data. Remarks include operationally significant weather conditions within a five mile radius of the airport (e.g., thunderstorms, precipitation, obstructions to vision when visibility is 3 miles or less, fog banks). Standard weather observation contractions are used.

DECODED REPORT:
Hometown Municipal Airport, observation at 1755 UTC, AWOS report. Measured ceiling 2000 feet overcast. Visibility 1 mile variable. Temperature 36 degrees (F), dewpoint 34 degrees (F), wind from 200 degrees true at 15 knots gusting to 25 knots, altimeter setting 29.90 inches. Precipitation accumulation during past hour 0.10 inch. Visibility variable between 1/2 and 2 miles. Wind direction variable from 170 degrees to 230 degrees true. Observer reports light rain (R-) and fog (F).

KEY TO METAR (NEW AVIATION ROUTINE WEATHER REPORT) OBSERVATIONS

TYPE OF REPORT	STATION DESIGNATOR, TIME OF REPORT	WIND	VISIBILITY,	WEATHER AND OBSTRUCTIONS TO VISIBILITY	SKY CONDITIONS	TEMPERATURE / DEWPOINT	ALTIMETER SETTING	REMARKS
METAR	KSEA 1250Z	08032G45KT	1/2SM R32L/1200 FT	TSRA	SCT008 OVC012CB	15 / 08	A2995	RMK RETSB24RAB24

VISIBILITY:

Visibility is reported in statute miles with "SM" appended to it.

Examples:

7SM - seven statute miles

15SM - fifteen statute miles

1/2SM - one half statute mile

Runway Visual Range (RVR), when reported, is in the format: R(runway)/(visual range)FT. The "R" identifies the group followed by the runway heading, a "/", and the visual range in feet (meters in other countries).

Example:

R32L/1200FT - runway 32 left visual range 1200 feet

WEATHER:

The weather as reported in the METAR code represents a significant change in the way weather is currently reported. In METAR, weather is reported in the format:

Intensity, Proximity, Descriptor, Precipitation, Obstructions to visibility, or Other

Intensity - applies only to the first type of precipitation reported. A "-" denotes light, no symbol denotes moderate, and a "+" denotes heavy.

Proximity - applies to and reported only for weather occurring in the vicinity of the airport (between 5- and 10 miles of the center of the airport runway complex). It is denoted by the letters "VC."

Descriptor - these seven descriptors apply to the following precipitation or obstructions to visibility:

TS - thunderstorm
DR - low drifting
SH - shower(s)
MI - shallow
FZ - freezing
BC - patches
BL - blowing

Precipitation - there are eight types of precipitation in the METAR code:

RA - rain
GR - hail (> 1/4")
DZ - drizzle
GS - small hail/snow pellets
SN - snow
PE - ice pellets
SG - snow grains
IC - ice crystals

Obstructions to visibility - there are eight types of obstructing phenomena in the METAR code:

FG - fog (vsby < 5/8 mile)
PY -spray
BR - mist (vsby 5/8 - 6 mi)
SA - sand
FU - smoke
DU - dust
HZ - haze
VA - volcanic ash

Note: Fog (FG) is reported only when the visibility is less than five eighths of a mile otherwise mist (BR) is reported.

Other - there are five categories of other weather phenomena which are reported when they occur:

SQ - squall
SS - sandstorm
DS - duststorm
PO - dust/sand whirls
FC - funnel cloud/tornado/waterspout

Examples:

TSRA - thunderstorm with moderate rain

SKY CONDITION:

The sky condition as reported in METAR represents a significant change from the way sky condition is currently reported. In METAR, sky condition is reported in the format: Amount, Height, (Type), or Vertical Visibility

Amount - the amount of sky cover is reported in eighths of sky cover, using the contractions:

SKC - clear (no clouds)
SCT - scattered (1/8 to 4/8's of clouds)
BKN - broken (5/8's to 7/8's of clouds)
OVC - overcast (8/8's of clouds)

Note: A ceiling layer is not designated in the METAR code. For aviation purposes, the ceiling is the lowest broken or overcast layer, or vertical visibility into an obscuration. Also, there is no provision for reporting thin layers in the METAR code.

Height - cloud bases are reported with three digits in hundreds of feet.

(Type) - if towering cumulus clouds (TCU) or cumulonimbus clouds (CB) are present, they are reported after the height which represents their base.

Examples:

SCT025TCU BKN080 BKN250 - scattered tower ing cumulus at 2,500 feet, broken clouds at 8,000 feet, broken clouds at 25,000 feet.

SCT008 OVC012CB - scattered clouds at 800 feet, overcast cumulonimbus cloud at 1,200 feet

SKC - clear, no clouds

Vertical Visibility - total obscurations are reported in the format "VVhhh" where VV denotes vertical visibility and "hhh" is the vertical visibility in hundreds of feet. There is no provision in the METAR code to report partial obscurations.

Example:

1/8SM FG VV006 - horizontal visibility one eighth of a mile in fog, vertical visibility six hundred feet.

TEMPERATURE/DEWPOINT:

Temperature and dewpoint are reported in a two-digit form in degrees Celsius. Temperatures below zero are prefixed with an "M."

Examples:

15/08 - temperature 15 degrees, dewpoint 8 degrees

00/M02 - temperature zero degrees, dewpoint minus 2 degrees

REMARKS:

Remarks are limited to reporting operationally significant weather, the beginning and ending times of certain weather phenomena, and low-level wind shear of significance to aircraft landing and taking off. The contraction "RMK" precedes remarks. The contraction "RE" is used to denote recent weather events.

DECODED REPORT:

Routine observation report for Seattle, Washington at 1250 UTC. Wind from 080 degrees at 32 knots with gusts to 45 knots. Visibility 1/2 statute mile, runway 32 left visual range 1,200 feet, thunderstorm with moderate rain. There are scattered clouds at 800 feet and over cast cumulonimbus clouds at 1,200 feet. Temperature 15 degrees Celsius, dewpoint 8 degrees Celsius, altimeter setting 29.95 inches of mercury.

Remarks: recent weather event, thunder- storm began 24 minutes past the hour, rain began 24 minutes past the hour.

HYDROLOGIC CYCLE

It is useful at the end of this chapter to look at the broader role of clouds and precipitation in global weather. Although they can cause a long list of problems for the pilot, they have many benefits. For example, clouds and precipitation are part of a natural worldwide cycle that replaces water lost from the earth's surface. The cycle begins with the loss of moisture from the earth by evaporation, transpiration (loss of water vapor from plants), and sublimation. The winds carry the water vapor great distances. Clouds form through condensation and deposition processes and, finally, some of those clouds produce precipitation, returning moisture to the earth to replenish supplies. This is the so-called hydrologic cycle which describes the moisture budget of the earth and the atmosphere. (Figure 6-28)

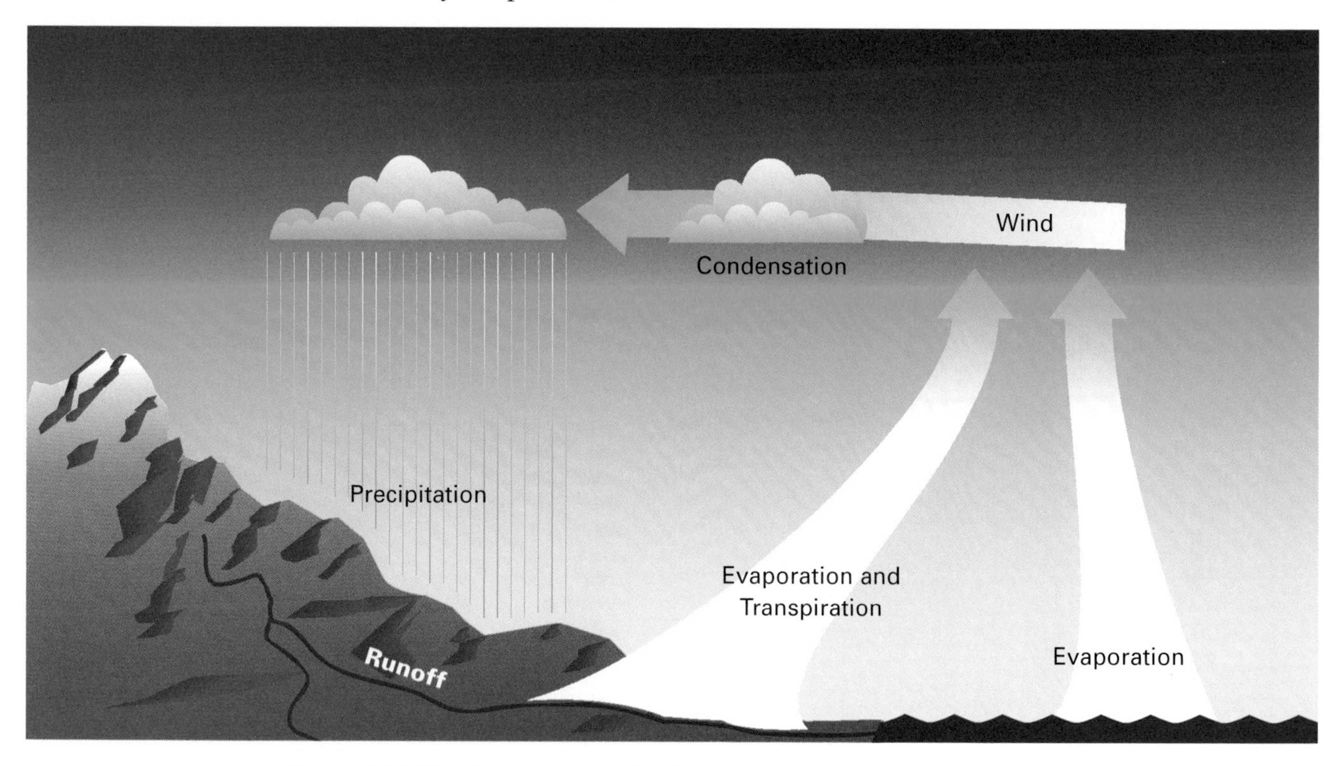

Figure 6-28. The hydrologic cycle. On and below the surface of the earth, moisture is stored as ice and snow, and as water in lakes, rivers, ground water, and the oceans. Water vapor removed from the earth's surface may be transported far from its source by the wind. Clouds form and produce precipitation which returns the moisture to the earth's surface.

SUMMARY

In this final chapter of Part I — Aviation Weather Basics, you have learned some important details about H_2O and its three states in the atmosphere. Changes of state and the associated latent heat exchanges have important effects on cloud formation and dissipation. You should now understand that there are important differences between the way clouds are formed and the processes by which precipitation is produced. On the very practical side, you should know the basic elements of cloud and precipitation observing and reporting. Your ability to recognize the 10 basic cloud types and the types and characteristics of precipitation give you valuable observational tools with which you can deduce much information about the state of the atmosphere and its likely effect on flight.

In the next part of the book, you will apply your knowledge of all of the basic physical processes gained thus far to understand how atmospheric storms and other circulations arise, and to determine their structures and future behavior.

KEY TERMS

Advection Fog
Atmospheric Moisture
Black Ice
Blizzard
Boiling
Broken (BKN)
Capture
Ceiling
Celestial Dome
Change of State
Clear (CLR)
Cloud
Cloud Amount
Cloud Height
Cloud Layer
Clouds of Vertical Development
Coalescence
Condensation
Condensation Level
Condensation Nuclei
Conditionally Unstable
Conditional Instability
Contact Cooling
Deposition
Dew
Dewpoint
Drizzle
Evaporation
Fallstreaks
Freezing
Freezing Drizzle
Freezing Rain
Frost
Frostpoint
Ground Fog
Hail
Heavy Snow Warning
High Clouds
Hydrologic Cycle
Ice
Ice Crystal Process
Ice Pellets
Instrument Flight Rules (IFR)
Instrument Meteorological Conditions (IMC)
Latent Heat
Low Clouds
Low IFR (LIFR)
Marginal VFR (MVFR)
Melting
Middle Clouds
Obscured (X)
Overcast (OVC)
Partial Pressure
Partially Obscured (-X)
Phase Change
Pilot Report (PIREP)
Precipitation
Prevailing Visibility
Psychrometer
Radiation Fog

Rain
Rainshadow
Rain Showers
Relative Humidity
Runway Visibility (RVV)
Runway Visual Range (RVR)
Saturated Adiabatic Process
Saturated Adiabatic Lapse Rate
Saturation
Saturation Vapor Pressure
Scattered (SCT)
Sensible Heat
Sky Condition
Snow
Snow Depth
Snow Grains
Snow Showers
Steam Fog
Sublimation
Supercooled Water Droplets
Temperature-Dewpoint Spread
Thin (-)
Trace
Tower Visibility
Upslope Fog
Vapor Pressure
Vertical Visibility
Virga
Visual Flight Rules (VFR)
Visual Meteorological Conditions (VMC)
Water
Water Equivalent
Water Vapor
White Dew

CHAPTER QUESTIONS

1. It is often observed that relative humidity reaches a maximum near sunrise and a minimum in the afternoon. Why?

2. If a saturated parcel is descending, say in the middle of a downburst, is the rate of heating less than or greater than 3°C per 1,000 feet? Why?

3. Is there any truth in the adage, "too cold to snow"?

4. You are in a pressurized cockpit that undergoes rapid decompression. Fog forms suddenly, then dissipates. Explain.

5. You are standing next to your airplane preparing for a night flight. It is overcast and the visibility is very good. Rain starts abruptly. You notice the droplets are quite large. The rain stops after a minute or so. What can you say about flying conditions at cloud level?

6. You place a gallon can of water on a burner until it comes to a boil. You remove the can from the burner and cap it. After awhile, the can begins to collapse. Explain.

7. Observe and record sky conditions at the same time, every day for a week. Obtain official weather observations from a nearby airport and compare them with your observations.

8. You walk out to your aircraft for preflight just before sunrise. It is parked in the open. It has been clear all night. There is no moisture on the ground, but you find a thin layer of ice on your wing. Explain.

9. Decode the following weather reports.

ORD RS 1556 M7OVC11/2R+F 995/62/61/3107/980
FOD SA 2255 AWOS M5BKN10OVC 3/4 32/32/0412/980
BNA SA 2154 7SCT 250SCT 6FK 128/60/59/2504/991 JFK RS 1853 W5X1/2F 180/68/64/1804/006/R04RVR22V30 TWR VSBY 1/4
METAR KSFO 1435Z VRB02KT 3SM MIBR SKC 15/12 A3012

Part II

Atmospheric Circulation Systems

Part II
Atmospheric Circulation Systems

Part II uses your knowledge of "weather basics" to develop an understanding of circulations that occur within the atmosphere. These circulations produce temperature, wind, and weather changes that you must understand in order to plan and carry out safe and efficient flights.

When you complete Part II, you will understand how circulation systems of all sizes develop, move, and dissipate. You will know how they produce their characteristic global, regional, and local weather patterns. This knowledge will prove invaluable when you use observed and predicted weather information to anticipate flight conditions. Further, it will serve as an important background for Part III, Flight Hazards.

(Part II view of an ET cyclone from space, courtesy of NASA)

Chapter 7 Scales of Atmospheric Circulation

Introduction

Weather is not a random occurrence. Every weather event is the result of the development of some sort of atmospheric circulation. In this context, circulation means a more or less organized movement of air. The word eddy is often used in the same sense. The motion in a given circulation or eddy may be vertical, horizontal, or both. A very important characteristic of the atmosphere is that circulations occur with many different dimensions, ranging in size from the earth itself to turbulent eddies as small as your hand. In this chapter, we formally introduce the concept of scales of atmospheric motion to help you organize our study of various atmospheric weather phenomena. We then apply this idea to the examination of two important circulations of very different sizes: the general circulation and the monsoon circulation. When you finish this chapter, your knowledge of these two macroscale circulations will provide you with important background for the study of smaller scale circulations that are embedded within the general and monsoon circulations.

Section A

SCALES OF CIRCULATIONS

A common way to study any system is to separate it into its component parts. Whether you are dealing with an airplane or the atmosphere, a complicated combination of parts becomes more understandable when you see what each of those parts does, how they fit into the whole, and how they interact to do what the system is designed to accomplish.

With regard to the atmosphere, if we measure the weather in a particular geographical region, the picture is often complicated, because many different physical processes are contributing to the total weather picture. To simplify things for better understanding, we want to be able to separate these processes.

One way of doing this is to consider the total circulation of the atmosphere as the sum of a number of individual circulations. The individual circulations are the parts of our system. By studying the characteristics of each of them in isolation, the total picture will become more understandable.

Scales of circulations refer to the sizes and lifetimes of individual circulations. In your own experience, you have seen many examples of these. For example, the sea breeze develops during the day, reaches its maximum strength in the afternoon, and dies out at night. You might say it has a lifetime of about a half a day. Sea breezes typically extend 10 to 100 n.m. across the coast from the ocean side to the land side. This range can be taken as a characteristic spatial dimension. To summarize, the sea breeze has a "time scale" of about 12 hours and a "space scale" of 10 to 100 nautical miles.

Another familiar circulation is the dust devil. It has a typical time scale of a few minutes and a space scale of 5 to 100 feet (diameter of the circulation).

Figure 7-1 shows the approximate space and time scales of a number of atmospheric disturbances that are critical for aviation. The diagram is separated into three segments labeled with the words microscale, mesoscale, and macroscale. These are rather broad meteorological terms that are frequently used to classify atmospheric circulations. Mesoscale refers to horizontal dimensions of 1 to 1,000 nautical miles. Macroscale is greater, microscale is less.

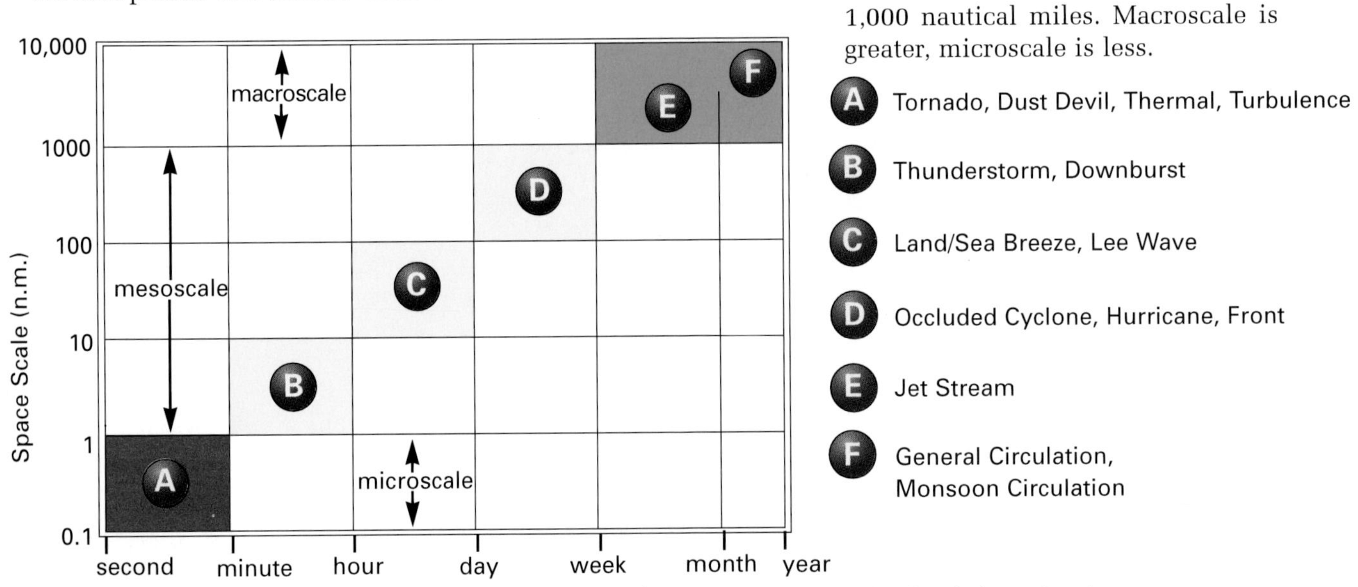

Figure 7-1. Horizontal dimensions and lifetimes of a selection of atmospheric circulations. Precise numbers for space and time scales cannot be given for each phenomenon because of variations caused by such things as local terrain, season, and larger scale weather systems. However, the range of the possible scales for each atmospheric circulation is clearly limited.

Figure 7-1 is useful not only as a summary of atmospheric circulations, but also to help you develop the idea of embedded circulations. At any one time, several circulations may be present, with smaller ones embedded in, and often driven by, larger scale circulations. A good example is a macroscale cyclone associated with a front which produces a number of mesoscale thunderstorms, one of which generates a microscale tornado. The concept of embedded circulations will prove very useful in your interpretation of current and forecast weather conditions. (Figure 7-2)

Figure 7-2. GOES global satellite image showing cloud patterns produced by circulations with a variety of scales.

Section B

THE LARGEST SCALE CIRCULATIONS

In the following paragraphs, we use the concepts of scale and thermal circulation to explain how global winds are affected by the equator-to-pole temperature gradient, the earth's rotation, the continents, and seasonal changes in solar radiation.

THE GENERAL CIRCULATION

The general circulation refers to the wind system that extends over the entire globe. The horizontal scale of this circulation is approximately 10,000 n.m. (macroscale). Aside from long term climatological changes, the time scale of the global circulation is one year. This is the period it takes the circulation system to go through a complete cycle of seasonal changes.

To help you understand the general circulation, it is helpful to begin with a simplified version. Consider an idealized earth with a smooth surface (no surface friction) and no land-sea differences. Let the earth rotate in its usual direction (towards the east), but much slower than the real earth. In this case, and as you would expect from our previous discussion of a thermal circulation, the equator-to-pole temperature gradients create pressure gradients. Surface high-pressure areas are located over each of the cold poles and a surface low pressure region is found around the warm equator. These features cause surface air to move from the poles toward the equator. The reverse occurs aloft where equatorial air moves toward the poles. (Figure 7-3) Each of these vertical circulation systems is called a circulation cell, or simply, a cell.

The simple general circulation cell that develops with slow rotation is similar to the thermal circulation (sea breeze) cell with an important exception. Because the scale of the global circulation is many times larger, Coriolis force has an important modifying effect. Surface winds in the Northern Hemisphere are deflected to the right and become northeasterly. In the Southern Hemisphere, surface winds become southeasterly. Winds aloft have the opposite directions.

We now increase the rotation of our idealized earth to its normal rate of one rotation every 24 hours. The resulting wind circulation becomes more involved, but much more realistic. In the remaining discussion, we will concentrate on the Northern Hemisphere pattern. Just remember the

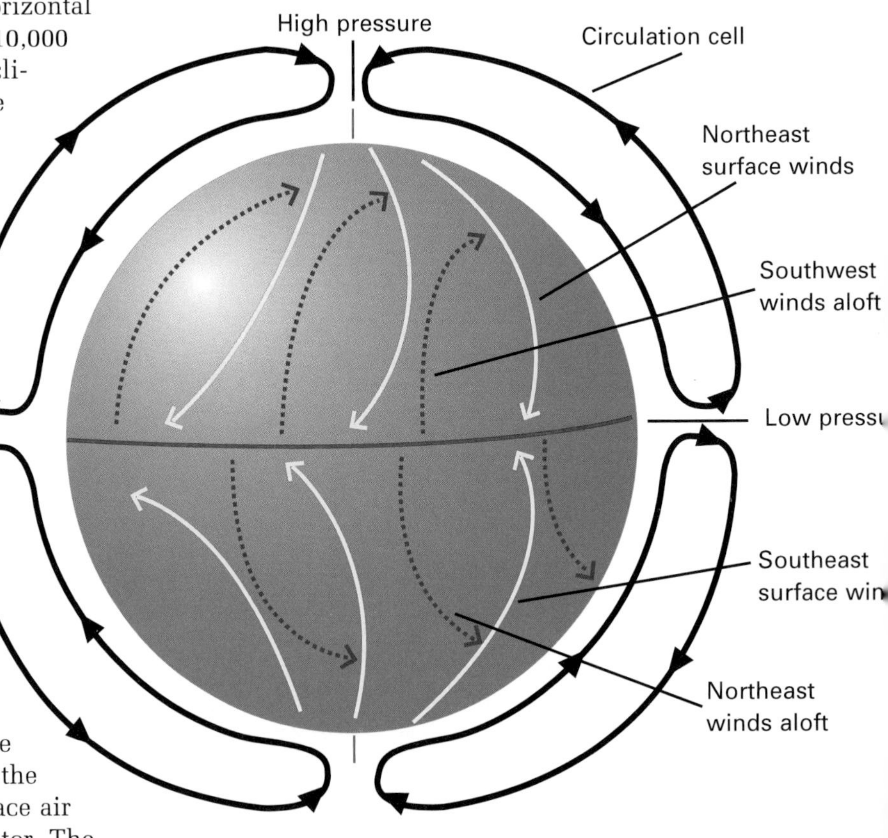

Figure 7-3. A slowly rotating earth has only one circulation cell in each hemisphere. Surface winds are indicated by solid arrows. Note how winds are changed from a strictly north-south direction by the Coriolis force. A cross section showing the vertical structure of the circulation cells in each hemisphere is shown on the eastern edge of the globe.

Southern Hemisphere circulation is a mirror image because Coriolis force acts in the opposite direction.

With the faster rotation rate, the single cell circulation breaks up into three cells. We find that air still rises at the equator and flows toward the pole aloft, but that branch of the circulation reaches only 30°N, where the air sinks. At the surface, we again find northeasterlies. (Figure 7-4) This cell is called the Hadley Cell for an 18th century scientist who first proposed a model of the general circulation.

In the highest latitudes, a Polar Cell has developed. It is defined by air rising near 60°N and sinking over the pole. Coriolis force causes the cold surface winds in the polar cell to be northeasterly, and winds aloft to be southwesterly.

In the latitude belt between 30°N and 60°N, the faster rotation and strong north-south temperature gradient in midlatitudes favors the development of smaller scale eddies in that region. We will examine these eastward-moving disturbances in the next chapter. Their influence on the general circulation is to cause the average surface winds to be southwesterly in this latitude belt, and to remain westerly up through at least the tropopause. These average winds define a midlatitude circulation cell called the Ferrel Cell. It is also named for an early investigator of the general circulation.

The three-cell circulation generates some important and well-known features in the surface wind pattern. (Figure 7-4) These include the steady, northeasterly trade winds between the equator and 30°N; the prevailing westerlies between 30°N and 60°N, and the polar easterlies north of 60°N.

Other important surface features of the general circulation are found in the surface pressure distribution. The low pressure area near the equator is called the "Doldrums." Because of the convergence of trade winds from both hemispheres into that area, it is also known as the Intertropical Convergence Zone (ITCZ). The instability and

Figure 7-4. A three-cell circulation develops in each hemisphere of a smooth, homogeneous earth rotating at one revolution per 24 hours. A cross section showing the vertical structure of the circulation cells in each hemisphere is shown on the eastern edge of the globe. The related surface winds are shown with the yellow arrows. Note that the wind pattern in the Southern Hemisphere is a mirror image of the pattern north of the equator.

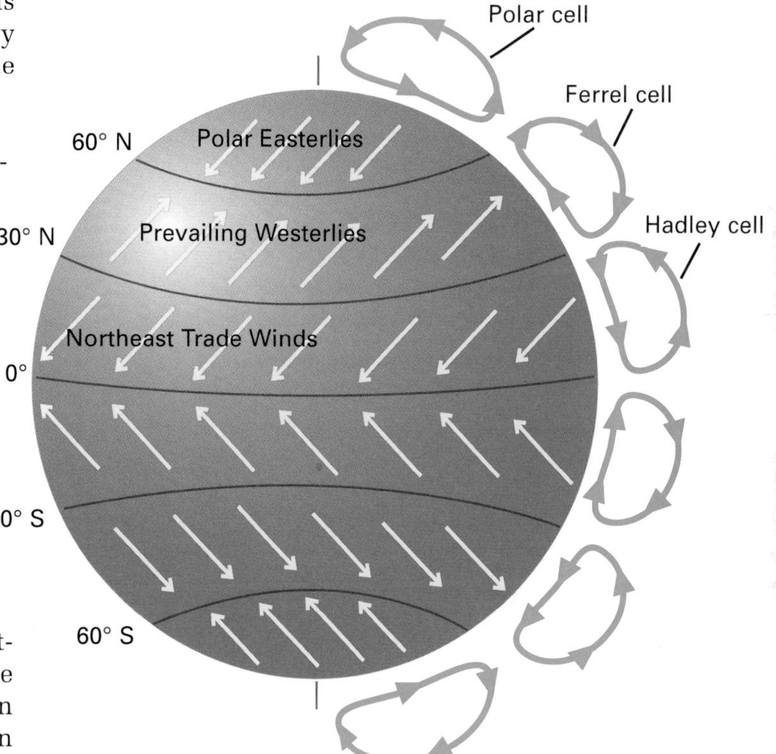

large moisture content of the air in the ITCZ, make it a favorite area for the development of thunderstorms. (Figure 7-5)

Air sinks in a region of diverging surface winds that correspond with a subtropical high pressure near 30°N. This part of the general circulation is known as the horse latitudes. Cloud formation is suppressed and precipitation is typically low in these areas.

Near 60°N, pressures are low and surface winds converge, bringing warm airmasses from tropical regions into contact with cold airmasses from polar regions. The line separating the airmasses at this location is called the Polar Front. It is another region of cloudiness and precipitation.

Finally, there are two areas of sinking air and diverging winds in high-pressure systems near the poles. As with the horse latitudes, precipitation is very low in these areas. The ground remains snow covered despite the low precipitation because of the very low temperatures.

THE MONSOON CIRCULATION

Our discussion to this point has centered on an idealized general circulation on an idealized planet. The real picture is different because of the existence of oceans and continents. To understand how the actual pattern of global winds develops, we must bring in the monsoon circulation.

The monsoon circulation or, simply, the monsoon, is a macroscale wind pattern that undergoes a seasonal reversal in direction. The low-level winds of the "wet" monsoon of summer flow from the ocean to the continent. The "dry" monsoon flow is in the opposite direction (the continent to the ocean). A rough measure of the scale of a monsoon is 5,000 n.m. or about the size of a continent.

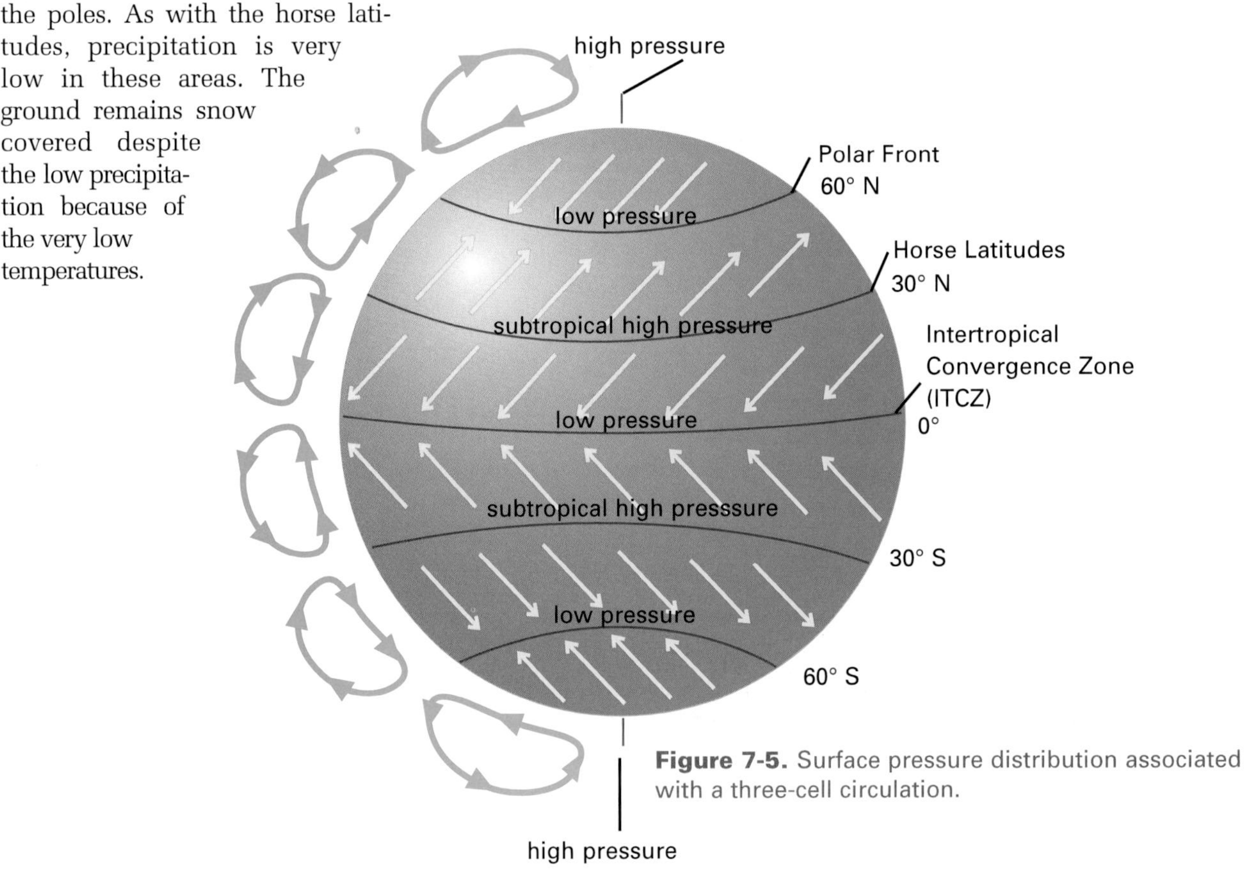

Figure 7-5. Surface pressure distribution associated with a three-cell circulation.

Figure 7-6. The summer and winter monsoon circulations for an idealized continent. These diagrams show the surface circulations caused by low sea level pressure over the continent in summer and high pressure in winter.

In order to understand how the monsoon works, we will look at another idealized situation. This time, we ignore the influence of the general circulation and consider a single continent with a simple shape in the Northern Hemisphere. (Figure 7-6) The monsoon has the characteristics of a thermal circulation. In the summer, the continent is much warmer than the surrounding ocean and the sea level pressure is lower over the land. Moist ocean winds sweep inland at that time. However, because the scale of the circulation is so large, Coriolis force is also important. With the added affect of friction, winds spiral counter-clockwise into the continental low.

In winter, the picture is reversed. The continent is cold relative to the surrounding ocean. High pressure prevails over the land and cool, dry surface winds spiral clockwise outward from the anticyclone. As you would expect, the circulations are opposite for a Southern Hemisphere continent.

On the real earth, the monsoon circulation is embedded in the larger general circulation. Additionally, continents vary in size, shape, and latitude. The results of these factors are that the monsoon is very well defined in some geographical areas (Southeast Asia), but is only barely noticeable in others (Europe). In the next section, we examine the effects of both the general circulation and monsoon circulations over the real earth.

Section C

THE GLOBAL CIRCULATION SYSTEM

Most global climatological wind charts are based on a monthly or seasonal average of the worldwide winds. This averaging process eliminates circulations with smaller time scales leaving what is called the global circulation system. It is a combination of the general and monsoon circulations. To illustrate, the average surface wind patterns for January and July are given in figure 7-7. The average sea level pressure has been repeated from Chapter 3 to emphasize the relationships between pressure and wind.

In both January and July, the underlying general circulation is apparent over the oceans where the prevailing westerlies, the trade winds, subtropical high-pressure regions, and ITCZ can be seen. (Figure 7-7) This is particularly true in the Southern Hemisphere where there is much less land area (less monsoon effect).

There is a strong seasonal variation in the global circulation pattern. In January, the Icelandic and Aleutian lows, which indicate the average position of the polar front, are stronger and farther south than in July. The subtropical highs are also farther south in January, but they are weaker than in July. The ITCZ tends to move northward in July in some parts of the world, and southward in January. For example, in July, the ITCZ is located in the north of the India, while in January, it is just south of the equator.

Figure 7-7. Global circulation for January and July. Average wind directions are indicated with arrows. Solid lines are mean sea level pressure in millibars.

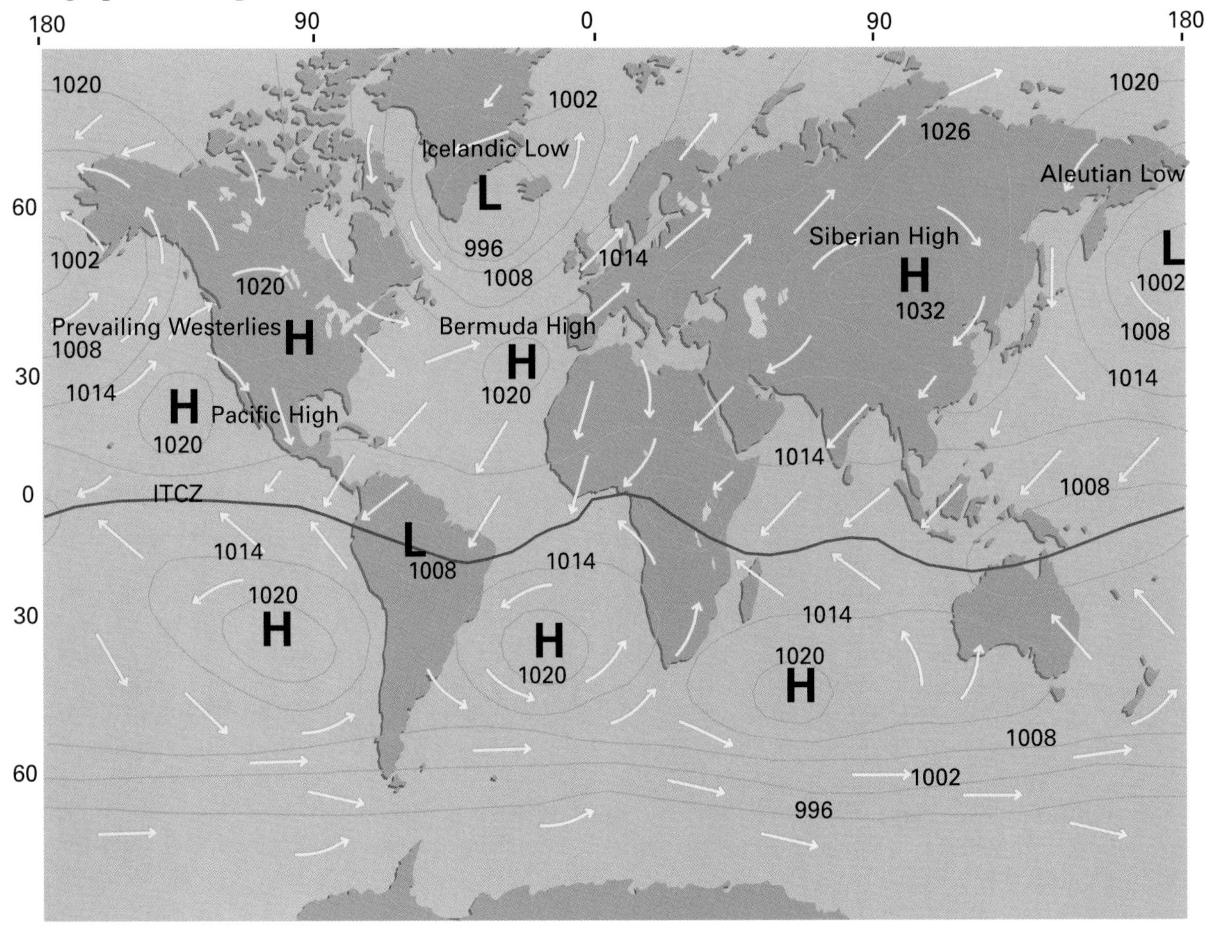

In the vicinity of nearly all of the continents, the influence of the monsoon becomes evident. Nowhere is it so obvious as over Southeast Asia. (Figure 7-7) This is due to the location of the Asian landmass to the north of a very warm ocean. There is well-defined cyclonic inflow into a low pressure area over Asia in July and anticyclonic outflow from a cold continental high-pressure system in December. The Asian monsoon influence is so pervasive that winds over the Indian Ocean become southwest in the summer, rather than northeast, as would be expected when considering only the general circulation.

Monsoon winds also develop over Africa, Australia, and some parts of North and South America, especially in the lower latitudes and usually in combination with the seasonal shift of the ITCZ. Their strengths depend on the shape and size of the continents and the temperatures of the surrounding oceans.

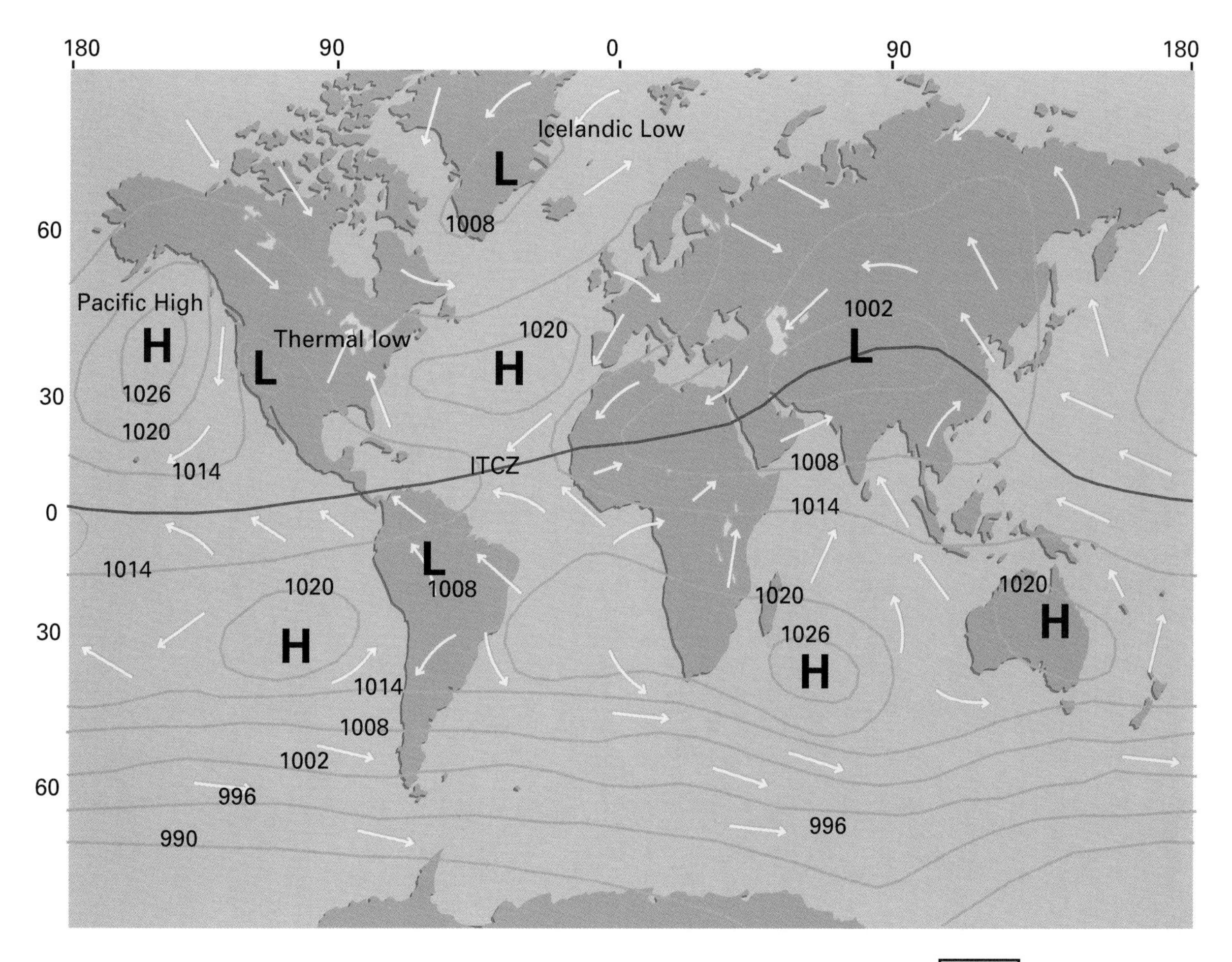

July

THE GLOBAL CIRCULATION ALOFT

The global circulation patterns aloft are far simpler than what you have just seen for the surface. Figure 7-8 shows average January and July 500 mb (18,000 feet MSL) charts with height contours and wind directions superimposed. Macroscale features of interest include wintertime cyclones over Western Siberia and the Canadian Arctic. These Northern Hemisphere lows are upper-level extensions of the Aleutian and Icelandic lows seen in the surface wind and pressure patterns. The subtropical highs seen near the surface are also identifiable aloft where they are closer to the equator. These features are particularly noticeable in the summer, as shown in figure 7-8.

In figure 7-8, there is also a well-defined wave structure in the contour and westerly wind patterns, especially in the Northern Hemisphere. For example, in winter, there are wave troughs along the east coasts of both Asia and North America and a weaker trough over Europe. The three waves in figure 7-8 are examples of the largest scale wave disturbances that occur in the atmosphere. Appropriately, these are called long waves. They tend to move much more slowly than the wind.

Long waves can be viewed as large scale disturbances embedded in the basic westerly flow around the globe. The airflow through the upper-level waves causes storms and cold air to move to lower latitudes in the vicinity of the wave troughs and carries warm air to higher latitudes in the

Figure 7-8. Average 500 mb heights and wind directions for January and July.

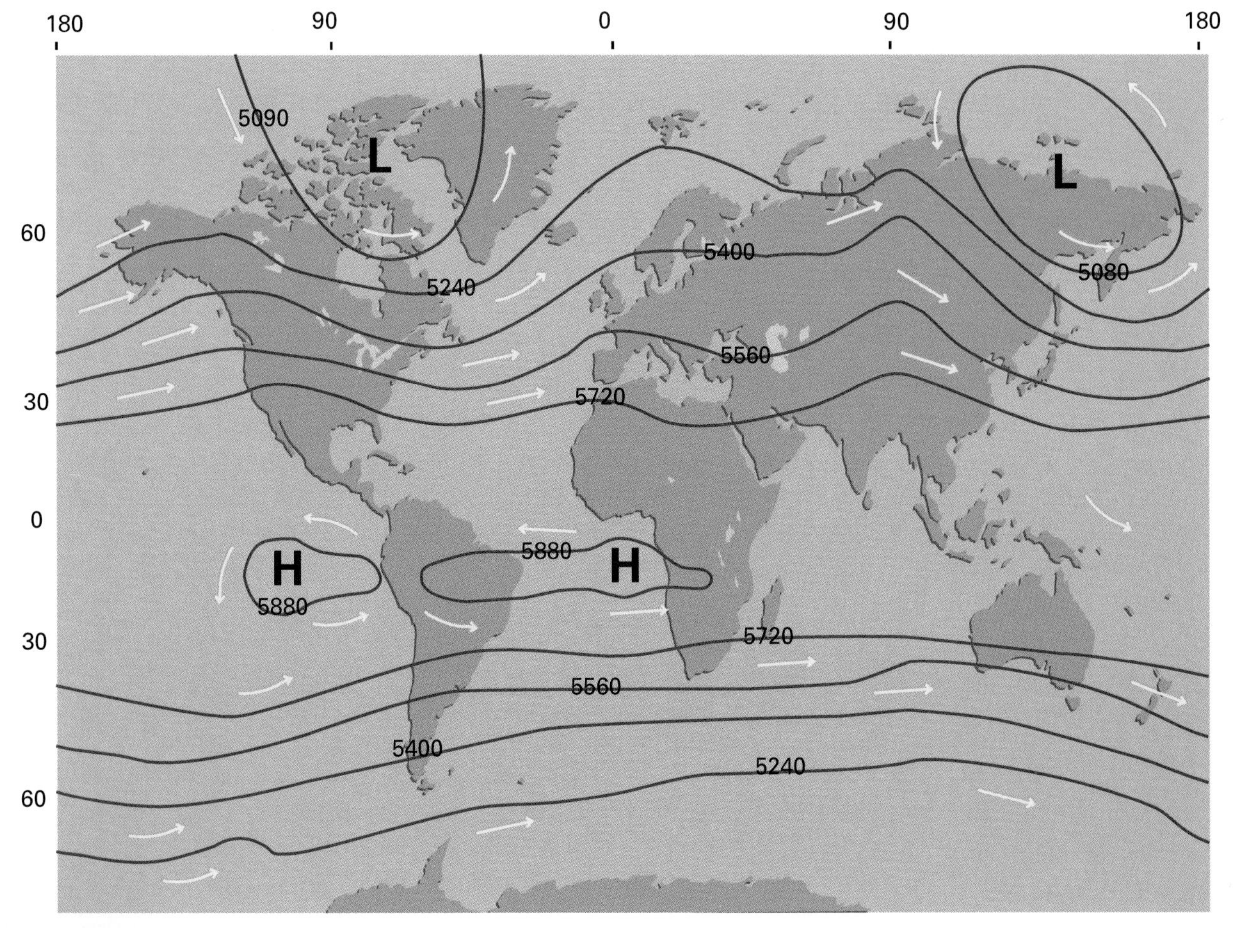

wave crests (ridges). They are necessary links in the heat exchange between equator and poles.

An important characteristic of the flow patterns shown in figure 7-8 are the prevailing westerly winds in middle latitudes, which strengthen (tighter packing of contours) in winter and weaken in the summer. In almost all areas of the globe, the upper-level westerly winds are stronger and exist across a broader latitude belt than in the surface pattern.

Very important upper-air features known as jet streams are often embedded in the zone of strong westerlies. A jet stream is a narrow band of high speed winds that reaches its greatest speed near the tropopause (24,000 to 50,000 feet MSL). Typical jet stream speeds range between 60 knots and about 240 knots. Jet streams are typically thousands of miles long, a few hundred miles wide, and a few miles thick.

On the average, two jet streams commonly occur in the westerlies, the polar front jet stream and the subtropical jet stream. Figure 7-9 shows the approximate positions of the two jet streams in the Northern Hemisphere. As the name implies, the polar front jet stream can be found near the latitude of the polar front. Although it exists year round, the polar front tends to be higher, weaker, and farther north in the summer. The subtropical jet stream is found near 25°latitude. It reaches its

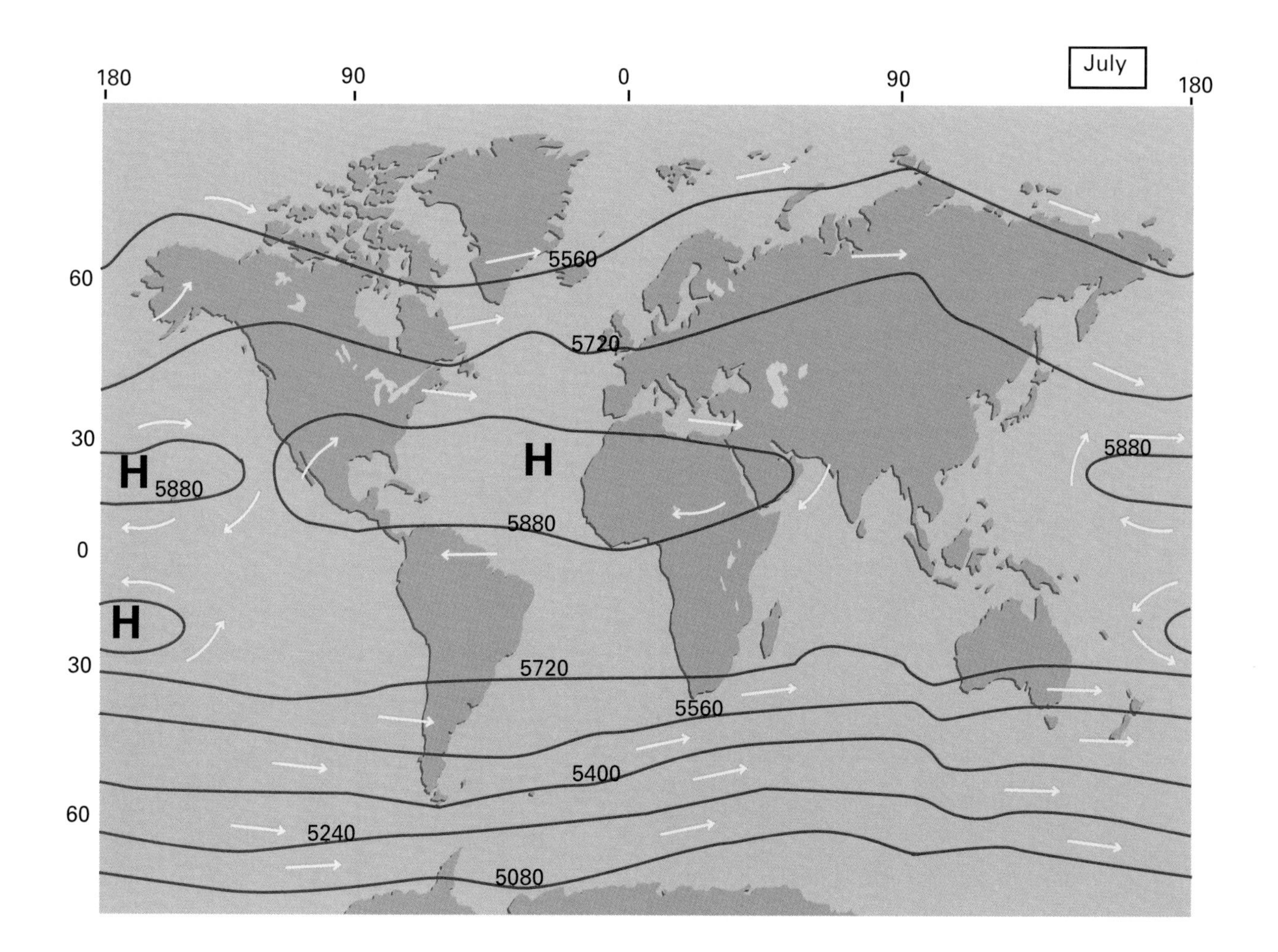

The strength and location of the jet stream is normally weaker and farther north in the summer.

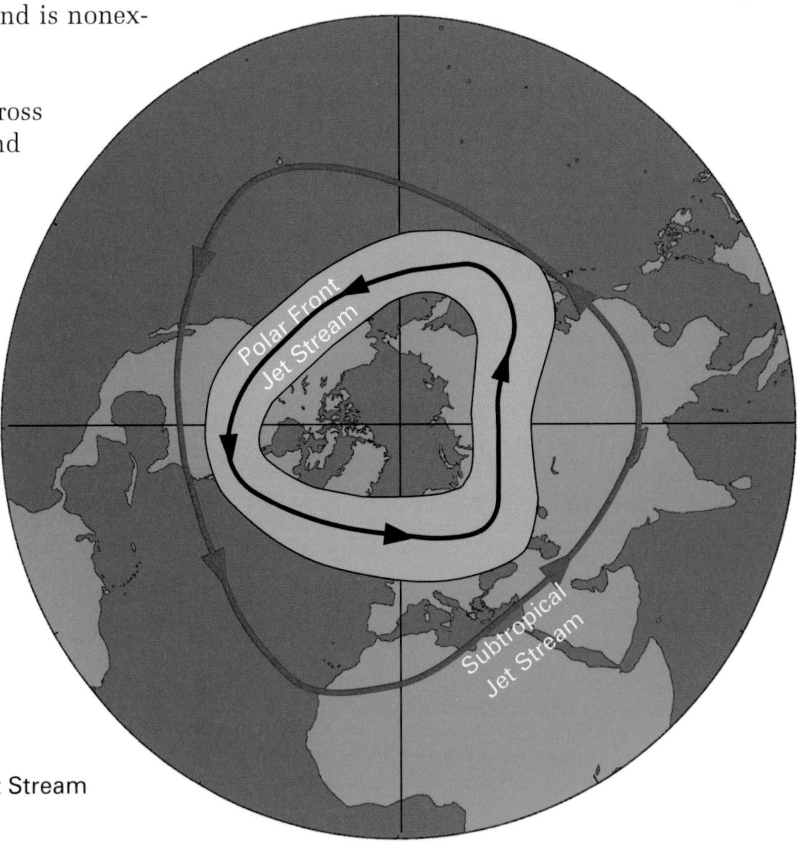

Figure 7-9. Approximate locations of polar front jet stream and subtropical jet stream near tropopause level in winter. The polar front jet is enclosed within a broad zone, because its position varies widely from day to day. Similar conditions are found in the Southern Hemisphere.

greatest strength in the wintertime and is nonexistent in the summer.

An equator-to-pole atmospheric cross section showing the polar front and subtropical jet streams is given in figure 7-10. Notice that the tropopause slopes upward from polar to tropical regions as described in an earlier chapter; but, upon closer examination, we find that the tropopause is not continuous. There is a break at the location of each jet stream. If you stand with the wind at your back, a distinctly higher tropopause occurs on the right side of each jet stream and a separate, lower tropopause occurs on the left. This structure is reversed in the Southern Hemisphere.

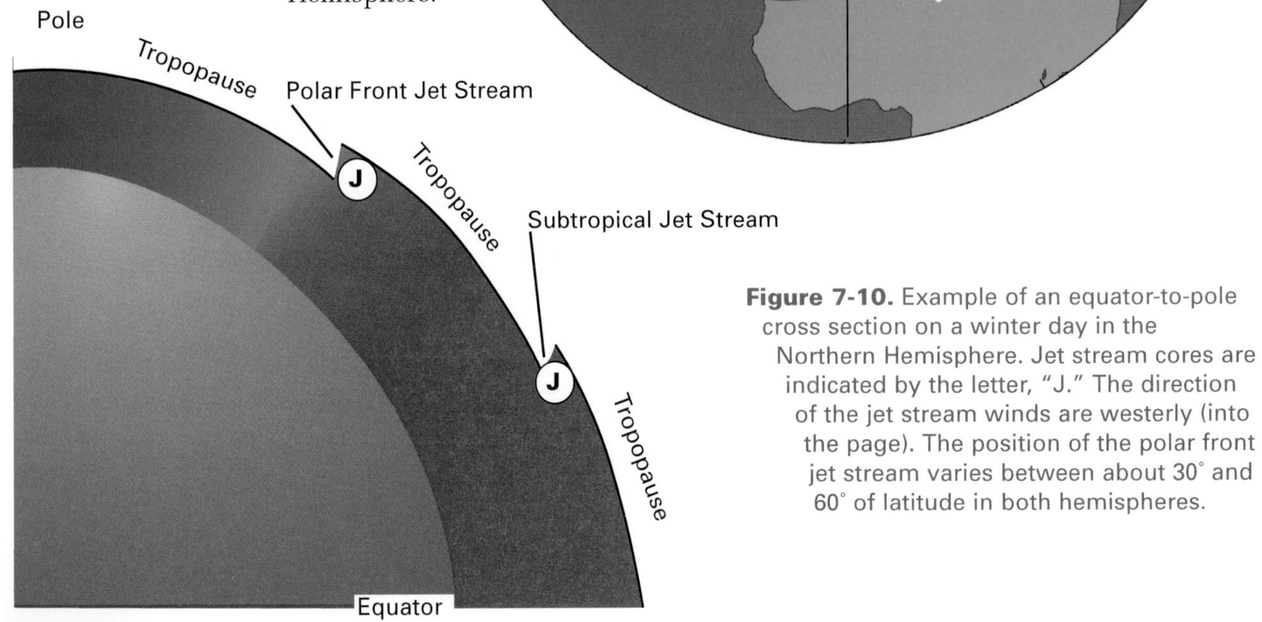

Figure 7-10. Example of an equator-to-pole cross section on a winter day in the Northern Hemisphere. Jet stream cores are indicated by the letter, "J." The direction of the jet stream winds are westerly (into the page). The position of the polar front jet stream varies between about 30° and 60° of latitude in both hemispheres.

Section D

THE GLOBAL CIRCULATION AND CLIMATOLOGY

Climatology is the study of the average conditions of the atmosphere. Although an in-depth examination of climatology is not the purpose of this text, your understanding of the global circulation has provided you with some basic climatological background. You will find this information useful in a number of ways. First, it will help you to better understand the processes that produce the weather. Second, it will aid you in the deduction of average weather conditions for some distant destination from simple climatological charts. Finally, it will help you to understand the basic results of current research on climate change. In the following paragraphs, we consider some brief examples.

Your knowledge of global winds is very useful for explaining something as complicated as the unequal distribution of precipitation around the world. Arctic deserts, rainforests in the Pacific Northwest, desert canyons in Hawaii: all of these can be explained on the basis of your knowledge of the global circulation. (Figure 7-11)

In figure 7-11, the high precipitation near the equator is the result of large amounts of tropical moisture, convergence, and upward motions in the ITCZ. In both the Arctic and the subtropics (especially noticeable over North Africa and the desert southwest of the U.S.), downward motions forced by divergence of the winds in semipermanent high-pressure regions minimize precipitation. Additionally, the Arctic atmosphere has low temperatures and, therefore, small amounts of water vapor.

Examples of the interaction of topography and the winds of the global circulation can be seen in western North America where a large area of low precipitation is found on the east side of the Rockies and several other north-south mountain ranges. Since westerlies dominate the middle latitudes, the east side of those mountains are subjected to downslope motion. In contrast, precipitation is enhanced by upward motions on the west side of the mountains.

An extreme example of the contribution of orographic lifting to precipitation is found along the southern edge of the Himalayas where the combination of the wet monsoon winds and orography causes large precipitation amounts. Cherrapunji, India, is in this area and has received over 1,000 inches of rain in a single year.

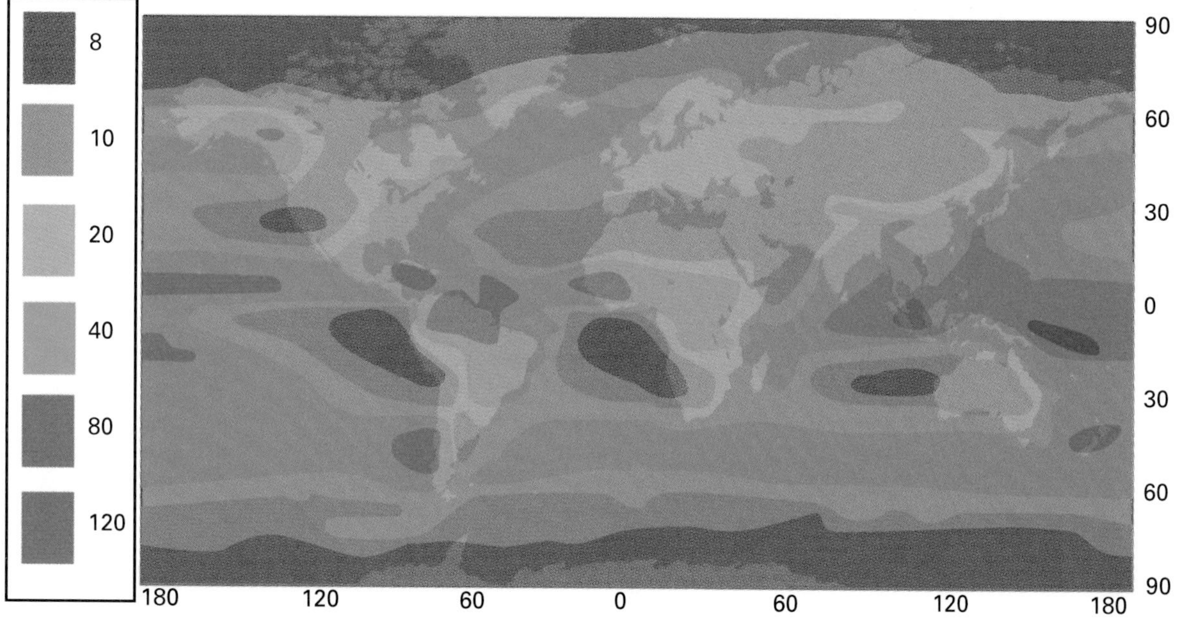

Figure 7-11. Annual average world precipitation.

A similar example is found on Mt. Waialeale on the island of Kauai, Hawaii. Waialeale protrudes into the steady, moist trade winds which produce an average rainfall of 460 inches at that location. Interestingly, not far downwind of Waialeale is Waimea canyon, one of the driest spots in the Hawaiian Islands, yet another example of orographic effects.

The scales of motion discussed thus far, and considered in future chapters, are associated with time periods of one year and less. Weather changes over longer periods do occur. For example, as you have probably noticed, some very wet years are followed by dry years. An occasional very warm or cold winter may occur between a couple of average winters. These "interannual" variations have been related to a variety of possible causes; such as, changes of ocean water temperatures (El Nino), long term oscillations in equatorial winds, fluctuations in solar output, and the interception of solar radiation by long-lived plumes from volcanic eruptions.

Much longer time scales of climate variation have also been identified. These include changes due to a gradual variation in the earth's tilt from 22° to 24.5° and back (cycle length: 41,000 years). This variation in tilt affects day length and seasonal changes. There is also a very gradual change in the shape of the earth's orbit around the sun from nearly circular to elliptical (cycle length: 100,000 years). This change affects the amount of energy received at the earth's surface and the length of the seasons. Finally, there is a "wobble" in the axis of the earth (cycle length: 23,000 years). This causes a change in the time of year that the earth is tilted toward or away from the sun. Scientists generally believe that these fluctuations have been responsible for the ice ages during the last 2,000,000 years.

Finally, there are scales of climate variations that are being driven by the impact of pollution caused by industrial growth, urbanization, and the demands of the earth's rapidly growing population. The ozone hole and the greenhouse effect are but two concerns. The long term influences and time scales of these and other man-made climate modifications are not well known. Currently, computer simulations of the global climate are being used to examine the impact of these influences.

SUMMARY

This chapter can be viewed as a transition between aviation weather basics and some useful applications of those basics. The concept of "scales of motion" has been introduced as a learning and organizing device. The observed state of the atmosphere is usually due to the effects of one or more individual circulations. By separating the variety of atmospheric disturbances according to their space and time scales, they become easier to understand. The interpretation of current and predicted weather also is much easier when you have an appreciation of the types and scales of the disturbances involved.

The scale approach has been applied in this chapter to describe the causes and characteristics of the general circulation and the monsoon. In combination, these largest circulation systems account for the average global winds, and help us explain many of the characteristics of global climatology.

KEY TERMS

Circulation
Circulation Cell
Climatology
Eddy
Embedded Circulation
Ferrel Cell
General Circulation
Global Circulation System
Hadley Cell
Horse Latitudes
Intertropical Convergence Zone (ITCZ)
Jet Stream
Long Waves
Macroscale
Mesoscale
Microscale
Monsoon
Monsoon Circulation
Polar Cell
Polar Easterlies
Polar Front
Polar Front Jet Stream
Prevailing Westerlies
Scales of Circulations
Subtropical Jet Stream
Trade Winds

CHAPTER QUESTIONS

1. You are responsible for advising an oceanic research group on the operation of an aircraft flight to do a low-level photo survey of the Atlantic between Panama and Gibraltar. You will be flying at about 1,000 feet MSL at an airspeed of 170 knots. Flights will be made one day a week for a year. There will be a full crew and scientific equipment going one way. The direction is arbitrary. The crew and equipment will return by commercial airline. It is up to you to minimize the cost of the missions (fuel, crew duty time, etc.). On the basis of average conditions in January and July, what would your meteorological advice be?

2. A minimum time track (MTT) is not necessarily the shortest path between two locations, but it is the fastest. Your company aircraft flies a daily, round trip flight from London to New York for a period of one year. The airspeed is 300 knots at an altitude of 18,000 feet MSL. On the basis of January and July conditions, give a rough estimate (draw a map) of how the average annual MTTs for the out and return flights would look? Discuss your reasoning. Would your answers change for flights at 300 mb? For an aircraft flying at Mach 3? If so, how?

3. You want to fly a balloon across the Atlantic. For technical reasons you must fly below 5,000 feet MSL. You want dependable (steady) winds. Where and when should you attempt your crossing? Discuss.

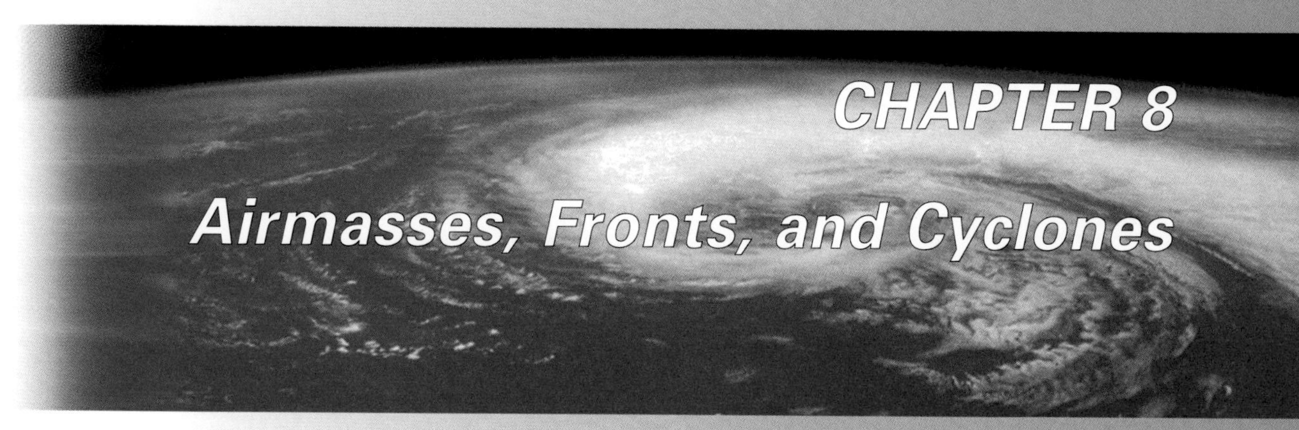

CHAPTER 8

Airmasses, Fronts, and Cyclones

Introduction

The general circulation and monsoon discussed in the previous chapter have very large horizontal dimensions and long time scales. Another notable characteristic is that seasonal; large-scale features, such as, highs and lows, tend to remain stationary. In this chapter, we look at another collection of circulations that are different in at least three ways: they are smaller in size, have shorter life times, and they have significant movement. These are extratropical cyclones and tropical cyclones known as hurricanes. They are the "weather makers" that have significant effects on aviation activities. Many of the flight hazards discussed in Part III are related to cyclones.

When you complete Chapter 8, you will be familiar with the causes and structure of extratropical cyclones and tropical cyclones, and the weather they produce. You will also have been introduced to a conceptual model of each type of cyclone, which will prove to be invaluable in the interpretation of meteorological observations, analyses, and forecasts.

Section A

EXTRATROPICAL CYCLONES

An extratropical cyclone is a macroscale low-pressure disturbance that develops outside the tropics. Extratropical cyclones draw their energy from temperature differences across the polar front, so they are also known as frontal lows or frontal cyclones. They move from west to east as macroscale eddies embedded in the prevailing westerlies. As shown in a later section, these disturbances distort the polar front into a wave shape; therefore, they are also referred to as wave cyclones and frontal waves.

In some circumstances, extratropical cyclones may simply be called "lows." We will use this term carefully because it can be ambiguous when taken out of context. Other, very different low-pressure systems develop in the tropics and elsewhere. These have different scales, different structures, different weather, and different behavior than extratropical cyclones.

Frontal lows have much shorter time scales than either the general circulation or the monsoon. They are much easier to identify on weather charts for a given day and time rather than on monthly or seasonal average charts. Figure 8-1 shows the scale relationship of a frontal low to the global circulation. On any given day, there are always several of these lows around the globe in various stages of development, as shown in figure 8-2.

THE POLAR FRONT MODEL

The important characteristics of the development and structure of a frontal low are represented by the polar front model. The origins of this model date from research begun by Norwegian meteorologists about the time of WWI. Since that time, the model has been expanded and improved with better observations and understanding of the atmosphere.

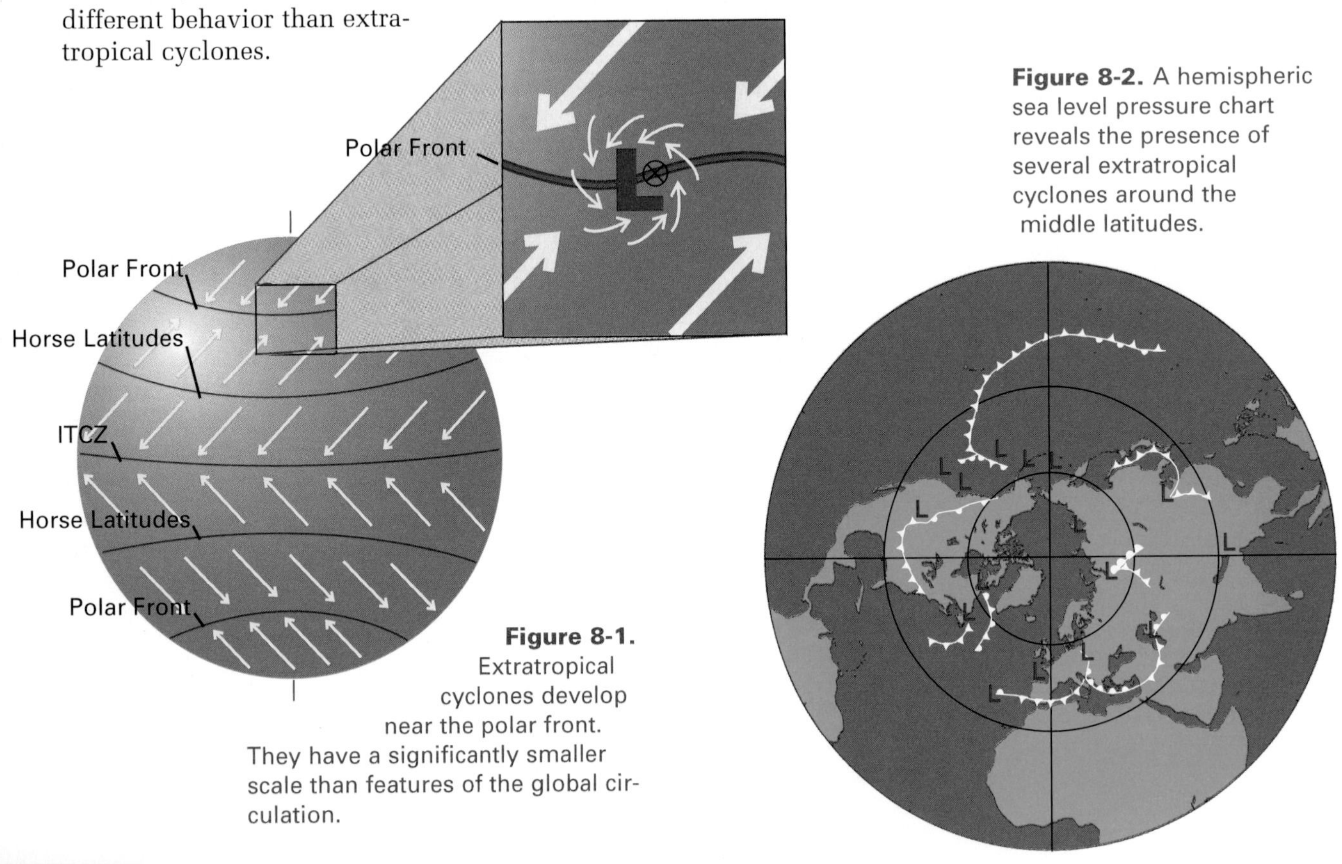

Figure 8-1. Extratropical cyclones develop near the polar front. They have a significantly smaller scale than features of the global circulation.

Figure 8-2. A hemispheric sea level pressure chart reveals the presence of several extratropical cyclones around the middle latitudes.

The modern polar front model has both a surface and an upper-air component. The surface model describes the structure and behavior of fronts and airmasses in the lower atmosphere. The upper air part of the model deals with the associated development of troughs, ridges, tropopauses, and jet streams. Both surface and upper air components contribute to unique cloud and weather patterns during the life cycle of the frontal low. We start our discussion with an examination of the surface components of the cyclone and work upward.

Arctic air only reaches the lower 48 states in the winter and initially has temperatures at or below 0°F. Maritime tropical air is common over Florida during the colder months and spreads northward in the spring and summer. It has dewpoint temperatures of 60°F or more.

AIRMASSES

An airmass is a large body of air that has fairly uniform temperature, stability, and moisture characteristics. Typical airmasses are about 1,000 n.m. across. In terms of time scale, it is not unusual for airmasses to be identifiable over periods of several days to more than a week after they leave their area of origin.

Airmasses develop in regions where surface conditions are homogeneous and winds are light. This allows the air to adapt to the temperature and moisture properties of the surface. The locations of such regions are usually near the centers of semipermanent high-pressure systems over the snow and ice fields of polar regions and over the subtropical oceans. It follows that an airmass is generally identified by its airmass source region; that is, by the geographical area where it develops. Common airmass types are Arctic (A), Polar (P), and Tropical (T).

Once an airmass leaves its source region, it is also classified according to its temperature relative to the ground over which it is moving. A cold airmass is colder than the ground and a warm airmass is warmer than the ground over which it is moving.

As polar and tropical airmasses move away from their source regions they are further identified by their moisture content. They are classified as to whether their recent trajectories were over land (continental) or over water (maritime). Figure 8-3 shows source regions and trajectories for North American airmass types.

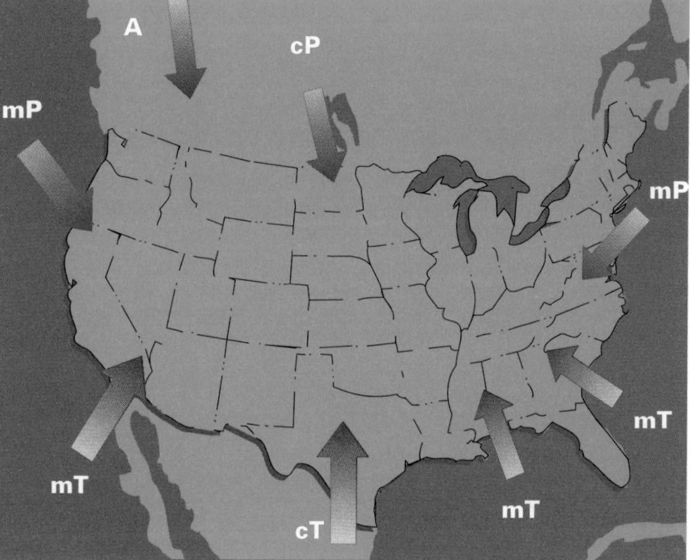

Figure 8-3. North American airmass source regions. Note standard airmass abbreviations: arctic (A), continental polar (cP), maritime polar (mP), continental tropical (cT), and maritime tropical (mT).

Airmasses undergo modification as they move away from their source region. If they move quickly, that modification will be small. For example, a fast-moving arctic airmass that moves through the prairie provinces of Canada will remain extremely cold as it penetrates the Central Plains of the U.S., because it has not had time to adjust to the new surface temperature conditions. In contrast, if airmasses move slowly and/or over great distances, they undergo large modifications. For example, a very cold airmass that moves from its source region over snow and ice often undergoes substantial modification as it moves over a large body of open water, such as the Pacific. (Figure 8-4)

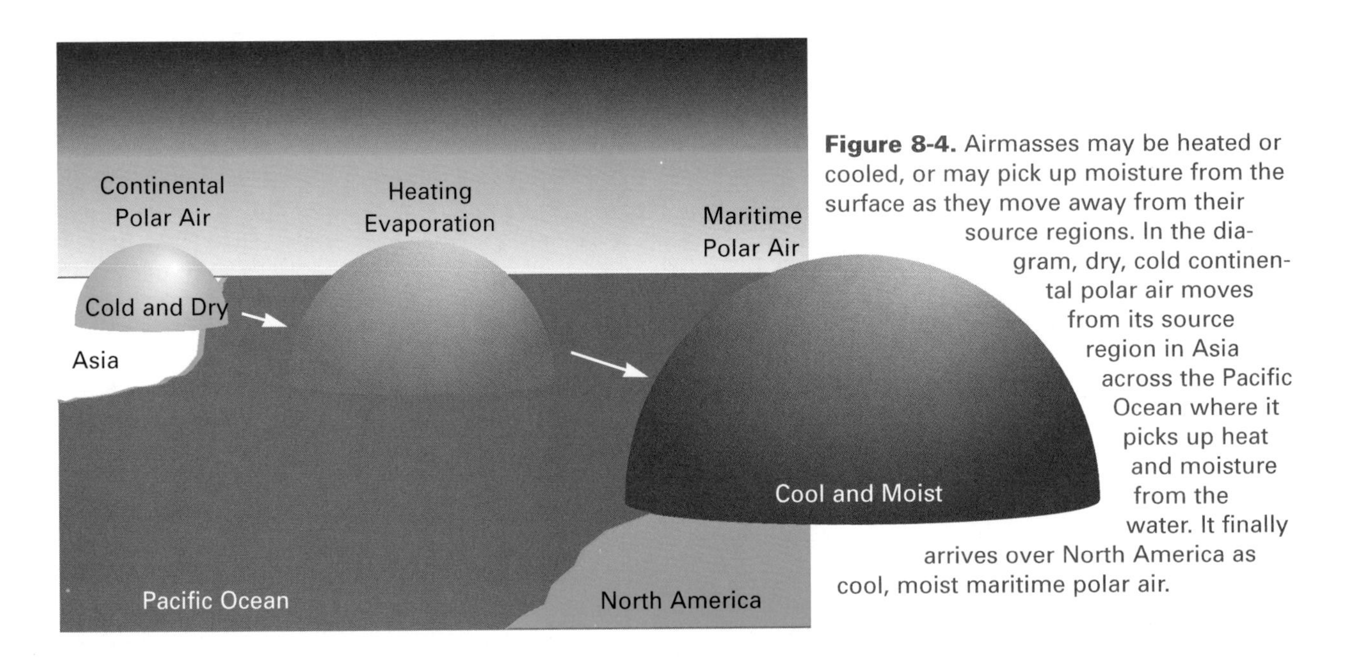

Figure 8-4. Airmasses may be heated or cooled, or may pick up moisture from the surface as they move away from their source regions. In the diagram, dry, cold continental polar air moves from its source region in Asia across the Pacific Ocean where it picks up heat and moisture from the water. It finally arrives over North America as cool, moist maritime polar air.

Another example of airmass modification is seen when a polar airmass moves from the eastern U.S. over the warm Gulf Stream along the Atlantic Coast. The airmass rapidly increases its moisture and temperature. Similarly, an arctic airmass moving slowly southward over the Great Plains will warm up rapidly during the day, especially if the skies are clear and the ground is not snow covered. On a smaller scale, a cold airmass crossing the Great Lakes in fall, before the formation of substantial amounts of ice, often becomes so moist that heavy snow showers occur in the airmass as it moves downwind from the Great Lakes.

A slow-moving, cold airmass that is unstable during the day because of surface heating will become stable at night due to cooling from below. This can result in a marked change in weather conditions from day to night. For example, a change from daytime convection with cumulus clouds and good visibilities to nighttime fog or low stratus clouds with poor visibilities may occur.

When an airmass is stable, it is common to find smoke, dust, haze, etc., concentrated at the lower levels, with resulting poor visibility.

FRONTS

Airmasses tend to retain their identifying characteristics for long periods, even when they are in close contact with another airmass. Because two airmasses with different characteristics do not mix readily, there is often a distinct boundary between them. As you know from your earlier reading about the causes of vertical motions, that boundary is called a front. Fronts are hundreds of miles long and have lifetimes similar to those of airmasses. As we will see, they are classified according to their movement.

One of the most easily recognized discontinuities across a front is a change in temperature.

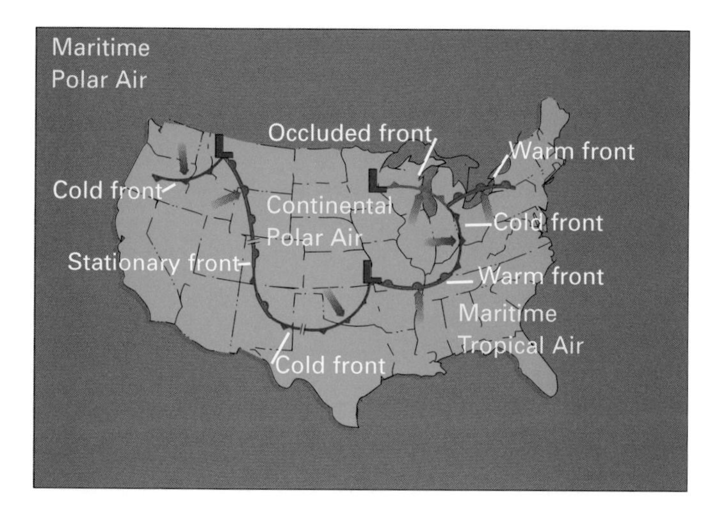

Figure 8-5. Typical front depictions. The name and direction of movement for fronts and airmasses are indicated.

FRONTAL CLASSIFICATIONS

Fronts are assigned a name according to whether the cold airmass is advancing (cold front) or the warm airmass is advancing (warm front). If the airmasses show no appreciable movement, the front is designated a stationary front (or quasi-stationary front). Later, we will examine a situation where a cold front overtakes a warm front; the result of which is termed an occluded front.

When airmasses meet, their relative motion frequently leads to the lifting of moist air (frontal lifting). When sufficient moisture is present, fronts are often locations of clouds and precipitation. For forecasting purposes, meteorologists find it useful to identify and track fronts. Examples of identifying symbols used to indicate fronts on surface analysis charts are shown in figure 8-5. You should become familiar with these useful "roadsigns" that indicate the locations of possible aviation weather problems.

FRONTAL SLOPES

Whenever two contrasting airmasses come into contact, the more dense, cold air wedges under the warm air so that the front always slopes over the cold airmass. (Figure 8-6)

The frontal slope refers to the ratio of the altitude of the top of the cold air at some point in the cold airmass to the horizontal distance of that point on the surface from the nearest edge of the airmass. It is defined in figure 8-6. For the types of macroscale fronts described in this chapter, slopes of fronts are usually in the range 1:50 to 1:500. The understanding of the slopes of fronts is important because major changes in pressure, wind, temperature, and weather occur as a front passes you on the ground or as you fly through a front.

It turns out that cold fronts have steeper slopes (about 1:50 to 1:100) than warm and stationary

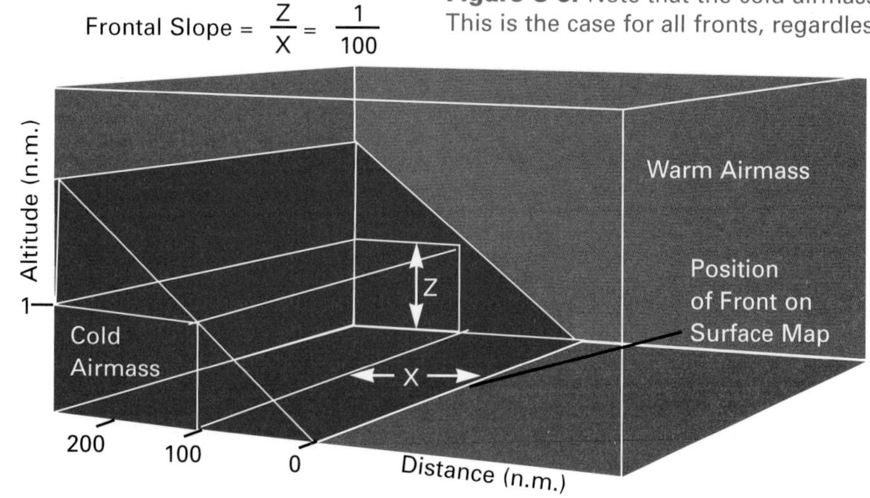

Figure 8-6. Note that the cold airmass wedges under the warm airmass. This is the case for all fronts, regardless of their classification. If this is a warm front, it would be moving from right to left. A cold front would be moving from left to right. The slope of the front is the ratio of the altitude (Z) of the frontal surface to the distance (X) from the surface position of the front. The slope of this front is 1:100 since X = 100 n.m. and Z = 1 n.m. This slope is typical of a cold front.

fronts (about 1:200 or less). (Figure 8-7) Some rapidly moving cold fronts are steeper than 1:50 in the lowest 1,000 or 2,000 feet. Occluded fronts may have the slope of either a warm or cold front, depending on the type of occlusion. This is clarified in the section on cyclone structure.

Although fronts have a distinct three-dimensional structure, they are not usually identified on upper air charts as they are on surface analysis charts. Therefore, if your flight will pass through a front, you should be aware that, because of the slope of the front, the frontal position aloft will be different than its surface position. You can anticipate this by remembering a simple rule:

> The frontal position aloft is always on the cold air side of the frontal position at the surface, regardless of the type of front.

FRONTAL ZONES

Another important concept to understand is that a front is not a thin line as shown on a chart, but is actually a narrow frontal zone through which there is a rapid transition of conditions from one airmass to the other. The width of the transition zone may be as little as 0.5 n.m. to more than 100 n.m. It is usually narrower near the ground than aloft. (Figure 8-8)

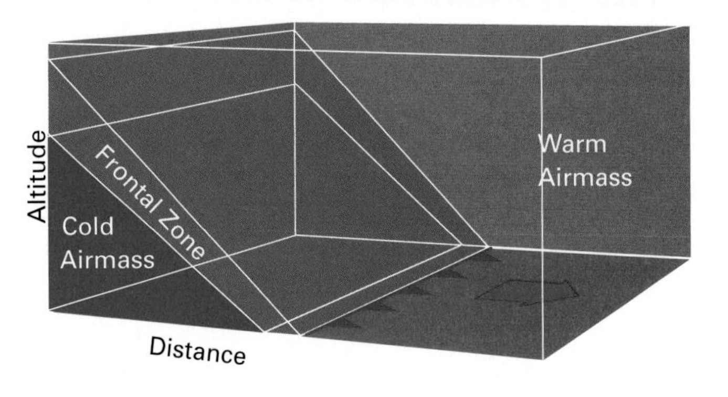

Figure 8-8. Perspective diagram of a cold front showing the frontal zone on the cold side of the line that indicates the front on a surface analysis chart. The front is moving from left to right.

On conventional surface analysis charts, single lines with appropriate symbols are used to represent the positions of the fronts. Because frontal zones are often the location of strong shears and turbulence, you should always be aware of their existence and location relative to the indicated frontal position. Remember,

> A frontal zone is always located on the cold side of the frontal position designated on the surface analysis chart, regardless of type of front.

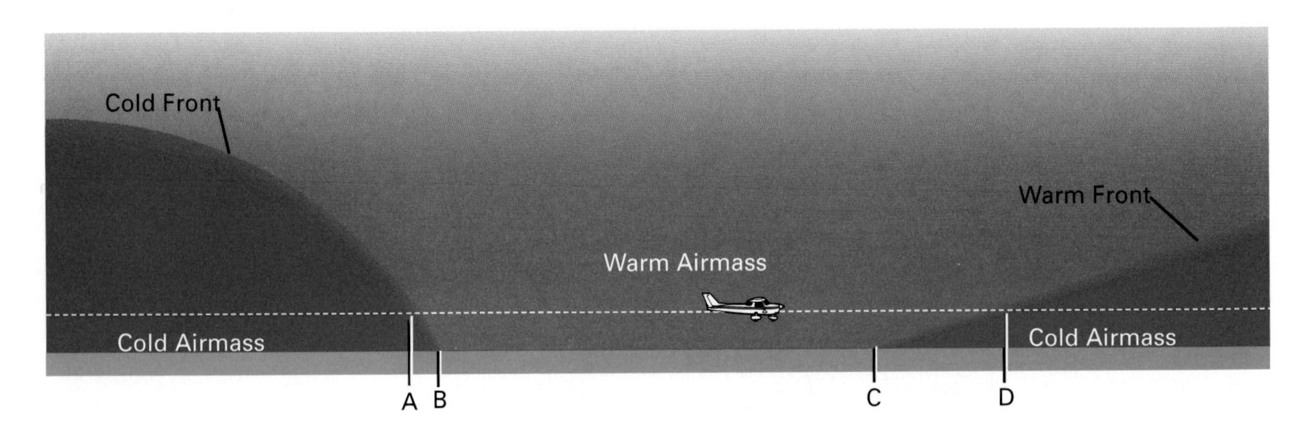

Figure 8-7. Cross sections through warm and cold fronts. For illustration purposes, the slopes are greatly exaggerated in this diagram. Both fronts are moving from left to right. The slope of a cold front is steeper than the slope of a warm front. The dashed line indicates the flight of an aircraft through both fronts. Notice the points of intersection along the flight path (A and D) versus the surface frontal positions (B and C).

EXTRATROPICAL CYCLONE STRUCTURE AND DEVELOPMENT

Extratropical cyclones are not accidental disturbances. They develop for a reason. As we saw in earlier chapters, there is an excess of solar energy received at the equator and a deficit at the poles. The temperature gradient caused by these differences is concentrated in the polar front. If that temperature gradient becomes excessive at some place along the polar front, a disturbance will occur to reduce the gradient. The disturbance in this case is the extratropical cyclone. In the following description of its development and structure, it will become very clear how these frontal lows accomplish the task of mixing warm air toward the pole and cold air toward the equator to reduce the temperature gradient.

With regard to aviation, frontal cyclones can produce almost every weather flight hazard that you can think of, ranging from clear air turbulence to icing and wind shear. The importance of understanding these large scale circulations is that their dangers are not impossible to avoid. With a thorough knowledge of the structure of frontal lows, their associated fronts, and with good planning, you can often circumnavigate the problem areas while taking advantage of favorable winds.

CLIMATOLOGY

Frontal lows develop in areas of the globe where favorable conditions exist. For example, in winter, locally strong temperature gradients are found along some coastlines where cold continents are next to very warm oceans. This is the case for the U.S. just off the coast of the Gulf of Mexico and along the East Coast. When fronts move into these areas, the development of frontal lows is common. (Figure 8-9)

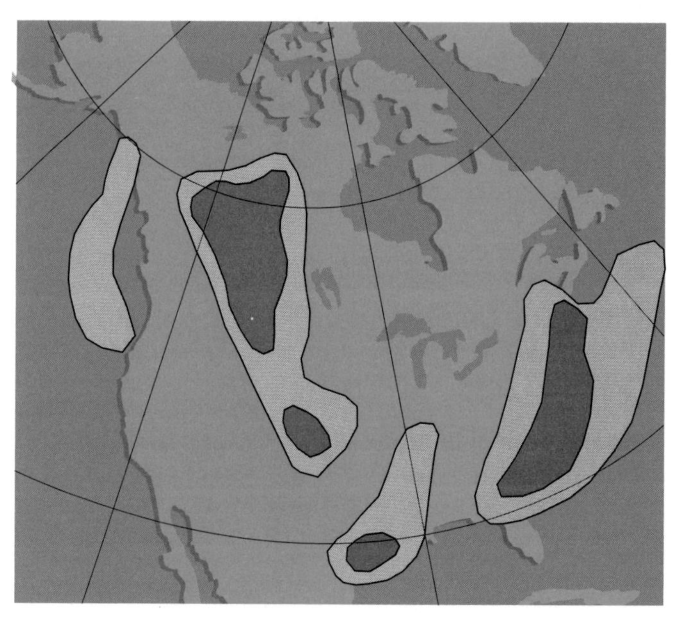

Figure 8-9. Frequent areas of cyclone development in North America are shaded. Regions of highest frequencies are dark red.

Vertical motions provided by large mountain chains and latent heat derived from moist air during condensation can also enhance cyclone development (cyclogenesis). These two processes frequently work together to produce frontal lows on the east slopes of the Rocky Mountains. (Figure 8-9)

SURFACE STRUCTURE AND DEVELOPMENT

The development of a frontal cyclone progresses through a distinctive life cycle. The incipient stage is shown in figure 8-10. Before the development begins, a stationary front is present in the area. The frontal zone is characterized by a change in wind speed and/or direction (wind shear) from the warm side to the cold side.

A variation which will always occur when flying across a front is a change in the wind direction.

As the cyclone development begins, pressure falls at some point along the original stationary front, and counterclockwise (Northern Hemisphere) circulation is generated. At this point, the cyclone is in the incipient stage; this is also called the wave cyclone stage because the previous stationary front has been distorted into a wave shape in response to the developing circulation.

The wave cyclone moves northeastward at 15 to 25 knots, pushing warm air northward ahead of it, and bringing cold air to the south into the wake of the cyclone. (Figure 8-10) Notice that the initial stationary front has been replaced by warm and cold fronts. The triangular region of warm air between the fronts and to the south of the cyclone is called the warm sector.

Both of the fronts in this idealized model lay in troughs of low pressure. This structure provides some useful indicators of the approach and passage of fronts.

Wind shear is the change of wind speed and/or wind direction over a distance. Shears may be vertical, horizontal, or both. A large change in wind over a short distance corresponds with strong shear. Wind shear is associated with a number of atmospheric disturbances. The strongest shears occur with mesoscale and microscale circulations, such as a microburst. Small scale wind shears are included in Part III. When discussing larger scale disturbances such as cyclones or fronts, we often specify whether the shear is cyclonic or anticyclonic. Cyclonic wind shear means that changes in the wind speed or direction correspond with what you would find as you cross a low-pressure area. Anticyclonic wind shear is what you would expect when crossing a high-pressure area. The wind shear across a front (which often lies in a trough of low pressure) is cyclonic.

Wave cyclones do not necessarily develop beyond the incipient stage. These stable waves simply move rapidly along the polar front, finally dissipating.

Figure 8-10. Development of a frontal low. Diagrams are labeled chronologically. Arrows represent circulation around the cyclone. Isobars are indicated on diagram 2 for reference. These diagrams show conditions during the first 12 hours after the development begins (incipient stage).

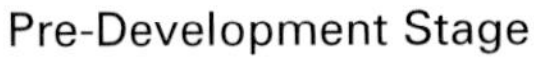

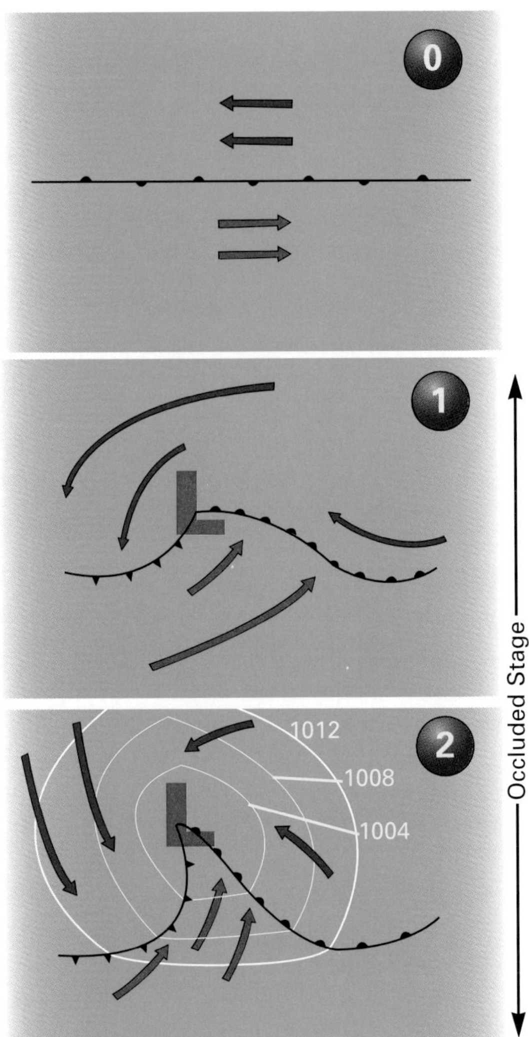

1. As a front approaches a given location, the pressure falls. As the front passes, the pressure rises. This is more noticeable with a cold front than a warm front.

2. The sharp change in pressure gradient across a front corresponds with an equally sharp wind shift, an example of a cyclonic wind shear.

The winds change rapidly from a south or southwesterly direction just ahead of the cold front (in the warm sector) to northwesterly in the cold air. With the warm front, wind directions change from southeasterly in the cold air to southwesterly in the warm air.

Because a frontal zone has a distinct slope, wind shears are experienced during flight through the front, whether the penetration is made horizontally (level flight) or vertically (climb or descent). The vertical wind shear through a cold front is often visible from the ground. Just after a cold front passes, low clouds in the cold air will be moving from the northwest, while middle and high clouds move from the southwest.

> The approach of a warm front is indicated by southeasterly winds, falling pressures, and a gradually lowering ceiling. The progression of cloud types as the front approaches is cirrus (CI), cirrostratus (CS), altostratus (AS), and nimbostratus (NS) with stratus (ST), fog, poor visibilities, and continuous precipitation.

As the cyclone progresses eastward, the central pressure continues to fall. This is an indication that the cyclone is deepening. The winds around the cyclone increase in response to the greater pressure gradient.

About 12 hours after the initial appearance of the frontal low, the cold airmass trailing the cyclone is swept around the low and overtakes the retreating cold air ahead of the cyclone. The warm sector air is pushed aloft by this occlusion process and the cyclone enters what is called the occluded stage. (Figure 8-11) As noted previously, the combined frontal structure is called an occluded front. The central pressure of the low falls below 1,000 mb, and the cyclone slows down appreciably. The storm reaches its greatest intensity within about 12 hours after occlusion.

Figure 8-11. Development of an occluded cyclone. Diagrams labeled 3 and 4 represent the early occluded stage (12 to 24 hours after development began) when the cyclone reaches its greatest intensity. Diagram 5 shows conditions a few days later when the system is dissipating. Fronts in the vicinity of the occlusion have also weakened.

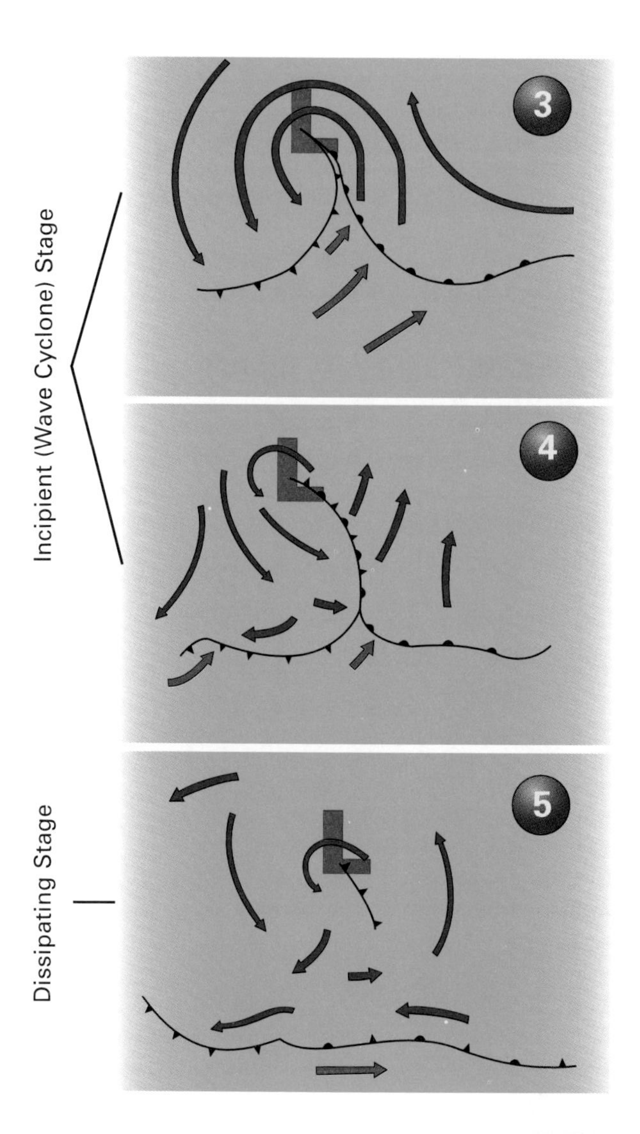

When a cyclone is in the wave cyclone stage, its direction of movement will be parallel to the geostrophic wind in the warm sector.

The central pressure begins to rise (the cyclone is filling) as the frontal low enters the dissipating stage of its life cycle. (Figure 8-11) The weakening of the cyclone begins 24 to 36 hours after the initial formation of the disturbance and lasts for another few days. The weakening is understandable if you recall that the low draws its energy from the temperature gradient across the front. In the occlusion process that gradient is destroyed, so the cyclone, now located entirely in the cold air, dies from the lack of an energy source.

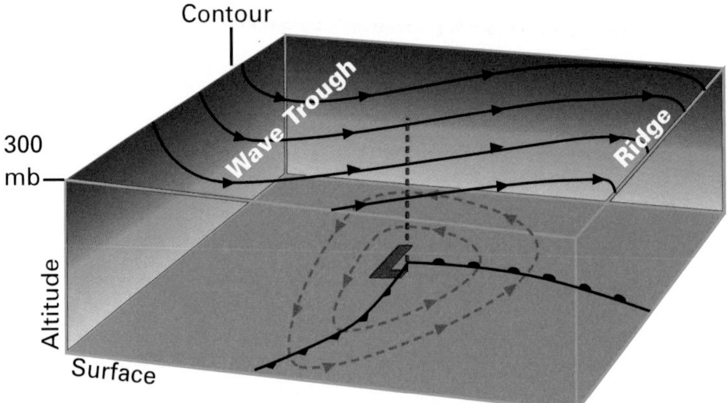

Figure 8-12.
Perspective diagram of 300 mb and surface charts for the case of a developing wave cyclone. Arrows have been added to the isobars and contours to show the wind directions at both levels.

Although the emphasis here is on the extratropical cyclone and its associated fronts, the frontal low development is usually accompanied by a well-defined, but shallow, cold high-pressure system behind the cold front. The center of this anticyclone is found near the center of the coldest air.

UPPER AIR STRUCTURE AND BEHAVIOR

Upper level troughs and ridges were first introduced in Chapter 3 as features commonly encountered in the middle and upper troposphere and stratosphere. In Chapter 5, very large macroscale troughs aloft were identified as important characteristics of the wintertime global circulation along the east coasts of Asia and North America.

The upper-level troughs which correspond to developing frontal lows are smaller scale than the long waves discussed earlier (1,000 n.m. vs. 3,000 n.m.). Also called short wave troughs, these disturbances move toward the east much more rapidly than long wave troughs, averaging about 600 n.m. per day.

The development of an extratropical cyclone often begins aloft before there is evidence at the surface. When surface development begins, the upper trough must be just upwind of the surface low so development can proceed in an efficient manner. The ideal arrangement is shown in figure 8-12.

The position of the short wave trough to the west of the surface low allows mass to be removed above the low by strong winds ahead of the trough. This causes the pressure to fall at the surface and the deepening of the surface low. Two useful rules of thumb regarding a developing cyclone are:

1. There is a good chance of the development of a frontal low when an upper-level, short-wave trough moves to within 300 n.m. of a stationary front at the surface.

2. The east side of an upper-level, short-wave trough is the bad weather side.

Once the surface cyclone begins to develop, the upper-air system develops with it. There are no longer separate surface and upper air disturbances, but a single cyclone that extends from the surface through the tropopause.

Both the upper trough and the surface cyclone deepen through the wave cyclone stage. About the time of occlusion, a closed cyclonic circulation, indicated by at least one circular contour, develops at about 10,000 feet MSL (700 mb) and below. Occasionally, a closed low will also be found at 18,000 feet MSL (500 mb) and above. At the time of the occlusion, or soon after, the upper cyclone becomes centered over the surface cyclone. (Figure 8-13)

Figure 8-13. An occluded cyclone at the surface corresponds with a closed low aloft. Compare with figure 8-12.

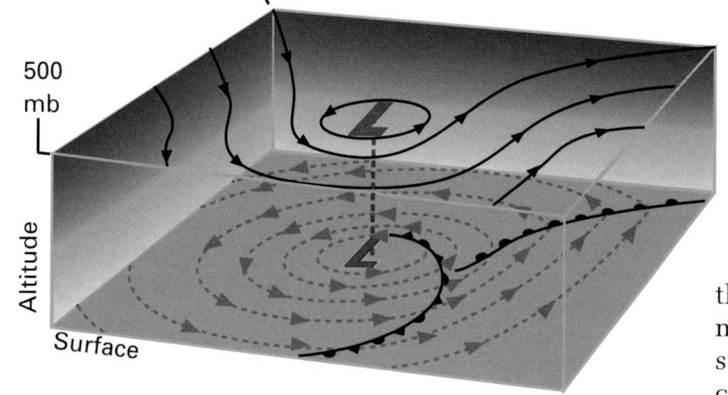

While flying cross-country in the Northern Hemisphere, if you experience a continuous left crosswind which is associated with a major wind system, you are flying toward a low-pressure area and generally unfavorable weather conditions.

An important temperature feature of the extratropical cyclone is the location of the cold air. Aloft, the cold air is found near the center of the trough. At the surface, the low is initially located on the boundary between the warm and cold airmasses, until occlusion occurs; then, the cold air reaches the center of the surface cyclone.

At every stage of development of the extratropical cyclone, the cyclone is stronger (as measured by the wind speeds) aloft than at the surface. The greatest wind speeds are in the jet streams near tropopause level.

In the troposphere, the polar front jet stream is on the edge of the coldest air; that is, it parallels fronts with the coldest air to the left of the wind, looking downstream. When an extratropical cyclone develops at jet stream levels, the segment of the jet stream that is found around the upper trough intensifies as the trough develops. In the early stages of some cyclones, this jet streak of high winds may be more obvious than the upper-level trough. (Figure 8-14)

CLOUDS AND WEATHER PATTERNS

In moist areas around the frontal low, broad layers of clouds and precipitation are produced where upward motions occur. Rising air is generated at low levels by fronts and by the converging winds around the cyclone. Low-level cloudiness may also be produced in the warm and cold airmasses if there is sufficient moisture. The processes are contact cooling in stable air (warm airmass) or surface heating and convection in unstable air (cold airmass).

At higher levels, upward vertical motions are generated through a deep layer over the surface low, because mass is being removed by the winds at jet stream level more rapidly than it is being

Figure 8-14. A jet streak refers to a segment of a jet stream where the winds are stronger than either upstream or downstream of that segment. The jet streak lies in a well-defined tropopause break. The tropopause is significantly lower over the cold tropospheric air in the trough than it is in the warm ridges that flank the cyclone.

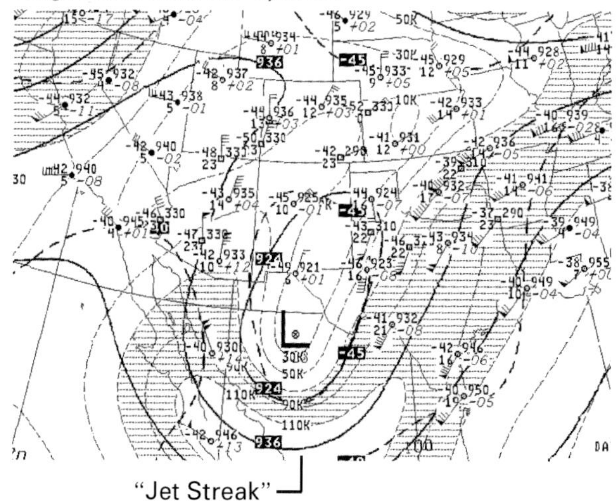

replaced near the surface. As noted in the previous section, this process is what causes the low to deepen. With adequate moisture, the resulting upward motions produce widespread cloud layers.

Frontal Clouds and Weather

At low levels, clouds and weather tend to concentrate in the center of the cyclone and near the fronts and troughs. An idealized pattern is shown in figure 8-15.

Clouds and weather caused by frontal lifting depend to a large degree on frontal type. Cross sections through those fronts, and their associated clouds and weather are shown in figures 8-16 and 8-17.

In the ideal picture, the cold air lifts unstable air along the relatively steep cold front. This leads to deep convection, with a narrow line of CU and CB clouds and associated shower activity.

Moist, warm, and stable air along the warm front moves over the retreating wedge of cold air in

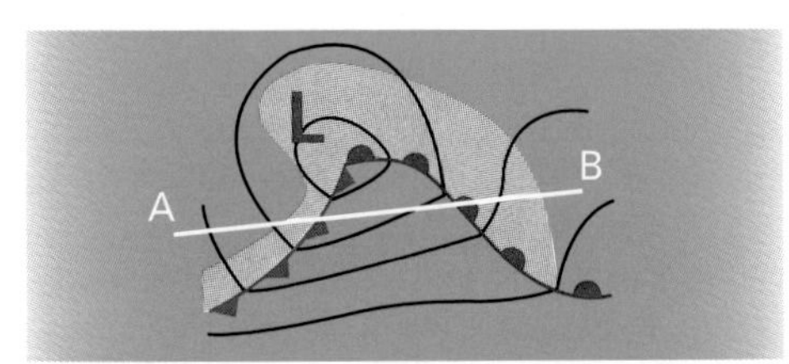

Figure 8-15. The distribution of clouds and precipitation in and around a developing extratropical cyclone. Jet stream cirrus and airmass cloudiness are not shown.

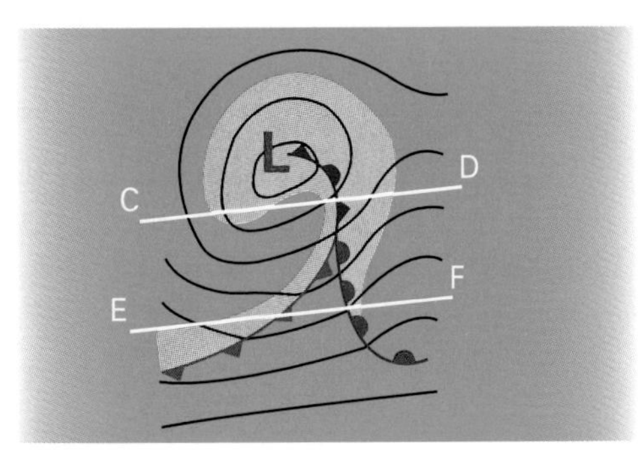

Figure 8-16. Idealized cross section showing the structure of cold front (left) and warm front (right), and associated clouds and weather in the vicinity of a cyclone in the wave stage of development along line A-B in figure 8-15.

what is frequently described as a "gentle, upglide motion" or overrunning. The result is that the warm front produces broad, deep layers of stratiform (NS, AS, CS) clouds and steady, continuous precipitation. In the stable, cold air below the sloping frontal zone, falling precipitation saturates the air; low ceilings with low clouds, ST, and low visibilities with fog are common. The situation is made worse in winter when freezing precipitation occurs.

The warm front cloudiness described above is modified when the warm, moist air moving over the front becomes unstable. This produces areas of convection. Thunderstorms embedded in otherwise stratiform cloudiness in the vicinity of a warm front represent a serious flight hazard.

When a cold front overtakes a warm front in the occlusion process, the resulting occluded

Overrunning also occurs when moist, southerly flow intersects a stationary front east of the Rockies in the winter. Under these conditions, a region of bad flying weather with low clouds, poor visibilities and, occasionally, freezing rain and icing occurs over a distance of 400 n.m. or so on the cold air side of the front.

front combines cold and warm frontal cloudiness, depending on the type of occlusion. In a warm front occlusion (figure 8-17, left), the warm front remains on the ground, because the cold air ahead of the warm front is much colder than the air behind the cold front. In this case, the cold front moves up the warm frontal boundary. Because of the flat slope of the warm front, the passage of this upper cold front with its showers and thunderstorms may precede the passage of the surface occluded (formerly warm) front by some distance. In comparison, the passage of a cold front occlusion (cold front remains on the ground) is followed fairly quickly by the passage of the upper warm front. (Figure 8-17, right) Both cases are subject to the hazard of embedded thunderstorms, especially during the warmer months of the year.

Indicators of an approaching warm front are steady precipitation with stratiform clouds.

The in-flight hazard most commonly associated with warm fronts is precipitation-induced fog.

In a cold front occlusion, the air ahead of the warm front is warmer than the air behind the overtaking cold front.

When the cold airmass following a cold front is moist and unstable, it is characterized by cumuliform clouds and showery precipitation.

A ridge or high-pressure area is characterized by downward motion.

Behind the cold front, cold air moving over warm ground or warm water may become unstable. Visibilities are good except where the instability is so great that post-frontal showers occur. Convective cloud clusters in the cold air are often visible in satellite images over oceanic areas where moisture is plentiful. As the ridge or anticyclone centered in the cold air moves into the area, downward motion is dominant. Clouds and precipitation are suppressed and clear weather prevails.

A Space View of An Extratropical Cyclone

Images of clouds from weather satellites are becoming more accessible every day. However, the space view of clouds is a far different perspective than the view from the ground or the cockpit. To provide some interpretive guidance, a schematic satellite view of typical cloud patterns for a fully developed occlusion is presented in figure 8-18.

Warm Front Occlusion

Warm

Cool

Cold

C

P

D

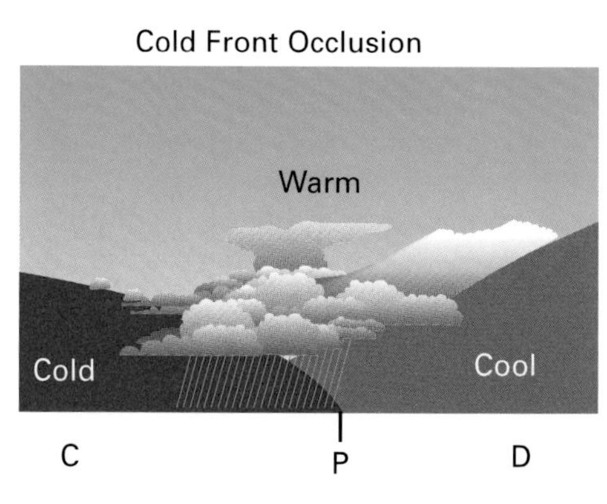

Figure 8-17. Idealized cross sections showing frontal cloudiness and weather in the vicinity of an occluded front. In both cases fronts and airmasses are moving to the right. These diagrams are two possible vertical cross sections along line C-D through the occluded front on figure 8-15. Point P corresponds with the position of the occluded front on the ground. Note that a cross section along E-F in figure 8-15 would be identical to the diagram in figure 8-16.

From the images in diagrams A and B, we see three primary cloud features that support our earlier contention that the east side of a trough aloft is the bad weather side. A low-level, cold frontal cloud band is found to the south of the cyclone and east of the trough aloft. Also on the east side of the upper trough are a comma cloud composed of middle clouds, and a broad, curved band of jet stream cirrus located just to the right of the jet axis. These cirrus clouds are the highest clouds near the cyclone, and they will be the brightest (coldest) in an infrared satellite image. They can usually be identified by the very sharp edge to the cirrus band near the jet stream axis. Other cirrus clouds are also found near the comma cloud. These are lower clouds, not to be confused with jet stream cirrus.

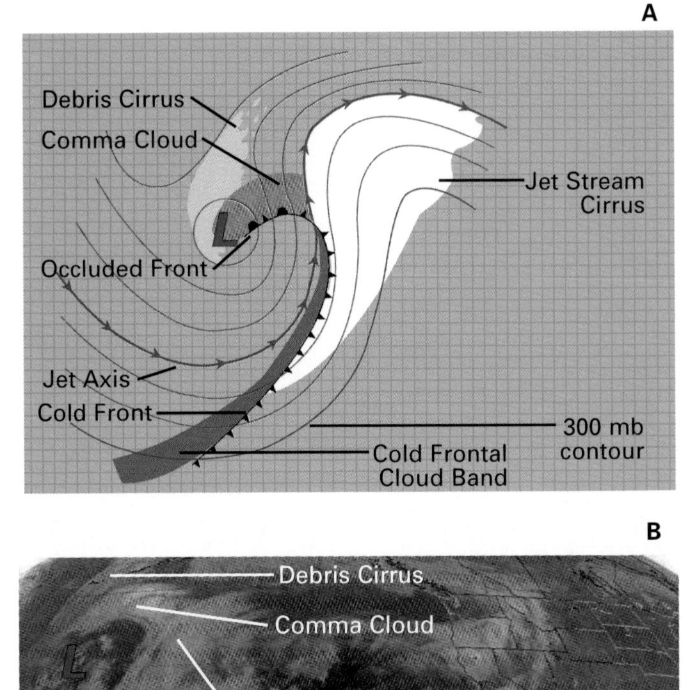

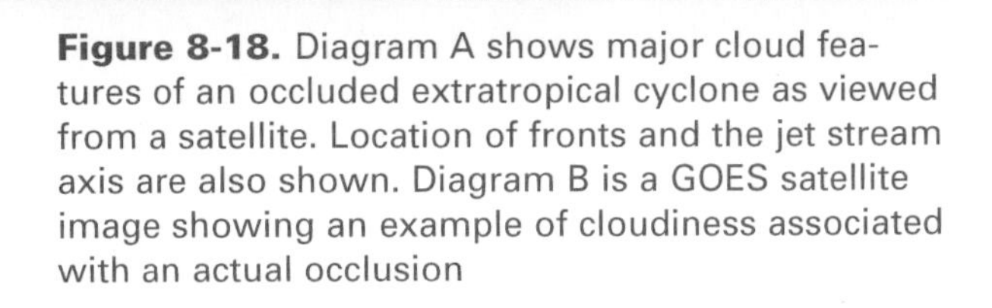

Figure 8-18. Diagram A shows major cloud features of an occluded extratropical cyclone as viewed from a satellite. Location of fronts and the jet stream axis are also shown. Diagram B is a GOES satellite image showing an example of cloudiness associated with an actual occlusion

As shown in figure 8-18, the cold frontal clouds can usually be seen from space, but jet stream cirrus covers much of the low level cloudiness in the vicinity of the warm front. Note also that the comma cloud is a good indicator of the location of the upper trough or low. In the early stages of extratropical cyclone development, the comma cloud may be separated (upstream) from the cold frontal cloud band.

The idealized polar front model is a useful guide for interpreting observed weather conditions, for understanding the surface analysis chart, and for anticipating future weather conditions. But keep in mind that it does not precisely describe the structure and behavior of every frontal low that occurs. For example, in some oceanic cases, the cyclone deepens significantly, but the occluded front does not develop as previously described. Also, warm fronts in occluded cyclones approaching the west coast of the U.S. are very difficult to locate and are often not shown on surface analysis charts. Over the North American continent, prominent mountain ranges and the very warm Gulf of Mexico often cause distortions in the various stages of wave cyclones. In some cases, wave cyclones that develop on the east slopes of the Rockies seem to be occluded from their initial appearance.

These variations should not be difficult to deal with if you remember that all frontal cyclones typically retain their identities for several days, allowing you to identify variations from the idealized polar front model and adjust your flight planning accordingly.

Section B

HURRICANES

All cyclones that develop in the atmosphere are not exactly like extratropical cyclones. In many cases, the only similarity is that the winds blow cyclonically around a low-pressure area. Otherwise, behavior, structure, and energy sources are all different. Perhaps the best example of such a contrast is the tropical cyclone.

A tropical cyclone is a mesoscale, cyclonic circulation that develops in the tropical easterlies. In its most intense form, it becomes a hurricane with strong convection, exceptionally strong winds, and torrential rains. In this section, we briefly examine the climatology, structure, and behavior of hurricanes.

CLIMATOLOGY

The term "tropical cyclone" covers a number of similar tropical disturbances which are classified according to their maximum sustained wind speeds: tropical disturbance (< 20 knots); tropical depression (20 to 34 knots); tropical storm (35 to 64 knots); and hurricane (> 64 knots). Hurricane-strength tropical cyclones are known by other local names, depending on the geographical location. For example, in the Western Pacific, they are called typhoons.

There is a large intensity range of tropical cyclones beyond the threshold of the hurricane definition; hurricanes with winds in excess of 100 knots are not uncommon. For this reason, the Saffir-Simpson scale was developed to rate the damage potential of individual hurricanes. (Figure 8-19)

When any tropical storm reaches hurricane strength, it is named from a list selected by international agreement. Recent examples are

Figure 8-19. Saffir-Simpson scale of damage potential of hurricanes. A *storm surge* is an abnormal rise of water due to a tropical cyclone.

Catagory	Central pressure (millibars)	Winds (knots)	Storm Surge (feet)	Damage
1	≥ 980	64 - 82	4 - 5	damage mainly to trees, shrubbery, and unanchored mobile homes
2	965 - 979	83 - 95	6 - 8	some trees blown down; major damage to exposed mobile homes; some damage to roofs of buidings
3	945 - 964	96 - 113	9 - 12	foliage removed from trees; large trees blown down; mobile homes destroyed; some structural damage to small buildings
4	920 - 944	114 - 135	13 - 18	all signs blown down; extensive damage to roofs, windows, and doors; complete destruction of mobile homes; flooding inland as far 10 km (6 mi.); major damage to lower floors of structures near shore
5	< 920	> 135	> 18	severe damage to windows and doors; extensive damage to roofs of homes and industrial buildings; small buildings overturned and blown away; major damage to lower floors of all structures less than 4.5m (15 ft.) above sea level within 500 meters of shore

Hurricane Hugo that struck the coast of South Carolina in 1989, and Hurricane Andrew that devastated southern Florida in 1992. Favored regions of development, some typical cyclone trajectories, and other local names are presented in figure 8-20.

The tropical disturbances that influence the continental U.S. are mainly produced in the Atlantic, the Caribbean, and the Gulf of Mexico. The areas of the U.S. that are most vulnerable to hurricanes are the Eastern Seaboard and the Gulf Coast. Hurricanes occur most frequently in the late summer and early fall.

Occasionally, the remnants of a dying tropical cyclone cross into the southwest U.S. from the Pacific coast of Mexico. However, the great majority of storms produced in the Eastern Pacific move northwestward, occasionally threatening the Hawaiian Islands, but more often they die over colder waters in the North Pacific.

DEVELOPMENT AND BEHAVIOR

As shown in figure 8-20, tropical cyclones develop within about 1,200 n.m. of the equator, over areas of very warm water (79°F or greater). Each begins its life cycle as a poorly organized tropical disturbance. If conditions are favorable, it moves through the successively stronger stages of a tropical depression and tropical storm. A relatively small number of tropical cyclones continue to intensify to hurricane strength. During the period of development, the hurricane progresses to the west or northwest at about 10 knots.

Since a hurricane draws its energy from warm water, it dissipates when it moves over cold water or over land. In some areas of the globe, (for example, the Atlantic) hurricanes move out of the easterlies, gradually turning poleward on the western side of the subtropical high. (Figure 8-20) As the storms reach higher latitudes during this recurving process, they weaken, finally moving rapidly back toward the east. When some dying hurricanes reach the polar front, or at least an area of strong temperature gradient, they undergo redevelopment as strong extratropical cyclones.

STRUCTURE AND WEATHER

When a tropical cyclone reaches hurricane strength, the storm is several hundred miles in diameter. Figure 8-21 gives different views of two hurricanes, one from a satellite and one from weather radar. The horizontal scale (diameter) of

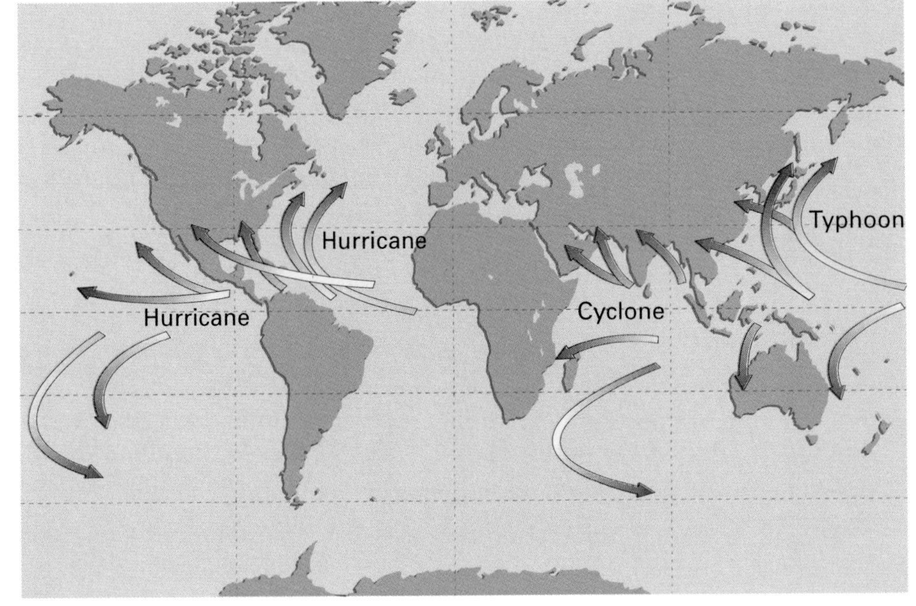

Figure 8-20. Common tropical cyclone tracks.

the hurricane cloud signature is typically 300 n.m. or more. The clouds observed by satellite often cover a larger area than the main part of the storm. This happens because the cirrus spreads out in a high level outflow region of an upper level anticyclone. The horizontal scale of the radar signature of a hurricane is much smaller than that seen in the satellite image because the radar doesn't observe the cloud structure. Rather, it senses the precipitation regions of the hurricane. (Figure 8-21) A still smaller portion of the storm is the diameter of the region of hurricane strength winds; that distance is often less than 50 nautical miles. These are approximate scales. There are considerable variations from storm to storm, depending on overall strength.

Major features of the hurricane environment are the eye, the eye wall, and cloud bands spiraling into the storm. (Figure 8-22) The hurricane eye is the circular, nearly cloudfree region approximately 10 to 20 n.m. in diameter that is located in the center of the storm. It is the warm core of the hurricane, a region of relatively light winds, and the lowest sea level pressure. As a matter of fact, the lowest sea level pressure in the world was observed in the center of a hurricane (870 mb in Typhoon Tip in 1979).

The eye wall is the cloudy region embedded with cumulonimbus (CB) clouds immediately adjacent to the eye. It is the region of strongest winds and most intense convection. Because hurricanes occur in the tropics where the tropopause is very high, it is not unusual for the CB cloud tops to extend to 50,000 feet MSL or higher in the eye wall and elsewhere in the storm.

The rainbands that spiral into the storm are also lines of convergence characterized by CB clouds and shower activity. They measure from a few nautical miles to about 30 n.m. wide and are spaced 60 n.m. to 200 n.m. apart.

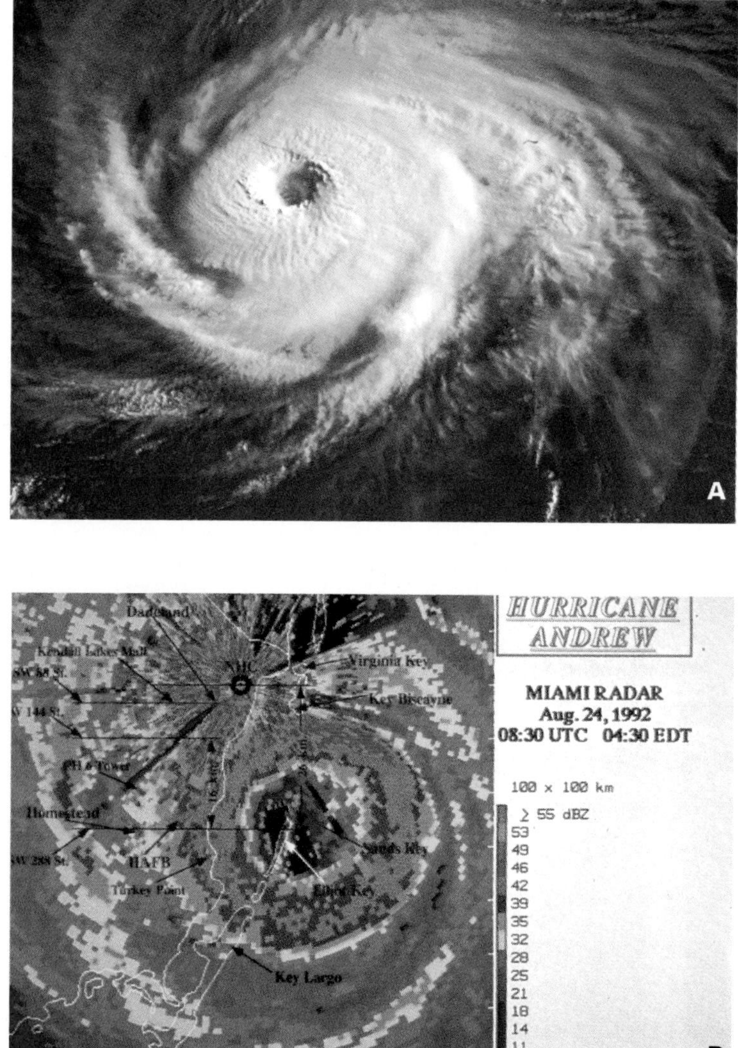

Figure 8-21. Hurricanes have distinctive signatures when viewed from space (diagram A) and distinctive radar signatures (diagram B) that reveal the eye and spiral band structure. (Hurricane photograph courtesy of NASA. Radar signature courtesy of National Center for Atmospheric Research/University Corporation for Atmospheric Research/National Science Foundation.)

As shown in figure 8-22, strong winds flow into the intense low-pressure area at the surface. High up in the storm (40,000 feet MSL or so), there is a change in the circulation. The top of the storm is marked by weak winds and outflow, and an anticyclone. This structure is a key characteristic of tropical cyclones: winds are strongest near the surface and weaken aloft, especially above 18,000 feet MSL. The same cannot be said about the convection associated with thunderstorms. More details on the structure of thunderstorms (CB) is given in the following chapter.

Although a hurricane may appear to be roughly symmetrical when viewed from a satellite, this is not true when considering the associated weather. Looking in the direction of hurricane movement (Northern Hemisphere), the strongest winds occur on the right side and the worst weather usually occurs in the right front quadrant of the hurricane. The movement of rainbands also can affect local winds, and precipitation intensity depending on the velocity and width of the bands. Although the strong, and often damaging, hurricane winds receive the greatest attention when a hurricane is described, the greatest damage from a hurricane is usually associated with coastal flooding caused by the storm surge.

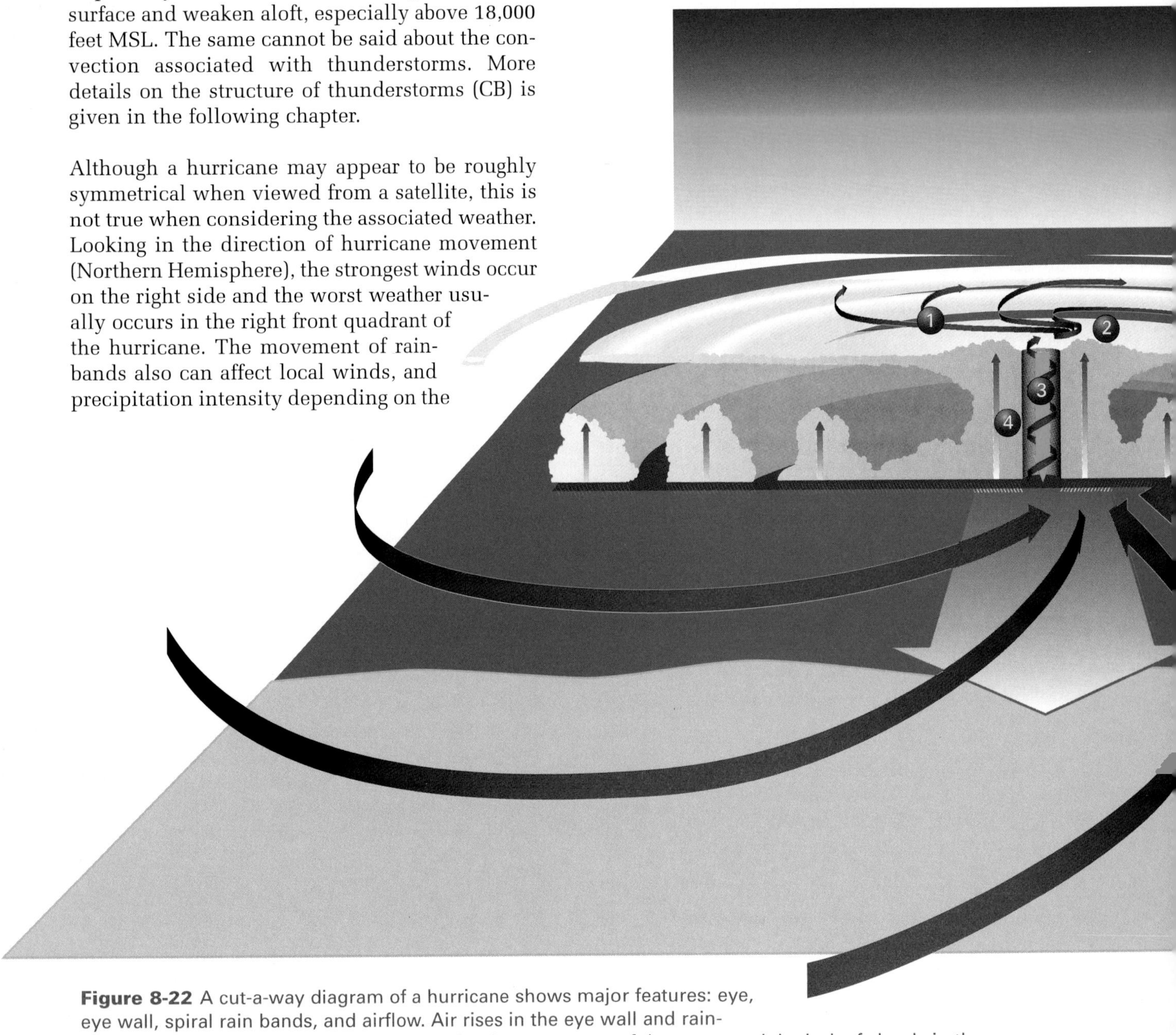

Figure 8-22 A cut-a-way diagram of a hurricane shows major features: eye, eye wall, spiral rain bands, and airflow. Air rises in the eye wall and rainbands. Strong sinking in the eye accounts for the warm core of the storm and the lack of clouds in the eye.

	Charleston Naval Station, SC	Charleston (City Site), SC	WSO Charleston Airport, SC
Date	September 21-22	September 21	September 22
Anemometer Height	118 feet	25 feet	20 feet
Peak Gust	137 m.p.h. @ 11:30-11:45 p.m., 9/21	108 m.p.h. @ 11:40 p.m.	98 m.p.h. @ 12:59 a.m.
Max. Sustained Speed	NA	87 m.p.h. @ 11:30 p.m.	78 m.p.h. @ 1:03 a.m.
Max. Mean Speed	15-minute, 74 m.p.h. @ 1:00 a.m., 9/22	NA	10-minute, 59 m.p.h. @ 1:10 a.m.

Figure 8-23. Surface wind observations from Charleston, South Carolina during Hurricane Hugo, which passed to the north of the city. The eye of Hugo crossed the coast about 1000 UTC, September 22, 1989.

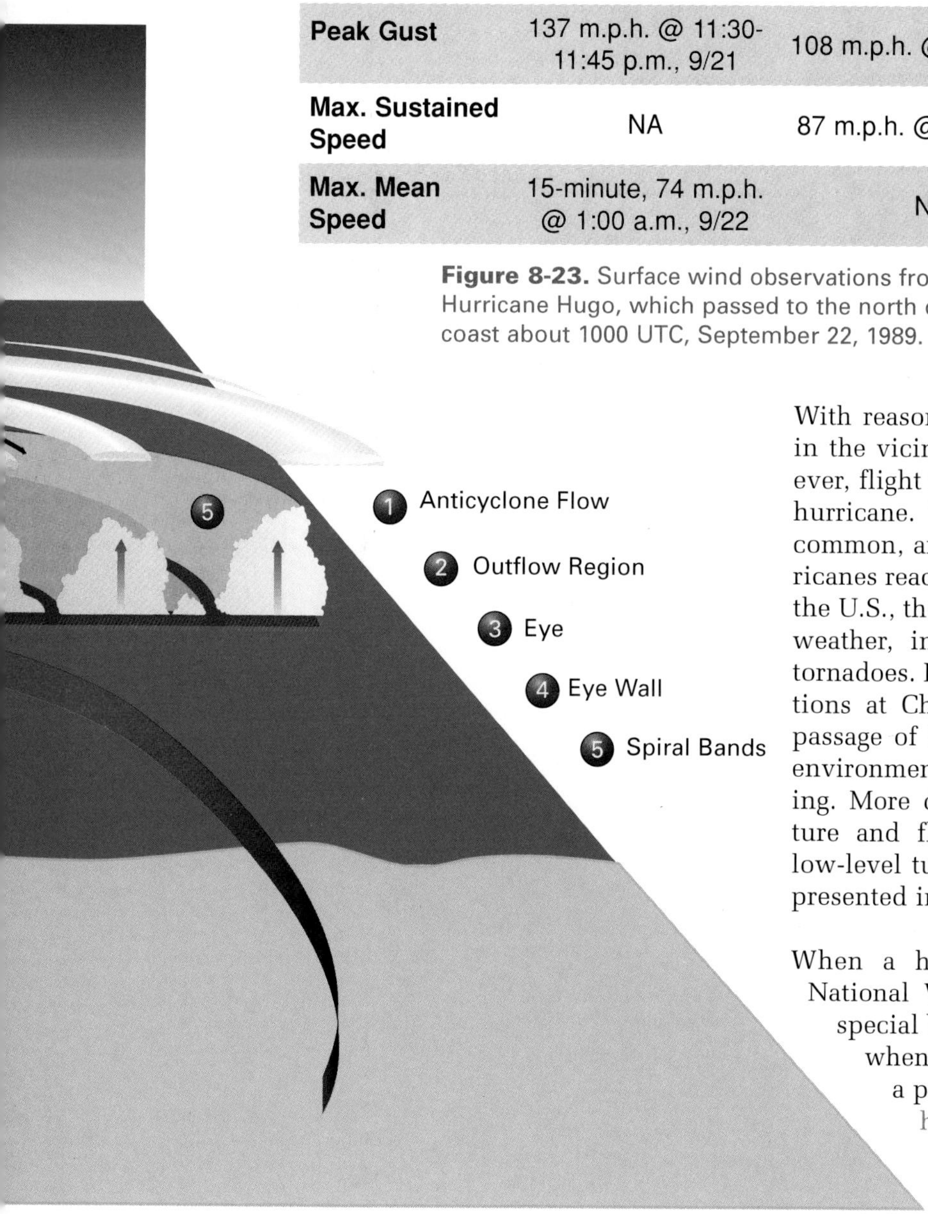

1 Anticyclone Flow
2 Outflow Region
3 Eye
4 Eye Wall
5 Spiral Bands

With reasonable caution, flight is often possible in the vicinity of an extratropical cyclone; however, flight is never advisable in the vicinity of a hurricane. Winds are strong, thunderstorms are common, and rainshowers are heavy. When hurricanes reach land along the East or Gulf coasts of the U.S., they frequently set off severe convective weather, including intense thunderstorms and tornadoes. Figure 8-23 lists surface wind observations at Charleston, South Carolina during the passage of Hurricane Hugo. Clearly, it is not an environment for commercial or recreational flying. More detailed information about the structure and flight hazards of thunderstorms and low-level turbulence caused by strong winds are presented in the next chapter and in Part III.

When a hurricane approaches an area, the National Weather Service issues one or more special bulletins. A hurricane watch is issued when hurricane conditions are expected in a particular area within a day or more. A hurricane warning is issued when the arrival of those conditions is expected within the next 24 hours.

SUMMARY

Extratropical cyclones are important large scale disturbances that move eastward in the middle latitudes. These extratropical cyclones develop in response to large horizontal temperature gradients. They form on the polar front and involve the movement of large airmasses and fronts near the earth's surface, as well as the development of troughs and jet streaks aloft. Unique wind, cloud, and precipitation patterns evolve during the life cycle of the extratropical cyclone; the main features are captured by the polar front model. In sharp contrast to extratropical cyclones, smaller, highly destructive hurricanes develop and move westward in low latitudes. They draw their energy from warm waters and die over cold waters or land.

KEY TERMS

Airmass
Airmass Source Region
Anticyclonic Wind Shear
Closed Low
Cold Airmass
Cold Front
Cold Front Occlusion
Comma Cloud
Cyclonic Wind Shear
Cyclogenesis
Deepening
Dissipating Stage
Extratropical Cyclone
Eye Wall
Filling
Front
Frontal Cloud Band
Frontal Cyclone
Frontal Low
Frontal Slope
Frontal Wave
Frontal Zone
Hurricane
Hurricane Eye
Hurricane Warning
Hurricane Watch
Incipient Stage
Jet Streak
Jet Stream Cirrus
Occluded Front
Occlusion Process
Occluded Stage
Overrunning
Polar Front Model
Rainbands
Short Wave Troughs
Stationary Front
Storm Surge
Tropical Cyclone
Tropical Depression
Tropical Disturbance
Tropical Storm
Warm Airmass
Warm Front
Warm Front Occlusion
Warm Sector
Wave Cyclone
Wave Cyclone Stage
Wind Shear

CHAPTER QUESTIONS

1. You are trying to land at an airport before the arrival of the weather associated with a cold front passage. The front is approaching the airport at a speed of 10 knots. You have been in the cold airmass for almost your entire flight and you manage to fly through the front into warm air at 3,000 feet AGL, about 5 n.m. from the airport. Your airspeed is 120 knots. Will you reach the airport before the front does? Discuss.

2. Obtain a U.S. surface analysis chart with a well-defined cold front to the east of the Rockies. Identify the frontal zone.

3. Why are extratropical cyclones farther south and stronger in the winter than in the summer?

4. A typical short-wave trough line is just crossing the west coast of the U.S. If it is moving directly eastward, when will it cross the Continental Divide?

5. Why does the front aloft precede the surface front by a large distance in a warm front occlusion and follow the surface front at a shorter distance in a cold front occlusion?

6. A sea level pressure of 888 mb was reported in Hurricane Gilbert in September, 1988. If your altimeter was set at 29.92 inches of mercury with that pressure, what would your altimeter read if you were at sea level?

7. Several years ago, when a typhoon was approaching an airfield on a Pacific island, it was determined that the eye of the storm would actually cross the base where several large aircraft were parked in the open. There was no hanger space and although it was too late to fly the aircraft to safety, minimum aircrews were placed in the aircraft. Why? Discuss.

8. Make a table that contrasts the following characteristics of extratropical cyclones and hurricanes.
    1. Geographical region of development
    2. Initial direction of movement
    3. Energy source
    4. Altitude of greatest intensity
    5. Stages of development
    6. Scale
    7. Temperature structure
    8. Weather

9. List the indicators of a cold front approach and passage.

CHAPTER 9 Thunderstorms

Introduction

In this chapter, we continue our scale approach to the understanding of atmospheric circulations. Moving to smaller scales, our consideration is now the mesoscale phenomenon known as a thunderstorm. The thunderstorm is one of the most spectacular atmospheric circulations, and one that you must respect as a pilot. It can be bright, loud, violent, and dangerous in many ways. As with our study of macroscale circulations, we will begin with an idealized model of the thunderstorm.

When you complete this chapter, you will understand thunderstorm structure and behavior as well as the wide variety of microscale phenomena that are frequently produced by a thunderstorm. You will also become familiar with larger mesoscale and macroscale circulations that provoke thunderstorms and organize them into lines and clusters.

A thunderstorm is always a threat to aircraft operations. A wise pilot will be sure he or she understands the what? why? and where? of thunderstorms for safety reasons.

Section A

THE ROOTS OF CUMULUS CLOUDS

Observing a growing cumulus cloud provides clear evidence that the air in the cloud is going upward, quickly! Our study of thunderstorms concentrates on this buoyant, saturated air. However, we must keep in mind that the air in cumulus clouds or cumulonimbus clouds originally comes from the boundary layer. So-called dry convection is a common process within a few thousand feet of the ground. Since thunderstorms are rooted in dry convection, a discussion of their development would be incomplete without an examination of the important properties of convection below the cloud base.

When the ground becomes much warmer than the air above it, the lapse rate in the lowest layer often becomes "superadiabatic" (LR>3C°/1,000 ft.). Under these absolutely unstable conditions, any air that receives the slightest vertical displacement, such as by a wind gust, will rapidly move in the direction of that displacement. When this process begins in the boundary layer, the air motion becomes organized into discrete "bubbles" of warm air rising from the ground. These bubbles are but another atmospheric circulation with a distinct range of scales and life cycles. They are more frequently called thermals, which are the cloudless roots of cumulus clouds.

Individual thermals have horizontal dimensions of a few hundred to a few thousand feet, lifetimes of five to ten minutes, and vertical speeds of a few hundred feet per minute to about 2,000 f.p.m. Thermals can develop day or night, as long as the ground is warmer than the overlying air. These conditions are often met in a cold airmass as it moves over warm land, such as following a cold frontal passage. Thermals are common in the daytime under clear skies and during the warmer months of the year.

The size and strength of thermals varies greatly, depending on how much warmer they are than their surroundings. This, in turn, depends on the surface heating. Thermals tend to be smaller and weaker in the morning than they are in the afternoon. Daytime sources of thermals are those surfaces that heat up more rapidly than surrounding areas. Favorable surfaces include dry fields, paved roads, parking lots, and runways. Elevated terrain is also a producer of thermals, because it is often warmer than the surrounding air at the same altitude. High, bare hills generate thermals earlier in the day and for a longer period than nearby valleys. Unfavorable thermal sources are forested areas, cool bodies of water, irrigated fields, and ground dampened by rain.

Close to the ground, thermals are elongated, plume-like structures of rising warm air, perhaps 100 feet across and a few hundred feet long. When winds are strong, thermals close to the ground become chaotic and difficult to identify. Otherwise, thermal plumes are often indicated by birds soaring over flat ground and looping smoke plumes.

Occasionally, the wind is partially blocked by an obstruction, such as a stand of trees or a small hill. Air sweeping around the sides of the obstruction causes eddies to form downwind. If a thermal happens to form in the same area as the eddies, it will rotate. As the thermal rises, it stretches vertically and shrinks horizontally. This causes a faster rotation. You see the same effect when a spinning skater pulls his arms in to his body. The result of the spin-up of a thermal is a vortex known as a whirlwind or dust devil, not to be confused with more violent tornadoes or waterspouts. Wind speeds up to 20 knots are not unusual within a dust devil; however, extremes of 50 knots have been reported. (Figure 9-1)

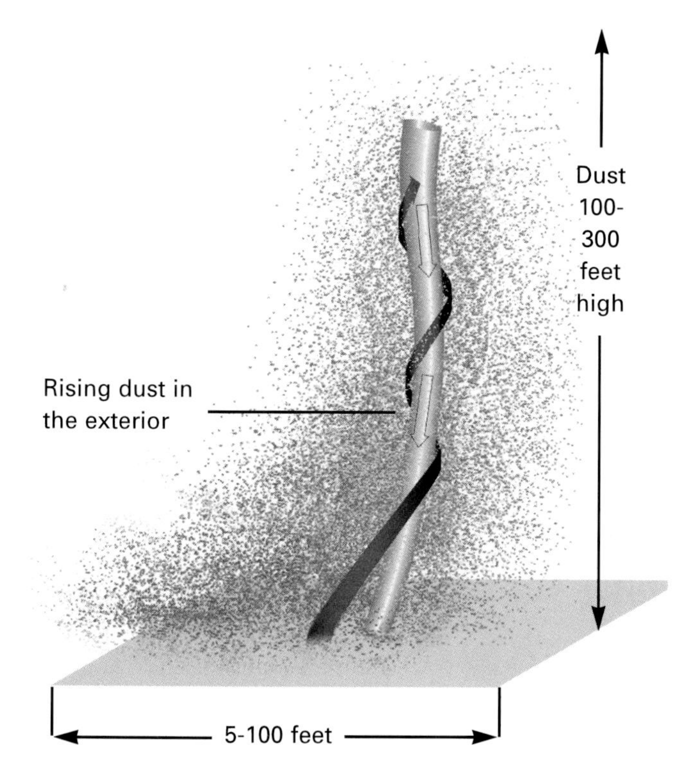

Figure 9-1. A dust devil is a product of extreme surface heating and light winds.

Dust devil formation is favored under light wind conditions over very hot surfaces, such as over barren desert areas in summer and in the early afternoon. Once formed, they have no preferred direction of rotation. They move with the speed and direction of the average wind in the layer that they occupy.

Dust devils are typically 5 to 100 feet in diameter and have lifetimes of 4 minutes or less, although extremes of hours have been reported. Dust lifted by the vortex often reaches altitudes of 100 to 300 feet AGL; although, the vortex itself may extend to a higher level. Over desert areas in the summer, dust devils occasionally reach altitudes of several thousand feet AGL. Remember, dust devils are a product of dry convection; they do not require the presence of CB or CU to form.

As ordinary thermals rise, they develop a distinct internal circulation. An idealized model of a thermal is shown in figure 9-2. Superimposed on the overall rising motion of the thermal is a microscale circulation cell that is best described as an elongated vortex ring. Extending upward from the ground, the vortex ring has a relatively narrow core of upward motions surrounded by a broad region of weaker sinking motions. The horizontal dimension of the thermal also grows with altitude as outside air is mixed into the circulation.

The shape of the thermal is idealized in figure 9-2. Although the updraft is the dominant feature, a real thermal twists and distorts as it rises due to internal temperature differences and external influences, such as wind shear. Furthermore, as thermals grow with altitude, they often merge.

Figure 9-2. A perspective diagram of an idealized thermal. The broad, vertical arrow indicates the warm air entering the thermal near the ground and the subsequent ascent of the thermal. The circular ribbons show the vortex ring that rises with the thermal. This circulation is similar to that of a smoke ring.

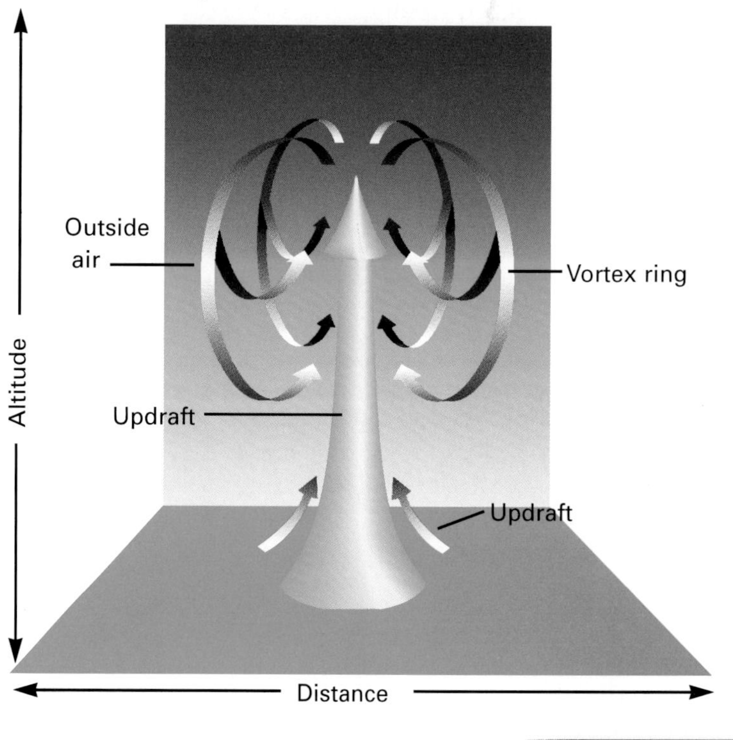

The vortex-ring circulation draws outside air into the thermal as it rises. This reduces the temperature of the thermal and, therefore, its buoyancy. The rise of the thermal is often halted when it reaches a stable layer where it become colder than its surroundings. This happens frequently in the vicinity of a large, high pressure region. In this case, boundary layer air becomes unstable because of daytime surface heating, but thermals may only rise to 2,000 feet or so because of the presence of an elevated stable layer. If clouds form in this process, they are of the fair weather cumulus variety, as shown in figure 9-3.

Figure 9-3. Fair weather cumulus clouds which have formed at the tops of rising thermals. Notice the flattened cloud tops, indicative of the presence of an elevated inversion. The clouds are often bigger than an individual thermal, because thermals tend to expand and merge as they rise.

Section B

CLOUDY CONVECTION

In its most common use, the term cloudy convection refers to saturated air that is rising because it is warmer than its surroundings. We also include in this definition saturated air that descends because it is colder than its surroundings. Therefore, cloudy convection includes all of the various forms of cumulus clouds and their updrafts and downdrafts. In order to better describe and explain thunderstorms, we begin by briefly examining the general process of convective cloud growth.

CLOUD GROWTH

The distinct appearance of cumuliform clouds reflects not only the saturation, but also the instability of the convective updraft. The characteristic flat bases of the clouds occur at the altitude where the rising, unstable air first reaches saturation and is called the convective condensation level. Above the convective condensation level, upward moving air cools at the saturated adiabatic lapse rate. You should recall that this rate is less than the dry adiabatic lapse rate because of the continued release of latent heat as the cloud forms. For this reason, an unstable updraft often becomes more unstable when the cloud begins to appear.

As shown in figure 9-4, above the convective condensation level, the cloudy updraft continues to accelerate upward. Clear air sinks around the cloud to compensate for the upward motion. The sinking is usually much weaker than the updrafts, typically taking place over a much broader area than that of the cloud. The updraft is similar to a jet of fluid pointed vertically. The strong shears between the updraft and the weak downdrafts outside the cloud produce turbulent eddies marked by cauliflower-like protuberances on the edges of the cumulus cloud.

The cloudy updraft continues its upward acceleration until it reaches the equilibrium level; that is, the altitude where the updraft temperature is equal to the temperature of its surroundings. Above that level, the air continues to rise, but decelerates because it is cooler than its surroundings. The top of the cloud occurs at the level where the updraft speed decreases to zero. Because the cloud top is colder than its surroundings, it finally collapses and spreads out around the equilibrium level.

The upper limit of convective clouds depends strongly on the presence or absence of stable layers above the convective condensation level. A strong tropospheric inversion may stop the vertical growth of cumulus clouds at any stage. So-called fair weather cumuli have limited vertical growth and last only a few minutes because they encounter a strong inversion aloft. Therefore, the bases of the clouds and the equilibrium level are close together.

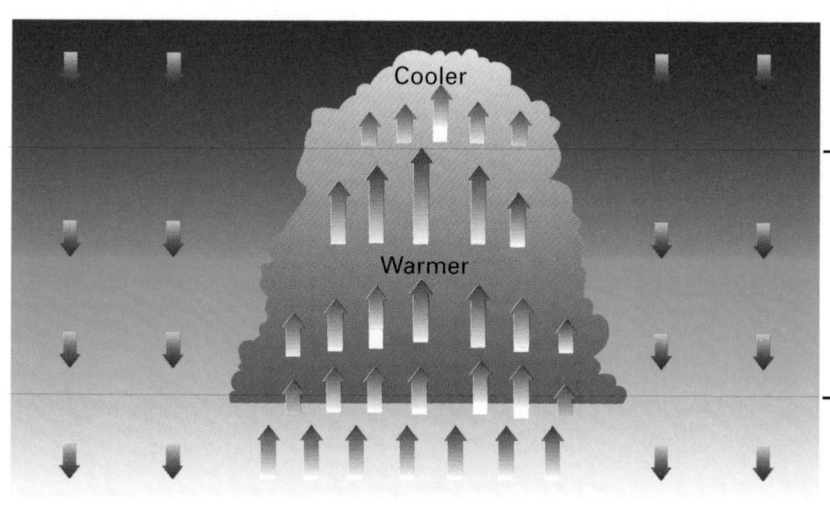

Figure 9-4. Schematic cross section of a growing cumulus cloud. Lengths of arrows are proportional to speeds of updrafts and downdrafts.

Another example of limited cloud development is often found in a cold airmass behind a cold front. Although the convection may be strong in these circumstances, the vertical development of the clouds is frequently limited by the stable layer near the top of the shallow, cold airmass. (Figure 9-5)

DOWNDRAFTS

When the altitude of the cumulus cloud exceeds the freezing level, there is a rapid growth of cloud particles by the ice crystal process. At some point in this process, the updraft is no longer strong enough to support the weight of the large particles. They begin to fall, dragging air downward. This is the beginning of the precipitation-induced downdraft. These internal downdrafts are much stronger than the sinking motions outside the cloud.

The downward vertical motions are strengthened where unsaturated air outside the cloud is mixed across the boundaries of the cloud. Evaporation further cools the downdraft, increasing the negative buoyancy. As snow turns to rain at lower levels, melting also contributes to the cooling of the air and the intensity of the downdraft. As we will see in the next section, the precipitation-induced downdraft is a major component of the life cycle of the thunderstorm.

Strong upward currents in clouds enhance the growth rate of precipitation.

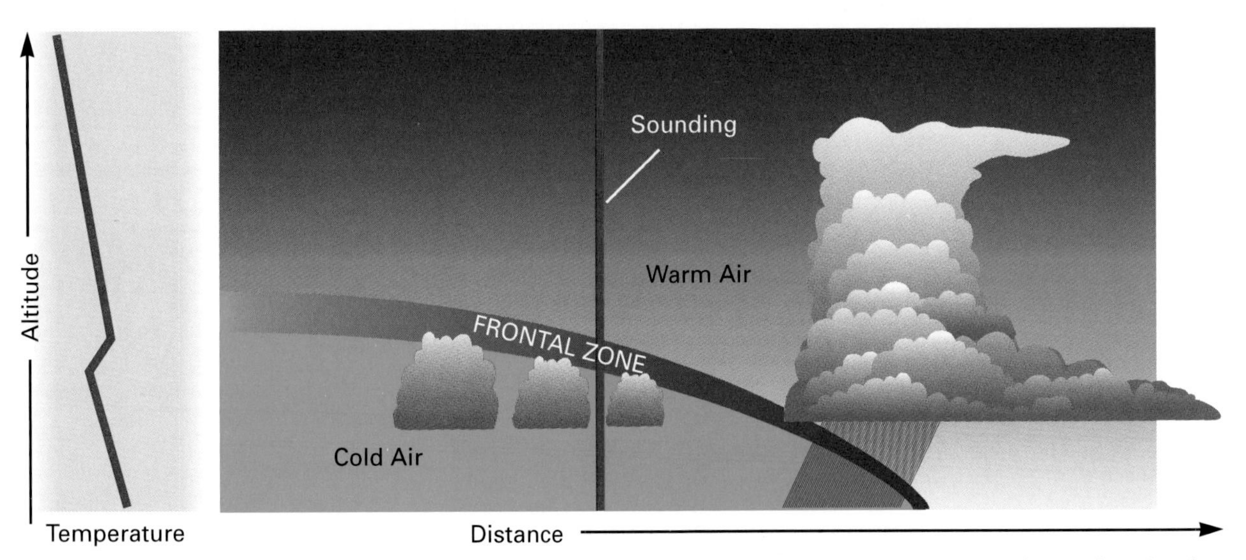

Figure 9-5. In contrast to cumulonimbus cloud formation along a cold front, the growth of cumulus clouds beneath the front in the cold airmass is often limited by the stable layer that corresponds with the frontal zone at the top of the cold air. The sounding on the left indicates the temperature along the heavy vertical line through the front.

Section C

WEATHER RADAR

Before describing the life cycle of a thunderstorm, it is helpful to briefly examine a useful thunderstorm detection tool: weather radar. In general, radar, for Radio Detection and Ranging, is an instrument that uses electromagnetic radiation to detect objects and determine their distance and direction from the radar site. The objects, or radar targets, must be composed of matter that scatters or reflects electromagnetic energy in the frequencies of radio waves.

Radar consists of a transmitter and receiver, usually in the same antenna. Electromagnetic energy from the transmitter is focused and emitted in a narrow beam by the antenna. Targets such as airplanes, buildings, and atmospheric particles reflect, scatter, and absorb the energy. A small fraction of the radiation reflects back toward the radar antenna/receiver where it is intercepted. The received signal constitutes a radar echo. By monitoring the time it takes for the transmitted energy to travel from antenna to target and back, the distance or slant range from the antenna to the target can be determined. The direction of the target is simply determined by noting the direction the antenna is pointing when the echo was received. (Figure 9-6)

Doppler radar has the capability to determine velocity of a target toward or away from the radar by measuring the frequency difference between the transmitted and received radiation. The returning frequency is lower than the transmitted frequency if the target is moving away from the radar, and higher if the target is moving toward the radar. You hear a similar doppler effect when the pitch of the whistle of a train decreases as the train passes your location.

Weather radar operates at specific frequencies, or wavelengths, of electromagnetic radiation that are sensitive to scattering by particles. Most

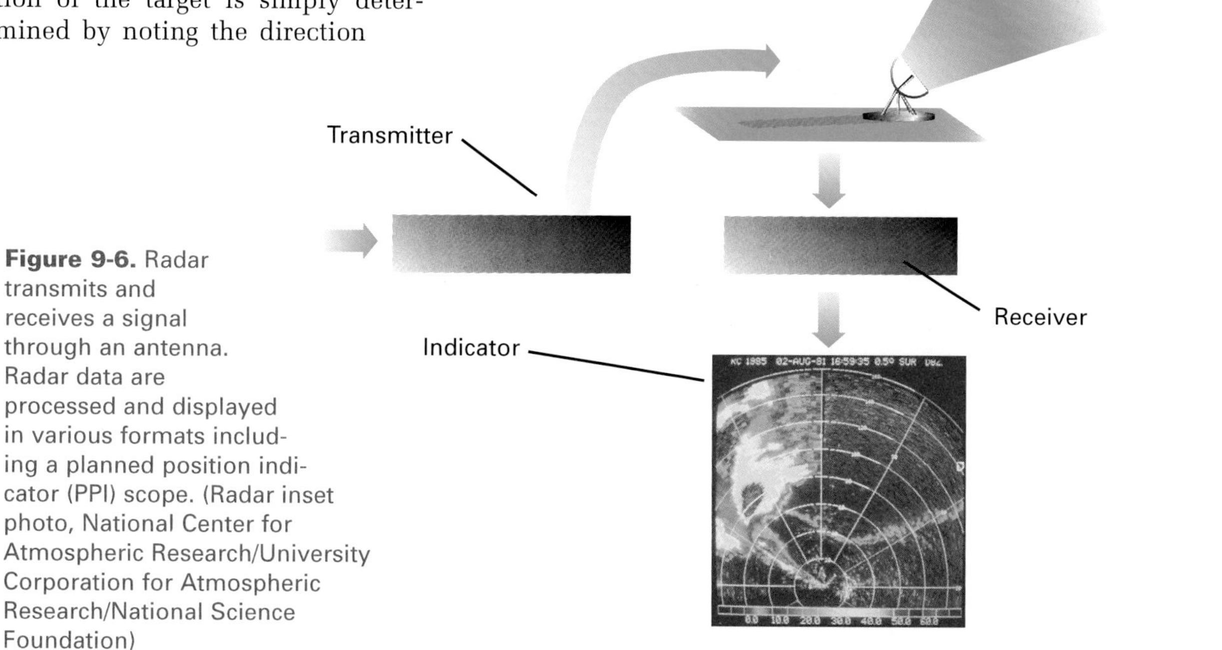

Figure 9-6. Radar transmits and receives a signal through an antenna. Radar data are processed and displayed in various formats including a planned position indicator (PPI) scope. (Radar inset photo, National Center for Atmospheric Research/University Corporation for Atmospheric Research/National Science Foundation)

weather radars detect relatively large particles, such as rain and snow, better than cloud particles. Some precipitation types provide more substantial targets than others. For example, large water droplets and wet hail give better echoes than snow or drizzle.

Currently, the United States weather radar network is being upgraded with the next generation weather radar system (NEXRAD). This is a more powerful doppler radar system. The sensitivity and narrow beam of the new radar systems actually allow the monitoring of some clear air echoes caused by targets as small as dust or insects. The NEXRAD system can determine winds out to a distance of about 60 n.m., and detailed severe weather features out to distances of 130 nautical miles.

Airborne weather radar and many conventional, nondoppler, ground-based radars do not have the power to detect extremely small targets. Cloud particles are generally too small to cause significant scattering in these cases. This means that the cloud boundaries are larger and cloud tops are generally higher than those shown by the radar echoes. Precipitating cloud bases are difficult to detect by radar because the echo often extends to the ground with the precipitation.

As a radar signal travels away from its source, it undergoes a process known as attenuation. This is a weakening of the signal that occurs when the signal is absorbed, scattered, or reflected along its path. Precipitation is an efficient attenuator of weather radar signals. For example, in heavy rain, a radar signal may be partially or totally absorbed by the target (the rain) in the foreground, so that targets in the background cannot be seen very clearly, if at all. This is a critical feature for airborne radar. Your decision to fly across a line of echoes is usually determined by your ability to "see" across to the other side. This may not be possible with a very strong echo in the foreground.

Weather is observed continuously by the previously mentioned network of ground-based weather radars maintained by the National Weather Service (NWS), the Department of Defense (DOD), and the Federal Aviation Administration (FAA). In Chapter 6, you were introduced to a radar intensity or VIP scale that gave an indication of precipitation rate. The same scale is used to rate the intensity of radar echoes. In this broader application, the VIP level is a measure of the overall strength of the convective circulation cell associated with the precipitation. The highest VIP levels occur with the most intense thunderstorms. In Part III, the relationship of VIP level to a specific turbulence intensity is discussed. An example of the display of VIP information available every hour from the national radar network is shown on a weather radar summary chart in figure 9-7.

The contoured VIP levels on radar summary charts must be used carefully. The contours only tell you, in broad terms, the intensity of the echo in a particular area. As you will see in the next section, thunderstorms have short lifetimes, often less than the time between radar summary charts. For that reason, a cautious interpretation is recommended by the NWS and FAA. The maximum reported VIP level should always be used to estimate the severity of the storm.

You should also be aware that surface weather radar signals can often be blocked by mountainous terrain, or a radar station may simply be inoperative at a particular time (see legend in figure 9-7). The result is that some thunderstorm cells may be missed. Other modes of thunderstorm detection such as surface observations, satellite observations, and the monitoring of lightning discharges should also be used in these cases.

The radarscope provides no assurance of avoiding instrument weather conditions.

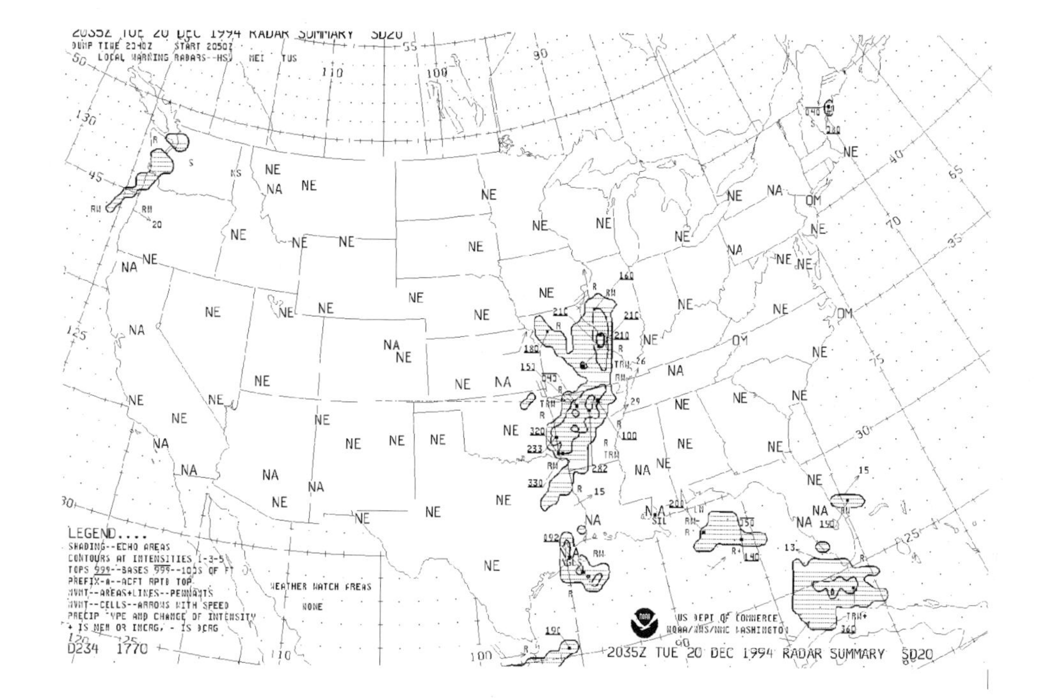

Figure 9-7. Weather Radar Summary Chart. Contours represent VIP levels 1, 3, and 5. The Legend for weather radar summary charts is included in the Appendix.

Information that is provided on the radar summary chart and not shown on other weather charts includes lines and cells of hazardous thunderstorms.

Section D

THUNDERSTORM STRUCTURES

Based on surface observations, a thunderstorm is defined as a local storm produced by a cumulonimbus cloud, and always accompanied by lightning and thunder. It typically produces strong wind gusts, heavy rain, sometimes hail, and occasionally tornadoes. It is usually of short duration, rarely over two hours for a single storm. On the basis of flight experience, thunderstorms are also characterized by significant turbulence, icing, and wind shear.

Special weather observations are taken to mark the beginning and end of a thunderstorm, and to report significant changes in its intensity. Besides the standard coded information about sky condition, weather, visibility, pressure, temperature, and wind, evidence of the presence of thunderstorms is also found in the remarks section of surface weather reports. An example follows.

> 1655 RS E30 BKN 160 BKN 300 OVC 1/4TRW+A 82/68/3020G32/982/T B35 OHD MOVG E OCNL LTGCCCG RB43 AB50
>
> Decoded Remarks: Thunderstorm began at 35 minutes past the hour. Thunderstorm is overhead moving east with occasional lightning, cloud-to-cloud and cloud-to-ground. Rain began at 43 minutes past the hour and hail began at 50 minutes past the hour.

As we examine the growth and structure of thunderstorms in the next sections, we will stress the relationship between the visible characteristics of thunderstorms and their internal structure. In this regard, it is not too early to be reminded of a practical and very important rule of thumb related to thunderstorms:

> If a convective cloud reaches the cumulonimbus stage, it should be considered a thunderstorm, whether or not any other evidence of thunderstorm activity is present.

THUNDERSTORM TYPES

There are two basic thunderstorm types: an ordinary thunderstorm, frequently described as an airmass thunderstorm, and a severe thunderstorm. The severe thunderstorm may also be called a squall line thunderstorm. A severe thunderstorm has a greater intensity than an airmass thunderstorm, as defined by the severity of the weather it produces: wind gusts of 50 knots or more, hail three-quarters of an inch or more in diameter, and/or strong tornadoes.

The basic component of any thunderstorm is the cell. In the initial stages, this is the updraft region of the growing thunderstorm. Later in the thunderstorm development, it includes the precipitation-induced downdraft. A thunderstorm may exist as a single cell, multicell, or supercell storm. A single-cell airmass thunderstorm lasts less than one hour. In contrast, a supercell severe thunderstorm may last two hours.

A multicell storm is a compact cluster of thunderstorms. It is usually composed of airmass thunderstorm cells in different stages of development. These cells interact to cause the duration of the cluster to be much longer than any individual cell.

AIRMASS THUNDERSTORM

The life cycle of a single-cell airmass thunderstorm is illustrated in figure 9-8. The cycle is divided into three stages: cumulus, mature, and dissipating.

CUMULUS STAGE

When atmospheric moisture and instability are sufficient, the evolution of the airmass thunderstorm begins. In the cumulus stage, an important change occurs in the nature of convection. There is a marked increase in the scale of the circulation. The size of the updraft region becomes larger than the size of any of the individual

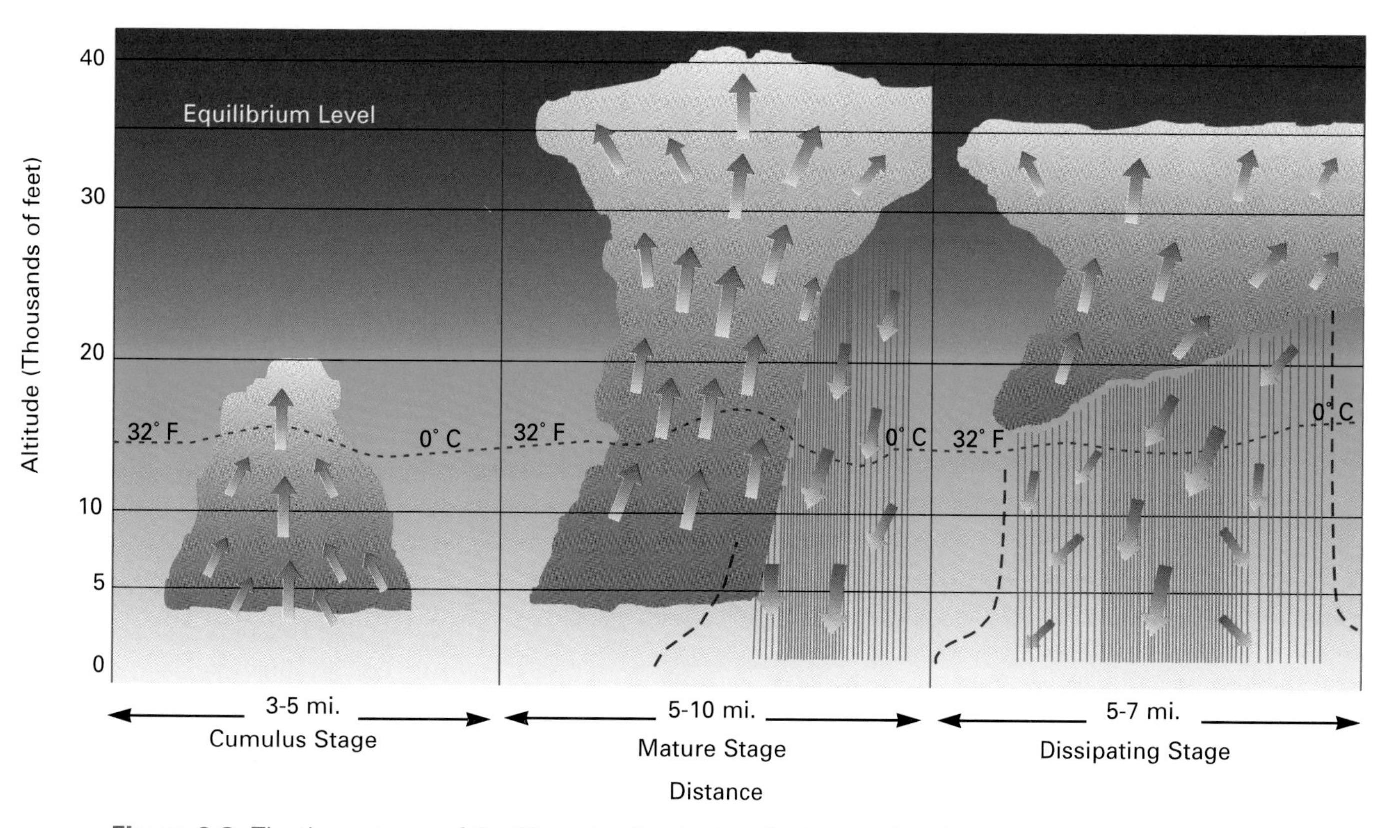

Figure 9-8. The three stages of the life cycle of a single-cell, airmass thunderstorm are the cumulus, mature, and dissipating stages.

thermals that are feeding the region. This can often be seen in a field of cumulus clouds in which one particular cloud begins to grow more rapidly than the others. Around the developing cloud, the smaller clouds will clear out as the air starts sinking in response to the larger scale updraft.

In this first stage of development of the thunderstorm, air initially rises throughout the cloud. The upward growth is much greater in some portions of the cloud than in others, and the cloud grows in an unsteady succession of upward bulges as thermals arrive at the top. These can be seen as turrets on the top of the cloud in figure 9-9.

During the cumulus stage, the convective circulation grows rapidly into a towering cumulus (TCU) cloud which typically grows to 20,000 feet in height and three to five miles in diameter. The cloud reaches the next stage of development in about 15 minutes.

Figure 9-9. Towering cumulus clouds characteristic of the cumulus stage of an airmass thunderstorm.

As the cloud continues to grow, precipitation begins to develop, initiating a downdraft within the cloud late in the cumulus stage.

A continuous updraft is normally associated with the cumulus stage of a thunderstorm.

MATURE STAGE

As shown in figure 9-8, the mature stage begins when the precipitation-induced downdraft reaches the ground. Lightning and thunder begin as the thunderstorm cell grows to about 5-10 miles in diameter.

The circulation of the thunderstorm cell is well organized in this stage. The relatively warm updraft and the cool, precipitation-induced downdraft exist side by side. The downdraft reaches its greatest velocity near the base of the cloud, while the updraft reaches its maximum speed near the equilibrium level in the upper part of the cumulonimbus cloud.

The top of the mature cell often reaches into the stratosphere. The cumulonimbus cloud which is characteristic of this stage is easily identified by the appearance of its top. The highest portion of the cloud develops a cirriform appearance because of the very cold temperatures and the strong stability of the stratosphere. Vertical motions are damped and the cloud spreads out horizontally, finally forming the well-known anvil shape. (Figure 9-10) When the anvil top forms, it points in the direction of the winds at the top of the thunderstorm. This is approximately the direction that the storm is moving.

Although the beginning of the mature stage of the airmass thunderstorm cell is usually indicated by the arrival of precipitation and wind gusts at the ground, there are exceptions. For example, a developing cumulus cloud may produce a shower and the associated downdrafts, but not reach the mature stage. In this case, lightning and thunder do not occur.

An indication that downdrafts have developed and that the thunderstorm cell has entered the mature stage is when precipitation begins to fall from the cloud base.

Another exception to the model of the mature stage is found in the arid regions of the western U.S., especially during summer. Because of the low humidity at the surface, air must rise a great distance to reach condensation. Thunderstorm bases can be as high as 10,000 feet AGL or more.

Thunderstorms reach their greatest intensity during the mature stage.

Figure 9-10. A cumulonimbus cloud indicates that the thunderstorm has reached at least the mature stage.

In these high-based storms, lightning and thunder occur, but the precipitation often evaporates before reaching the ground. In this case only a veil of precipitation known as virga is observed immediately below the cloud base. (Figure 9-11) This combination of lightning and gusty winds in the absence of precipitation is often a cause of forest fires. Also, despite the lack of rain, the associated downdraft and gusty winds can still produce flight hazards. These will be considered in Part III.

If the precipitation-induced downdraft is exceptionally strong and small, it may be classified as a microburst, producing dangerous wind shear conditions on landing and takeoff. Furthermore, as the rain reaches the ground below the thunderstorm, the cool downdraft spreads out, causing horizontal gusts which often extend beyond the edges of the thunderstorm cell. These topics are also discussed in Part III with other aviation weather hazards.

DISSIPATING STAGE

Thirty minutes or so after it begins, the single-cell airmass thunderstorm reaches the dissipating stage. As shown in the right hand panel of figure 9-8, precipitation and downdrafts spread throughout the lower levels of the thunderstorm cell, cutting off the updraft. Since the source of energy for thunderstorm growth is the supply of heat and moisture from the surface layer, the cut-off of the updraft spells the end of the storm. With no source of moisture, the precipitation decreases and the entire thunderstorm cloud takes on a stratiform appearance, gradually dissipating. Because the anvil top is an ice cloud, it often lasts longer than the rest of the cell. It is helpful to remember that an observation of an anvil will not be conclusive as to the stage or even the presence of a thunderstorm.

Although the lifetime of a single-cell airmass thunderstorm is less than an hour, odds are that you have encountered thunderstorms that have lasted much longer. How can this be? In such cases, the explanation is that you were actually observing a multicell thunderstorm or a supercell thunderstorm, both of which last longer and affect larger areas than an airmass thunderstorm.

Figure 9-11. Virga indicates precipitation falling from the base of a cumulonimbus cloud, but not reaching the ground. An invisible downdraft will often continue to the ground below the virga.

In the life cycle of a thunderstorm, the dissipating stage is dominated by downdrafts.

MULTICELL THUNDERSTORM

The life cycle of any one of the cells of a multicell thunderstorm is much like any airmass thunderstorm. However, the life cycle of the multicell cluster is much different. Cell interaction produces more cells, thus sustaining the life of the cluster. (Figure 9-12)

The key to the long life of the multicell is the development of a thunderstorm gust front. It is the sharp boundary found on the edge of the pool of cold air that is fed by the downdrafts and spreads out below the thunderstorm. The main updraft for the multicell is located just above the gust front at low levels, slanting upward toward the back of the storm. A shelf cloud often indicates the rising air over the gust front.

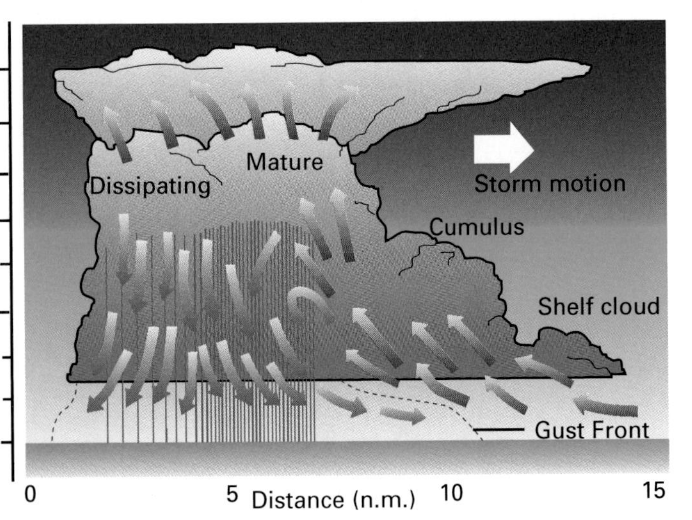

Figure 9-12. Multicell thunderstorm with cells in various stages of development. Small arrows indicate airflow. The vertical lines on the left indicate precipitation. The boundaries of the cool, downdraft air are shown by dotted lines below the thunderstorm base. Note, the right hand boundary is the gust front.

The updraft enters the storm from the direction in which the storm is moving. New cells develop toward the front of the thunderstorm as the gust front lifts unstable air. At the same time, the gust front is maintained by a supply of cool air from the precipitation-induced downdrafts of several mature and dissipating cells at the rear of the storm. This process of cell regeneration explains why the multicell thunderstorm influences a larger area and lasts longer than a single-cell airmass thunderstorm.

The strongest part of the gust front usually moves several miles ahead of the thunderstorm; that is, in the direction the cells are moving. When a gust front passes it is much like a mesoscale cold front. There is a wind shift, the winds are strong and gusty, the pressure increases, and the temperature decreases.

A rough estimate of the movement of a thunderstorm cell is given by the speed and direction of the 10,000-foot (700 mb) wind.

An outflow boundary is the remnant of a gust front that continues to exist long after the thunderstorms that created it have dissipated. On some occasions, outflow boundaries generated by thunderstorms late in the day have been observed to continue moving throughout the night, often covering well over one hundred miles. New convection may develop along an outflow boundary as it moves into unstable areas or intersects fronts or other outflow boundaries.

Multicell storms vary widely in intensity. They can produce severe convective weather when they are organized into mesoscale convective systems. These structures will be discussed in the final section of this chapter.

SUPERCELL THUNDERSTORM

While the multicell thunderstorm occasionally produces severe weather, the supercell thunderstorm almost always produces one or more of the extremes of convective weather: very strong horizontal wind gusts, large hail, and/or tornadoes. This difference in severity is due primarily to differences in thunderstorm structure.

The supercell storm can occur almost anywhere in middle latitudes, but the favored area is in the southern Great Plains of the U.S. in spring. This is because the supercell requires extreme instability and a special combination of boundary layer and high level wind conditions that are most frequently found over Texas, Oklahoma, and Kansas at that time of year.

The internal structure of the supercell is more complicated than either the single-cell or multicell thunderstorms. This complication is, in fact, what makes the supercell large, intense, and long-lasting. Specifically, the supercell forms in an environment that tilts and twists the thunderstorm updraft. In order to illustrate the details of this airflow, it is necessary to use a three-dimensional model of the storm. (Figure 9-13)

The updraft enters the storm with low-level flow from the southeast. The air rises in a strong, steady updraft which slants upward toward the back of the thunderstorm, in this case, toward the northwest. Overshooting tops indicated by bulges on the top of the anvil show the location of the updraft. Under the influence of the strong westerly winds at upper levels, the former updraft, now mainly horizontal, twists toward the east where it exits the thunderstorm through the anvil.

A major precipitation-induced downdraft occurs north of the updraft where the rainshaft can be seen in figure 9-13. Another downdraft (not visible) also spreads around to the west of the updraft as the supercell develops.

The gust fronts caused by these two downdrafts are indicated by a dashed line at the surface. The "flanking line," which parallels the gust front to the southwest of the updraft, is composed of growing cumulus towers.

There are two important differences between the structures of the single cell and the supercell. First, the supercell is much larger and, second, its updrafts and precipitation downdrafts remain separated. The change in the wind direction from surface southerlies to upper westerlies is the key to this structure. Because of this vertical shear, the precipitation-induced downdraft occurs where it cannot interfere with the updraft as happens in the airmass cell. This structure allows the supercell to develop a strong, steady updraft. This is the reason why a supercell lasts longer and reaches a much greater size and intensity than single-cell airmass thunderstorms. Also, this explains why some aviation weather manuals refer to a severe thunderstorm as a "steady-state" thunderstorm.

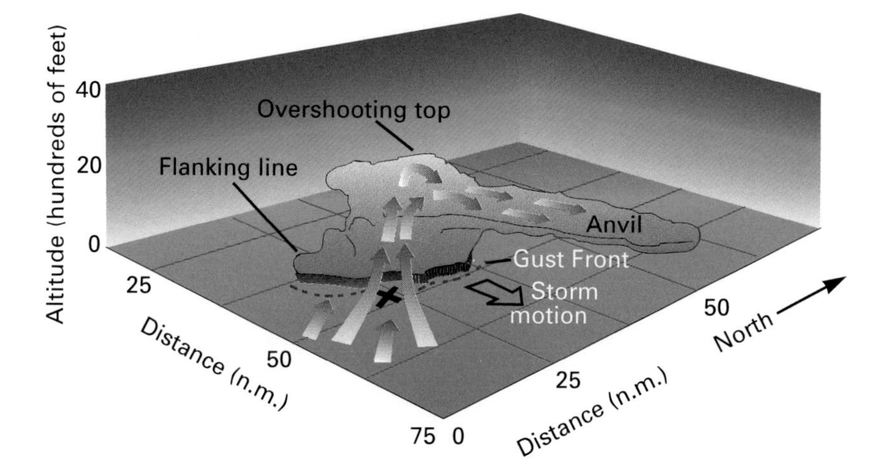

Figure 9-13. Perspective diagram of a supercell thunderstorm. For clarity, only the updraft is shown. The direction of movement of the supercell is toward the east in this example.

The horizontal separation of vertical drafts in a supercell can be better appreciated when observing the storm from the ground. Consider an observer looking at a supercell from the southeast (point X in figure 9-13). The view reveals the cloud features shown in figure 9-14. In addition to the anvil, the main storm tower, and the flanking line are also seen.

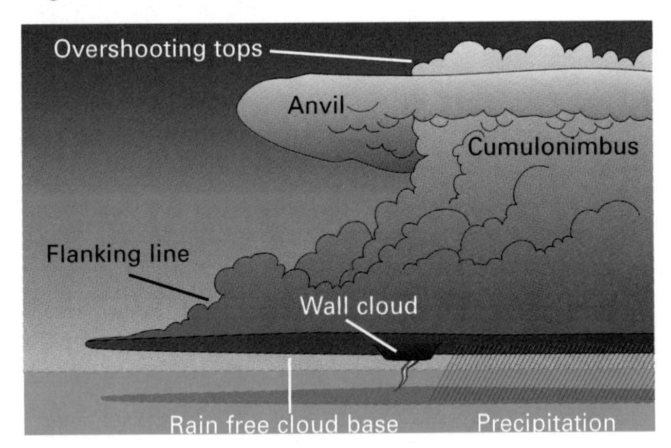

Figure 9-14. This is the ground view of the supercell thunderstorm at point "x" in figure 9-13.

In figure 9-14 below the cloud to the northwest (right side), the downdraft area is indicated by precipitation. To the southwest (left side) is the rain-free cloud base of the flanking line. The area of the storm immediately to the west (middle) is very close to the main updraft. As illustrated in the diagram in that area, a portion of the rain-free base of the clouds may appear lower in what is called a wall cloud. Significant rotation of the cloud is often seen. In fact, it is the location in the severe supercell where the strongest tornadoes occur. The bulges that appear under the anvil are known as mammatus.

Supercells may occur in isolation or in mesoscale convective systems with multicells and other supercells. These large lines and clusters of thunderstorms are discussed in a later section.

TORNADOES

A tornado is a violently rotating column of air which is found below cumulonimbus clouds. A tornado that does not reach the surface is called a funnel cloud. Tornadoes vary widely in intensity with the strongest, most damaging tornadoes (wind speeds exceeding 180 knots) usually associated with severe thunderstorms. Most tornado diameters range from 300 feet to 2,000 feet, although extremes of one mile have been reported. Because of the large scale weather conditions under which tornadoes are created in the U.S., they commonly move from southwest to northeast at a typical speed of 30 knots. Tornado lifetimes average only a few minutes, but unusual cases of over three hours have been documented. (Figure 9-15)

The strongest tornadoes are most often associated with severe thunderstorms. As described in the previous section, the supercell has an internal circulation that promotes the rotation favorable for tornado formation in the vicinity of the wall cloud.

Much of the damage associated with a tornado is caused by the presence of one or more suction vortices. These are small (about 30 feet in diameter), very intense funnels that rotate within the larger funnel of the tornado. These phenomena have been documented photographically and from damage surveys. Professor T. Fujita of the University of Chicago has developed a tornado intensity scale based on damage and wind speed. (Figure 9-16)

Figure 9-15. Tornado below the base of a cumulonimbus cloud. (NOAA)

Figure 9-16. The Fujita scale of tornado intensity.

Scale	Category	MPH	Knots	Expected Damage
F0	Weak	40-72	35-62	Light: tree branches broken, sign boards damaged
F1		73-112	63-97	Moderate: trees snapped, windows broken
F2	Strong	113-157	98-136	Considerable: large trees uprooted, weak structures destroyed
F3		158-206	137-179	Severe: trees leveled, cars overturned, walls removed from buildings
F4	Violent	207-260	180-226	Devastating: frame houses destroyed
F5		261-318	227-276	Incredible: structures the size of autos moved over 300 feet, steel-reinforced structures highly damaged

Tornadoes also occur with nonsevere thunderstorms, although they are usually weaker. A tornado that occurs over water is called a waterspout. It is weaker than most tornadoes over land. Fair weather waterspouts are very common near the Florida Keys between March and October. These vortices form over warm water near developing cumulus clouds. In general, fair weather waterspouts are very weak, short-lived, and slow-moving.

Near gust fronts and the edges of downbursts, tornado-like vortices known as gustnadoes sometimes occur. These phenomena are more like intense dust devils caused by the strong horizontal wind shear and strong updrafts.

A cold air funnel is a weak vortex that occasionally develops with rain showers and nonsevere thunderstorms. These vortices rarely reach the ground. In fact, cold air funnels do not appear under conditions usually associated with severe weather. For example, outbreaks commonly occur in a cold airmass, a half day or so after the passage of a cold front. Cold air funnels are more frequent in the spring and fall.

HAIL

The water droplets observed in rainshowers from convective clouds are notable by their size; that is, they are much bigger than the droplets that fall as rain from nimbostratus clouds. One of the main reasons for the large droplet size is that the upward motions in cumulus and cumulonimbus clouds are much stronger than in nimbostratus clouds. Another product of strong upward motions is hail, which was first described as a precipitation form in Chapter 6.

Large hail with diameters greater than three-quarters of an inch creates a danger to life and property on the ground as well as in the air. An understanding of its formation is useful. When ice particles grow to precipitation sizes through the ice crystal process, we expect to see snow, at least in the subfreezing environment in which it forms. However, this is not always the case in a thunderstorm where there are often very large numbers of supercooled water droplets. As snow collides with water droplets, the droplets freeze. This process, known as accretion, produces a larger particle which becomes the nucleus of a hailstone. If the updrafts in the thunderstorm are strong enough to keep the particle suspended in the cloud, it may grow to significant sizes. In a thunderstorm with a tilted updraft, hail may be thrown out of the storm near the top, only to fall back into the updraft. If the updraft is strong

Hail is most likely to be associated with cumulonimbus clouds.

enough, the hailstones are again carried up through the thunderstorm, growing still larger until they are so large that they finally fall to the ground. (Figure 9-17)

Figure 9-17 illustrates why hail has been observed at all levels throughout thunderstorms (up to about 45,000 feet AGL) as well as in clear air outside the clouds. These trajectories also explain why hailstones often have layers of alternating clear and opaque ice. The layered structure reflects varying temperatures and concentrations of supercooled water droplets encountered by the hailstone along its path through the thunderstorm.

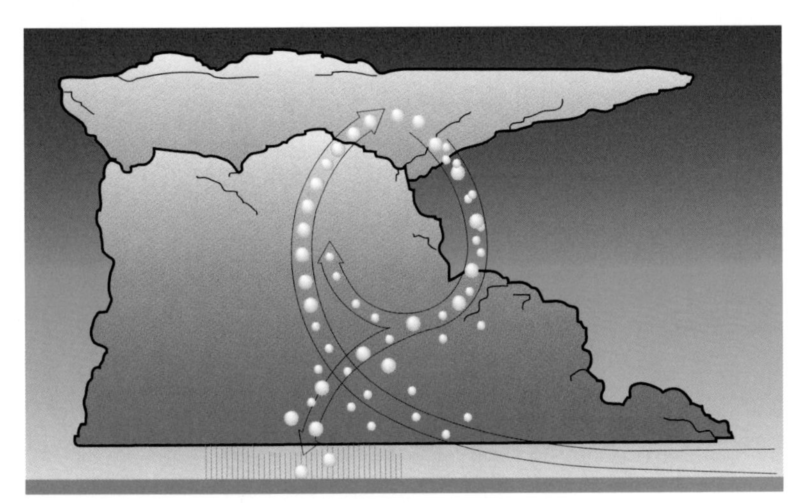

Figure 9-17. One way hail can grow to large sizes is by being recirculated through the storm.

LIGHTNING

Lightning is the visible electric discharge produced by a thunderstorm. It occurs in several forms, including in-cloud, cloud-to-cloud, cloud-to-ground lightning, and occasionally, between the cloud and clear air. A branched lightning stroke is called forked lightning. Lightning that occurs within a cloud, illuminating it diffusely, is often described as sheet lightning. Lightning that occurs far in the distance, when no thunder is heard, is called heat lightning.

Whatever the form of lightning, a lightning discharge involves voltage differences of about 300,000 volts per foot. Air along the discharge channel is heated to more than 50,000°F causing the rapid expansion of air and the production of a shock wave that moves away from its source, finally reaching your ear as thunder. Since the flash of lightning travels at the speed of light (186,000 miles per second) and.the shock wave travels with the speed of sound (about 1,000 feet per second), the difference of the arrival times of these two thunderstorm indicators becomes greater the farther you are from the thunderstorm.

As with any electrical discharge, lightning requires a charge separation. One way this can come about is for large and small particles (water droplets, ice crystals, hail) to develop opposite charges and then become separated by gravity or by convection. There are a number of ways by which charging can be accomplished before the charged particles are separated, although it is not clear which are the most important processes. For example, particles can become charged through collision, through a transfer of ions when a warm hailstone comes in contact with cold ice crystals, and a variety of other ways including freezing and splintering. In any event, the heavier, negatively charged particles end up in the lower part of the cloud with lighter, positively charged particles at the top. This distribution induces a positive charge on the ground because opposite charges attract one another. (Figure 9-18)

Lightning is always present in (and near) a thunderstorm.

When the charge difference between particles becomes large enough, (300,000 volts per foot or more) lightning occurs. The lightning stroke is actually a series of events which begins with a nearly invisible stepped leader that carries electrons from the base of the cloud to the ground,

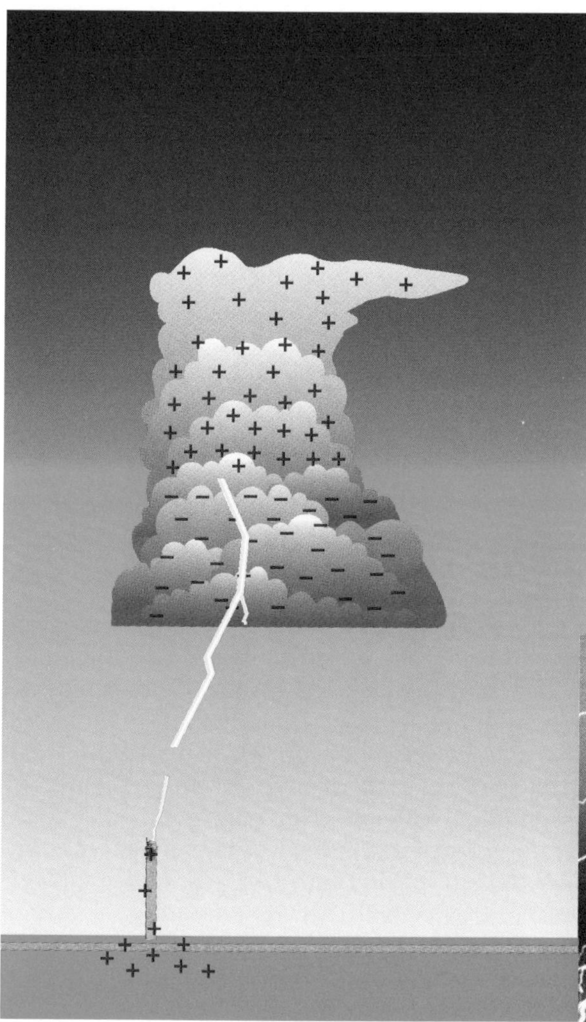

Figure 9-18. Charge separation in a convective cloud causes a region of positive charge at the top of the cloud and negative charge in the lower half. The photo inset shows the visible lightning discharge or return stroke. (National Center for Atmospheric Research/University Corporation for Atmospheric Research/National Science Foundation)

creating an ionized channel for the subsequent discharge. Close to the earth, the stepped leader is met by an upward-moving positive charge. A bright return stroke occurs, marking the route of the positive charge along the original path of the stepped leader, back up into the cloud. The initial discharge is often followed by several so-called dart leaders and further return strokes. These individual events are so fast that the eye cannot resolve them.

High points on the earth's surface, such as mountain tops, tall buildings, treetops, antennas, and steeples are particularly vulnerable to lightning strikes because the lightning discharge favors the path of least resistance to the earth's surface. The closer the positively charged object is to the cloud, the more vulnerable it is to lightning strikes. In areas where there is a high frequency of thunderstorms, many buildings have lightning rods to carry the discharge harmlessly into the ground.

A typical thunderstorm produces three or four lightning flashes per minute. In general, severe thunderstorms will produce more. Although there is great variability in the number of cloud-to-ground lightning strokes from storm to storm, it should be kept in mind that, only 10 to 25 percent of all lightning strokes are cloud-to-ground. That is, most discharges from a thunderstorm are within or between clouds. In-flight lightning strikes are a definite hazard that will be discussed in Part III.

Section E

THUNDERSTORM ENVIRONMENT

Thunderstorms and the phenomena that develop in their presence – downbursts, gust fronts, and tornadoes – are themselves embedded in larger scale circulations that produce conditions favorable for thunderstorm development. In this section, we consider that environment and attempt to answer the questions, where, when, and why are thunderstorms likely to occur?

REQUIREMENTS FOR DEVELOPMENT

Two basic requirements must be met for the formation of thunderstorms: the air must have large potential instability, and there must be a source of initial lift. A layer of air that is potentially unstable is not only conditionally unstable, but it has a high moisture content. In other words, if it is lifted a sufficient distance, significant convection and thunderstorms will occur. If the air is too stable, no amount of moisture or lifting can cause thunderstorms to develop. Initial lift is the minimum amount of vertical displacement necessary to release the potential instability.

The conditions necessary for the formation of cumulonimbus clouds are a lifting action and unstable, moist air.

Macroscale and mesoscale circulations in which thunderstorms are embedded provide potential instability by bringing in warm, moist air at low levels. Instability can also be caused by bringing in cold air aloft because this increases the lapse rate. Initial lift is provided by surface heating, orography, fronts, low-level convergence, and upper-level divergence. It follows that thunderstorms are favored in geographical areas that are close to moisture sources and sources of lift. Severe thunderstorms have stricter requirements that include very high potential instability, plus a unique wind shear that provides the thunderstorm with the tilt and rotation needed to produce supercells.

Whether they contribute to airmass or severe thunderstorms, these influences organize thunderstorms into distinctive patterns. If you understand the relationships between thunderstorm occurrence, larger scale circulations, and geography, then you should be able to understand and anticipate thunderstorm development. In the next few paragraphs, we examine some of the more common of these relationships. A useful starting point is thunderstorm climatology.

CLIMATOLOGY

Figure 9-19 shows the number of thunderstorm days in the contiguous U.S. by season. Summer is by far the most active season. The supply of potentially unstable air from the Gulf of Mexico at this time of year accounts for many of the thunderstorms which occur from the Rocky Mountains eastward. The high number of thunderstorm days in the maritime tropical environment of Florida is caused by sea breeze convergence, coupled with daytime heating. The high frequency of thunderstorms over the southern Rockies is due to a good supply of moisture, orographic lifting and, again, daytime heating. An increase in thunderstorm activity in the very hot desert southwestern U.S. occurs during July and August as moist air reaches that area.

Thunderstorms are much less common on the West Coast of the U.S., because the sinking air around the Pacific high pressure region increases the stability of the air, especially in summer. Also, the water along the West Coast is much colder than that found in the Gulf of Mexico. This increases the stability of the air at lower levels. Because of this, some locations along the coast actually experience fewer thunderstorms in summer than in winter.

Thunderstorm activity reaches its peak in the afternoon in most areas because of the influence

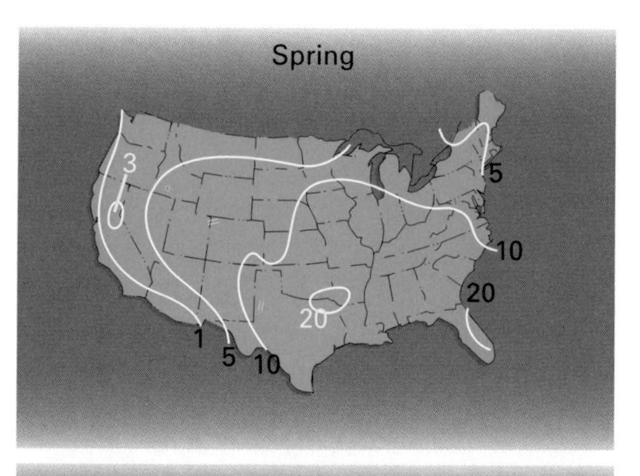

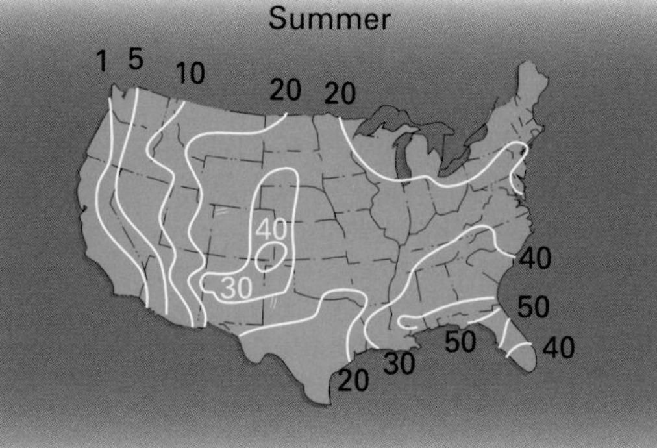

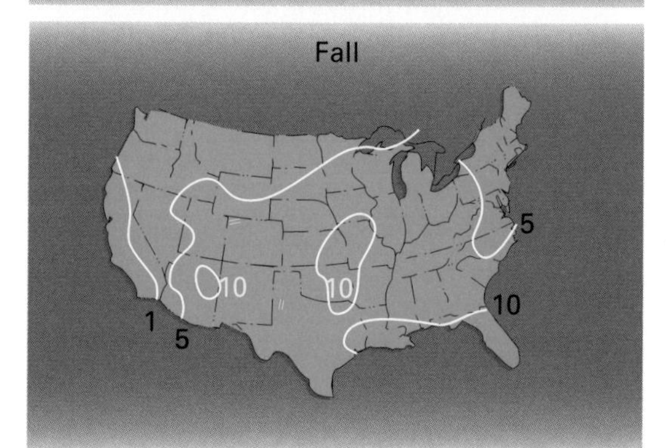

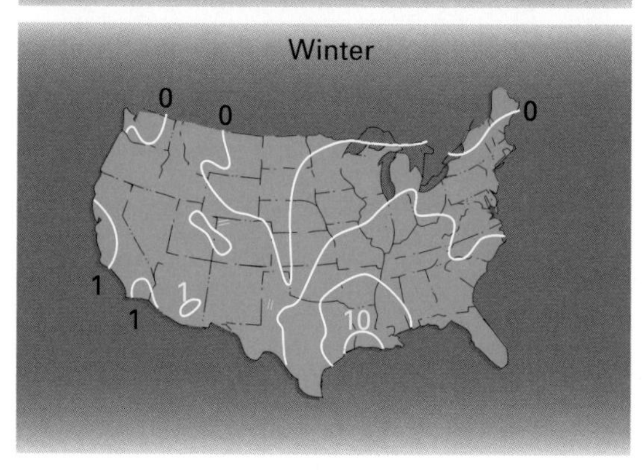

Figure 9-19. Average number of thunderstorm days in each season.

of solar heating. However, the Great Plains region of the U.S. has a nighttime maximum in thunderstorms during the summer. This is caused, in part, by strong southerly winds that frequently develop in a narrow band from south Texas into Oklahoma between 650 feet and 1,500 feet AGL during nighttime hours. This low-level "jet" carries potentially unstable air northward from the Gulf of Mexico.

The highest frequency of severe thunderstorms (supercells) occurs in the springtime in the area between the Rockies and the Mississippi and from Texas to the Dakotas. The primary region of activity is northern Texas, Oklahoma, and southern Kansas. That area is particularly conducive to severe weather because of its proximity to the warm, moist air from the Gulf of Mexico and the tendency for extratropical cyclones (a source of initial lift) to develop just east of the Rockies.

Another climatological feature of thunderstorms is that their tops tend to be lower in winter than in summer, and lower at high latitudes than at low latitudes. Cumulonimbus cloud bases, on the other hand, are lower in the moist environment of the eastern U.S. and higher over the drier west.

Thunderstorm climatology provides a useful "first guess" of where and when thunderstorms are likely to develop, given the right conditions. But caution should be used in its application. Day-to-day variations from the average picture may be large. In the next few paragraphs, we examine some of these variations.

INSTABILITY PATTERNS

A practical approach to the evaluation of the potential instability requirement for thunderstorms is to use a stability index. One of the most common is the lifted index. It is simply the difference between the observed 500 mb temperature and the temperature that a parcel of air would have if lifted from the boundary layer to the 500 mb level. The lifting process takes into account any condensation and latent heat release that occurs so that the temperature difference at

500 mb also reflects the influence of moisture. If the observed 500 mb temperature is colder than an air parcel lifted to that level, then the lifted index is negative (unstable) and thunderstorms are likely. The relationship between the lifted index and thunderstorm severity is shown in figure 9-20.

Lifted Index	Chance of Severe Thunderstorm
0 to -2	Weak
-3 to -5	Moderate
< = -6	Strong

Figure 9-20. The lifted index numbers on the left are related to the thunderstorm severity on the right.

The difference found by subtracting the temperature of a parcel of air theoretically lifted from the surface to 500 millibars and the existing temperature at 500 millibars is called the lifted index.

Keep in mind that the lifted index evaluates thunderstorm severity, not the probability of occurrence. A negative lifted index doesn't mean that a thunderstorm is likely to occur. If it does occur, the lifted index shows how strong it will be. Also, note that airmass (nonsevere) thunderstorms can occur when the lifted index is slightly positive.

Another stability index, the K index, has proved useful in determining the probability of occurrence of airmass thunderstorms. Similar to the lifted index, the K index is determined from a current sounding. It is defined as:

$$K = \overset{A}{(T_{850} - T_{500})} + \overset{B}{D_{850}} - \overset{C}{(T_{700} - D_{700})}$$

T is the air temperature and D is the dewpoint temperature in degrees Celsius at the pressure levels indicated by the subscripts, 850 mb, 700 mb, 500 mb. The K index may seem a bit complicated at first glance, but it has a simple physical interpretation.

Term A is a measure of the lapse rate between the 850 mb and 500 mb levels. If it is a large positive number, then air is less stable. Term B measures the moisture at 850 mb. If the 850 mb dewpoint temperature is high, then there is a large possible contribution to instability by the release of latent heat when that air is lifted to saturation. Term C, including the minus sign, simply measures the dryness of the air at 700 mb. When the difference between the air temperature and the dewpoint temperature is large, the air at that level is dry. The contribution to instability is negative.

It follows that when the atmosphere has plenty of moisture and a large lapse rate, the K index will be large and the probability of airmass thunderstorms will be high. Figure 9-21 shows the relationship between K and airmass thunderstorm probability.

K Index	Thunderstorm Probability (%)
<15	near 0
15 to 20	20
21 to 25	20 to 40
26 to 30	40 to 60
31 to 35	60 to 80
36 to 40	80 to 90
>40	near 100

Figure 9-21. The K index numbers on the left are related to the probability of airmass thunderstorm occurrence on the right.

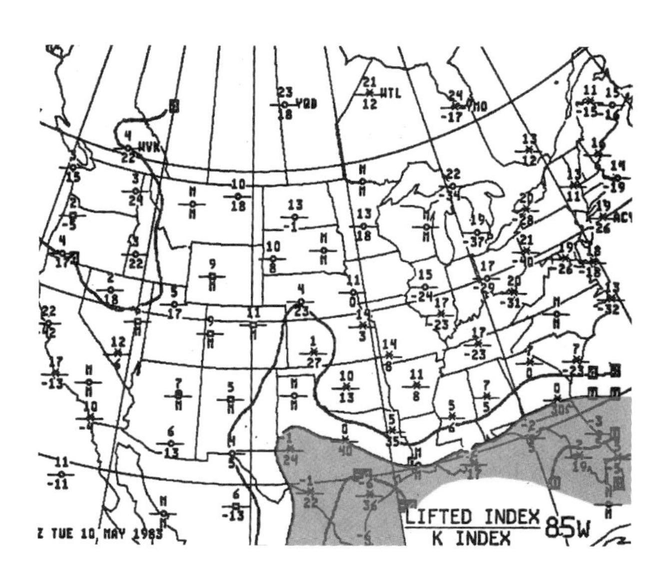

Figure 9-22. A portion of a stability panel from a composite moisture chart. Areas with lifted index less than zero are shaded. A K index of 28 corresponds with a 50 percent probability of air-mass thunderstorms. Note only the lifted index is contoured.

The lifted and K indices are computed twice per day for all soundings made across the U.S., Canada, and northern Mexico. Stability charts which display both indices are prepared and are regularly available. An example is shown in figure 9-22.

The regions of thunderstorm activity are located where critical stability index values and adequate initial lifting coexist. In many cases, the initial lift occurs in distinct lines or areas.

THUNDERSTORM LINES

Multicell thunderstorms often form along lines much longer than the diameter of any single storm. The processes by which wind conditions along lines can produce upward motions (initial lift) were introduced in Chapter 5. They include frontal lifting and surface convergence. Such lines may contain nonsevere and/or severe thunderstorms.

MACROSCALE FRONTS

A cold front is often the location of a line of abrupt lifting because it is fast moving and has a relatively steep slope. If the potential instability is sufficient, a line of frontal thunderstorms will be present. Thunderstorm production along fronts is particularly efficient when cold air is present aloft. (Figure 9-23)

The types of thunderstorms that form along cold fronts are multicell thunderstorms which occasionally become severe, and severe supercell thunderstorms. The lifted index and VIP level give indications of thunderstorm type and intensity.

Not all cold fronts produce thunderstorm activity. For example, some cold fronts may be dry (cloudless), while others will produce mainly nimbostratus with steady rain, or showers without thunderstorms. Because of these differences, it is important to examine other information than the surface weather analysis chart to determine if thunderstorms are associated with a front. Weather radar data, satellite imagery, and surface weather data provide many useful clues as to thunderstorm presence.

Figure 9-23. Fast moving cold front and resultant abrupt lifting can create a line of thunderstorms.

Some aviation meteorology literature refers to a type of frontal thunderstorm, which is described as being more severe and longer lasting than the single-cell airmass thunderstorm. The use of this terminology does not differentiate between what are now identified as multicell and supercell thunderstorms. It also omits reference to the organization of thunderstorms into areas as well as along lines. This text does not use the term frontal thunderstorm.

Thunderstorms may be aligned along warm fronts when unstable air overruns the wedge of retreating cold air at lower levels. Over land, this occurs more frequently in the warmer months of the year. It is a particularly troublesome situation, because the thunderstorms are often embedded in stratiform clouds and cannot be seen from the ground or from the air, unless you are above the cloud deck. This problem is also likely in occlusions.

SQUALL LINES

A squall line, or instability line, is a broken or continuous line of nonfrontal thunderstorms. It ranges from about one hundred to several hundred miles in length. Depending on the degree of instability and the wind variation through the troposphere, thunderstorms along squall lines may be ordinary multicell, supercell, or a mixture.

A squall line frequently develops along or just ahead of a cold front in the warm sector of a cyclone. Once multicell thunderstorms form, their ability to regenerate new cells helps to maintain the line. A squall line generally moves across the warm sector in the direction of the winds at 500 mb. When a squall line approaches and passes a particular location, the effect is similar to an idealized cold front. Examples of the appearance of a squall line on a surface analysis chart, a radar summary chart, and a view from space are given in figure 9-24.

Embedded thunderstorms are thunderstorms that are obscured by massive cloud layers and cannot be seen.

The most severe weather conditions, such as destructive winds, heavy hail, and tornadoes are generally associated with squall lines.

Squall lines are nonfrontal and often contain severe, steady-state thunderstorms.

Squall lines most often develop ahead of a cold front.

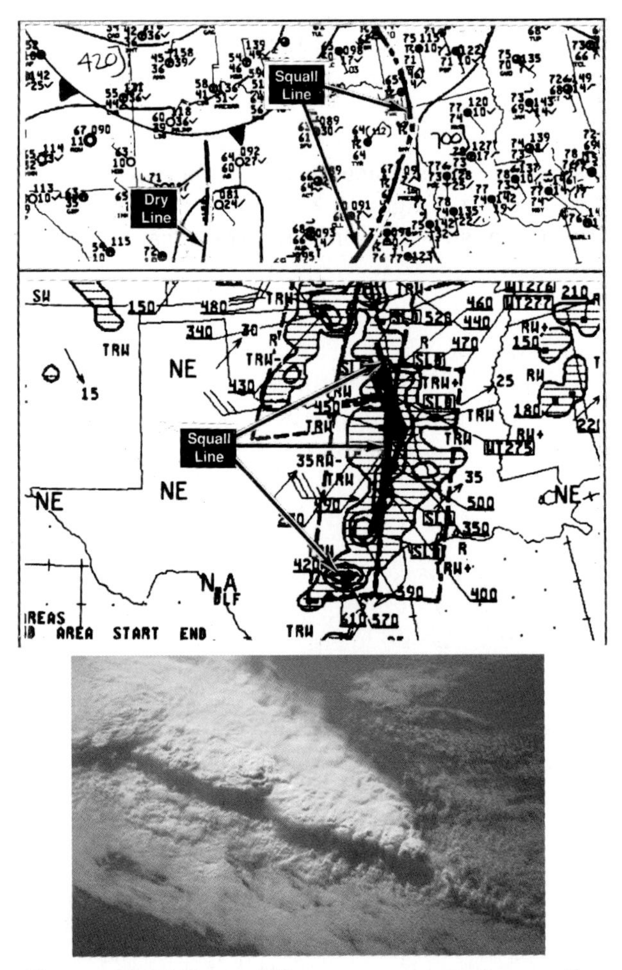

Figure 9-24. Three different examples of a squall line. Top: Section of surface analysis chart. Middle: Radar summary chart showing alignment of echoes along a squall line. Bottom: Photograph of a squall line from space. (Photo courtesy of Lunar and Planetary Institute, NASA Photograph)

OTHER MESOSCALE LINES

There are several other mesoscale phenomena that are known to contribute to the formation of thunderstorms along narrow bands. For example, in Chapter 11, you will see the lifting that is generated at a sea breeze front. If potentially unstable air is present, thunderstorms form in a line along that boundary. Similarly, outflow boundaries from thunderstorms produce favorable conditions for the formation of thunderstorm lines when those boundaries move into unstable regions.

In west Texas during late spring and early summer, there often exists a north-south boundary between moist tropical air flowing northward from the Gulf of Mexico and dry air over the higher terrain to the west. The moisture boundary, which is called a dry line, is apparent in the distribution of surface dewpoint temperatures in the top chart of figure 9-24. Weather radars in this area often show a long, narrow clear air echo known as a "fine line." This is where the moisture content of the air changes rapidly from one side of the dry line to the other. Dry lines are often the initial location of eastward moving squall lines.

When orographic lifting of potentially unstable air occurs, often with the added help of surface heating, thunderstorms will be aligned along mountain ranges. In some cases these lines will remain stationary as thunderstorm cells keep forming in the same location. This condition can lead to local flooding.

In all cases of thunderstorms along lines, thunderstorm development is enhanced when there is the presence of an upper-level short wave or upper cyclone in the area. Divergence near jet stream level provides upward vertical motions, and cold air aloft contributes to instability.

THUNDERSTORM CLUSTERS

Under certain circumstances, thunderstorms in middle latitudes develop into large clusters that are more circular than linear. These include both macroscale and mesoscale clusters which may contain nonsevere and/or severe thunderstorms.

MACROSCALE CLUSTERS

There are many macroscale upper air circulations that can cause thunderstorm outbreaks over areas of thousands of square miles. A few of the more common patterns are described here.

Large areas of thunderstorms are often caused when a developing upper level disturbance moves over an unstable area. The disturbance in this case is usually a cold trough or a low in the upper troposphere. As viewed from a meteorological satellite, the associated thunderstorm cluster often appears as a comma-shaped cloud mass below and slightly ahead of the upper disturbance. This pattern happens more often over oceans because of the availability of large amounts of moisture. (Figure 9-25)

Figure 9-25. Comma-shaped region of convective activity below and slightly ahead of an upper-level short wave trough.

As we saw in the previous chapter, when the upper air disturbance shown in figure 9-25 intensifies further in the vicinity of a surface front, an extratropical cyclone with a surface low pressure region and warm and cold fronts frequently develop. Thunderstorms are then distributed along fronts or squall lines as described earlier.

Macroscale regions of thunderstorms observed to develop without any obvious frontal structure include those associated with old occluded cyclones, cold lows aloft, and tropical cyclones. For example, a macroscale cold low aloft will bring widespread thunderstorms to the southwestern U.S. during the warmer months. The same region is subjected to widespread thunderstorm activity whenever a dissipating tropical cyclone moves across the area from the west coast of Mexico. Hurricanes, of course, often bring macroscale thunderstorm activity when they reach land.

MESOSCALE CONVECTIVE COMPLEXES

Mesoscale convective complexes are nearly circular clusters of thunderstorms that develop primarily between the Rockies and the Appalachians during the warmer part of the year. Heavy rains and severe weather are not unusual. The complexes, as viewed by satellite, are typically a few hundred miles in diameter.

Mesoscale convective complexes typically develop in late afternoon as a result of the interactions and merging of smaller groups of thunderstorms. They generally move eastward, reaching their maximum development about midnight, then weaken in the early morning hours. An example of a mesoscale convective complex is shown in the satellite image of figure 9-26.

Figure 9-26. A large mesoscale convective complex covers Missouri and portions of Kansas, Oklahoma, and Arkansas. The area of major activity appears larger than it actually is due to the blow off of cirrus anvils from the thunderstorms.

SUMMARY

The thunderstorm, by itself, is a distinct mesoscale atmospheric circulation that begins as cloudless convection in the boundary layer and develops through a great depth of the atmosphere in a very short period of time. Thunderstorms have a range of structures. Some of these support the generation of new thunderstorms as well as the development of very intense and long-lived severe thunderstorms. It has become clear that in order to really understand thunderstorms, it is also necessary to know about a variety of related circulations. These range from fronts and squall lines to individual thermals, and from downdrafts and gust fronts to tornadoes and suction vortices. Considering that thousands of thunderstorms occur over the earth's surface every day, and that a single thunderstorm may produce lightning, very heavy rainshowers, hail, strong winds, low visibilities, wind shear, turbulence, and icing, it is not surprising that they have a substantial impact on aircraft operations. Details of these hazards are examined in Part III.

KEY TERMS

Accretion
Airmass Thunderstorm
Attenuation
Cloudy Convection
Cold Air Funnel
Convective Condensation Level
Cumulus Stage
Dart Leaders
Dissipating Stage
Doppler Radar
Dry Convection
Dry Line
Dust Devil
Equilibrium Level
Funnel Cloud
Gust Front
Gustnadoes
Hail
Initial Lift
K Index
Lifted Index
Lightning
Mammatus
Mature Stage
Mesoscale Convective Complex
Multicell
NEXRAD
Outflow Boundary
Potential Instability
Precipitation-Induced Downdraft
Radar
Radar Echo
Radar Summary Chart
Radar Target
Return Stroke
Severe Thunderstorm
Shelf Cloud
Single Cell
Squall Line
Stepped Leader
Suction Vortex
Supercell
Thermals
Thunderstorm
Tornado
Towering Cumulus (TCU)
Virga
Vortex Ring
Waterspout
Weather Radar

CHAPTER QUESTIONS

1. The cluster of cells that makes up a multicell storm will not necessarily move in the same direction as the individual cells. Why?

2. As you complete the preflight inspection of your aircraft, you notice the lightning and rain south of the airport. You can't really see any movement of the storm. The airport winds are from the south. The airport is isolated and uncontrolled. You have no radar or other supplementary meteorological information. Discuss your options.

3. Convection also occurs at high levels producing altocumulus and cirrocumulus clouds. Surface heating obviously doesn't play a role. How do they form?

4. On a particular summer day, you notice that thermals are exceptionally strong. Fair weather cumulus clouds form, but there are no thunderstorms. Why?

5. Decode the following weather reports.

6. Why is the main downdraft in a thunderstorm cold?

7. In the morning, the K index is 32, and the lifted index is -3. What are your conclusions about thunderstorm occurrence in the afternoon?

8. Occasionally, aircraft will experience strong turbulence in a part of a convective cloud that does not show up on the radar. Give some reasonable explanations.

9. Do some research and find out what a storm scope is, how it works, and how it compares with airborne radar.

10. Thunderstorm tops often penetrate thousands of feet across the tropopause, past the equilibrium level into the stable stratosphere. How can the air continue to rise under these conditions?

MDW RS 1856 M8BKN1 TRW 990/63/61/3215G32/980/TSTM N MOVG EWD

ICT SP 1944 3SCT M8BKN 200VC 11/2TRW 132/72/64/0214G25/992/R01VR30V50 T SW MOVG E RB25

FOD SA 0055 AWOS M5 OVC 1/2 70/68/3325G30/992/P110/WND 30V36/WEA: TRW+ T OVHD

METAR
SPECI KCVG 2228Z 28024G36KT 3/4SM +TRSA BKN008 OVC020CB 28/23 A3000 RMK RETSB24RAB24

Introduction

Local winds refer to a variety of mesoscale circulations other than thunderstorms. These circulations fall into two broad catgories. Thermally driven local winds are caused by local differences in radiation heating or cooling. They are most noticeable when large scale wind systems are weak or absent. Externally driven local winds are produced when strong winds interact with the local terrain. They are most noticeable when strong, large scale wind systems are present. The phenomena discussed under these categories all have the potential to produce important flight hazards. It is imperative for you to know what their causes and characteristics are, and how to identify them and avoid their worst consequences. When you complete this chapter, you will have gained this knowledge and, additionally, you will have condensed it into some useful conceptual models of land, sea, mountain, valley breezes; mountain waves; and downslope winds.

Section A

THERMALLY DRIVEN LOCAL WINDS

Thermally driven local winds include sea and land breezes, mountain and valley breezes, and slope circulations. Because they depend on radiational heating and cooling, these winds commonly develop in middle latitudes in the warmer part of the year. They may also develop in lower latitudes during any season as long as there are no effects of larger-scale circulations, such as extratropical or tropical cyclones. The dependence of thermally driven local winds on radiational heating and cooling causes directions and intensities of the circulations to be linked closely to the time of day.

SEA BREEZE

In Chapter 4, the concept of thermal circulation was introduced to help you understand how horizontal temperature gradients cause pressure gradients, which in turn, cause the wind to blow. The sea breeze was used as a brief example.

To review, daytime heating along coastlines brings land areas to higher temperatures than nearby water surfaces. The pressure over the land falls, establishing a horizontal pressure gradient. Often, this pressure difference is so small that you can't observe it in the isobar patterns on a surface analysis chart. However, it is large enough to cause cool air to begin moving across the coastline toward land in late morning. This is the sea breeze.

In the sea breeze, and other circulations of this scale and smaller, Coriolis force is usually much less important than the horizontal pressure gradient force and frictional forces. Therefore, the wind tends to blow directly from high to low pressure.

The sea breeze continues to intensify in the afternoon, reaching typical speeds of 10 to 20 knots in middle or late afternoon and decreasing thereafter. A well-developed sea breeze is usually 1,500 to 3,000 feet deep and capped by a weaker, deeper, and oppositely directed return flow aloft. The combined sea breeze and return flow are called the sea breeze circulation (Figure 10-1)

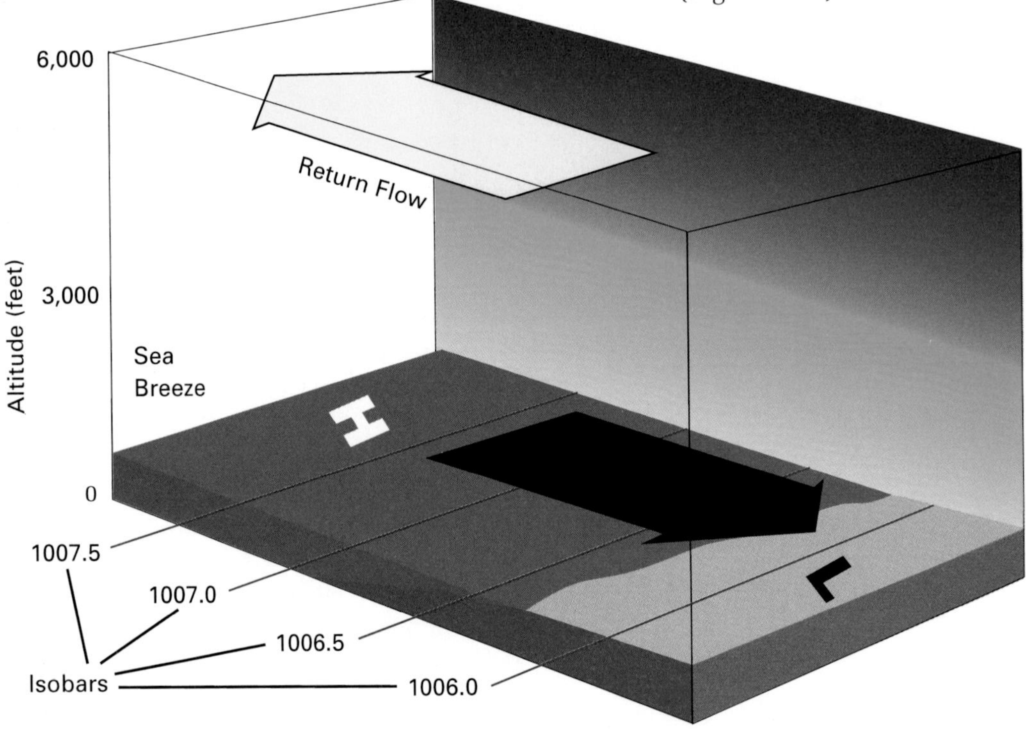

Figure 10-1. Sea Breeze Circulation. During the daytime, a low-level sea breeze flows from land to sea in response to a pressure gradient caused by the heating of the land. The sea level pressure pattern is shown by isobars labeled in millibars. Aloft, the pressure gradient is reversed and a return flow is directed from land to sea.

The sea breeze circulation is sometimes made visible by large differences in visibility at various altitudes. For example, over some urbanized coastlines, clear marine air moves inland at low levels with the sea breeze, while the return flow aloft is made visible by the offshore movement of polluted urban air.

SEA BREEZE FRONT

The boundary between the cool, inflowing marine air and the warmer air over land is often narrow and well defined. This feature is known as the sea breeze front. The frontal location is often identifiable by differences in visibilities between the moist and dry airmasses, or a broken line of cumulus clouds along the front. In locations where there is large conditional instability, the sea breeze front is marked by a line of thunderstorms. (Figure 10-2)

The sea breeze front moves inland more slowly than the winds behind it. Frontal speeds vary over a wide range (2 to 15 knots) depending on macroscale wind conditions and terrain. In some areas, the inland movement of the front is limited to a few miles by coastal mountains. In contrast, over regions with broad coastal plains, the front can move a hundred miles or more inland during the course of the day.

Certain coastline shapes and a favorable distribution of coastal mountains and hills promote the convergence of sea breezes. For example, sea breezes often converge from both sides of the Florida peninsula, producing lines of thunderstorms. Near coastal hills or mountains, gaps in the terrain allow the sea breeze front to move farther inland in those areas. Also, isolated hills may split the sea breeze into two parts which then move around the barrier and converge inland. (Figure 10-3)

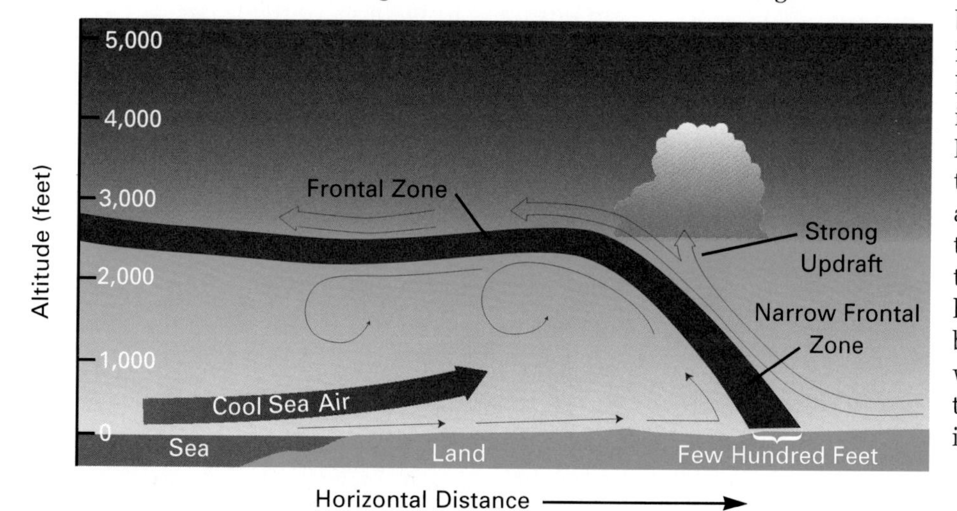

Figure 10-2. Cross section through a sea breeze front. The see breeze is blowing from left to right. The front will be marked by a line of clouds only when adequate moisture is present.

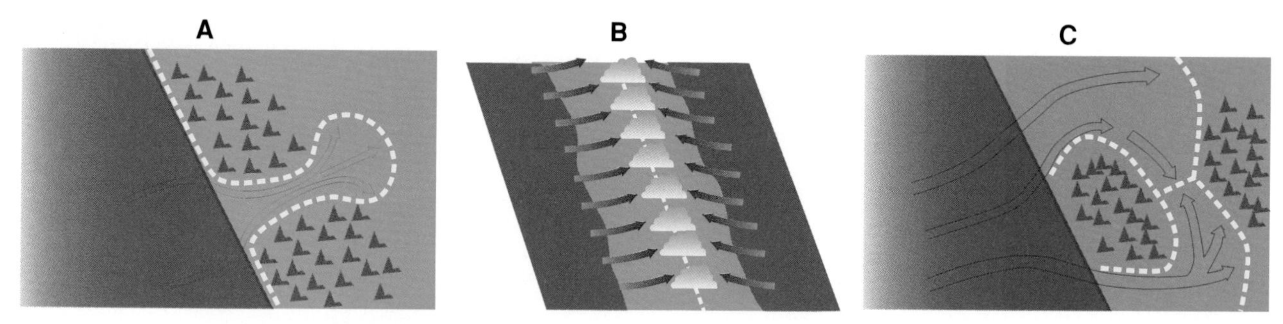

Figure 10-3. Examples of effects of topography on the sea breeze. The yellow dashed lines represent the sea breeze front. In diagram A, the sea breeze penetrates farther inland through a gap in the terrain. In diagram B, sea breezes from opposite sides of a peninsula converge. In diagram C, the sea breeze spreads around hills on a coastal plain and converges in a valley on the opposite side.

LAND BREEZE

A few hours after sunset, the land surface near a coastline has cooled much more rapidly than the nearby water surface. When the land becomes colder than the ocean, the pressure gradient across the coast reverses, so that the lower pressure is offshore. The low-level flow which begins to move from land to sea under the influence of this pressure difference is called the land breeze The land breeze circulation is also made up of a ground level breeze and an opposite return flow aloft, the reverse of the sea breeze circulation. (Figure 10-4)

The land breeze continues to strengthen throughout the night, reaching its greatest intensity about sunrise. Because of the strong stability typical of nighttime conditions over land, the depth of the land breeze circulation is considerably less than that of the sea breeze. It normally reaches only a few hundred feet above the surface. The land breeze is usually weaker than the sea breeze with typical maximum speeds of about five knots. Some exceptions may occur when cold air moves down the slope of a mountain range located along the coast. Details of such drainage winds are given in a later section.

Because a land breeze front occurs over water, it is not as well documented as the sea breeze front. However, in many coastal areas, a convergence line is found offshore at night. The distance of the land breeze front from the coast varies widely between locations, from less than 5 n.m. to over 100 nautical miles. The actual distance depends on the strength of the land breeze. If the water is warm, the convergence zone may be identified by a line of convective clouds or thunderstorms.

Noticeable land and sea breeze circulations are not restricted to ocean coastlines. For example, lake and land breezes are generated by the Great Lakes, as well as by smaller bodies of water such as Lake Tahoe and the Salton Sea in California. Wind speeds generated by such circulations are proportional to the area of the water surface and to the land-water temperature difference. They are stronger and more frequent in the summer.

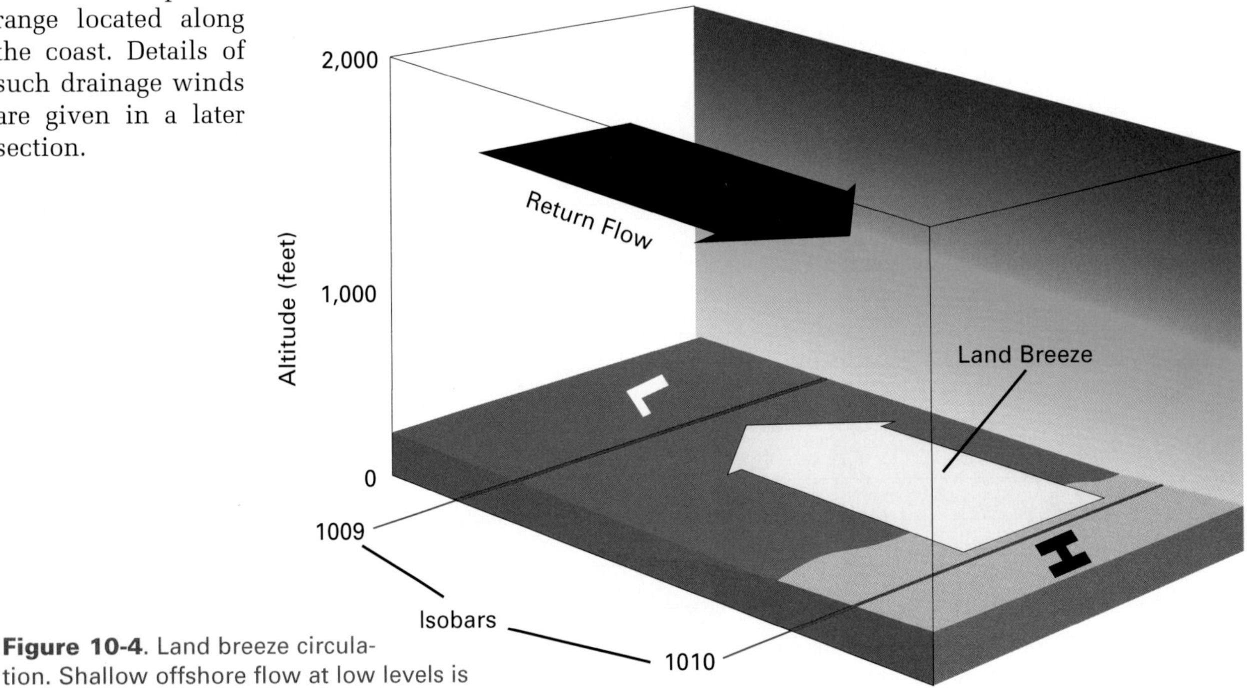

Figure 10-4. Land breeze circulation. Shallow offshore flow at low levels is capped by a weaker onshore return flow aloft.

VALLEY BREEZE

In mountainous areas that are not under the influence of large scale wind circulations, the daytime winds tend to be directed toward higher terrain. This primarily warm season circulation leads to rising air over mountains or hills and sinking air over nearby lowlands. It is often marked by greater thermal activity and cumulus clouds over the highlands than over the valleys.

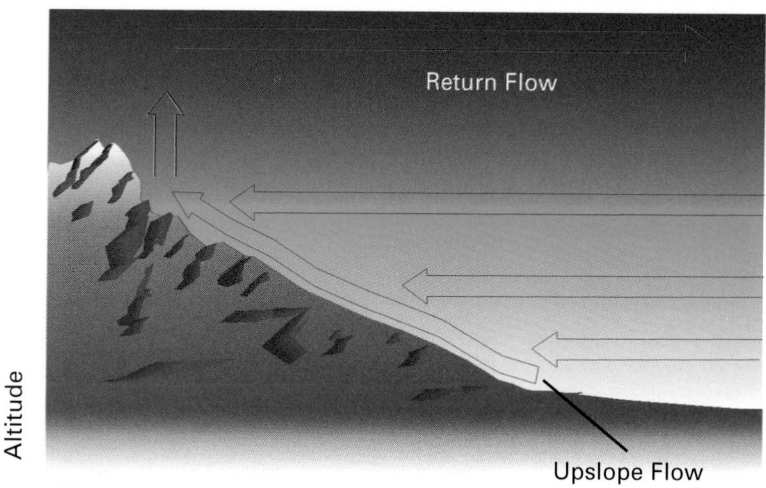

Figure 10-5. Upslope circulation. In the daytime, cooler air moves toward the warm slopes at and below ridge-top level while, aloft, there is a weak return flow.

These flow patterns occur because the hills and mountains are heated to temperatures that are warmer than air at the same level over nearby valley areas. Because of the horizontal temperature gradient, a horizontal pressure gradient develops with the lower pressures over the mountains. Below the peaks, air responds by flowing toward the slopes of the warmer mountain. The hillside deflects the air, producing an anabatic or upslope wind. As expected, a return flow is found above the mountain. (Figure 10-5)

If the mountainside is part of a valley, the upslope flow may be part of a larger scale valley breeze which is also directed toward higher terrain. Above the mountain tops, a weak return flow known as an anti-valley wind is found. Together with the valley breeze, this is called the valley breeze circulation. (Figure 10-6)

The precise time that an upslope or valley flow begins depends on local sunrise. This time is determined not only by latitude and time of year, but also by the depth and orientation of the valley. In many valleys, upslope flow begins on one side of the valley while the other side is still in shadow. Similarly, in the afternoon, the upslope

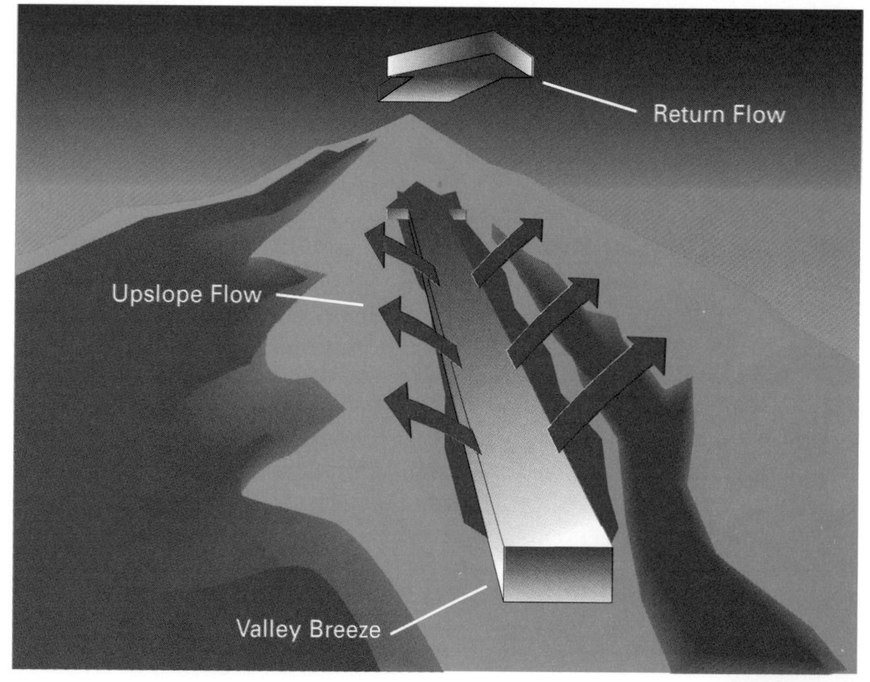

Figure 10-6. The daytime valley breeze is similar to an upslope circulation, but on a larger scale. At low levels, air flows up the centerline of the valley and toward the warm slopes. Aloft, a return flow is directed down the valley.

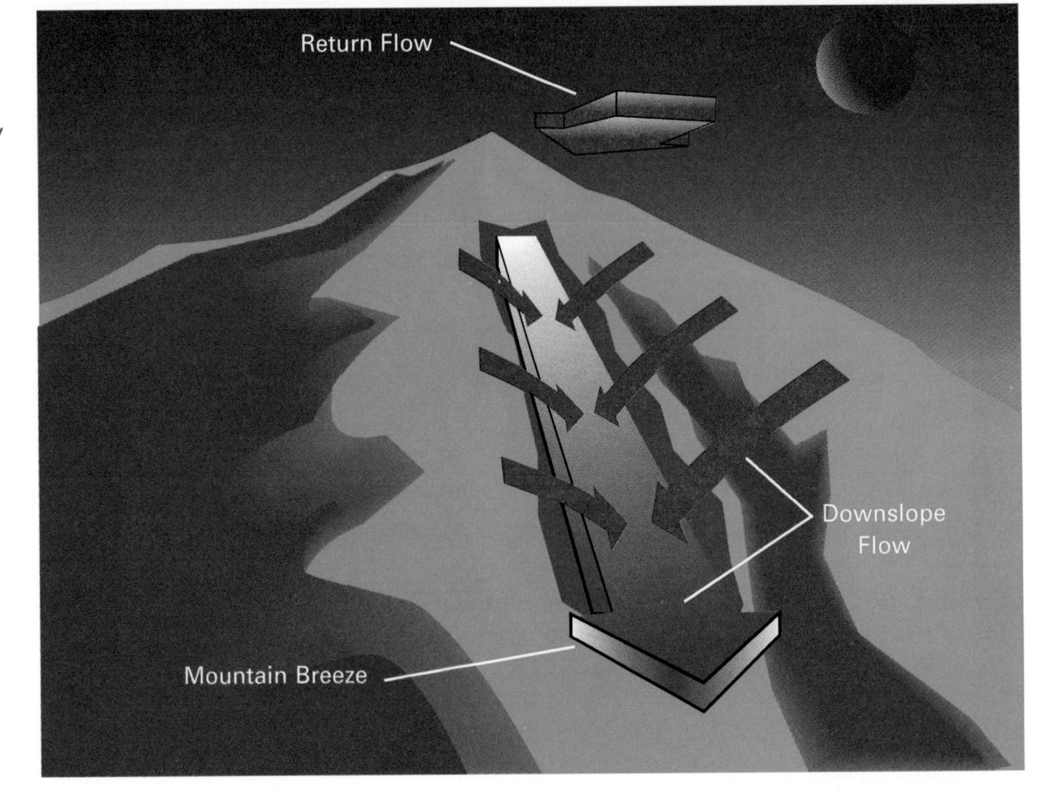

Figure 10-7. Mountain breeze circulation. As the ground cools at night, air flows away from the higher terrain. Winds are directed downslope along the mountain slopes and valley floor.

circulation persists longer on sun-facing slopes. When the slopes become very warm, instability and convection make upward motion even stronger. Usually upslope and valley breezes begin a few hours after sunrise reaching maximum speeds in middle afternoon. Typical up-valley wind speeds reach 5 to 20 knots, with the maximum winds occurring a few hundred feet above the surface.

Strong winds in the atmosphere above the mountain peaks may disturb valley and slope circulations depending on the direction and speed of the upper winds and the orientation and depth of the valley. When winds above the mountains exceed about 15 knots, the valley circulation may be significantly modified. It may strengthen, weaken, or even reverse direction as strong winds aloft are carried down into the valley through mechanical mixing and/or by afternoon convection.

Over snow-covered terrain, heating of the high slopes is small due to the reflection of solar radiation by the snow. Therefore, over bare slopes, the daytime upslope flow does not usually extend beyond the snow line. However, if the hillsides are tree covered, the flow may be different. When the tree tops are snow free, they absorb solar radiation, warming the air and producing a well-defined upslope circulation despite the presence of snow on the ground.

MOUNTAIN BREEZE

At night, when the high terrain cools off, the air over the mountains becomes cooler than the air over the valley. The pressure gradient reverses and katabatic, or downslope winds develop along the hillsides. On the larger scale of the valley, a mountain breeze blows down the valley with a return flow, or anti-mountain wind, above the mountain tops. This configuration is known as the mountain breeze circulation. (Figure 10-7)

As with the valley breeze, the size, depth, orientation of the valley, and the steepness of the slopes determine the intensity of the mountain

A. Sea Breeze/Upslope

Figure 10-8. In diagram A, a combined sea breeze/upslope flow with heating over the island produces cloudiness inland. In diagram B, the flow is reversed. The combined land breeze and downslope winds keep the island clear of clouds but converge offshore to produce clouds.

B. Land Breeze/Downslope

breeze circulation. Prior to sunrise, speeds of 5 to 15 knots are common. It is not unusual to find greater speeds, sometimes exceeding 25 knots, at the mouth of the valley.

In locations where mountains are found along coastlines, the sea breeze enhances the upslope or valley breeze. Similarly, land breeze and mountain breeze effects are combined to produce significant offshore winds. These effects are well developed on tropical islands where thermally driven circulations dominate. Because of the availability of moisture, the daytime circulation is marked by convective clouds over the island, while the nighttime flow produces a ring of cloudiness over the ocean around the island. (Figure 10-8)

COLD DOWNSLOPE WINDS

Just as upslope and valley breezes are intensified by warm air rising in convective currents; cold, dense air sinking down the slopes strengthens the downslope and mountain breezes. Gravity is effective in causing the air to move downward as long as the air near the slopes remains cooler than the air away from the hills at the same level. Small scale flows of this type are called drainage winds. These are one of several types of air flows identified as cold downslope winds.

Drainage winds are very shallow. They can begin before sundown; that is, as soon as a slope becomes shaded. Their flow is similar to water, following the natural drainage patterns of the terrain. The cold air typically pools at the bottom of the slope, unless it is caught up in a larger scale mountain breeze.

Drainage flows are often very weak because of the offsetting effect of adiabatic heating as the air moves downslope. However, in some situations strong winds can develop. Anywhere permanent ice and snow fields occur in mountainous areas, drainage flows occur day or night in the absence of strong, large-scale winds. This effect is especially noticeable in the warmer part of the year. A shallow layer of cold, dense air flows rapidly down the surface of the glacier. Gravity accelerates this glacier wind as it moves downslope, so the strongest winds occur at the lower end, or toe, of a glacier. The maximum speeds depend on the length and steepness of the glacier and the free-air temperature. Extreme examples of such winds are found along the coast of Antarctica.

An example of an extreme cold, downslope wind is the bora. It develops along the coast of the former Yugoslavia in the winter. The terrain in that area rises rapidly from the Adriatic Sea to about 2,000 feet AGL. Shallow, cold airmasses moving from the east literally spill over the steep mountain slopes. In extreme cases, the cold air reaches the coast with speeds in excess of 85 knots. Many cold, downslope winds in other geographical areas are also called bora, or described as bora-like. For example, during winter in North America, a very cold airmass associated with high pressure will occasionally spread over the mountains of British Columbia, Idaho, and western Montana. These conditions cause strong, cold winds to flow down the east slopes of the Northern Rockies in Alberta and Montana.

Section B

EXTERNALLY DRIVEN LOCAL WINDS

When large-scale circulations cause airflow across rugged terrain, numerous mesoscale circulations develop over and downwind of the mountains. Two of the most important are mountain lee waves and warm downslope winds.

MOUNTAIN LEE WAVES

When a stable airstream flows over a ridgeline, it is displaced vertically. Downwind of the ridge, the displaced air parcels are accelerated back to their original (equilibrium) level because the air is stable. They arrive at the equilibrium level with some vertical motion and overshoot it, only to repeat the oscillation as they are swept downstream. The mesoscale wave pattern that they follow is known as a mountain wave or mountain lee wave. It is a particular form of an atmospheric gravity wave. These phenomena are so-named because, in a stable atmosphere, gravity (through stability) plays a major role in forcing the parcels to return to, and oscillate about their equilibrium level. (Figure 10-9)

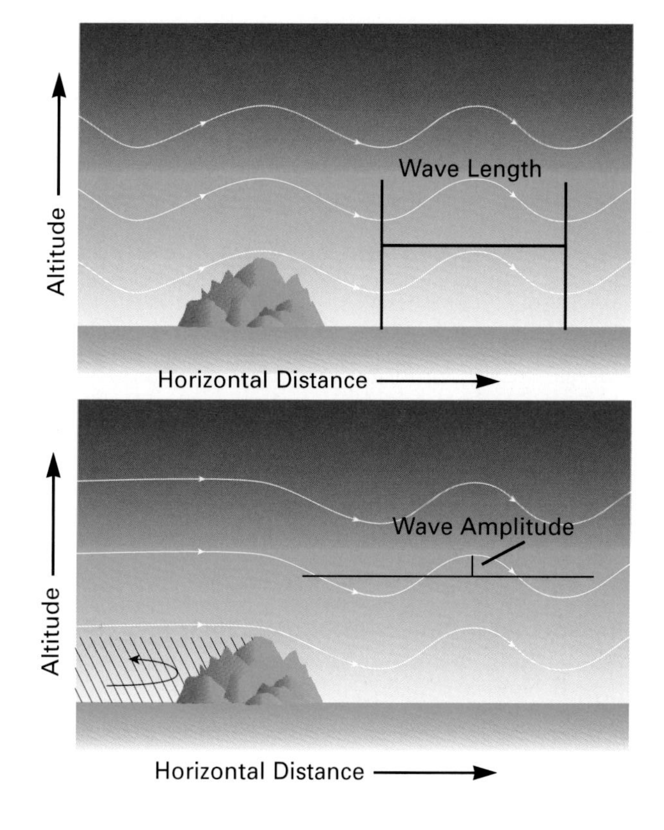

Figure 10-9. Two situations which contribute to the formation of lee waves downwind of a mountain ridge. In the top diagram, stable air flowing from left to right is lifted by the ridge. In the bottom diagram, the airstream is blocked at low levels in front of the mountain ridge, and stable air from aloft descends on the lee side of the mountain. In both cases, the vertical displacement of the stable air by the mountain causes lee waves.

Because the mountain which initially displaces the air doesn't move, lee waves tend to remain stationary despite the fact that the air moves rapidly through them. Lee waves are relatively warm in the wave troughs where stable air has descended, and cold in the crests where stable air has risen. Once established in a particular locale, mountain lee wave activity may persist for several hours, although there are wide variations in lifetime.

The formation of mountain lee waves requires movement of stable air across a mountain range.

Lee waves are important phenomena because they produce vertical motions large enough to affect aircraft in flight. Furthermore, they are often associated with turbulence, especially below mountaintoplevel and near the tropopause. The flight hazards associated with lee waves are discussed in detail in Part III. In the remainder of this section we will briefly describe the general features of lee waves.

Under typical mountain wave conditions, wavelengths average about five nautical miles. However, they can vary widely from a few miles to more than 30 n.m., depending on stability and wind speed. The stronger the wind speed, the longer the wavelength of the lee wave; the greater the stability, the shorter the wavelength of the lee wave.

Figure 10-10. Lee wave activity is revealed by the "wash board" wave pattern lenticular clouds over the western United States.

Although the lee wave wavelength is an accurate measure of the horizontal scale of an individual mountain lee wave, it must be kept in mind that effective scale may be quite a bit larger. This is because a single ridge often sets up a train of several lee waves, and a mountainous area may be composed of many parallel ridges, all capable of producing lee waves. Therefore, it is not unusual for lee waves to cover a horizontal area of a few hundred nautical miles. (Figure 10-10)

The strength of a lee wave is indicated by the strength of the vertical motions it produces. In a weak lee wave, upward and downward air motions usually have vertical speeds of a few hundred feet per minute or less. Strong lee waves have vertical motions of 1,800 f.p.m. or more.

The vertical motion in the waves depends on wind speed, wavelength, and wave amplitiude. Wind speed determines how fast air parcels move through the wave pattern; for a given wave amplitude and wavelength, the faster the wind at mountain top level, the greater the vertical motions in the lee waves.

The amplitude indicates how far an air parcel will deviate from a horizontal path as it moves through a wave pattern; the larger the wave amplitude, the greater the vertical motions. Lee wave amplitudes may reach 4,000 feet or more in strong wave cases. Figure 10-11 illustrates the effect of wave amplitude on vertical speed.

If two waves have identical amplitudes and wind speeds, then the wave with the shorter wavelength will have the greatest vertical velocity. This effect is shown in figure 10-12.

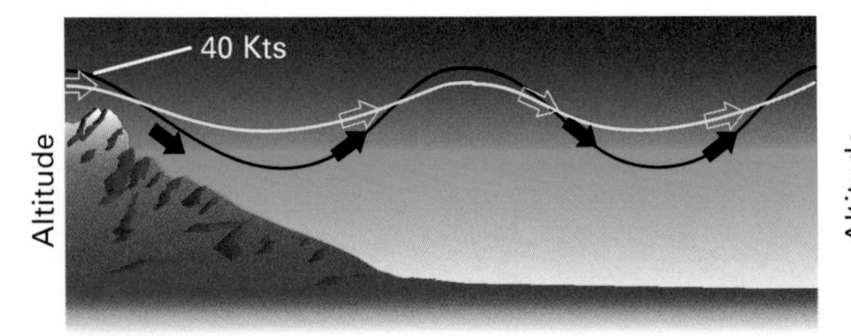

Figure 10-11. Two examples of the influence of amplitude on the vertical speed of the wind. The wind speeds and wavelengths are the same in both cases. The yellow line indicates the path of an air parcel through a lee wave with a relatively small amplitude. The black line shows a case where the amplitude is greater. The vertical speeds are larger in the second case.

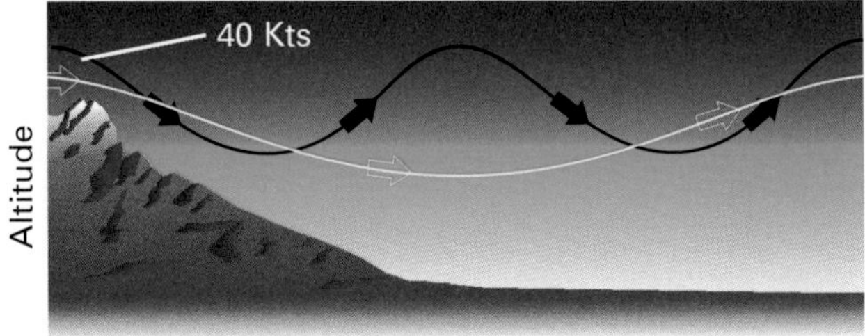

Figure 10-12. Two examples of the influence of wavelength on vertical motions. The wind speeds and amplitudes are the same in both cases. The yellow line indicates the path of an air parcel through a lee wave with a relatively long wavelength. The black line shows a case where the wavelength is shorter. The vertical speeds are larger in the second case.

Tall, relatively narrow mountains are more effective in producing strong lee wave amplitudes. For wide mountains, the steepness of the lee slopes, rather than the overall width of the mountain is important in determining lee wave strength. Two locations where exceptionally strong mountain waves are produced because of high, steep lee slopes are the Sierra Nevada range near Bishop, California, and the Rockies near Boulder, Colorado.

Winds nearly perpendicular to a ridgeline are more effective in the production of lee waves than winds nearly parallel to the ridgeline. Significant vertical motions will occur in lee waves if the winds perpendicular to the ridgeline exceed 20 knots at the top of the ridge and the lee wave wavelength exceeds 5 n.m.

High, broad ridges with steep lee slopes often produce large amplitude lee waves. This is especially true when the height of the terrain decreases 3,000 feet or more downwind of the ridge line.

THE LEE WAVE SYSTEM

All lee waves, regardless of their geographic location, produce certain common flow features and clouds. These characteristics are captured in the idealized model of the lee wave system shown in figure 10-13. The lee wave system is divided into two layers, an upper lee wave region where smooth wave flow dominates and microscale turbulence occasionally occurs, and a lower turbulent zone from the ground to just above mountain-top level.

This simple model is a valuable guide that helps you locate regions of wave action and turbulence from cloud observations and other visible indicators. The model also helps you deduce lee wave conditions from wind measurements and from macroscale airflow patterns shown on weather charts. For example, soundings taken during lee wave conditions show a number of similar features. Temperature soundings typically have a stable

The conditions most favorable to wave formation over mountainous areas are a layer of stable air at mountain top altitude and a wind of at least 20 knots blowing across the ridge.

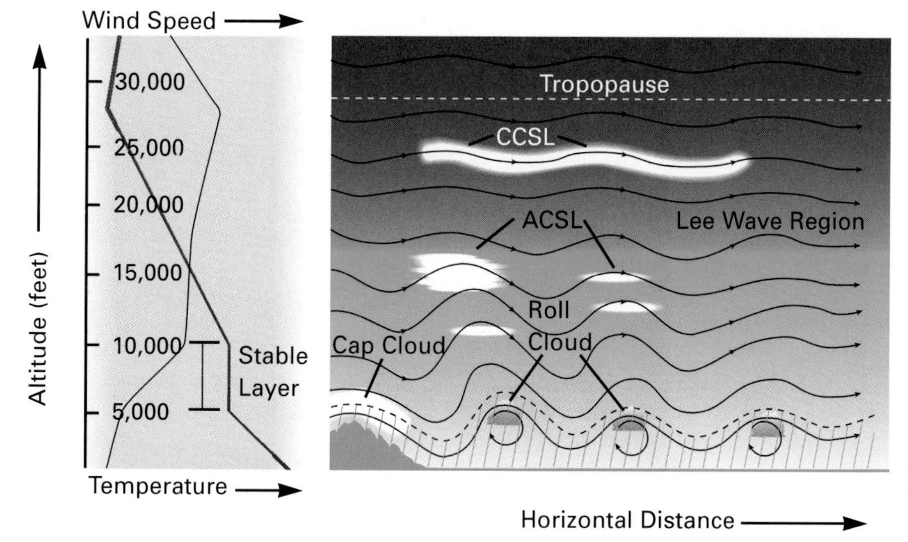

Figure 10-13. The lee wave system. Airflow through the lee waves is indicated by thin solid lines with arrows. The mountain is located on the left side of the diagram. The lower turbulent zone (shaded) and characteristic clouds in this area are shown. Temperature (red) and wind speed (black) soundings taken just upstream of the ridge are shown on the left.

layer near mountain top level with less stable layers above and below. Wind soundings display increasing wind speed with altitude.

In the lee wave portion of figure 10-13, the most intense lee wave is the first or primary cycle immediately downwind of the mountain. Successive cycles tend to have reduced amplitudes. Figure 10-13 shows that the lee waves have their greatest amplitudes within a few thousand feet above the mountains, decreasing above and below.

A major feature of the lower turbulent zone is the rotor circulation found under one or more of the lee wave crests. The altitude of the rotors is about mountain top level. The rotor under the first wave crest is the most intense and is usually the major source of turbulence in the lower turbulent zone, if not in the entire lee wave system.

The lower turbulent zone is generally characterized by strong gusty winds with the strongest surface winds along the lee slopes of the mountain. These features contribute to the dominance of turbulence in this layer. Related flight dangers are examined in the chapter on aviation weather hazards.

Crests of standing mountain waves may be marked by stationary, lens-shaped clouds known as standing lenticular clouds.

When there is adequate moisture, the lee wave system produces one or more of three unique cloud forms. These are presented schematically in figure 10-13 and shown photographically in figure 10-14. They are the cap cloud immediately over the mountain tops; the cumuliform rotor, or roll cloud associated with the rotor circulation; and the smooth, lens-shaped altocumulus standing lenticular (ACSL) or lenticular clouds in the crests of the lee waves. Higher lenticular clouds are sometimes reported as CCSL in aviation weather reports.

It is important to realize that clouds may be your only indication of the presence of lee waves. Because mountain waves are mesoscale phenomena, they escape detailed measurements in the regular network of weather stations.

Aside from pilot weather reports, the only direct evidence of lee wave activity comes from weather satellite

Figure 10-14. Cap cloud, roll cloud, and altocumulus standing lenticular cloud. These clouds are easily recognized by their proximity to mountains and unique shapes. They are useful indicators of lee wave activity.

images and occasional surface weather reports of lee wave clouds. Automated reports from unmanned stations have no such information. Some examples of weather reports which indicate the presence of lee waves are shown in figure 10-15.

As with most of the atmospheric models, wide variations from the average picture of the lee wave system may occur. For example, lee wave activity may be limited to the lower troposphere because the wind speed or direction changes radically with altitude, or if the stability weakens at higher levels, wave action is no longer possible. In other cases, waves may intensify with altitude, leading to intense wave action near the tropopause and in the stratosphere. Significant mountain wave activity has been observed at altitudes of more than 60,000 feet MSL.

The presence of lenticular, roll, and/or cap clouds indicate lee wave activity and the location of wave crests and rotor circulation. However, these observations should not be used to estimate the strength of the vertical motions or associated turbulence.

There are also horizontal variations in lee wave structure. Our model of the lee wave system is based on airflow across an idealized long ridge. In contrast, real mountain ridges are complicated by rugged peaks of different sizes and separations. Also, a typical mountainous area is more likely to be made up of several ridges which are not quite parallel and are irregularly spaced. This means that even when wind and stability conditions are generally the same across a particular geographical area, significant variations in wave characteristics may occur.

WARM DOWNSLOPE WINDS

We have already seen that very cold air flowing down the steep slope of a mountain can produce a strong, cold wind called a bora. In contrast, when a warm, stable airmass moves across a mountain range at high levels and descends on the lee side, it often produces a strong warm wind called a chinook.

A chinook, or foehn, is defined as a warm, dry, gusty wind that blows from the mountains. It often occurs under the same conditions that produce mountain waves, although it can extend much farther downwind of the mountains than the wave activity. Surface winds in a chinook are typically 20 to 50 knots, and extreme speeds near 100 knots have been measured. The strongest chinook winds and greatest warming occur closest to the mountains. In the winter, especially in high latitudes, temperature changes with the onset of the chinook can be very large; changes of 20F° to 40F° in 15 minutes have been widely documented.

Warm downslope winds are identified by many different local names. Although the terms, chinook and foehn, are used widely now to describe downslope winds in many geographical areas, they also come from specific regions. Chinook is usually applied to winds along the east slopes of the Rocky Mountains, while foehn is used to describe warm winds along the northern slopes of

GFA SA 2054 80 SCT 100 SCT E250 OVC 15 008/55/42/2428G40/956/ ACSL ALQDS

TPH SA 2253 120 SCT 30 062/40/30/1803/971/ CCSL W ROTOR CLD SW

DEN SA 1757 E100 BKN 250 OVC 20 020/50/38/2714G23/959/ ACSL SW-NW

Figure 10-15. Weather reports with remarks that indicate lee wave activity. The important remarks are not available from weather stations that are completely automated.

the Alps. Some other local names for warm, dry, gusty, downslope winds are Santa Ana (southern California), zonda (Argentina), Canterbury Northwester (New Zealand), and Mono Wind (central and northern California).

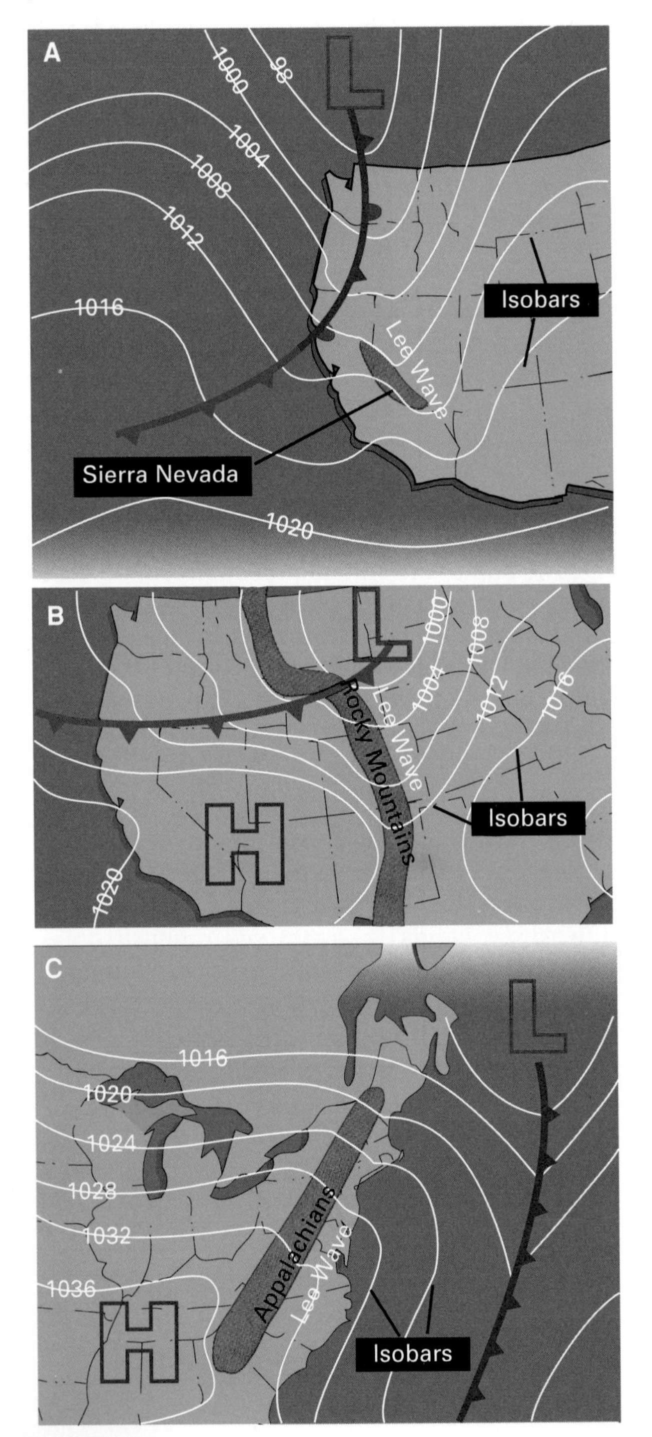

MACROSCALE WEATHER PATTERNS

Large scale wind systems conducive to the development of lee wave and chinook activity must satisfy two important requirements:

1. There must be a stable layer just above the mountain.
2. The wind speeds across the mountaintop must be at least 20 knots.

These conditions are frequently met when extratropical cyclones move across mountainous areas, bringing strong mountain top winds and widespread stable layers associated with fronts. Lee waves occur most often during the cooler months of the year when extratropical cyclone activity is greater.

In the western U.S., major mountain ranges run north-south or northwest-southeast. Therefore, mountain waves occur when fronts and upper air troughs approach the mountains from the west. After the fronts and troughs pass, winds shift and significant wave activity ceases. In contrast, the Appalachian Mountains are oriented northeast-southwest. Lee wave activity is favored after the fronts and troughs aloft pass and the winds become northwesterly. These conditions are illustrated in figure 10-16.

Figure 10-16. Three surface analysis chart diagrams from actual lee wave situations along the east slopes of the Sierra Nevada (A), the Colorado Rockies (B), and the Appalachians (C). The mountains are shown in gray. Notice in all cases that the sea level pressure is higher on the west side of the mountains and lower on the east. In A and B, the lee waves are prefrontal; while in C, they are postfrontal.

SUMMARY

Chapter 10 has shown how the simple concept of thermal circulation is used to explain the development and general features of sea and land breezes, mountain and valley breezes, and drainage winds. Such small-scale circulations can certainly affect flight conditions, but their presence is not immediately obvious on surface analysis charts. The information you have learned in this chapter will help you anticipate winds produced by local terrain and land-water differences.

Additionally, when large-scale circulation systems such as extratropical cyclones make their way across rugged terrain, interactions of their winds with mountains and hills produce other unique mesoscale circulations including mountain lee waves and warm downslope winds. These phenomena offer many more serious problems to pilots than do most thermally driven circulations. However, there is a problem. The conventional network of surface weather observing stations does not observe these mesoscale circulations very well. Therefore, there is not much detailed information available to the pilot to determine, for example, the location and strength of lee waves for flight planning and avoidance purposes. Your new knowledge of cloud indicators and large-scale patterns favorable for their development will prove exceptionally valuable in your analysis of the presence and intensity of externally driven local winds.

The completion of this chapter ends our formal consideration of atmospheric circulation systems. You have examined a wide spectrum of atmospheric phenomena, ranging from the macroscale general circulation to microscale dust devils. In Part III, we examine weather flight hazards and the role played by the circulations presented in Part II.

KEY TERMS

Atmospheric Gravity Wave
Cap Cloud
Chinook
Cold Downslope Wind
Downslope Wind
Drainage Wind
Externally Driven Local Winds
Glacier Wind
Land Breeze
Land Breeze Circulation
Lee Wave
Lee Wave Region
Lee Wave System
Lenticular Cloud
Local Winds
Lower Turbulent Zone
Mountain Breeze
Mountain Breeze Circulation
Mountain Wave
Primary Cycle
Roll Cloud
Rotor
Sea Breeze
Sea Breeze Circulation
Sea Breeze Front
Thermally Driven Local Winds
Upslope Wind
Valley Breeze
Valley Breeze Circulation
Warm Downslope Wind

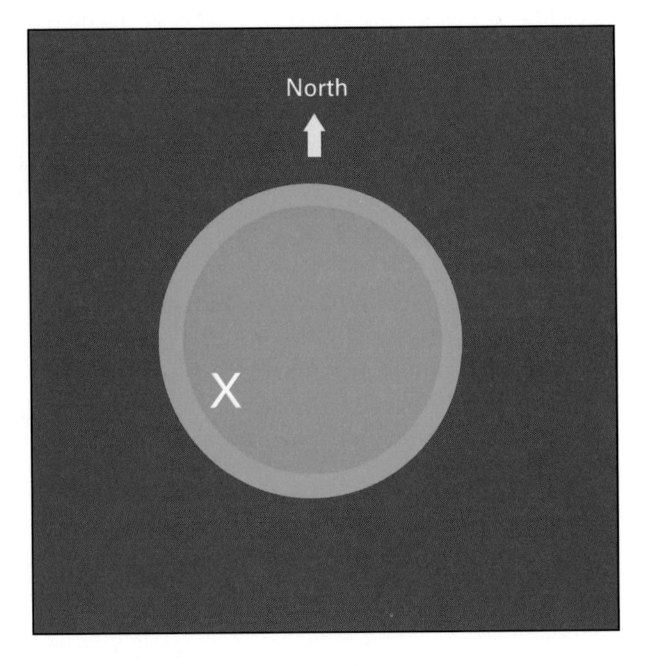

CHAPTER QUESTIONS

1. Why shouldn't you use observations of lenticular, cap, or roll clouds to estimate the strength of the vertical motions in the lee wave system. How should you use such observations?

2. The island shown to the left is located in the trade winds of the Northern Hemisphere. It is about 60 n.m. in diameter and is low and flat. An airport is located at the point marked "X." A pilot lands at the airport twice in a given 24-hour period. Both final approaches were from the southwest. The first landing was accomplished with a 25-knot headwind. The second landing also experienced a headwind, but it was less than 10 knots. At what times (24-hour clock) did the landings most likely occur? Explain.

3. (True, False) A moist airmass from the Pacific moves across the Rocky Mountains producing chinook conditions along the lee slopes. Clouds form during the passage of the air over the mountains, although there is no precipitation. The clouds dissipate as the air moves down the east slopes of the Rockies. The release of latent heat has contributed substantially to the warming of the air in the lee of the mountains. Explain.

4. You fly your aircraft well above the mountain peaks, upwind and downwind through a train of standing waves (lee waves). The wave conditions are exactly the same during both flights. You don't attempt to hold altitude, letting your aircraft "ride the waves." Draw cross sections along each flight track showing the airflow through the waves and the aircraft path. Be sure to indicate the flight direction.

5. A helicopter is carrying water to put out a fire along the east slopes of a ridge. Winds aloft are light. As would be expected, the flight is at low levels, so winds along the slopes are critical. It is afternoon

and, except for the smoke from the fire, the skies are clear. There are no significant large-scale wind systems in the area. As the pilot approaches the fire from the east, she notices that since her last run 15 minutes earlier, the fire has changed directions and is burning downhill. Give a reasonable explanation for the wind shift.

6. A dry, cool high pressure region has stagnated over Nevada in the early fall. For several days, the afternoon temperature at Tonopah, Nevada (elevation 5,425 feet MSL) has been 55°F. A dry, easterly wind is blowing over the beaches just west of Los Angeles. If the air in the latter region originated in Tonopah, what is its approximate temperature at the coast? Explain clearly, showing all of your work.

7. Sailplanes must often fly upwind in order to stay in the "up" portion of lee waves. Give a plausible explanation. (Note: there is more than one).

8. In some circumstances, a large forest fire can set up its own circulation so that oxygen-rich air is circulated into the fire, actually making it worse.

    1. Explain how the circulation develops.
    2. Do a little research to find some cases where such fire behavior has been doc umented.
    3. Describe each case briefly.

9. The chart given to the right is a simple topographic map that shows an airport at point "X" in a large valley. In this location, pilots claim that mountain waves occur over the valley both before and after the passage of a surface cold front and its associated trough aloft. Is this possible? Explain.

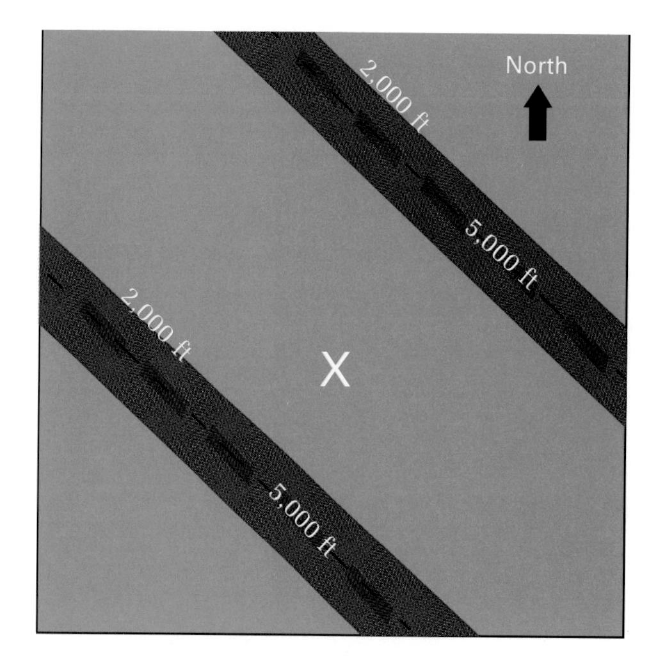

PART III

Aviation Weather Hazards

PART III AVIATION WEATHER HAZARDS

The circulations discussed in Part II produce a variety of flight hazards. Strong winds, low ceilings, turbulence, wind shear, icing, and lightning are just a few of the phenomena that can threaten safety of flight. The purpose of Part III is to closely examine several common weather hazards in order to give you better tools to anticipate and avoid them whenever possible.

(Thunderstorm photograph on previous page, copyright C. Melquist)

CHAPTER 11

Wind Shear

Introduction

Strong wind shear is a hazard to aviation because it can cause turbulence and large airspeed fluctuations and, therefore, serious control problems. It is a threat especially to aircraft operations near the ground because of the limited altitude for maneuvering, particularly during the takeoff and landing phases of flight. In this chapter, we examine wind shear and its causes. When you complete the chapter, you will know what wind shear is and its critical values. You will also know how, why, and where it develops in the vicinity of thunderstorms, surface-based inversions, and fronts.

Section A

WIND SHEAR DEFINED

Wind shear was defined briefly in Chapter 8 in connection with fronts. Because of the critical nature of wind shear, we will review and refine that definition.

A wind shear is actually a wind gradient. It is interpreted in the same sense as a pressure gradient or temperature gradient; that is, it is the change of wind over a given distance. Since wind is a vector, with both speed and direction, wind shear can involve a change in either speed or direction, or both.

Wind shear is best described as a change in wind direction and/or speed within a very short distance in the atmosphere.

In aviation, wind shear is a concern in all phases of flight. When considering its causes, it is convenient to visualize wind shear as being a horizontal wind shear (a change in wind over a horizontal distance) or vertical wind shear (a change in wind over a vertical distance) or a combination of both. Figure 11-1 shows examples of wind shear.

In diagram A, an aircraft is descending through air that is vertically sheared. In this particular case, the wind increases with altitude with no change in direction. Since the aircraft is flying into the wind, it experiences a decrease in head-

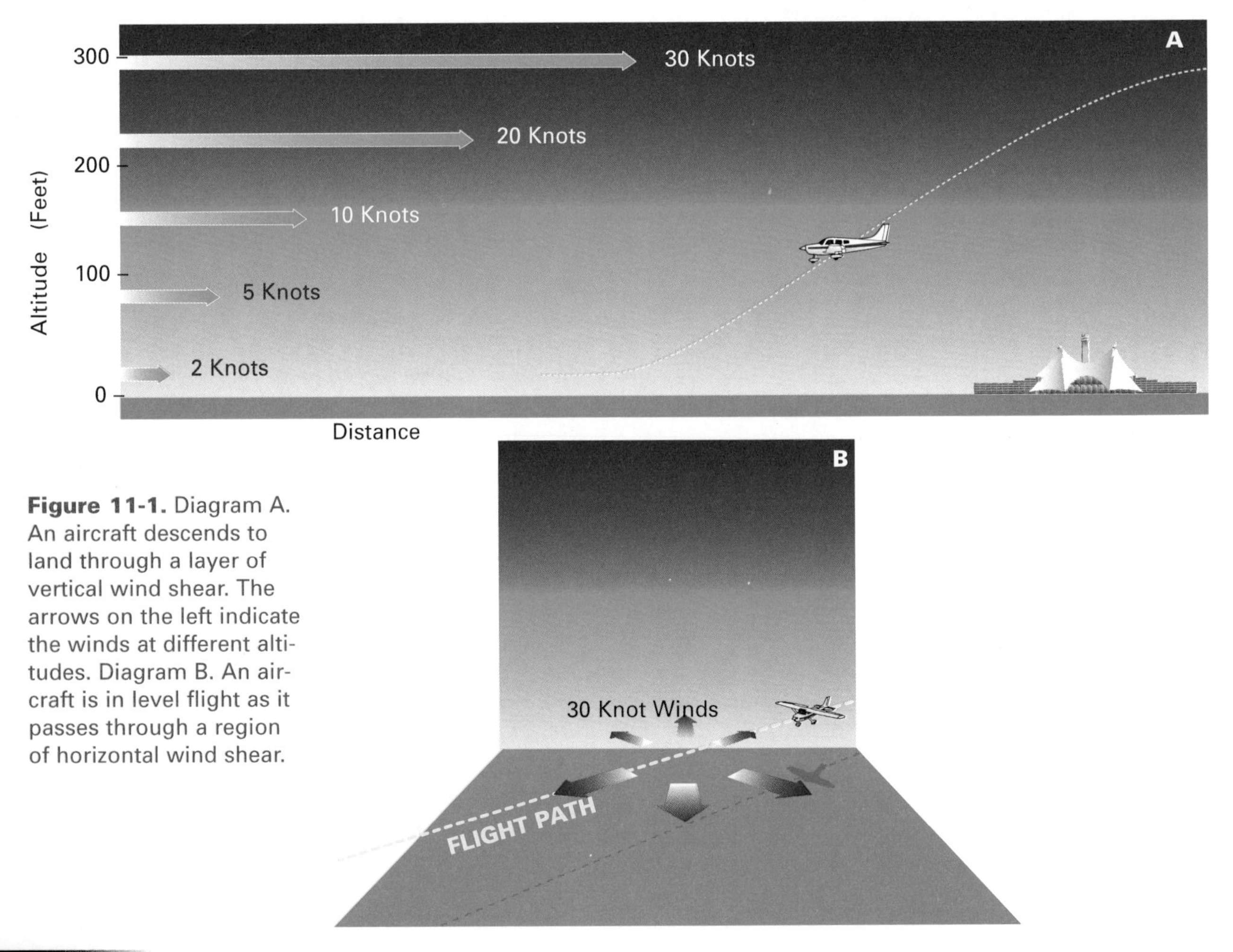

Figure 11-1. Diagram A. An aircraft descends to land through a layer of vertical wind shear. The arrows on the left indicate the winds at different altitudes. Diagram B. An aircraft is in level flight as it passes through a region of horizontal wind shear.

wind as it descends. In this example, the vertical wind shear is 30 knots per 300 feet (10 knots per 100 feet).

In diagram B of figure 11-1, an aircraft is in level flight toward a point where the wind direction changes 180°. As you will see, this is the type of pattern expected when penetrating a microburst. We must be careful in evaluating the shear in this situation. The difference in wind speeds (zero) along the flight path is misleading because of the reversal in wind directions. The meaningful shear is the change in headwind across the point of wind direction reversal (-30 knots). That decrease over a given horizontal distance, say from one end of the runway to the other, is the wind shear. Technically, the correct units of horizontal wind shear in this case are knots per nautical mile or the equivalent. However, wind shear information may also be given as a statement of the loss in headwind in knots, or as surface wind speeds and directions at two different locations along the runway.

If the wind direction changes over a given distance, the actual wind shear is always greater than the wind speed shear alone.

The two diagrams in figure 11-1 are idealized. In general, changes in wind speed and direction occur simultaneously with wind direction changes somewhere between 0° and 180°. Your concern as a pilot is how much the headwind or tailwind changes along your flight path, and in how short a distance the change occurs.

Low-Level Wind Shear (LLWS)	
LLWS Severity	**LLWS Magnitude (kts/100 ft)**
Light	< 4.0
Moderate	4.0 to 7.9
Strong	8.0 to 11.9
Severe	≥ 12

Figure 11-2. Severity categories of LLWS expressed in terms of vertical wind shear. Remember that LLWS applies to wind shear within 2,000 feet of the surface along the final approach path or along the takeoff and initial climbout path. According to these values, the vertical wind shear conditions in figure 11-1 correspond to strong LLWS.

Wind shear below 2,000 feet AGL along the final approach path or along the takeoff and initial climbout path is known as low-level wind shear (LLWS). The influences of wind shear on aircraft performance during landing and takeoff are well known. If the pilot of an aircraft encounters wind shear on approach and fails to adjust for a sudden decreasing headwind or increasing tailwind, the airspeed will decrease and the aircraft may undershoot the landing due to loss of lift. Similarly, a suddenly increasing headwind or decreasing tailwind on approach can cause an overshoot.

During departure under conditions of suspected low-level wind shear, a sudden decrease in headwind will cause a loss in airspeed equal to the decrease in wind velocity.

When wind shear is encountered on takeoff and the headwind decreases or the tailwind increases, the angle of climb and rate of climb will be lower. In critical situations, obstacles near the airport may not be cleared. During both takeoff and landing, a strong wind shear with a crosswind component may cause the aircraft to deviate from the centerline of the runway. Close to the ground, clearance of nearby obstacles may become difficult or impossible. Figure 11-2 assigns LLWS severity categories to various ranges of vertical wind shear.

A wind shear is considered significant when airspeed fluctuations of 15 to 20 knots occur.

Section B

CAUSES OF WIND SHEAR

In the last few years, the term wind shear often has been used to describe LLWS in the vicinity of convective precipitation, such as microbursts. There are not many experienced pilots who don't first think microburst when LLWS is mentioned. The reason that wind shear associated with convective precipitation receives so much attention is a simple one; it has been identified as the cause of a significant number of weather-related accidents with great loss of life and destruction of property. However, it is important to remember that microbursts are only one of the causes of wind shear. You will see in later sections that several other, quite different weather phenomena can also produce serious wind shear conditions.

An important characteristic of wind shear is that it may be associated with a thunderstorm, a low-level temperature inversion, a jet stream, or a frontal zone.

Critical wind shear for aircraft operations generally occurs on the microscale; that is, over horizontal distances of one nautical mile or less and vertical distances of less than 1,000 feet. These scales are so small that pilots may not have time to safely maneuver the aircraft to compensate for the wind change.

Although the dimensions of regions with significant shear are small, that shear is caused by circulations on scales which range from macroscale low pressure systems to microscale thermals. Some of the more important sources of wind shear are considered in the following paragraphs.

MICROBURSTS

The key to understanding development of wind shear below the bases of convective clouds is a good knowledge of the characteristics of a typical precipitation-induced downdraft. You were introduced to this phenomenon in Chapter 9. Its main features are summarized in figure 11-3.

Not all precipitation-induced downdrafts are associated with critical wind shears. However, there are two types of downdrafts that are particularly hazardous to flight operations because of their severity and small size. Professor T. Fujita, an atmospheric scientist from the University of Chicago, coined the term downburst for a concentrated, severe downdraft that induces an outward burst of damaging winds at the ground. He also

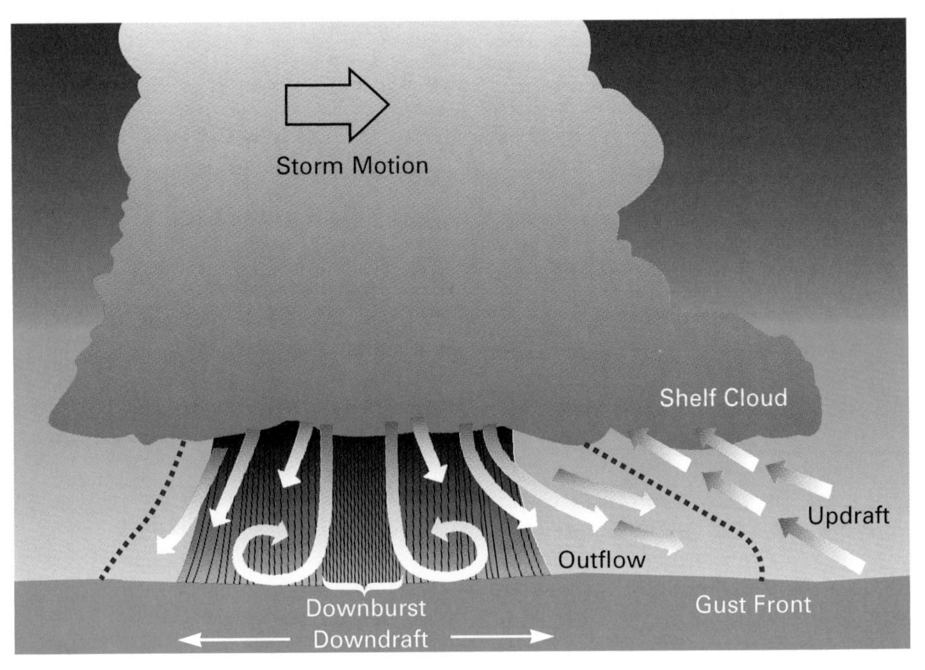

Figure 11-3. Conditions below the base of a multicell thunderstorm. Wind shear and turbulence are found within the main downdraft; in smaller, stronger downdrafts occasionally embedded in the main downdraft (downbursts and microbursts); and at the boundaries of the outflow (the gust front).

introduced the term microburst for a downburst with horizontal dimensions of 2.2 n.m. (4 km) or less. Because the term "microburst" is used more frequently in the general aviation literature to describe any precipitation-induced downdraft that produces critical wind shear conditions, we will use that term in subsequent discussions to avoid confusion.

Microbursts form by the same processes that produce the more common and less intense downdrafts; that is, by precipitation drag and cooling due to the evaporation and melting of precipitation particles. In a microburst, the downdraft intensifies with heavy rain and when dry air is mixed into the downdraft causing evaporative cooling and great negative buoyancy.

Microbursts may occur in airmass, multicell, and supercell thunderstorms. Isolated, single-cell storms present a greater hazard to aviation; because, they are common, small scale, rapidly developing, and they have strong outflows. Larger multicell storms are usually easier to avoid because they are less frequent, are larger scale, have longer lifetimes, and are often already identified as severe.

Don't land or take off in the face of an approaching thunderstorm.

A perspective view of a microburst is shown in figure 11-4. The microburst is characterized by a strong core of cool, dense air descending from the base of a convective cloud. As it reaches the ground, it spreads out laterally as a vortex ring which rolls upward along its outer boundary. You can interpret this pattern as an upside-down version of a thermal.

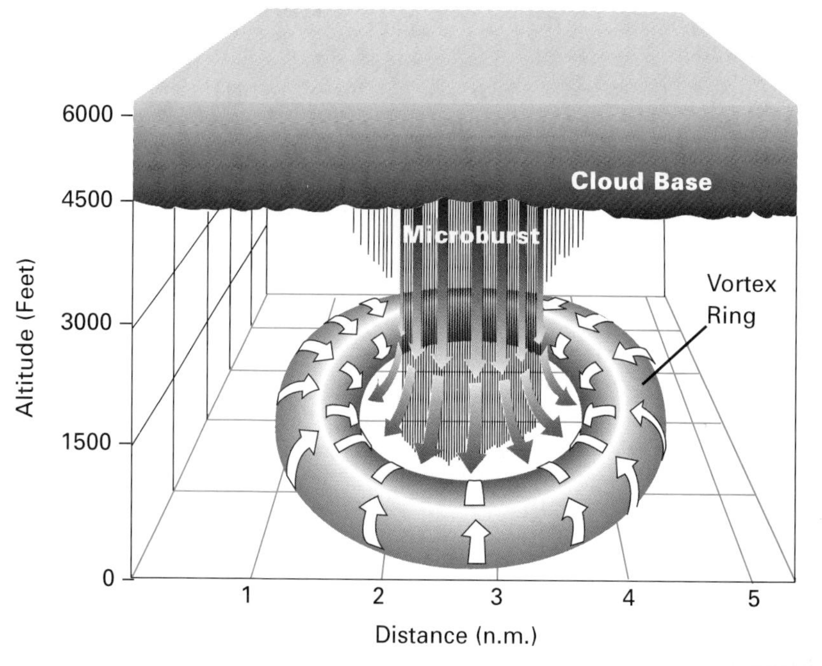

Figure 11-4. A symmetrical microburst. The arrows indicate airflow and the vertical lines indicate precipitation.

Typically, the microburst descends from the base of its parent cloud to the ground in a minute or so.

A cross section of an idealized microburst as it reaches the ground is shown in figure 11-5. The shear of the horizontal wind across the base of the microburst is apparent. You now see why this shear is so dangerous. Wind speeds are strong and the directions reverse 180° across the centerline of the microburst. Furthermore, the strong downward motions and heavy rain in the center of the downburst also reduce lift. All of this takes place in a very short period of time, near the ground, and frequently with low ceilings and visibilities.

An aircraft that encounters a headwind of 45 knots with a microburst may expect a total shear across the microburst of 90 knots.

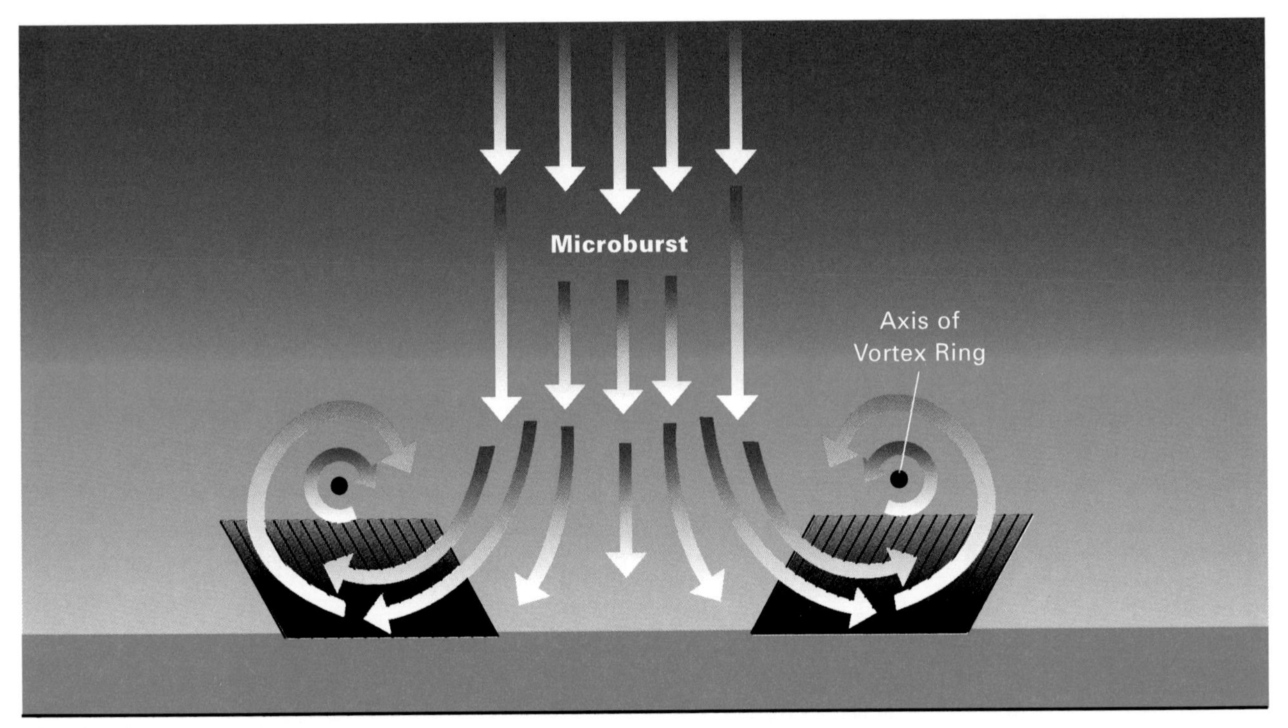

Figure 11-5. Microburst cross section. The flight hazards include the strong downdraft, often with heavy precipitation; gusty horizontal winds (shaded); strong horizontal wind shear from one side of the microburst to the other side; and turbulence in the vortex rings.

Within 100 feet of the ground, only a few seconds may be available for the recognition and recovery from wind shear associated with a microburst.

The lifetime of a microburst ranges from 5 to 30 minutes, once it reaches the ground. Most microbursts weaken significantly in only a few minutes. There is good evidence that some longer downburst events are a combination of successive microbursts a few minutes apart in the same location.

The duration of an individual microburst is seldom longer than 15 minutes from the time the burst strikes the ground until dissipation.

The peak outflow speed observed in an average microburst is about 25 knots. Winds in excess of 100 knots are possible. More critical is the change in wind speed across a microburst. An aircraft intersecting a typical microburst experiences an average headwind change of about 45 knots. This LLWS exceeds the capabilities of most light aircraft and is about the maximum that can be tolerated by heavy jet transports. The effect is illustrated in figure 11-6.

There are several variations in the formation and appearance of precipitation-induced downdrafts and microbursts. For example, a downdraft does not require a thunderstorm. As discussed previously, showers are common from cumulus clouds that do not reach the cumulonimbus stage. Therefore, microbursts also occur under these conditions.

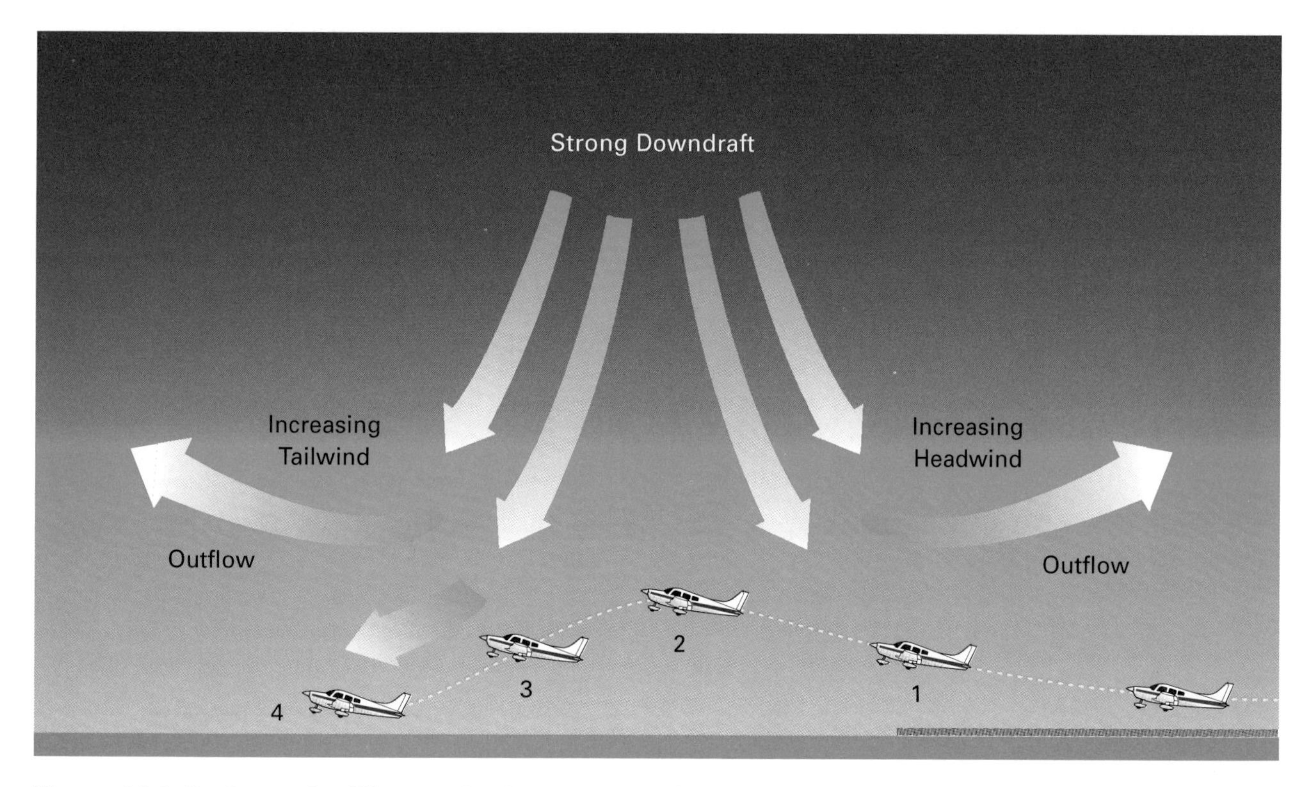

Figure 11-6. During a takeoff into a microburst, an aircraft experiences an increasing headwind (position 1), followed by a decreasing headwind and downdraft (position 2), and finally a tailwind (position 3). The most severe downdraft will be encountered between positions 2 and 3. Together with the loss of airspeed due to the tailwind, it can result in terrain impact or operating dangerously close to the ground (position 4).

Another variation on the ideal microburst model is that it may move and be distorted under the influence of the larger scale wind field in which the thunderstorm is embedded. Such "traveling" microbursts have stronger winds on the downwind side; that is, in the direction in which the microburst is moving.

When a shear from a headwind to a tailwind is encountered, while making an approach on a prescribed glide slope, the pilot should expect an airspeed and pitch attitude decrease with a tendency to go below glide slope.

There are several visual indicators of the presence of larger downbursts and microbursts. In humid climates, convective cloud bases tend to be low. These conditions produce "wet" downbursts or microbursts which are closely associated with a visible rainshaft. However, in dry climates, such as in the deserts and mountains of the western U.S., thunderstorm cloud bases are often high and the complete evaporation of the rainshaft can occur. In this case, a "dry" downburst or microburst may be produced. All that may be visible is virga at the cloud base and a characteristic dust ring on the ground.

Fortunately, dry downbursts occur mainly in the afternoon when these visible features can be identified. (Figure 11-7)

As the larger scale downdraft spreads out from one or more thunderstorms, strong shears persist in the gust front on the periphery of the cool air. Therefore, wind shear, including LLWS, may be found beyond the boundaries of the visible rainshaft.

If there is thunderstorm activity in the vicinity of an airport at which you plan to land, you should expect wind shear and turbulence on approach.

Because of the low-level wind shear hazards of downbursts, microbursts, and gust fronts, low-level wind shear alert systems (LLWAS) have been installed at many large airports around the U.S. where thunderstorms are frequent. This system continuously monitors surface winds at remote sites around the airport and communicates the information to a central computer. The computer evaluates the wind differences across the airport to determine whether a wind shear problem exists. Wind shear alerts are issued on the basis of this information.

Figure 11-7. Visible indicators of a wet downburst (or microburst) in diagram A, and a dry downburst in diagram B (facing page). (Dry downburst photograph, National Center for Atmospheric Research/University Corporation for Atmospheric Research/National Science Foundation.)

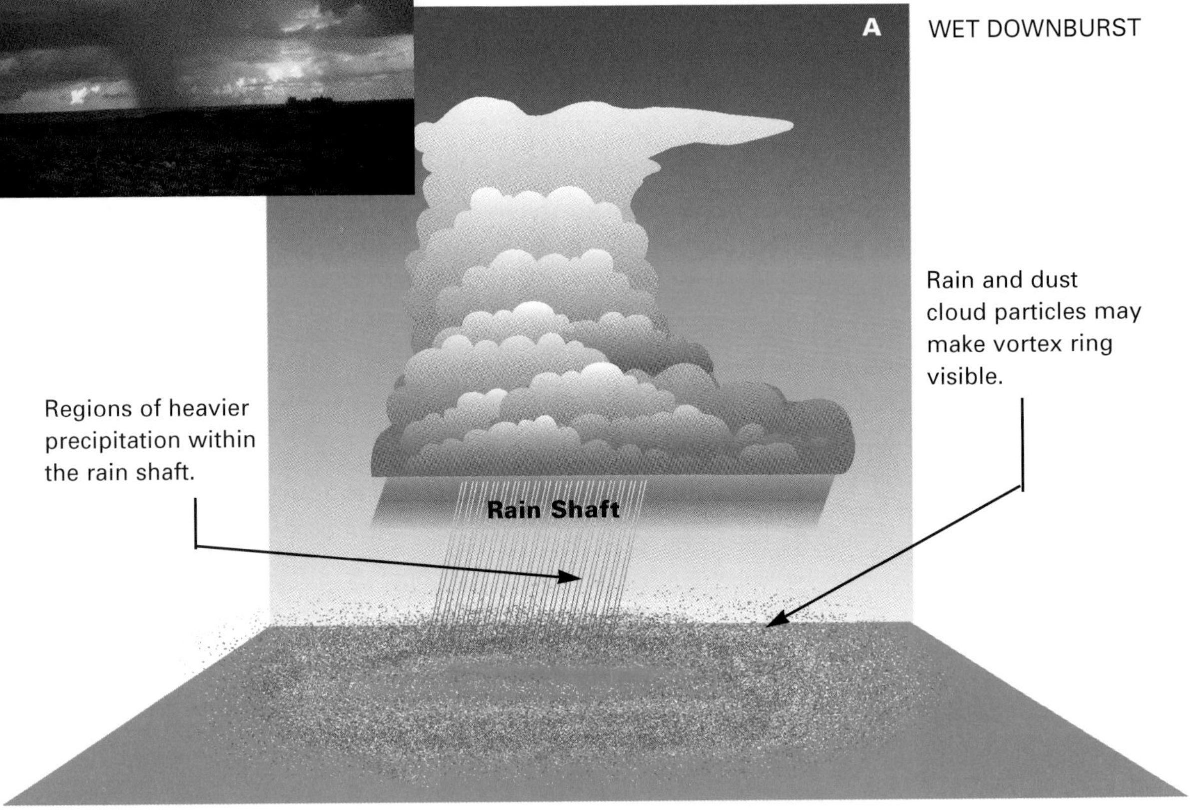

Additionally, terminal doppler weather radar (TDWR) systems are being installed across the U.S. at many vulnerable airports to provide more comprehensive wind shear monitoring. These radar systems have greater power and a narrower radar beam than conventional radar, providing greater resolution of thunderstorm details.

As we leave this brief discussion of microburst wind shear, it is important to note that a wide variety of information regarding microbursts and related flight techniques is contained in the FAA *Pilot Windshear Guide* and the FAA *Windshear Training Aid.*

Don't attempt to fly under a thunderstorm even if you can see through to the other side.

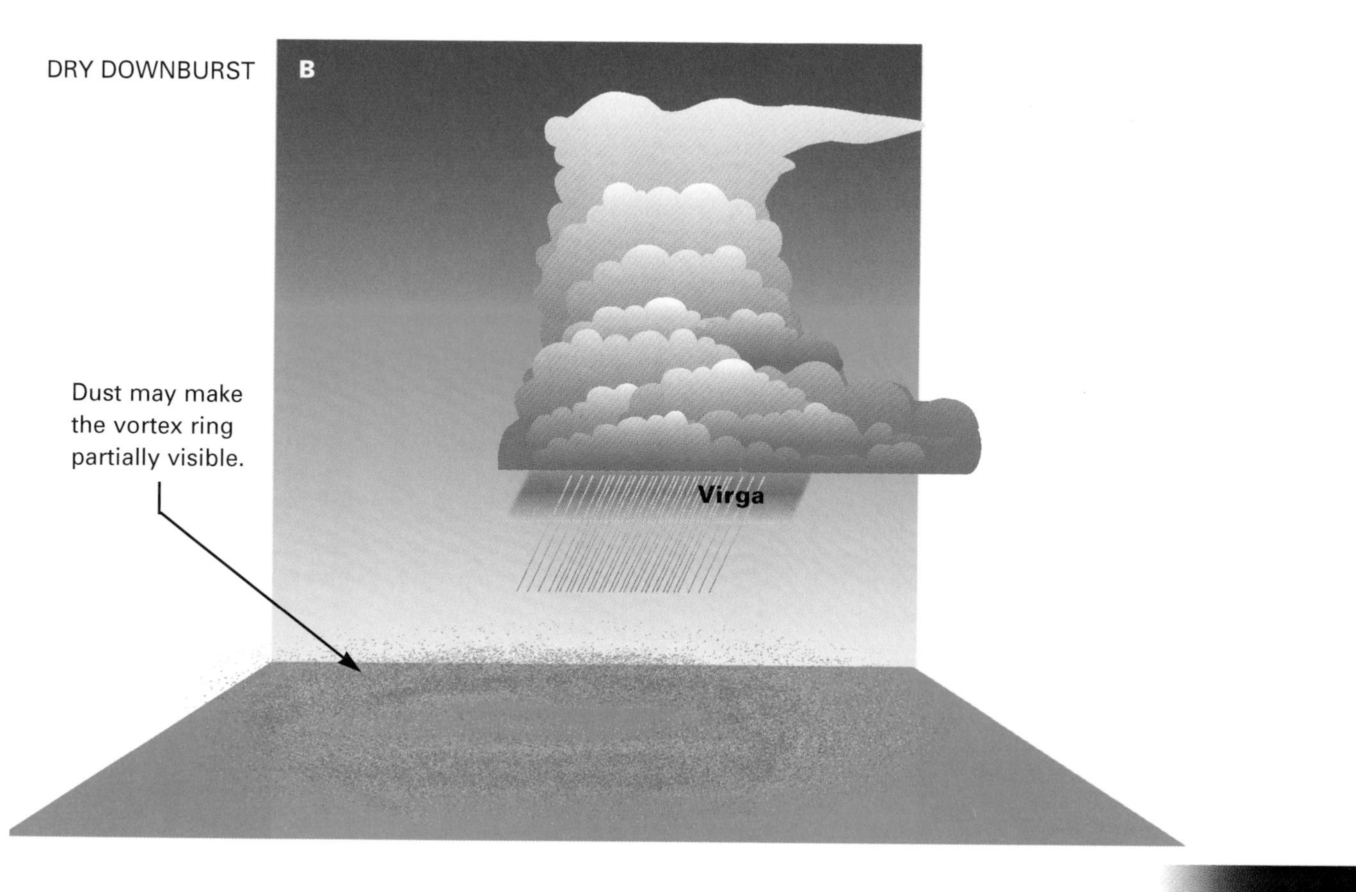

Reports of wind shear and other flight hazards are often available through pilot weather reports (PIREPs). This may be the only direct evidence of these phenomena. Pilots are encouraged to use and to report PIREPs. A PIREP is usually transmitted as an individual report, but it can be appended to a surface aviation weather report. An example of a PIREP and a PIREP code breakdown follows.

Elements	Explanations
UA or UUA — Type of Report	UA is routine PIREP; UUA is urgent PIREP
/OV — Location	In relation to VOR or route segment (station identifier, radial, DME)
/TM — Time	Coordinated Universal Time (UTC)
/FL — Altitude	Above mean sea level (MSL)
/TP — Type of aircraft	Example, CE172
/SK — Sky cover	Cloud bases and tops (both MSL), amount of coverage (scattered, broken, overcast)
/WX — Weather	Precipitation, visibility, restrictions to vision
/TA — Temperature	Degrees Celsius
/WV — Wind	Direction in degrees true, speed in knots
/TB — Turbulence	Light, moderate, severe, as appropriate
/IC — Icing	Trace, light, moderate, severe, as appropriate
/RM — Remarks	To clarify the report or for additional information

CODED PIREP:
UA/OV OKC 063064/TM 1522/FL 080/TP CE172/SK 020 BKN 045/060 OVC 070/
TA -04/WV 245040/TB LGT/RM IN CLR

DECODED PIREP:
Routine pilot report . . . 64 n.m. on the 63° radial from Oklahoma City VOR . . . at 1522 UTC . . . flight altitude 8,000 ft . . . type of aircraft is a Cessna 172 . . . base of broken layer at 2,000 feet with tops at 4,500 feet. Base of overcast layer at 6,000 feet tops at 7,000 feet . . . outside air temperature is minus four degrees Celsius wind is from 245° true at 40 kts . . . light turbulence and clear skies.

FRONTS AND SHALLOW LOWS

You already know from Part II that fronts are regions of wind shear. In fact, when we inspect the surface analysis chart, we often use the wind shift across a front as an identifying feature of its location. You should also recall that a front is a zone between two different airmasses and frontal wind shear is concentrated in that zone. Since the cold air always wedges under the warm air, the sheared frontal zone always slopes back over the cold air, regardless of the type of front. It follows that the sloping frontal zone contains both horizontal and vertical shear. (Figure 11-8)

Figure 11-8. Perspective view of a cold front. Broad arrows indicate winds. Wind shears through the sloping frontal zone are both vertical shears, along line V, and horizontal shears, along line H. The shears are concentrated in the frontal zone.

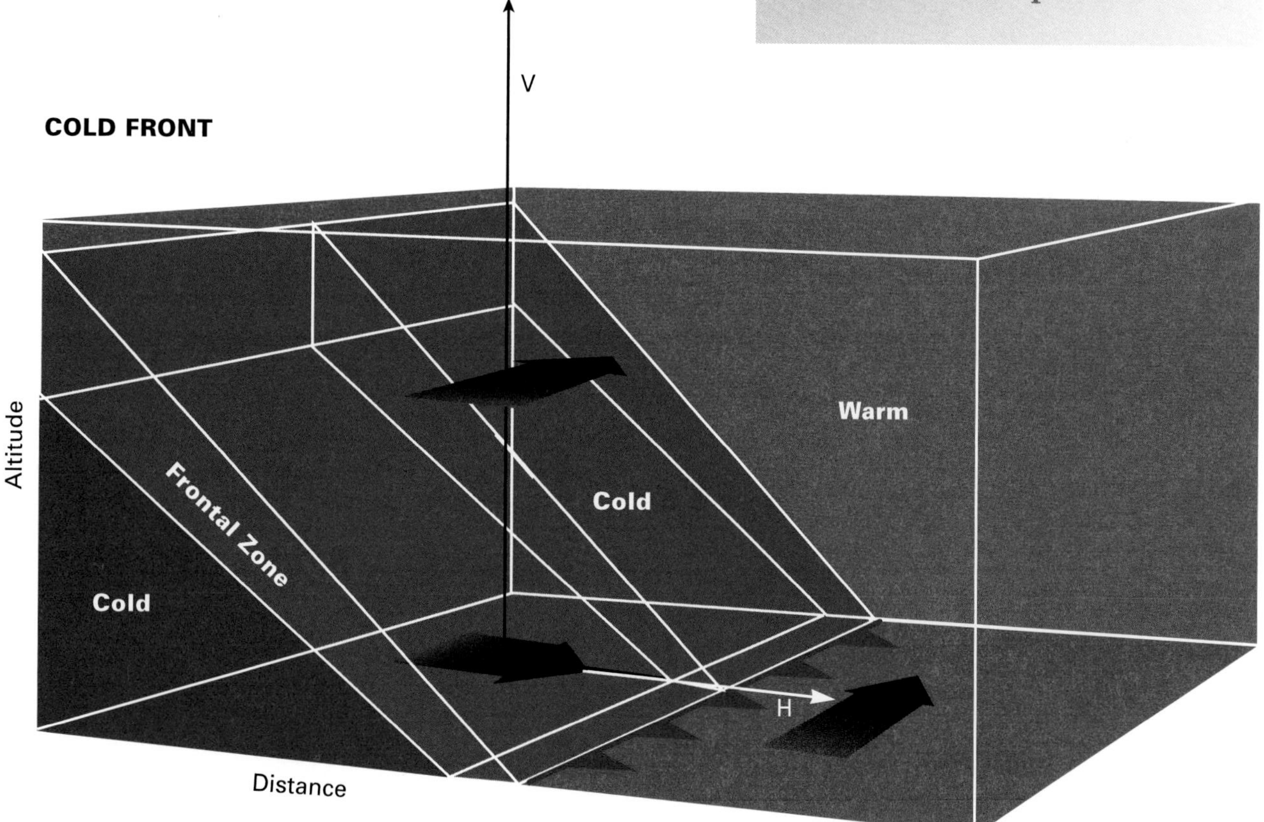

A frontal passage is reported at a weather station when the warm boundary of the frontal zone passes the station. Therefore, the onset of LLWS follows a cold frontal passage and precedes a warm frontal passage. Typical periods for critical LLWS with frontal passages are one to three hours after a cold front and up to six hours before a warm front. Wind shear with a warm front causes the most problems; because, it is often strong, lasts longer, and frequently occurs with low ceilings and poor visibilities.

The stronger the horizontal temperature change across the frontal zone, the stronger the wind shear.

With a warm front, the most critical period for LLWS is before the front passes.

Strong wind shear often occurs in shallow wave cyclones during the cooler part of the year, especially in the vicinity of the warm front. In contrast with occluded cyclones, wave cyclones in their initial stages of development may not extend to 700 mb (10,000 feet MSL). An example is shown in figure 11-9.

Wind shear is not limited to large scale fronts. It also occurs in the vicinity of mesoscale boundaries such as sea breeze fronts.

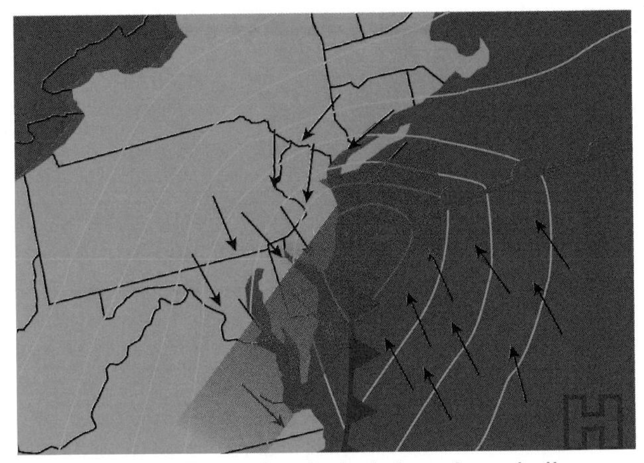

Figure 11-9. Example of wind shear in a shallow cyclone. Surface winds (thin arrows) are northeasterly ahead of the warm front, while just above the warm frontal zone, winds are strong southwesterly (broad arrow).

AIRMASS WIND SHEAR

Airmass wind shear occurs at night under fair weather conditions in the absence of strong fronts and/or strong surface pressure gradients. It develops when the ground becomes cooler than the overlying airmass as a result of radiational cooling. If the cooling is strong enough, a ground based, or surface inversion results. In this case, the temperature increases with altitude from the surface to an altitude of a few hundred feet. This layer is also known as a nocturnal inversion. (Figure 11-10)

If a temperature inversion is encountered immediately after takeoff or during an approach to a landing, a potential hazard exists due to wind shear.

The stability of the boundary layer in the nocturnal inversion hinders the mixing of faster-moving air down to the surface; therefore, surface winds tend to decrease at night. In contrast, winds above an altitude of a few hundred feet AGL often increase because they are insulated from the frictional influence of the surface by the nocturnal inversion. The result of these processes is that vertical wind shear increases through the nocturnal inversion. LLWS encountered during descent through the top of the nocturnal inversion can be particularly strong and unexpected.

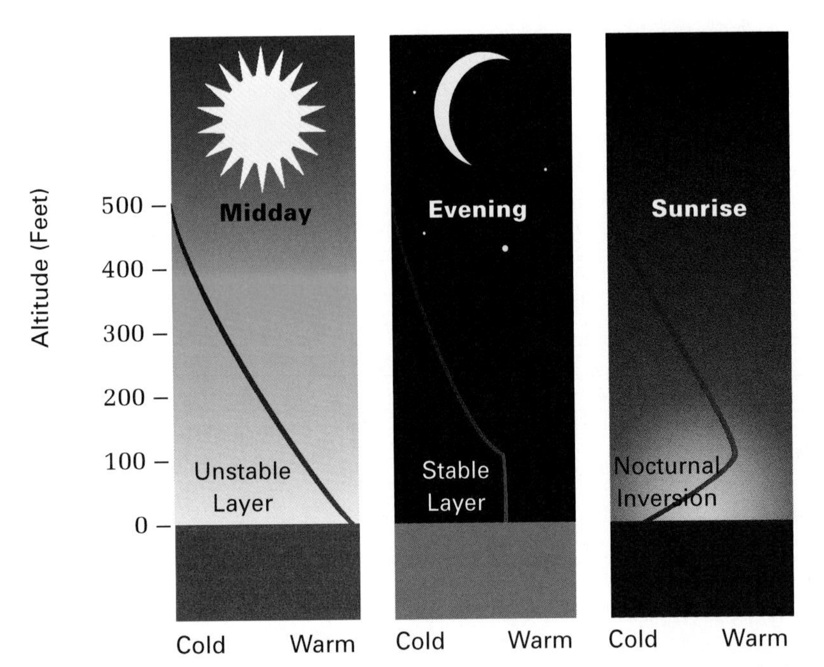

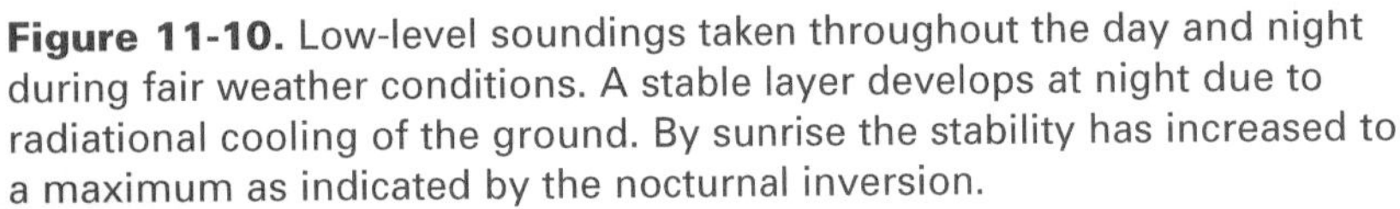

Figure 11-10. Low-level soundings taken throughout the day and night during fair weather conditions. A stable layer develops at night due to radiational cooling of the ground. By sunrise the stability has increased to a maximum as indicated by the nocturnal inversion.

After sunrise, mixing by convection destroys the nocturnal inversion. The connection between surface friction and the flow aloft is reestablished and the vertical shear weakens.

In winter, over regions of snow and ice, surface-based inversions are particularly strong and persist day and night. Caution is advised during the landing and takeoff phases, especially when winds above the inversion are strong.

A pilot can expect a wind shear zone in a temperature inversion whenever the wind speed at 2,000 to 4,000 feet above the surface is at least 25 knots.

ELEVATED STABLE LAYERS

In addition to fronts and nocturnal inversions, wind shears may be found in other stable layers in the free atmosphere, known as elevated stable layers. These are found over shallow, cool airmasses. In the warmer months, daytime convection concentrates the shear at the top of the airmass. In winter, these stable layers may extend all the way to the ground, especially over snow and ice fields.

After a cold airmass moves across a mountainous area, cold air will often remain trapped in the valleys as warmer air moves in aloft. A strong, elevated stable layer is typically found just below the mountain peaks. If strong winds are present above the mountains, there are large vertical wind shears in the stable layer; that is, between the weak, cold airflow in the valleys and warmer air flowing across the mountains. (Figure 11-11)

When a climb or descent through a stable layer is being performed, the pilot should be alert for a sudden change in airspeed.

Figure 11-11. Vertical wind shear is often found in an elevated stable layer that caps cold air trapped in a valley.

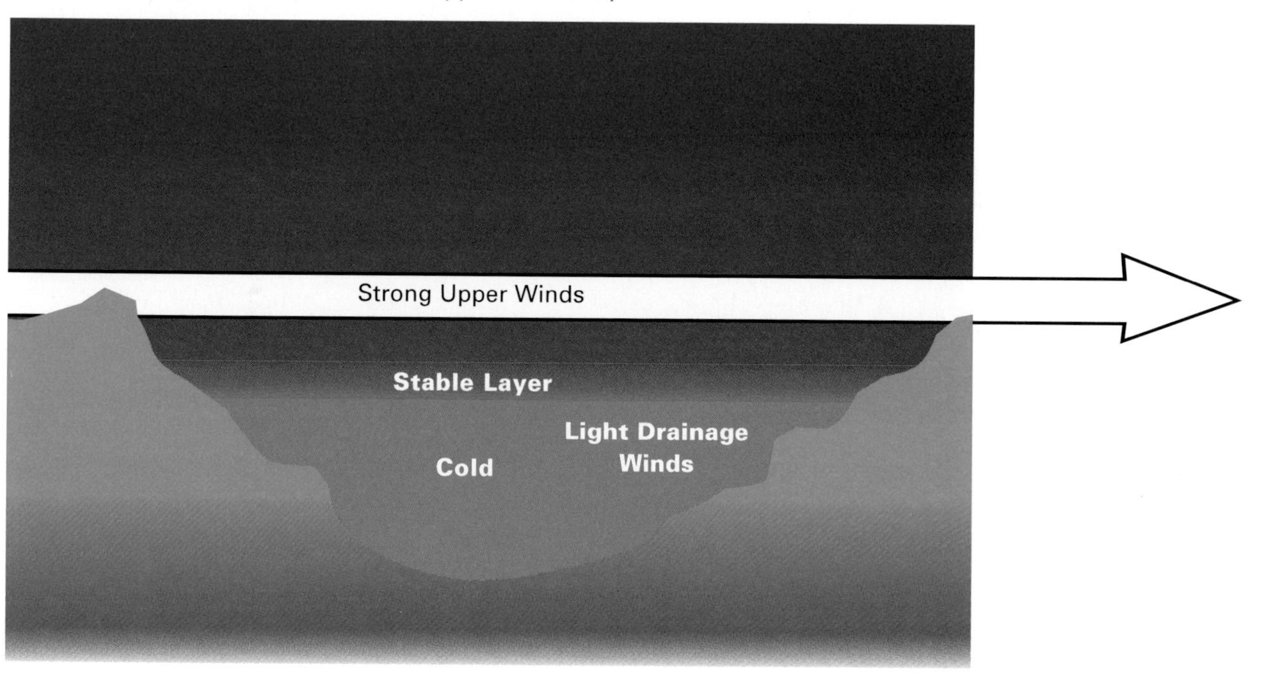

JET STREAMS

Certain patterns of upper level short wave troughs and ridges produce significant wind shear. The strongest shears are usually associated with sharply curved contours on constant pressure surfaces and/or strong winds. Regions near jet streams and within a few thousand feet of the tropopause have the highest probabilities of strong shears. Occasionally, the shear is strong enough to cause large airspeed fluctuations, especially during climb or descent. Since these sheared layers are also prime for clear air turbulence, this topic will be discussed in greater detail in the next chapter.

SUMMARY

Wind shear is one of the most serious low-level flight hazards in the atmosphere. Significant wind shear not only occurs with microbursts, but also with fronts, and nocturnal inversions. Wind shear is also found in elevated stable layers and near the jet streams. Failure to be aware of all causes and weather conditions that produce wind shear can lead to catastrophic results. An encounter with LLWS, in particular, is unforgiving because of the proximity of your aircraft to the ground. You now have some basic tools to recognize and, where possible, avoid potential wind shear conditions. These include useful conceptual models and rules of thumb. In the next chapter, you will become aware of a number of situations where wind shear and turbulence can be present at the same time in the same location.

KEY TERMS

Airmass Wind Shear
Downburst
Elevated Stable Layer
Frontal Wind Shear
Horizontal Wind Shear
Low-Level Wind Shear (LLWS)
Low-Level Wind Shear Alert System (LLWAS)
Microburst
Nocturnal Inversion
Terminal Doppler Weather Radar (TDWR)
Vertical Wind Shear
Vortex Ring
Wind Shear

CHAPTER QUESTIONS

1. The wind at 200 feet AGL is 330° at 15 knots. The surface wind is 240° at 15 knots.

    1. What is the wind speed difference over the 200-foot layer?
    2. What is the wind shear, magnitude only, between the surface and 200 feet AGL?
    3. What is the severity of the LLWS? Explain how you determined the severity.
    4. What is the severity of the LLWS in Diagram A of figure 11-1? Show all of your work.

2. You are taxiing out in preparation for takeoff. An isolated rain shower can be seen over the opposite end of the runway. There is no thunder or lightning.

    1. Should you take off?
    2. Why?
    3. If you elect not to take off, about how long will you have to wait until you can go? Explain.

3. At just about sunrise, a pilot is descending to land on an island. The airstrip is on the north shore of the island, between the ocean and a range of volcanic peaks. The prevailing winds are easterlies. On final approach from the west, strong LLWS is encountered at 100 feet. Explain why this happened and provide appropriate sketches.

4. The critical period for a low-level wind shear hazard is longer for a warm front than for a cold front. Why? There are two reasons. A sketch will help.

5. Perform the following experiment to simulate the structure and behavior of a downburst. You need an eye dropper of whole milk and a tall glass of water. Be sure the water is not moving. Place the end of the eye dropper close to the surface of the water and release a single drop. Make a sketch and describe the results.

    1. What will happen if you use skim milk?
    2. What does this say about the intensity of downbursts?

6. Decode the following PIREPs.

    1. UA/OV TOL/TM 2200/FL 240/TP UNKN/TB MDT CAT 180-240.

    2. . . . . /RM LLWS -15KT SFC-030 DURC RNWY 22 JFK.

Introduction

A characteristic of most naturally occurring fluids is that they contain some degree of turbulence. This means that you usually can find some part of the fluid where the velocities are fluctuating in a chaotic manner. The atmosphere is one of those fluids. The velocity fluctuations found within the atmosphere are often weak and barely noticeable in flight. Occasionally, however, the turbulence is of such a magnitude that passengers and crew are injured and the aircraft is damaged or destroyed.

The purpose of this chapter is to provide information that will help you avoid or at least minimize the effects of turbulence on your flight. When you complete this chapter, you will understand the basic types of turbulence and their causes, and you will know the large scale conditions under which turbulence occurs. Also, you will have learned some rules of thumb that will help you to anticipate and deal with the turbulence problem.

Section A

TURBULENCE DEFINED

Based on descriptions from pilots, crew, and passengers, aviation turbulence is best defined simply as

"bumpiness in flight."

It is important to notice that this definition is based on the response of the aircraft rather than the state of the atmosphere. This means that the occurrence of aviation turbulence can be the result of not only disorganized turbulent motions, but also organized small scale circulations. In addition, the magnitude of the bumpiness in flight depends on aircraft design and pilot reactions. For convenience, we will use the term turbulence for "aviation turbulence" except where the meaning is ambiguous.

AIRCRAFT AND PILOT RESPONSE

Since the identification of turbulence depends on the effects on the aircraft and pilot, we will begin our discussion by examining the nature of those effects. This will help you understand why certain atmospheric circulations are more closely associated with turbulence than others.

If the sizes of the atmospheric circulations (eddies) through which an aircraft is flying are large enough, the pilot has time to climb, descend, or divert in order to avoid any adverse effects of the circulations. On the other hand, if the eddies are sufficiently small, the aircraft will pass through them before they can have any significant influence. The circulations that cause turbulence fall between these two size ranges. For most aircraft flying today, the horizontal dimensions (or scales) of turbulence-producing eddies are 50 feet to 8,000 feet. If turbulent eddies were circular or spherical (oversimplifications), these dimensions would correspond to the eddy diameters.

Atmospheric motions produced by turbulent eddies are often referred to as turbulent gusts. In general, vertical gusts are more likely to have a larger impact on flight than horizontal gusts, because they change the angle of attack and lift. However, in certain phases of flight (landing and takeoff), and in some circulations (lee waves and rotors), horizontal gusts may be as important as vertical gusts.

The pilot's influence on aviation turbulence and the influence of turbulence on the pilot can arise in a number of ways. Turbulence of any intensity is, at best, uncomfortable. A pilot exposed to turbulent conditions for long periods of time will experience greater fatigue. When the frequency of aircraft shaking is very large (4-5 cycles per second), the pilot cannot read the instruments. If the frequency is near one cycle per four seconds, air sickness may result. All of these effects, together with experience and ability, affect the pilot's response to the turbulence. If the pilot (or autopilot) overreacts, control inputs may actually add to the intensity of bumpiness. The latter actions are known as maneuvering.

TURBULENCE MEASURES

How is turbulence characterized? More practically, if you fly through a turbulent area, what do you report? By far, the most important property of turbulence is its intensity. The most common turbulence reporting criteria are shown in figure 12-1. Turbulence intensity varies from light to extreme, and is related to aircraft and crew reaction and to the movement of unsecured objects about the cabin.

The turbulence scale in figure 12-1 has been used for many years and is the basis of most pilot weather reports. The criteria are highly subjective and are dependent on aircraft type, airspeed, and pilot experience. PIREPs of turbulence should be cautiously interpreted. For example, turbulence reports from large aircraft are often less intense than those from small aircraft in the same area at the same time.

Quantitative indications of turbulence intensity can be determined from the on-board measurements of g-load, airspeed fluctuations, and rate-of-climb. G-load (or gust load) is the force that arises because of the influence of gravity.

TURBULENCE REPORTING CRITERIA TABLE			
Intensity	Aircraft Reaction	Reaction Inside Aircraft	Reporting Term-Definition
Light	Turbulence that momentarily causes slight, erratic changes in altitude and/or attitude (pitch, roll, yaw). Report as Light Turbulence;* or Turbulence that causes slight, rapid and somewhat rhythmic bumpiness without appreciable changes in altitude or attitude. Report as Light Chop.	Occupants may feel a slight strain against seat belts or shoulder straps. Unsecured objects may be displaced slightly. Food service may be conducted and little or no difficulty is encountered in walking.	Occasional—Less than 1/3 of the time. Intermittent—1/3 to 2/3 Continuous—More than 2/3
Moderate	Turbulence that is similar to Light Turbulence but of greater intensity. Changes in altitude and/or attitude occur but the aircraft remains in positive control at all times. It usually causes variations in indicated airspeed. Report as Moderate Turbulence;* or Turbulence that is similar to Light Chop but of greater intensity. It causes rapid bumps or jolts without appreciable changes in aircraft altitude or attitude. Report as Moderate Chop.	Occupants feel definite strains against seat belts or shoulder straps. Unsecured objects are dislodged. Food service and walking are difficult.	NOTE: 1. Pilots should report location(s), time (UTC), intensity, whether in or near clouds, altitude, type of aircraft and, when applicable, duration of turbulence. 2. Duration may be based on time between two locations or over a single location. All locations should be readily identifiable. EXAMPLES: a. Over Omaha, 1232Z, Moderate Turbulence, in cloud, Flight Level 310, B707. b. From 50 miles south of Albuquerque to 30 miles north of Phoenix, 1210Z to 1250Z, occasional Moderate Chop, Flight Level 330, DC8.
Severe	Turbulence that causes large abrupt changes in altitude and/or attitude. It usually causes large variations in indicated airspeed. Aircraft may be momentarily out of control. Report as Severe Turbulence.*	Occupants are forced violently against seat belts or shoulder straps. Unsecured objects are tossed about. Food service and walking are impossible.	
Extreme	Turbulence in which the aircraft is violently tossed about and is practically impossible to control. It may cause structural damage. Report as Extreme Turbulence.*		
	* High level turbulence (normally above 15,000 feet MSL) not associated with cumuliform cloudiness, including thunderstorms, should be reported as CAT (Clear Air Turbulence) preceded by the appropriate intensity, or light or moderate chop.		

Figure 12-1. Turbulence reporting criteria. Standard turbulence symbols that are used on aviation weather charts are shown on the left side of the figure.

Normal gravity corresponds to a g-load of 1.0g. A change in g-load above or below the normal value is a rough measure of the intensity of the turbulence. For example, if an aircraft experiences a total g-load of +1.5g, it means that associated turbulence (or maneuvering) caused an excess load of +0.5g.

Airspeed fluctuations refer to the largest positive and negative airspeed deviations from the average during a turbulent event. For example, if your average airspeed is 140 knots with variations between 130 and 150 knots, you are experiencing fluctuations of ± 10 knots.

Until actually flying in an area of reported turbulence, the severity of the turbulence should be interpreted as being at or above the reported values. Never downgrade reported severities.

Rate of climb simply refers to the largest positive or negative values during a turbulent event. Figure 12-2 relates various categories of these variables to the turbulence reporting criteria.

The indicated rate of climb induced by a turbulent gust can only be used as a very rough estimate of the vertical gust speed, because it includes both the effect of the vertical gust and the motion of the aircraft. "Derived equivalent gust velocity" is a theoretical estimate of the vertical gust required to produce a given incremental change in gust load.

	Airspeed Fluctuation (kts)	G-load (g)	Derived Gust (f.p.m.)
Light	5 – 14.9	0.20 – 0.49	300 – 1199
Moderate	15 – 24.9	0.50 – 0.99	1200 – 2099
Severe	≥25	1.0 – 1.99	2100 – 2999
Extreme	—	≥2.00	≥3000

Figure 12-2. Quantitative measures of turbulence intensity. Values may be positive or negative.

Section B

TURBULENCE CAUSES AND TYPES

Aviation turbulence can be divided into four categories, depending on where the turbulence occurs, what larger scale atmospheric circulations are present, and what is producing the turbulence. They are:

1. Low-level turbulence (LLT)
2. Turbulence in and near thunderstorms (TNT)
3. Clear-air turbulence (CAT)
4. Mountain wave turbulence (MWT)

LOW-LEVEL TURBULENCE (LLT)

For operational purposes, "low-level turbulence" is often defined simply as turbulence below 15,000 feet MSL. We will be a little more specific in order to concentrate on turbulence causes. Low-level turbulence (LLT) is defined here as that turbulence which occurs primarily within the atmospheric boundary layer. Recall from Chapters 4 and 9 that the boundary layer is the lowest few thousand feet of the atmosphere; that is, where surface heating and friction influences are significant. LLT includes mechanical turbulence, thermal turbulence, and turbulence in fronts. Although wake turbulence may be encountered at any altitude, it is particularly hazardous near the ground, so it is also considered with LLT. For discussion purposes, turbulence which occurs near the ground in thunderstorms and in other moist convection is included with turbulence in and near thunderstorms. Turbulence which occurs near the ground in mountain wave conditions is included under mountain wave turbulence.

MECHANICAL TURBULENCE

Over flat ground, significant LLT occurs when surface winds are strong. This is called mechanical turbulence. It occurs because friction slows the wind in the lowest layers causing the air to turn over in turbulent eddies. (Figure 12-3) The turbulent eddies cause fluctuations (gusts) in winds and vertical velocities.

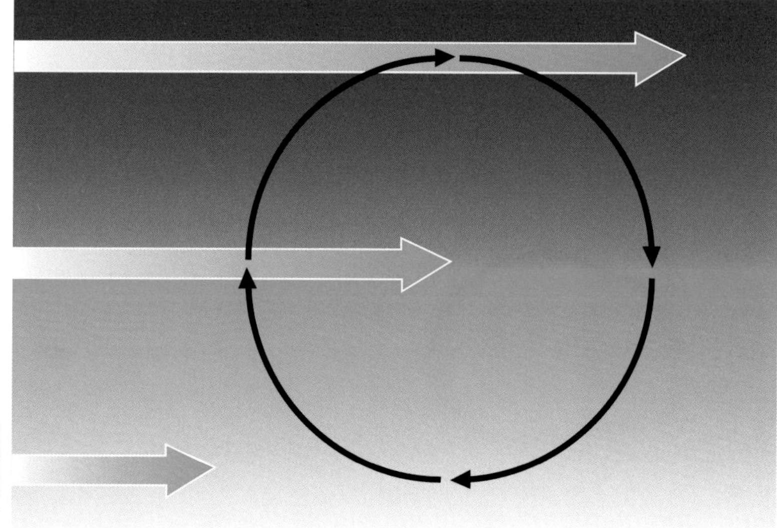

Figure 12-3. The broad arrows indicate sustained winds increasing with altitude. The circulation represents a turbulent eddy caused by surface friction. The observed wind field is a combination of the sustained wind and the eddy. Turbulent eddies mix stronger winds downward and weaker winds upward causing gustiness (LLT) and large fluctuations in wind shears. This effect increases with wind speed.

Turbulent eddies which are swept along by the sustained wind also cause rapid fluctuations in the wind shear near the ground. Fluctuating shears and turbulence contribute to rough approaches and takeoffs.

With strong winds over flat land, maximum surface wind gusts are typically 40% stronger than the sustained wind.

As the winds become stronger, the mechanical turbulence extends to greater heights above the ground. When surface wind gusts are 50 knots or greater, significant turbulence due to surface effects can reach altitudes in excess of 3,000 feet AGL.

When the sustained surface wind exceeds 20 knots, airspeed fluctuations of 10 to 20 knots will occur on approach. When the surface wind over flat land exceeds 30 knots, mechanically produced LLT will be moderate or greater.

The presence of obstructions such as buildings and stands of trees increase the effect of surface roughness and strengthen LLT and wind shear. (Figure 12-4) Typically, a trail of turbulent eddies is produced downwind of an obstacle with a sheared layer between the ground-based turbulent region and smooth flow aloft. This is generally referred to as a turbulent wake. When hangers and other large buildings are near an airport, there is a potential for additional control problems on takeoff or landing in strong winds.

The type of approach and landing recommended during gusty wind conditions is a power-on approach and power-on landing.

Hills can produce some very strong turbulent wakes with

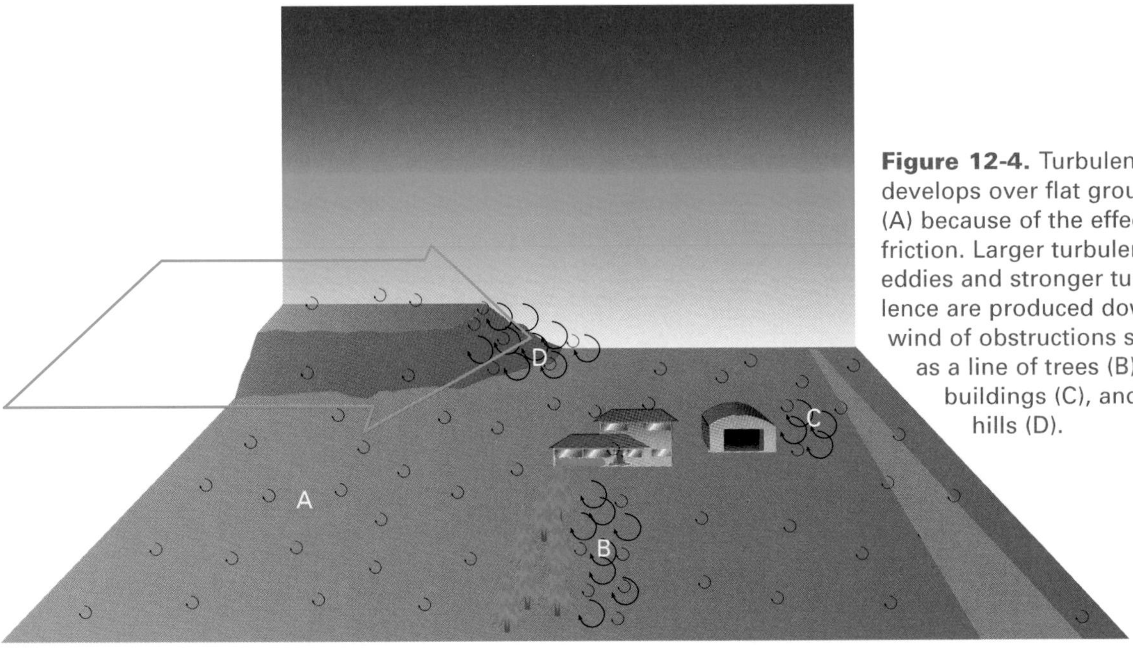

Figure 12-4. Turbulence develops over flat ground (A) because of the effect of friction. Larger turbulent eddies and stronger turbulence are produced downwind of obstructions such as a line of trees (B), buildings (C), and hills (D).

strong winds. In comparison with turbulence over flat ground, the turbulent eddies downwind of hills are larger because the obstructions that

When taking off from a valley, climb above the level of the highest peaks before leaving the valley. Maintain lateral clearance from the mountains, sufficient to recover if caught in a downdraft.

cause them (the hills) are larger. The resulting wind shears and turbulence are also stronger. The nature of the turbulence also depends on the shape of the topography. Steep hillsides encourage the flow to separate from the surface, producing eddies, LLT, and sheared regions.

Care must be taken when flying in a valley with strong crosswinds aloft. There is often a distinct upslope and downslope flow on either side of the valley, as shown in figure 12-5. However, in a very narrow canyon, turbulent gusts can be treacherous because of their strength and the limited space for maneuvering. (Figure 12-5)

As an airstream crosses a ridge line, wind speeds and wind shears are frequently greater near the peaks than at the same altitude over the nearby flatlands. This occurs because the depth of the airstream is reduced as it flows over the peaks. In order to move the same mass of air through the smaller depth, the air must speed up. This is similar to the increase in the speed of the current of a river when it narrows. Similarly, strong local winds with substantial LLT and wind shear are created when a broad airstream is forced to flow through a narrow mountain pass.

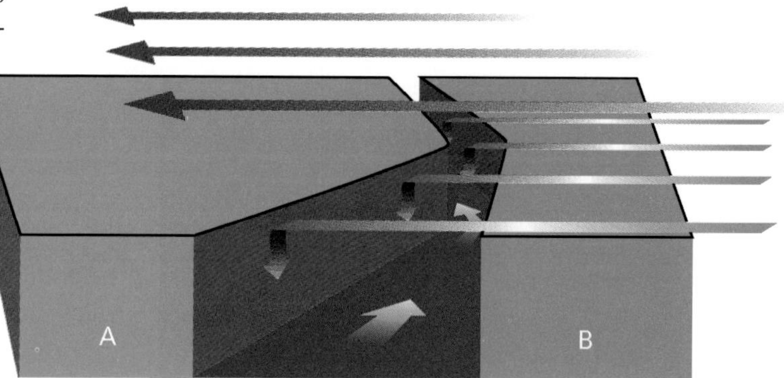

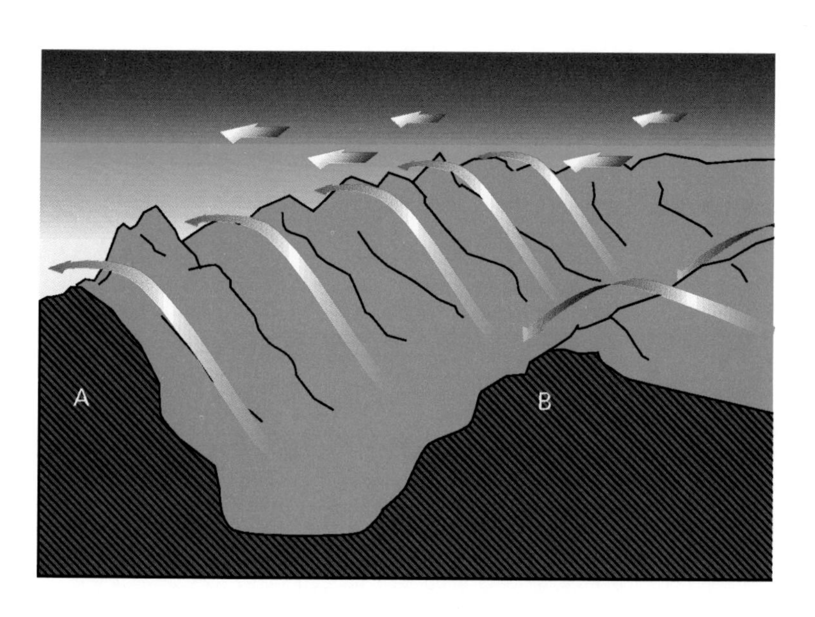

Figure 12-5. Cross valley winds on a gentle sloping valley (left) produce upslope winds on side A and downslope on side B. Conditions may be significantly different in a very narrow canyon with strong crosswinds (top). In this case, turbulence or downward motions occur on side A. Additionally, airflow within a narrow canyon is often turbulent at locations near the bottom, where there are sharp bends, and where side canyons intersect.

Strong winds due to this funneling effect may extend well downstream of the pass. (Figure 12-6)

For larger hills and mountains, stable airflow across peaks and ridges creates lee waves in addition to the effects described in the preceding paragraphs. As you will see in a later section, the lee wave system is very effective in producing widespread turbulence in certain layers of the atmosphere.

When there is a strong pressure gradient across a mountain range, flight through a mountain pass at low levels may expose the aircraft to strong winds (blowing toward the lowest pressure), strong shears, and turbulence.

The intensity of LLT increases with wind speed and steepness of the terrain. Over rough terrain, moderate or greater LLT is likely when sustained winds exceed 25 knots, and severe turbulence can be expected with winds of 40 knots or more.

THERMAL TURBULENCE

Thermal turbulence is LLT produced by dry convection in the boundary layer. As described in Chapter 5, it is typically a daytime phenomena that occurs over land under fair weather conditions. Solar radiation heats the ground generating convection at the bottom of the boundary layer. During the morning and early afternoon, the convection intensifies and deepens. It reaches a maximum in the afternoon, then gradually dies out as the earth's surface cools. Where cool air moves over warm land or water, thermal turbulence can occur any time of day or night.

As indicated by the name, thermals are the basic elements of thermal turbulence. You were introduced to them as the "roots" of cumulus clouds in Chapter 9. They are an important source of LLT. Thermals are rising warm "bubbles" of air. They typically develop near the ground where conditions are somewhat chaotic. Thermal plumes, narrow curtains of rising air, and dust devils are common close to the ground, especially when the ground is very hot. All of these are LLT sources.

As thermals move away from the ground, they gain upward speed, grow in size, and become more organized. They may be aligned in patterns depending on winds and terrain. Over flat terrain, in light winds, thermals are often arranged in a "honeycomb" pattern. With stronger winds (about 20 knots) they frequently form lines along the wind. With very strong winds, thermal patterns are chaotic. In this case, the combination of mechanical and thermal effects usually produces strong LLT.

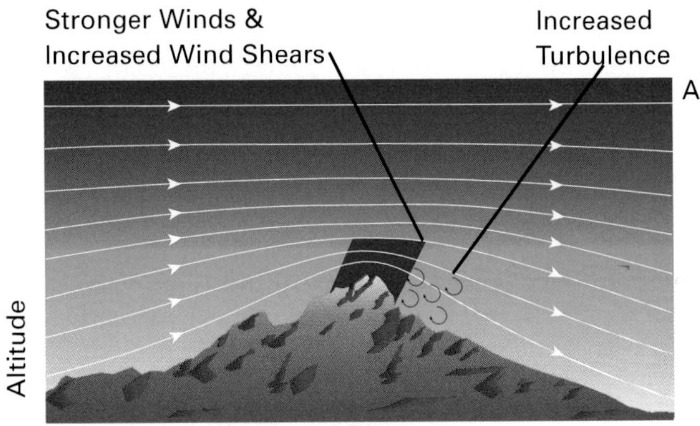

Figure 12-6. Cross section through a ridge line (A). Thin arrows indicate air flow. Winds and wind shears are stronger immediately over the mountain top (shaded). Winds, turbulence, and wind shears are stronger through and downwind of the mountain pass (B).

B

As you might suspect from our previous discussion of slope circulations, thermal turbulence is common over higher terrain. Thermals tend to be narrower and stronger over sun-facing slopes and LLT is stronger in these areas.

The characteristics of an unstable cold airmass moving over a warm surface are cumuliform clouds, turbulence, and good visibility. A stable airmass is most likely to have smooth air.

Glider pilots have long taken advantage of the upward motions in thermals to gain altitude and fly long cross-country distances. However, thermal sources of lift for slow-moving gliders are often sources of LLT for powered aircraft. An aircraft flying through dry convection is commonly exposed to turbulence intensities ranging from light to moderate. Typical upward gusts in thermals range from 200 to 400 f.p.m., with extremes of 1,000 to 2,000 f.p.m. reported. Flight through the boundary layer at midday in the summer will expose you and your aircraft to frequent (and uncomfortable) LLT due to thermals. It is not unusual for a PIREP from a low-level daytime flight in such conditions to read "CONT LGT-MDT TURBC"; that is, continuous light to moderate turbulence.

Relief from the continuous bumpiness of thermals can often be found by climbing into and above what is called the capping stable layer. This is a layer caused by a very slowly sinking motion aloft. It is typically associated with a macroscale high pressure region. The capping stable layer is at the top of the dry convection. As your aircraft climbs through it, there is a sudden cessation of turbulence. The height of the capping stable layer is typically a few thousand feet above the ground. Over desert terrain in the summer, the top of the dry convection can exceed 10,000 feet AGL.

The base of the capping stable layer is often visible as the distinct layer of haze and dust carried upward by thermals from below. If cumulus clouds are present, the haze layer is at the base of the clouds with the cloud tops extending into the lower part of the capping stable layer. Care must be exercised when descending into the convective layer from the smooth air above; the onset of turbulence is rapid and may cause problems for the unsuspecting pilot. (Figure 12-7)

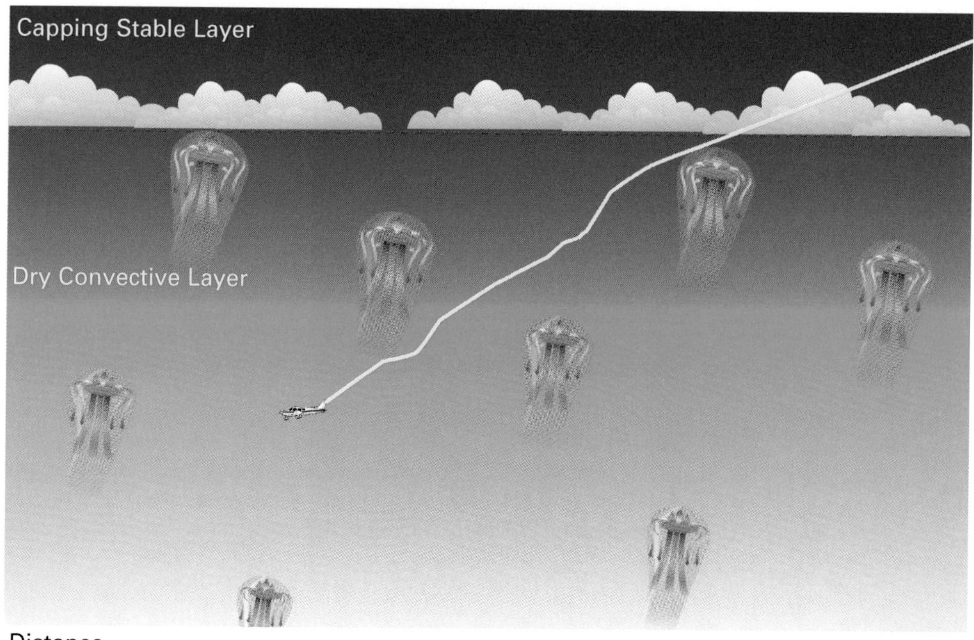

Figure 12-7. An aircraft descending from the capping stable layer through a layer of CU will encounter persistent LLT due to dry thermals in the convective boundary layer. Your only visual cue to the existence of LLT may be a CU layer, haze, or dust. The thermals may be tilted or otherwise distorted due to vertical wind shear.

Fair-weather LLT produced by thermals may be modified or interrupted by frontal passages, extensive cloud cover, and wet or snow-covered surfaces. Also, when instability is very large, moist convection and thunderstorms are often the result. The boundary layer is greatly modified under these conditions and LLT is primarily the result of thunderstorm activity.

Generally, over flat ground, any front moving at a speed of 30 knots or more will generate moderate or greater LLT. However, over rough terrain, all fronts should be assumed to have moderate or greater turbulence, regardless of their speed.

TURBULENCE IN FRONTS

Fronts are not only sources of wind shear but they also can produce moderate or greater turbulence. In the boundary layer, fast-moving cold fronts are usually steeper than at higher levels, and updrafts may reach 1,000 f.p.m. in a narrow zone just ahead of the front. When these conditions are combined with convection and strong winds across the front, LLT and wind shear can produce serious flight hazards over a broad area. (Figure 12-8)

Mesoscale frontal zones cause turbulence in the same manner as the macroscale fronts previously described. The intensity of the turbulence also depends on the strength and speed of the fronts. Examples of mesoscale fronts are the sea breeze front and the thunderstorm gust front. The gust front is discussed in more detail under the topic of turbulence in and near thunderstorms. Macroscale frontal zones found in the middle and upper troposphere are also sources of turbulence. These are discussed later in connection with jet streams.

WAKE TURBULENCE

As most pilots know, turbulent wakes are generated by aircraft in flight. This can be considered a form of mechanical turbulence. However, rather than the air blowing past an obstacle, the obstacle (in this case the wing of the aircraft) is moving through the air. The result is still the same, a turbulent wake is produced behind the obstacle. The term vortex or

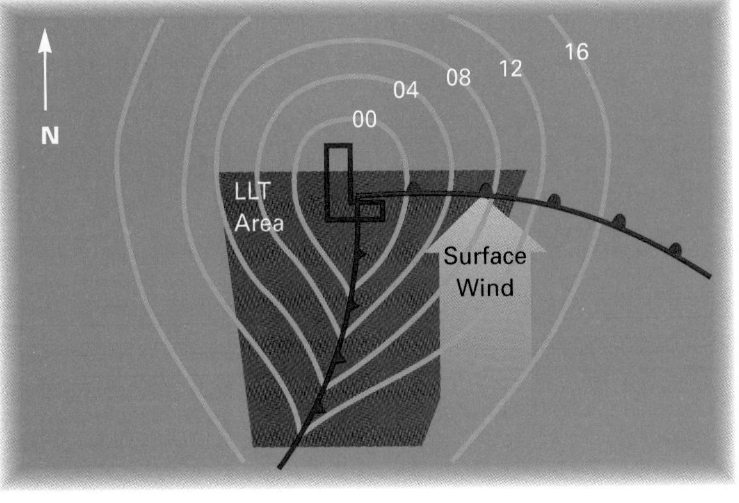

Figure 12-8. The primary LLT area associated with a typical wave cyclone is shaded. The area extends from about 200 miles behind the cold front through the region of strong southerly winds ahead of the front.

wake turbulence is applied to the turbulence that forms behind an aircraft that is generating lift. (Figure 12-9)

In contrast to other types of mechanical turbulence, wake turbulence is somewhat more predictable since all aircraft generate lift and lift is a requirement to generate wake turbulence. Because most aircraft have the same basic shape, the vortices they produce tend to be similar; however, they vary widely in intensity based on aircraft size, wing configuration, weight, and speed. Slow, heavy, large aircraft produce larger vortices and stronger turbulence than small aircraft. Wake turbulence is caused by high pressure air under the wing flowing toward lower pressure above the wing near the wingtip. This process creates two counter-rotating vortices that trail behind aircraft. Vortex generation begins near liftoff and ends at touchdown. As shown in figure 12-9, each vortex provides downwash inboard of the wingtips, and upward flow outboard.

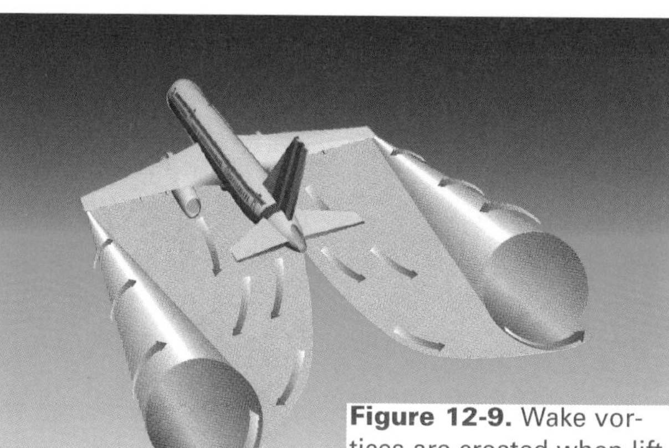

Figure 12-9. Wake vortices are created when lift is generated by the wing of an aircraft. A similar pattern is also created by the main rotor of a helicopter when it is producing lift.

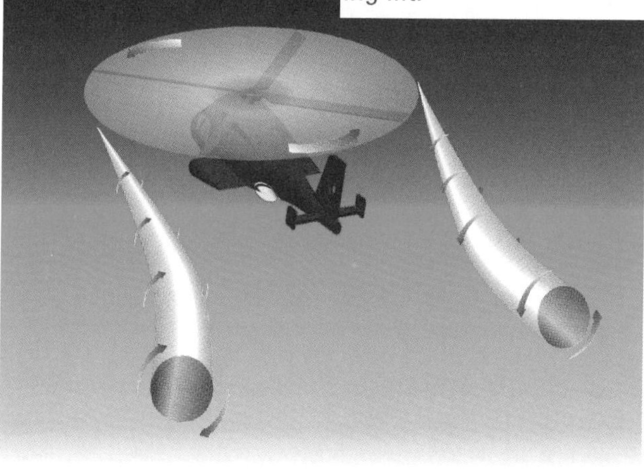

When fully formed at a distance of about two wingspans behind the aircraft, the vortices are typically 25 to 50 feet in diameter; their actual size depends on the wing dimensions. They tend to remain about three quarters of the wing span apart. The two vortices descend 800 to 900 feet below the aircraft within about two minutes and remain at that level until they dissipate. They weaken with distance from the generating aircraft. Atmospheric turbulence also hastens their dissipation.

The greatest vortex strength occurs when the generating aircraft is heavy, clean, and slow. Wake turbulence is near maximum behind a jet transport just after takeoff because of the high angle of attack and high gross weight.

If the vortices reach the ground, they will move outward from the aircraft at about five knots in calm wind conditions. If the wind is blowing, the net movement of the eddies will be the sum of the wind velocity and the "no-wind" motion of each vortex. Thus, a light crosswind could cause one vortex to remain nearly stationary over the runway.

The wind condition that prolongs the hazards of wake turbulence on a landing runway for the longest period of time is a light quartering tailwind.

The flight hazard of wake turbulence is obvious. A small aircraft departing or arriving too close to a large aircraft may encounter vertical and horizontal gusts which cannot be compensated for by any flight maneuver. This is especially true when a trailing aircraft enters directly into a vortex. The maximum gusts in the wake of the aircraft occur in the cores of the vortices. Gusts of more than 10,000 f.p.m. have been reported. Flight into the downwash region between vortices will also cause serious difficulties when the trailing aircraft is taking off or landing. Another hazardous situation occurs when an aircraft intersects the wake at a large angle. In this case, an aircraft would encounter a vortex, the downwash, and then another vortex in close succession.

When landing behind a large aircraft, the pilot should avoid wake turbulence by staying above the large aircraft's final approach path and landing beyond the large aircraft's touchdown point. When departing behind a heavy aircraft, the pilot should avoid wake turbulence by maneuvering the aircraft above and upwind from the heavy aircraft.

Wake turbulence is not restricted to "heavy" aircraft. Even light aircraft can cause significant wake turbulence. For example, there are cases where an aircraft involved in aerial spraying and flying near stall speed, has crashed after intersecting its own wake from a previous spraying swath. Specific flight procedures have been developed to avoid the effects of wake turbulence. The details of these have been published widely in FAA manuals, circulars, and other training media.

TURBULENCE IN AND NEAR THUNDERSTORMS (TNT)

Turbulence in and near thunderstorms (TNT) is that turbulence which occurs within developing convective clouds and thunderstorms, in the vicinity of thunderstorm tops and wakes, in downbursts, and in gust fronts.

A model of a developing thunderstorm was presented in Chapter 9. It also serves as the basic model for the discussion of TNT. Figure 12-10 shows the primary turbulence regions of a single thunderstorm cell in the mature stage.

The best advice that can be given about flight in or near thunderstorms is "Don't!" But if you inadvertently fly into a thunderstorm, set your power for turbulence penetration speed and maintain a wings-level flight attitude.

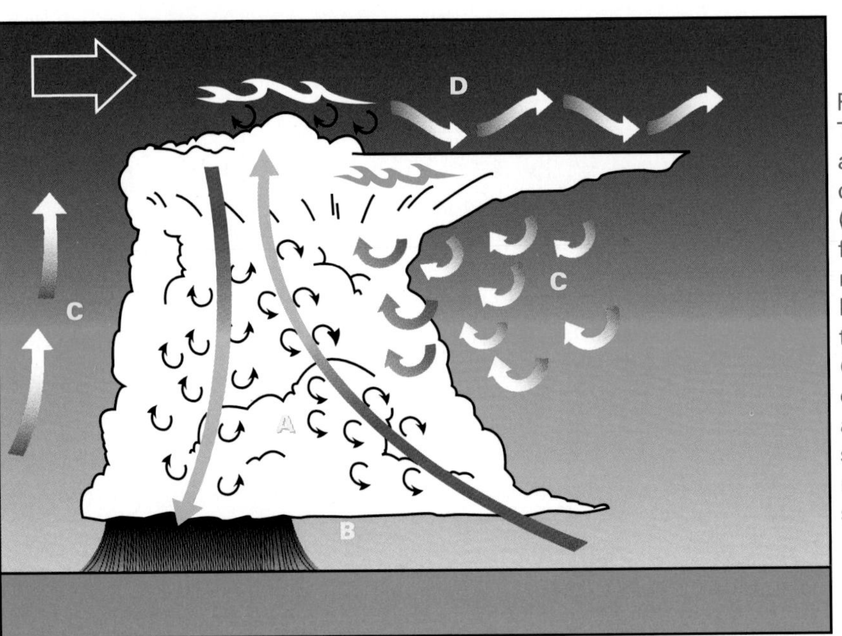

Figure 12-10. Turbulence in and near thunderstorms (TNT) occurs in four distinct regions: turbulence within thunderstorms (A), below thunderstorms (B), around thunderstorms (C), and near thunderstorm tops (D).

TURBULENCE WITHIN THUNDERSTORMS

Turbulence within the thunderstorm cloud boundaries is caused by the strong updrafts and downdrafts. The most frequent and, typically, the most intense TNT is found within the cloud (although turbulence below the cloud may have more disastrous consequences). Furthermore, it is made worse because it occurs in instrument meteorological conditions with heavy rain, lightning, and possible hail and icing. The combination of these hazards increases the chances of disorientation and loss of control, major factors in many fatal general aviation accidents in thunderstorms.

All thunderstorms should be considered hazardous, but if the thunderstorm top exceeds 35,000 feet MSL, it should be regarded as extremely hazardous.

Turbulence inside thunderstorms occurs on at least two different scales. The largest eddies have sizes comparable to the major updrafts and downdrafts. Small scale gusts are produced by strong shears on the edges of the vertical drafts.

Assume that moderate or greater turbulence is present while the cumulonimbus cloud continues to show some degree of organization. Also, beware of multicell thunderstorms. A nearby mature cell may be obscured by the clouds of a dissipating cell.

In the cumulus stage of thunderstorm development, the turbulence inside the storm is due to the updraft, which usually occupies less than half the cloud area. Updraft speeds increase from the base of the cloud to a maximum near the top of the cloud.

In the mature stage, updraft speeds accelerate through the depth of the storm, reaching a maximum in the upper part of the cell at the equilibrium level. This is often (but not always) near the tropopause. Because of the rapid rate of rise of the cloud tops, aircraft flying just below the tropopause are occasionally surprised with a strong burst of turbulence as the cloud top reaches the flight level.

Updraft speeds in the mature stage of the airmass thunderstorm normally vary from 400 to 1,200 f.p.m. near the base of the thunderstorm to 4,000 f.p.m. near the equilibrium level. Extreme vertical gusts of more than 10,000 f.p.m. have been reported in the strongest thunderstorms.

Although updrafts weaken above the equilibrium level, in intense thunderstorms, they may penetrate several thousand feet into the stratosphere before they are overcome by the stability. The strongest updrafts can often be identified by cumuliform bulges that extend above the otherwise smooth anvil top of the thunderstorm. These are called overshooting tops and they are evidence of very strong thunderstorms.

As expected, thunderstorm downdrafts are strongest in the areas of precipitation. Downdrafts typically reach their greatest intensities near the base of the thunderstorm. Extremes of near 5,000 f.p.m. have been reported.

Turbulence intensity increases with the development of the thunderstorm; that is, light and moderate intensities in the cumulus stage and moderate and severe (or worse) in the mature stage. When the thunderstorm cell begins to dissipate, turbulence within the thunderstorm weakens. However, in the absence of radar, a high degree of caution should be exercised in evaluating the turbulence potential of a dissipating cell. The exact point at which a mature cell becomes a dissipating cell is not well defined by visual observations. For example, early in the dissipating stage, turbulence in some locations of the thunderstorm is as intense as it is in the mature stage. Late in the dissipating stage, isolated patches of severe turbulence may still be present. Even with radar, it is extremely hazardous and not recommended to fly in a thunderstorm at any stage.

VIP	LEVEL	TURBULENCE
1	Weak	Light to Moderate possible
2	Moderate	Light to Moderate possible
3	Strong	Severe possible
4	Very strong	Severe likely
5	Intense	Severe with organized surface wind gusts
6	Extreme	Severe with extensive surface wind gusts

Figure 12-11. Video Integrator Processor (VIP) Levels with estimated turbulence intensity. The probability of lightning increases above VIP level 1, and the likelihood of large hail increases above VIP 4.

The radar intensity scale (VIP level) that is used with the radar summary chart and with most airborne radars is related to turbulence intensity as shown in figure 12-11.

TURBULENCE BELOW THUNDERSTORMS

The downdrafts, downbursts, and microbursts described in the last chapter define the primary turbulent areas below the thunderstorm. These phenomena produce intense turbulence as well as wind shear. Strong winds in the outflow from the downdraft generate mechanical turbulence. Turbulence also occurs on the edge of the downburst in the vortex ring, near the gust front and, of course, near any funnel clouds, tornadoes, and other tornado-like vortices. The combination of turbulence, wind shear, heavy precipitation, low ceilings, and visibilities make the area below a thunderstorm a very dangerous area.

Do not fly within 20 miles of a thunderstorm that is classified as severe or that has an intense radar echo (VIP 5 or greater).

Do not fly under the overhang of a cumulonimbus cloud.

TURBULENCE AROUND THUNDERSTORMS

Turbulence "around the thunderstorm" refers to that found outside the main region of convection. This includes turbulence in clear and cloudy air next to the main cumulonimbus cloud and turbulence in and over the anvil cloud.

For the most part, downdrafts in the clear air around airmass thunderstorms are a few hundred feet per minute or less. However, there are occasions when severe turbulence occurs in the clear air. The causes of some of these events are not well understood, and those that are understood are not very well measured. These uncertainties are the primary reasons why you must always maintain a substantial separation between a thunderstorm and your aircraft.

Turbulence is produced outside the thunderstorm when the cell acts as a barrier to the large scale airflow. Multicell and supercell thunderstorms move more slowly than the winds at upper levels. Under these conditions, part of the prevailing airflow is diverted around the thunderstorm, producing a variety of turbulent eddies. This effect is greater with strong thunderstorms and strong winds aloft.

A turbulent wake occurs under the anvil cloud downwind of the thunderstorm. This is one of the most hazardous regions outside of the thunderstorm and above its base. Sometimes identified as the region under the overhang (anvil), it is an area well known to experienced pilots and is a location of severe turbulence and possibly hail.

TURBULENCE NEAR THUNDERSTORM TOPS

Near the top of the thunderstorm, several circulations are possible. The cumuliform appearance of the overshooting tops is a warning that this region is a source of significant turbulence due to the convective currents. Additionally, the interaction of strong winds in the stable stratosphere with the updraft can produce vertical shears, turbulent eddies, and atmospheric gravity waves (similar to lee waves) over and downwind of the thunderstorm top. Flight near thunderstorm tops should be avoided wherever possible.

Do not fly in the anvil cloud. Your altitude should be 1,000 feet above the cloud for every 10 knots of wind at that level. For example, with a 50 knot wind, you should be 5,000 feet above the top. If this is above the ceiling of your aircraft, go around the thunderstorm.

CLEAR AIR TURBULENCE (CAT)

Clear air turbulence (CAT) is that turbulence which occurs in the free atmosphere away from any visible convective activity. CAT includes high level frontal and jet stream turbulence. Its name is derived from early experiences of pilots who encountered significant high level turbulence in clear skies; however, turbulence at high levels is not limited to cloudless conditions. We now know that the processes that produce CAT can also be present in nonconvective clouds. Nevertheless, the name remains "CAT."

In the discussion that follows, it is emphasized the CAT we are concerned with is "significant" turbulence; that is, the turbulence intensity is moderate or greater. Because we can't observe CAT very well, we often find it more convenient to describe it in statistical terms. For example, during a given flight anywhere in the atmosphere, an aircraft has about a 6 in 100 (6%) chance of encountering significant CAT. The chance of severe or greater CAT is less than 1 percent. The chances of CAT are higher in regions like the jet stream than in other areas, but not by much. A 10% probability of encountering significant CAT is considered large. Therefore, keep in mind as we discuss "favored" areas and "higher" frequencies for CAT that these are relative terms. On an absolute scale, the chances of significant CAT are almost always small.

The reason you must be concerned about CAT encounters is that severe and extreme incidents do occur, causing injuries and occasionally damage to aircraft. It is your responsibility and the responsibility of the meteorologist to minimize significant CAT encounters, whenever possible.

Some other descriptive characteristics of CAT are the following. CAT occurs more frequently

1. within a few thousand feet of the tropopause than either above or below.
2. over mountains than elsewhere.
3. in winter than in summer.

CAT is found near high level stable layers that have vertical wind shear. When the air parcels in the stable layer are displaced vertically, atmospheric gravity waves develop. These waves can have wavelengths from a few hundred feet to a mile or two. If the vertical shear is strong, it causes the wave crests to overrun the wave troughs, creating a very unstable situation. The air literally "overturns" in the waves, often violently. (Figure 12-12). The result is a layer of CAT. To keep these short gravity waves separate from the much longer mountain waves, we will call them shearing-gravity waves.

> The chances of significant CAT increase rapidly when the vertical wind shear exceeds 5 knots per 1,000 feet.

Jet streams, certain high level stable layers, and tropopauses are regions where vertical shears develop. This explains why these regions favor CAT. It also explains why the strength of vertical shear is used as a CAT indicator.

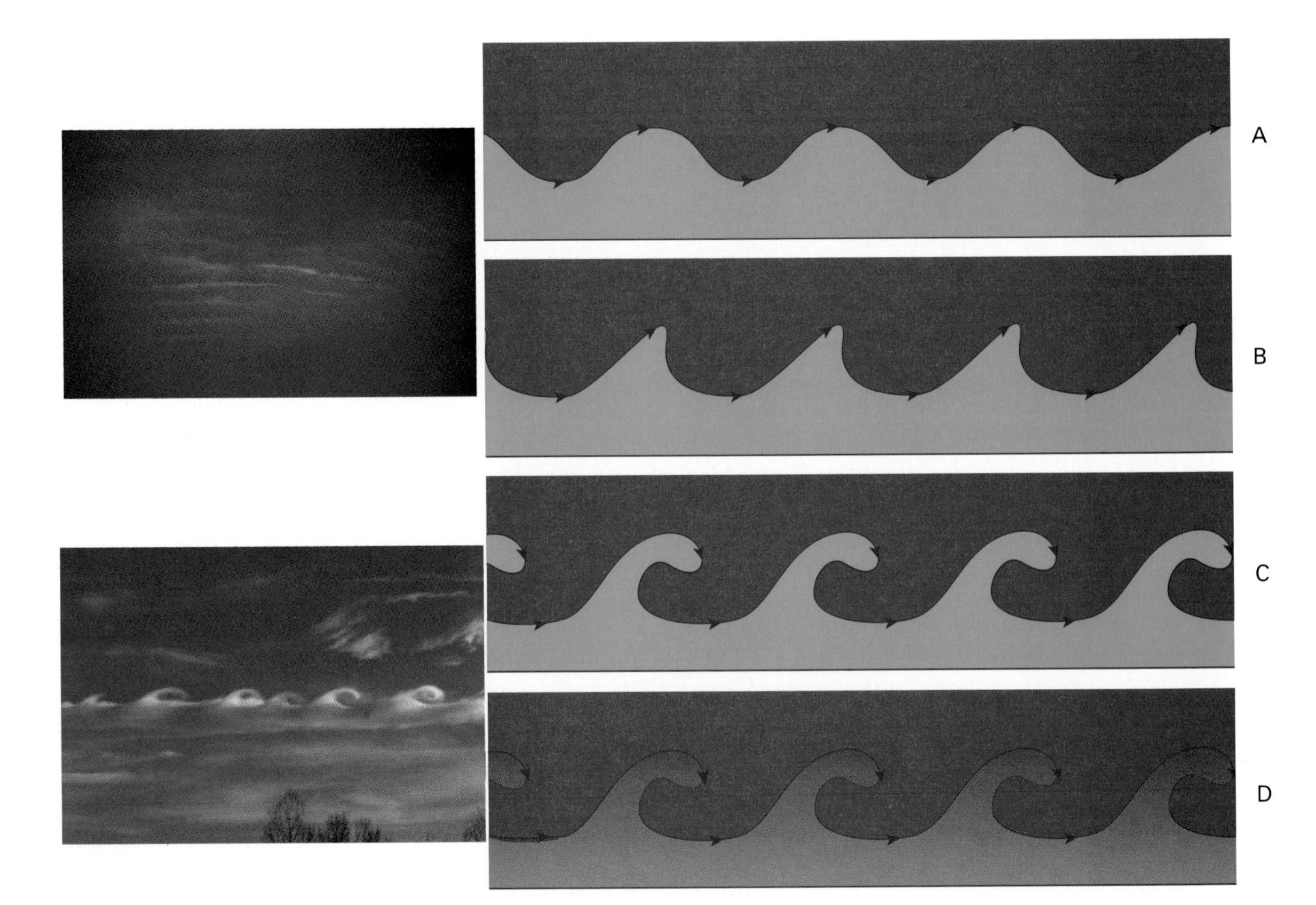

Figure 12-12. Vertical cross sections through shearing-gravity waves. Diagrams A through D are a time sequence showing how wind shear causes stable parcels to overturn as they move along a wavy trajectory. Large scale winds are indicated by horizontal arrows on the left. Wind speeds are greater in the wave crests than in the trough. It typically takes a few minutes to go from stage A to stage D. A photograph of the process is possible when the air is saturated. The "herring bone" pattern of billow clouds indicates wave crests in the upper photograph. In the lower right is a photograph of a wave in stage C or D. (Breaking Wave courtesy of National Center for Atmospheric Research/University Corporation for Atmospheric Research/National Science Foundation)

The jet stream, where about 2/3 of CAT occurs, is the focus of our model for CAT. Figure 12-13 is a vertical slice through an idealized jet stream. It shows favored layers for CAT occurrence around the jet stream core.

In the vicinity of the jet stream, there are two specific regions where CAT occurs most frequently. One is in the sloping layer below the jet core. This is a high-level frontal zone, also called a jet stream front. A jet stream front will be found with every atmospheric jet stream. The other favored layer for CAT is the sloping tropopause located on the warm side of the jet stream, but above the level of the jet core. The warm side of the jet stream is typically the side nearest the equator.

CAT tends to occur in thin layers, typically less than 2,000 feet deep, a few tens of miles wide and more than 50 miles long. CAT often occurs in sudden bursts as aircraft intersect thin, sloping turbulent layers.

Below the jet core, CAT is most likely in the jet stream frontal zone. On jet stream charts, the frontal zone is on the left side of the jet axis, looking downwind.

When a pilot enters an area where significant CAT has been reported, an appropriate action when the first ripple is encountered is to adjust airspeed to that recommended for rough air.

Because of the lack of good observations of CAT, forecasters use large scale weather patterns as important tools for predicting CAT. Based on known statistical relationships between CAT occurrences and certain contour and jet stream configurations, forecasters are able to identify large regions as potentially turbulent. The exact locations of CAT cannot usually be specified unless an aircraft happens to encounter and report the turbulence. Figure 12-14 shows some common high level patterns which favor CAT outbreaks. They include sharply curved troughs and ridges, and areas where jet streams converge and diverge.

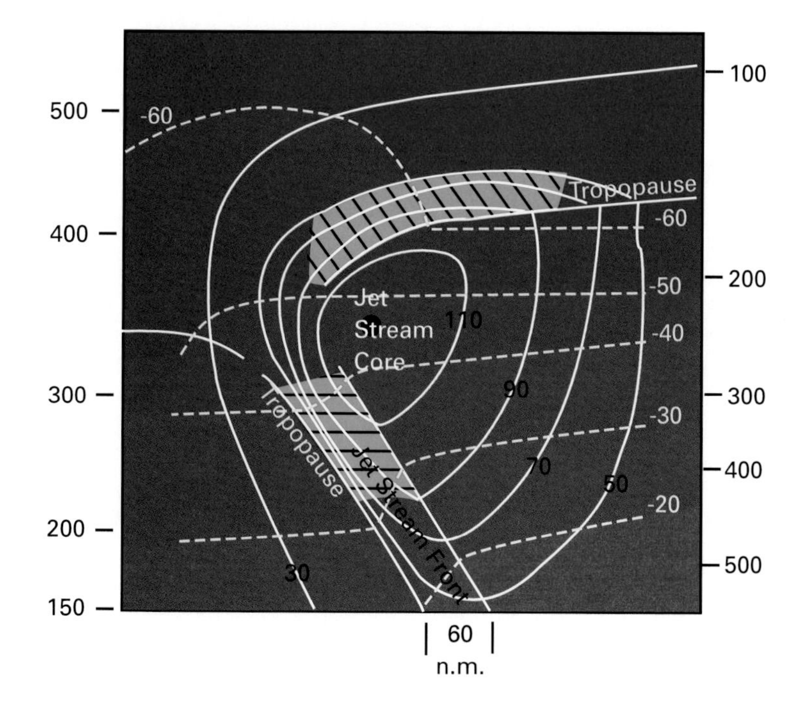

Figure 12-13. Idealized cross section through a jet stream. Thin solid lines represent wind speed in knots. The dashed lines show temperature in °C. Heavy solid lines are tropopauses and boundaries of the jet stream frontal zone. The most likely zones of significant turbulence are hatched. Note: The slope of the front is greatly exaggerated, and altitudes of various features may be significantly different depending on latitude and season.

A curving jet stream associated with a deep low pressure trough produces greater turbulence than a straight jet stream.

Significant CAT and wind shear in the vicinity of the jet stream is more likely when the speed at the core exceeds 110 knots.

MOUNTAIN WAVE TURBULENCE (MWT)

Mountain wave turbulence (MWT) is that turbulence produced in connection with mountain lee waves. It is responsible for some of the most violent turbulence that is encountered away from thunderstorms. Mountain wave turbulence occurs mainly in two well-defined regions of the lee wave system: near the tropopause and near the ground in the lower turbulent zone. (Fig 12-15)

It is useful to keep in mind that the intensity of MWT depends on the wind speed near the mountain peaks. The simple rule is the stronger the winds at mountain top level, the better the chances for turbulence.

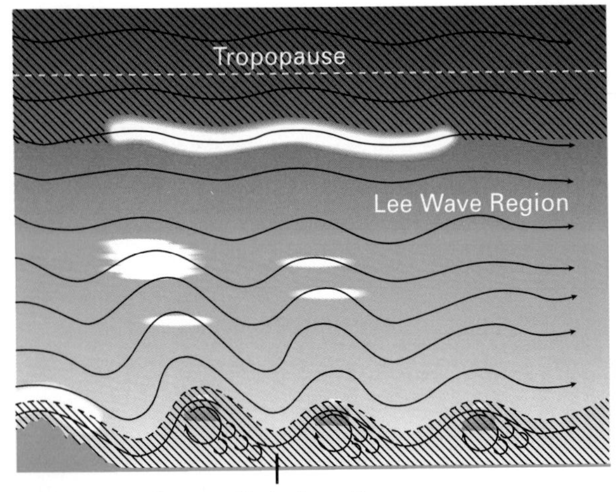

Figure 12-14. Jet stream patterns that are often related to CAT. Thick lines with arrows are jet stream axes. Shaded areas indicate potential CAT areas.

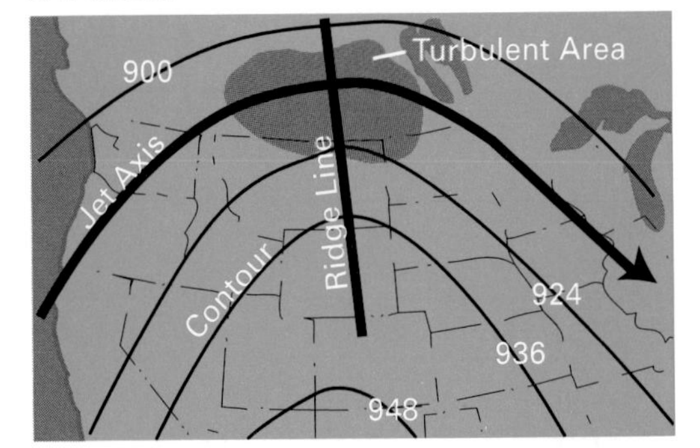

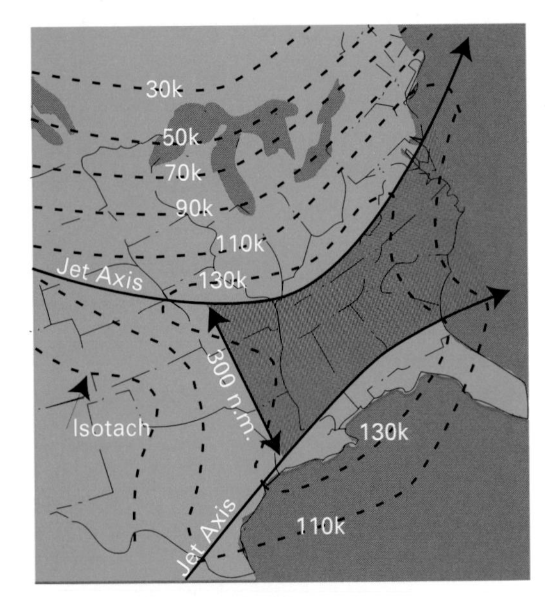

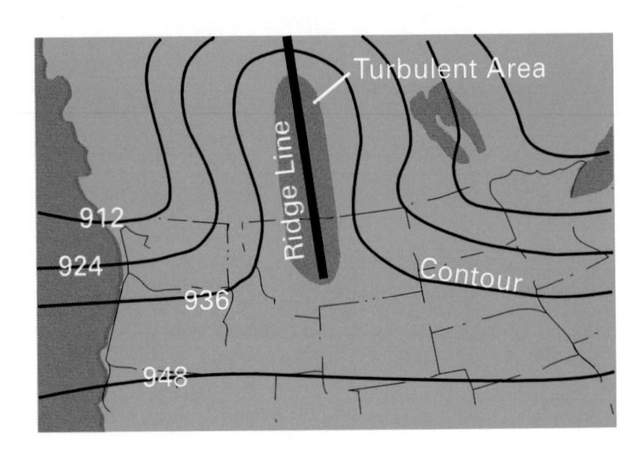

Fig 12-15. Cross section through the idealized lee wave system. Airflow is shown by thin lines with arrowheads. Favored regions for turbulence are shown with hatching.

When mountain top wind speeds are between 25 and 50 knots, expect moderate turbulence at all levels to 5,000 feet above the tropopause. The turbulence can extend 150 to 300 miles downwind of the ridge. When mountain top wind speeds exceed 50 knots, expect severe turbulence 50 to 150 miles downwind of the ridge, especially within 5,000 feet of the tropopause and below the top of the rotor.

LEE WAVE REGION

Lee waves are more often smooth than turbulent, but if turbulence does occur in the lee wave region, it is most likely to occur within 5,000 feet of the tropopause. This happens because the winds reach maximum speeds near the tropopause, with vertical shears above and below that level. Mountain lee wave activity strengthens the shear, promoting the development of shearing-gravity waves. The conditions for mountain waves and CAT are more favorable when a jet stream is present over a mountainous area. This helps to explain why high level turbulence is reported more frequently over mountains than elsewhere. MWT is usually strongest in the first wave cycle, just downwind of the mountain ridge. When lee waves have high amplitudes, the airflow is usually smooth; PIREPs often describe "...strong wave action..." rather than "...severe turbulence..." What the aircraft actually experiences depends not only on wave amplitude, but also on lee wavelength and aircraft speed.

Smaller scale turbulence in the vicinity of smooth lee waves is often made visible by the sawtooth appearance of shearing-gravity waves on lenticular clouds. Avoid ACSL with ragged edges.

To avoid mountain wave activity, change your route. If this is not possible, change your altitude away from the most likely layers of turbulence (near the tropopause).

The impact of lee waves on an aircraft is much different depending on whether the ridge line that generates the waves is approached upwind or downwind. An upwind approach will give you plenty of warning of the lee waves because the wave action typically increases as you get closer to the ridge. A downwind approach, however, immediately puts the aircraft in the primary cycle. There is little or no warning as the most intense part of the lee wave is encountered first.

Weak lee waves have updrafts and downdrafts of 300 to 900 f.p.m., while strong lee waves range from 1,800 to 3,600 f.p.m. Extreme vertical drafts of 5,000 to 8,000 f.p.m. have been reported. Caution: Altimeter readings may be inaccurate in strong lee waves.

Except for the highest mountains, the following procedure is recommended: When approaching a mountain wave area from the lee side during strong winds. Start your climb at least 100 n.m. away from the mountains. Climb to an altitude that is at least 3,000 to 5,000 feet above the mountain tops before you cross the ridge. The best procedure is to approach the ridge at a 45° angle to enable a rapid retreat in case turbulence is encountered. If you are unable to make good on the first attempt, and if your aircraft has higher altitude capabilities, you may want to make another attempt at a higher altitude. Sometimes you have to choose between turning back and diverting to another more favorable route.

In potential mountain wave areas, watch your altimeter, especially at night. Vertical motions in lee waves may be strong, resulting in large altitude excursions.

LOWER TURBULENT ZONE

The lower turbulent zone is the boundary layer in the lee wave system. Strong winds and wind shears produce widespread turbulence there. In the typical case the worst turbulence occurs along

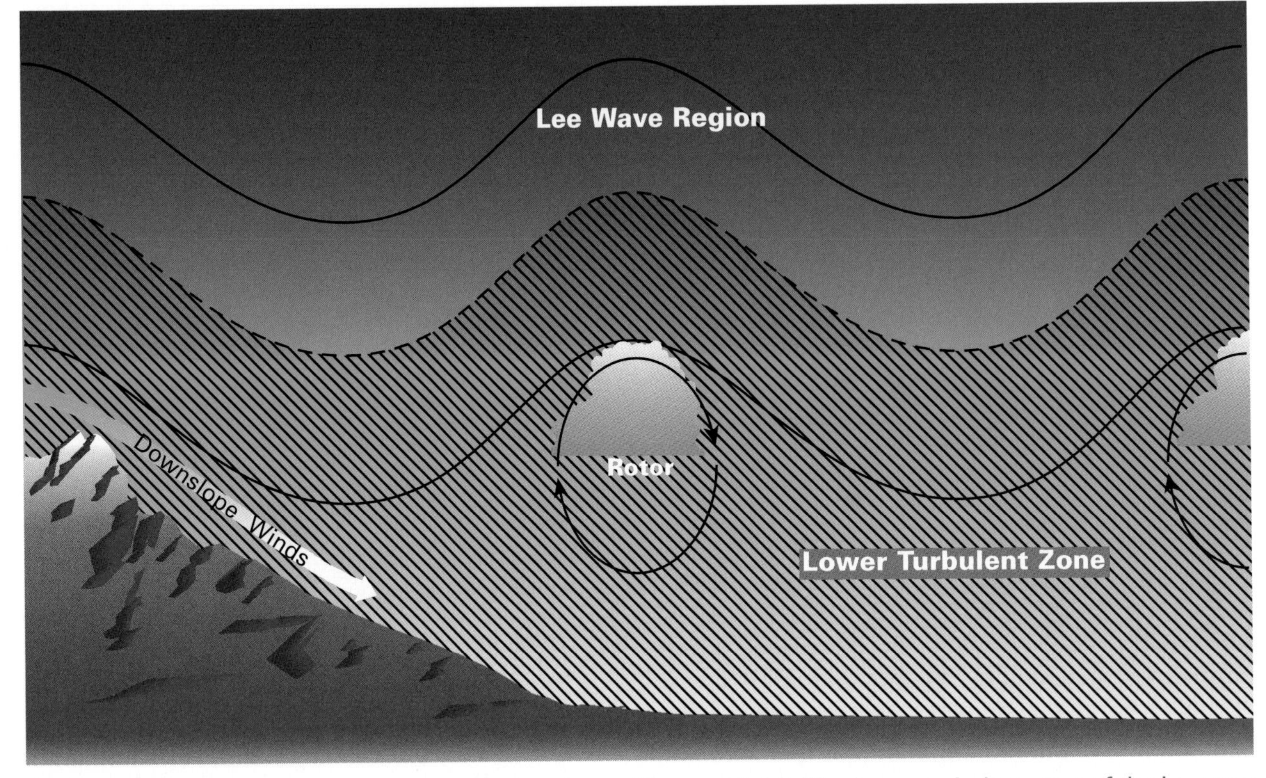

Figure 12-16. The lower turbulent zone of the lee wave system showing the locations of strongest turbulence.

the lee slopes of the mountain, below the first lee wave trough, and in the rotor. (Figure 12-16)

Close to the mountain, strong winds are directed downslope. The region of strong winds commonly extends part way down the lee slopes. Under strong lee wave conditions, strong, gusty surface winds can extend beyond the base of the mountains. (Figure 12-16) As you saw in Chapter 10, warm, dry, gusty winds blowing from the mountains are often identified by local names such as chinook.

Flying mountain passes and valleys is not a safe procedure during high winds. If winds at mountain top level are 25 knots or more, go high, go around, or don't go.

One of the most dangerous features of mountain waves is the turbulent area in and below rotor clouds

It follows from this description that you should avoid attempting a low-level flight across any substantial ridge when a mountain wave is present. If there is sufficient moisture, mountain wave clouds are useful indicators of such activity. In particular, the cap cloud indicates strong downward motion over the lee slopes. Remember, the absence of clouds does not guarantee the absence of MWT.

The greatest MWT typically occurs in rotor circulations which are found under the lee wave crests. The rotor associated with the first wave cycle downwind of the ridge is usually the most intense. The altitude of the center of the rotor circulation is about ridge top level. If present, the roll cloud is normally located in the upper part of the rotor. When strong and fully developed, a rotor is a closed circulation, producing a reversal of the winds in the lower levels.

The strength of the rotor is proportional to the strength of the lee wave. In particular, the rotor will be strong where the mountain top wind speed is strong, the lee slope is steep, and, where the mountain is high compared to the valley downwind. On some occasions, the rotor may exist only as a weak circulation or it may not be present at all. However, a conservative approach is always advised. When you suspect that lee waves are present, with or without roll clouds, you should assume that one or more strong rotors are also present.

If reported surface winds from a station on the lee slopes of a ridge are directed away from the ridge and exceed 20 knots, downslope winds and rotor activity should be suspected. Verification from other indicators should be sought: ACSL, roll cloud, cap cloud, blowing snow, or dust that is carried from the ground up into the rotor.

SUMMARY

Aviation turbulence is caused by a number of different atmospheric phenomena. In this chapter we have considered the four most common types: turbulence generated in the boundary layer (LLT), turbulence caused by strong convection (TNT), turbulence in the vicinity of the jet stream (CAT), and turbulence caused by mountain waves (MWT). You now know why and where this turbulence develops. You have some useful conceptual models which help you connect the various types of turbulence to the larger scale circulations in which it is embedded. Finally, you have learned some rules of thumb to aid you in turbulence avoidance. In Part IV, we will cover some of the aviation weather products that are available from the NWS and the FAA that help you identify turbulent areas.

KEY TERMS

Aviation Turbulence
Billow Cloud
Capping Stable Layer
Clear Air Turbulence (CAT)
Funneling Effect
G-load
Jet Stream Front
Low-Level Turbulence (LLT)
Maneuvering
Mechanical Turbulence
Mountain Wave Turbulence (MWT)
Overhang
Overshooting Tops
Shearing-Gravity Waves
Thermal Turbulence
Turbulence in and Near Thunderstorms (TNT)
Turbulence Reporting Criteria
Turbulent Gusts
Turbulent Wake
Wake Turbulence

CHAPTER QUESTIONS

1. You are approaching a north-south mountain range from the east. Mountain top is 10,000 feet MSL. Your altitude is 12,000 feet MSL. 500 mb winds are 340° at 25 knots. 700 mb winds are 270° at 35 knots. Discuss. Consider the same situation but this time your approach is from the west.

2. Why do both turbulence and hail often develop under the overhang of a CB?

3 You are diverting around a thunderstorm on the downwind side. The sky under the anvil is clear. If you fly under the anvil, it will save time and you might not have to stop for additional fuel. Discuss your options and risks of each option.

4. If you are flying at 30,000 feet MSL with a strong tailwind over the North Pacific and you encounter CAT, what would you do to get out of it? Explain.

5. The following are two rules of thumb for dealing with lee waves and rotors:

   "When there is a sustained loss of altitude while flying parallel to a ridge, rising air will often be found a few miles to the left or right of track. The exception is a downdraft close to the ridge. In that case, fly downwind."

   "In order to avoid rotors during arrivals and departures across rugged terrain in strong low level wind conditions, delay descent until `clear of the area. If necessary, pass over the airport and make descent from the other side."

   Draw clear and well-labeled diagrams of lee waves, rotors, and flight paths that show why this advice is good.

6. Interpret the PIREP at the bottom of the page.

7. The tower has just cleared you to land behind a 757. The reported wind is a seven knot crosswind.

   1. Do you see any potential hazards in this situation?
   2. If so, what are your options and the risks associated with each option?

UA/OVR MRB FL060/SK CLR/TB MDT/RM TURBC INCRS WWD.

CHAPTER 13
Icing

Introduction

When you fly through clouds or rain at temperatures near or below 0°C, you must consider the possibility of icing. Even when your aircraft is well-equipped with de-icing/anti-icing devices, they only provide partial protection. Aircraft icing is insidious. Unless you are observant and carefully select the environment in which you fly, icing can quickly become a factor and threaten the safety of flight. In this chapter, you will learn what causes icing, the forms it can take, how to report it, and under what meteorological conditions it is most likely to occur. When you complete the chapter, you should have a basic understanding of the icing threat and knowledge of how to avoid it or at least minimize the problem. You should never fly in clouds without first obtaining the appropriate training and an instrument rating.

Section A

ICING CHARACTERISTICS AND CAUSES

Icing refers to any deposit or coating of ice on an object. In this section, we describe the causes and types of aircraft icing, including frost.

ICING EFFECTS

When ice forms on an aircraft, it degrades performance in many ways. The aerodynamic shapes of lifting surfaces are distorted by the coating of ice. Weight and drag are increased, and lift is decreased. Icing on the propellers decreases thrust and ice in the carburetor or air intake decreases power. The stalling speed of the aircraft is increased. Control surfaces and landing gear may not operate properly if they are hindered by heavy icing.

An uneven distribution of ice on the propeller can cause damaging vibrations. In jet aircraft, chunks of ice breaking loose from the aircraft can be ingested into the engine, causing damage to compressor blades. Icing on the pitot-static system results in airspeed, altitude, and vertical speed indicator errors. Icing of radio antennas may adversely affect communications, while windscreen icing reduces visibility. Even ground operations can be adversely affected when runway ice/snow makes braking difficult.

ICING TYPES AND CAUSES

There are two categories of icing that are of importance to aviation. They are:

1. Structural icing.
2. Induction icing.

Understanding the causes of structural and induction icing requires that you recall some important meteorological concepts from previous chapters. First, ice or frost formation requires the presence of water or very moist air and some way of cooling the temperature below 0°C. Second, water droplets can exist at temperatures below 0°C (supercooled water droplets). Third, three of the most common ways to cool the atmosphere are by contact with a very cold surface, by adiabatic expansion, and by evaporation.

STRUCTURAL ICING

Airframe or structural icing refers to the accumulation of ice on the exterior of the aircraft. It includes ice on the wing and tail surfaces, propellers, radio antennas, windscreen, pitot tube, and static ports. The major effect of structural icing is loss of aerodynamic efficiency of the aircraft.

The primary cause of structural icing is the freezing of water droplets on the skin of the aircraft as it passes through a cloud. For this to happen, the aircraft must be cold (temperature below 0°C) and the cloud it flies through must contain supercooled water droplets. In these conditions, the droplets freeze when they strike the aircraft. Icing does not occur in clouds composed only of ice crystals.

This freezing process produces three different icing types, depending on the temperature and the number and size of droplets within the cloud: clear, rime, and mixed ice. If the droplets are large, they spread over the structure, freezing slowly and forming glaze or clear ice. Clear ice has a glossy surface and may be either clear or translucent. This is the most dangerous form of structural icing because it is heavy and hard, it adheres strongly to the aircraft surface, and it can greatly distort the shape of the airfoil.

If the droplets are small and freeze immediately when they strike the aircraft, they trap air and form brittle rime ice. Rime ice grows into the airstream from the leading edges of wings, struts, and other exposed parts of the aircraft. Its appearance is opaque and milky white with a porous texture.

Rime ice may also have serious effects on the aerodynamics of the aircraft wing, but it is generally lighter than clear ice and easier to remove with de-icing equipment. Mixed ice is a combination of rime and clear ice. It forms when snow or ice particles become embedded in clear ice, developing a very rough accumulation. (Figure 13-1)

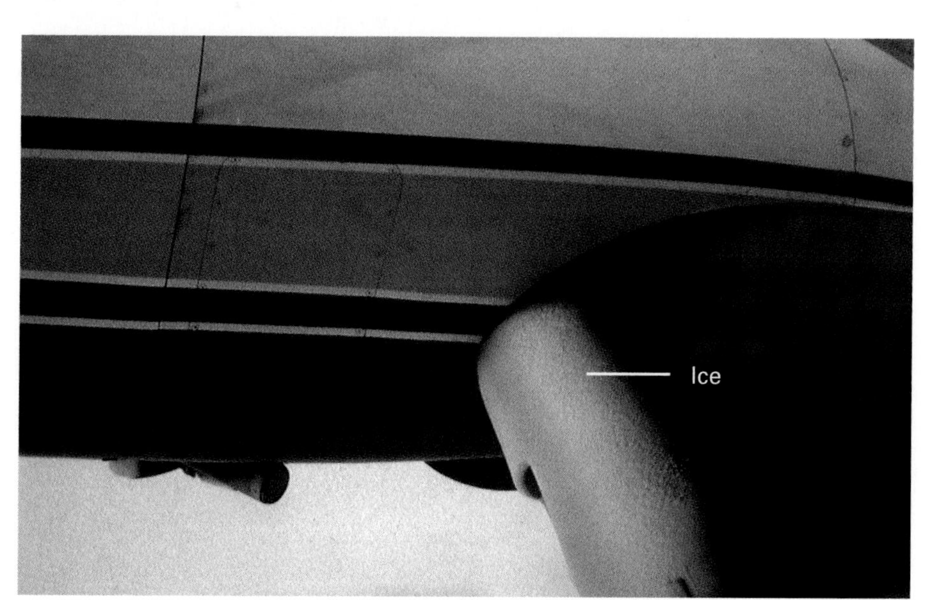

Figure 13-1. Examples of clear ice (above) on the outboard section of a wing, and rime ice (right) on a section of a wing between the nacelle and the fuselage. (Rime ice photograph courtesy of Wayne Sand; Clear ice photograph courtesy of National Center for Atmospheric Research/University Corporation for Atmospheric Research/National Science Foundation.)

Structural icing is reported in pilot weather reports (PIREPs) according to icing type and intensity. Intensity is categorized according to the rate of accumulation, the effectiveness of available de-icing/anti-icing equipment, and what actions a pilot must take to combat the accumulation of ice. De-icing equipment removes ice after it has formed, while anti-icing equipment prevents the formation of ice. (Figure 13-2)

Figure 13-2. Icing intensities, accumulation, and recommended flight procedure.

Intensity	Airframe Accumulation	Pilot Action
Trace	Ice becomes perceptible. Rate of accumulation of ice is slightly greater than the rate of loss due to sublimation.	Unless encountered for one hour or more, de-icing/ anti-icing equipment and/or heading or altitude change not required.
Light	The rate of accumulation may create a problem if flight in this environment for one hour.	De-icing/anti-icing required occasionally to remove/ prevent accumulation or heading or altitude change required.
Moderate	The rate of accumulation is such that even short encounters become potentially hazardous.	De-icing/anti-icing required or heading or altitude change required.
Severe	The rate of accumulation is such that de-icing/anti-icing equipment fails to reduce or control the hazard.	Immediate heading or altitude change required.

PIREPs are good sources of icing information:
UA/OV ABI/TM 1700/FL 080/TP PA60/SK 024
BKN 040 OVC /TA -11/IC LGT-MDT RIME 050-075

INDUCTION ICING

Induction icing affects the powerplant operation. It includes icing on air intakes and carburetor icing. The main effect of induction icing is power loss due to the blocking of the air before it enters the engine.

In a given icing environment, the potential for icing depends on the aircraft design and speed. Jet aircraft are usually the least susceptible to icing because of their excess thrust capabilities and the fact that they usually operate at high levels, out of the critical temperature range for icing. Many small reciprocating-engine aircraft, are highly susceptible because of their lack of icing protection and their frequent operation at low altitudes. Helicopters are extremely susceptible to icing on the rotor blades which provide both thrust and lift.

Ice develops on air intakes (for example, on screens and air scoops) under the same conditions favorable for structural icing. Carburetor icing occurs when moist air drawn into the carburetor is cooled to a dewpoint temperature less than 0°C. The cooling is caused by the adiabatic expansion of air in the carburetor and the vaporization of fuel.

Ice in the carburetor may partially or totally block the flow of the air/fuel mixture. In fact, carburetor icing can cause complete engine failure. Complicating the problem is the fact that cooling in the carburetor is so significant (it can decrease the temperature by 20°C or more), that outside temperatures may be well above freezing when such icing occurs. This can lead the pilot to misdiagnose the problem. The carburetor heat mechanism is essential for icing prevention. Applications and limitations of such devices are detailed in the pilot's operating handbook.

Carburetor icing is possible in a wide range of temperatures: 22°C (72°F) to -10°C (14°F).

In jet aircraft operations during taxi, takeoff, and climb, there are reduced pressures in the compressor air intakes. Air moving through the intakes is adiabatically cooled. Depending on the type of engine, ice may form when the outside air has high relative humidity and a temperature above 0°C, even though clouds or liquid precipitation are not present and the outside air temperatures are above 0°C. Air blockage and the reduction of engine performance may result. For a detailed description of this affect with a particular engine type, consult the aircraft flight manual.

Test data indicate that ice, snow, or frost having a thickness and roughness similar to medium or coarse sandpaper on the leading edge and upper surface of a wing can reduce lift by as much as 30 percent and increase drag by 40 percent.

FROST

As you previously learned, frost is a deposition product that forms when saturated air is cooled to its dewpoint and the dewpoint (frostpoint) is below 0°C. In this case, water vapor sublimates directly to ice. Frost typically has a white, feathery appearance, indicative of its ice crystal structure.

All frost should be removed from the aircraft prior to takeoff.

Frost forms on aircraft in the same manner that it forms on other objects. Typically, at night under clear skies, radiation heat loss reduces the temperature of the skin of a parked aircraft below the dewpoint, which is below 0°C.

The primary danger in frost is the roughness it adds to the aircraft surface. Although it may not look very threatening, especially in comparison to the bulk of clear ice, the added drag of a thin coating of frost slows the airflow which provides lift. Frost extends the takeoff roll and may make it difficult, if not impossible, to take off. A hard frost can increase the stalling speed by as much as 5 or 10 percent. An aircraft carrying a coating of frost is particularly vulnerable at low levels if it also experiences turbulence or wind shear, especially at slow speeds and in turns.

Frost may also occur in flight. This typically occurs when a cold aircraft descends or ascends into a cloudless warmer region that has a high relative humidity. The frost is often short-lived as the aircraft warms up, but as long as it is present, the problem of increased stalling speed exists.

Frost may prevent the airplane from becoming airborne at normal takeoff speed.

Section B

THE ICING ENVIRONMENT

The icing environment refers to the large scale setting in which structural icing occurs. It includes the temperatures, cloud types, and weather patterns which are conducive to icing.

TEMPERATURE

Anytime you fly in clouds with temperatures at or below 0°C, icing is possible. You should recall from the previous chapter dealing with the ice-crystal process of precipitation formation, that supercooled water droplets are found in clouds at temperatures down to about -40°C. This defines the limit for icing. Within the temperature range between 0°C and -40°C, the threat of icing decreases as the temperature decreases. Except within a cumulonimbus cloud, icing is not a great hazard at temperatures below about -20°C. Between 0°C and -20°C, there are three important temperature zones which are defined by icing type. These should be committed to memory. (Figure 13-3)

Outside Air Temperature Range	Icing Type
0°C to -10°C	Clear
-10°C to -15°C	Mixed Clear and Rime
-15°C to -20°C	Rime

Figure 13-3. Critical outside air temperature (OAT) ranges for icing. These ranges are approximate and often overlap by 5C°. Also, rime ice may occur at temperatures below -20°C.

CLOUDS

The greater the amount and size of the supercooled cloud droplets, the greater the icing potential. Droplets are larger where vertical motions within the cloud are large, and where temperatures are just below 0°C. Under these conditions, a droplet striking an aircraft will freeze more slowly, allowing the unfrozen part to spread back from the aircraft's leading edges. This explains the greater tendency for clear ice in the 0°C to -10°C temperature range.

An in-flight condition necessary for structural icing to form is visible moisture.

Since vertical motions are particularly strong in convection, the threat of significant clear icing is great in cumuliform clouds. Because of the cellular nature and vertical development of CU and CB, icing hazards in those clouds tend to be limited horizontally, but have a large vertical extent. Icing can be intense in the upper half of a cumulonimbus cloud, at least up to the level where the temperature falls below -20°C. As would be expected, clear icing is more likely in the lower part of the icing zone, with mixed and rime icing above. There have been reports of icing in CB at temperatures as low as -40°C.

Stratiform clouds have smaller droplets, so the tendency is toward rime and mixed icing. In contrast with clouds of vertical development; in layer type clouds, the vertical limit of icing conditions are typically only 2,000 or 3,000 feet thick and are rarely found more than 5,000

Avoid cumuliform clouds if at all possible. Clear ice may be encountered anywhere at altitudes above the freezing level. Most rapid accumulations are usually at temperatures from 0°C to -15°C.

feet above the freezing level. However, layers of icing in stratiform clouds can have a greater horizontal extent than icing layers in cumulus clouds.

Cirriform clouds do not usually present an ice hazard. An exception is found in the anvil cloud associated with a thunderstorm. Occasionally, icing will occur there as strong convective currents carry supercooled droplets to the top of the cloud despite the very cold temperatures.

FREEZING PRECIPITATION

In the cooler months of the year, it is not unusual to have a shallow layer of subfreezing temperatures at the ground with a warmer layer (temperature greater than 0°C) just above the layer. (Figure 13-4).

We saw in an earlier chapter that precipitation which forms as snow in cold clouds aloft will melt to rain in the warm layer, then refreeze to ice pellets as it falls through the cold layer. If the lower cold layer is shallow and in contact with a subfreezing surface, liquid precipitation freezes on contact; that is, it becomes freezing rain or freezing drizzle. If the skin temperature of your aircraft is at or below 0°C, it becomes an efficient collector of ice under these conditions. This situation occurs most often with warm fronts.

In stratiform clouds, you can lessen the chance of icing by changing your altitude to a level where either the temperatures are above freezing, or where the temperatures are colder than -10°C. An altitude change may also take you out of the clouds.

High clouds are least likely to contribute to structural icing on an aircraft.

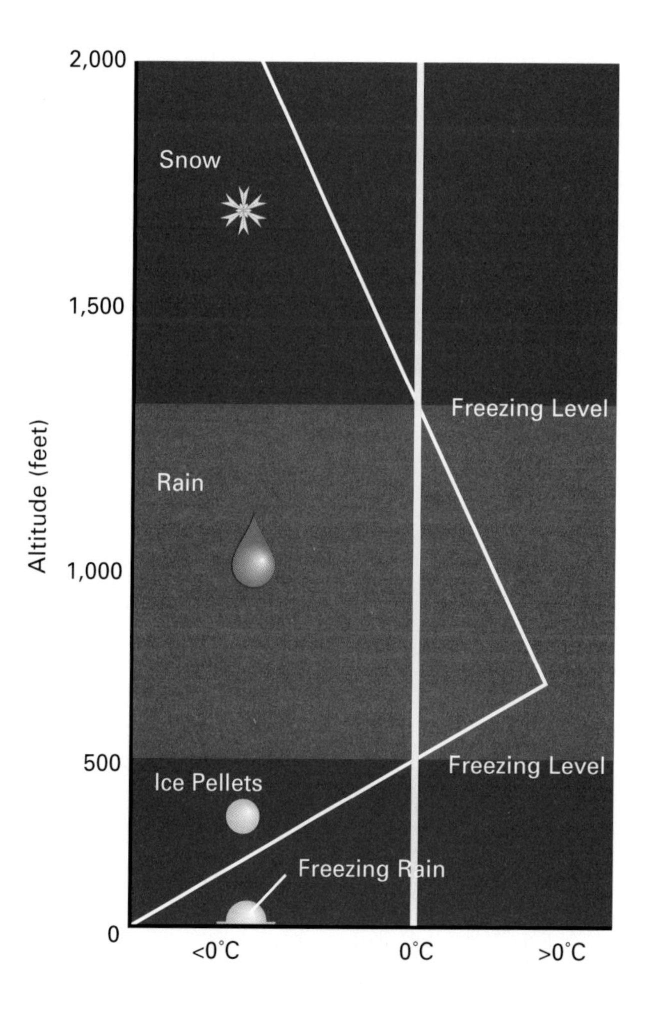

Figure 13-4. Sounding through a surface-based inversion. Red shading indicates temperatures greater than 0°C. Notice that there is a shallow layer of subfreezing temperatures near the surface. A snow particle will melt and refreeze on its way to the ground (ice pellets) or freeze on contact with the ground (freezing rain).

Structural ice is most likely to have the highest accumulation rate in freezing rain.

An important piece of practical information for the assessment of icing and freezing rain potential is the altitude of the freezing level; that is, the level in the atmosphere where the temperature is precisely 0°C. More often, there is only one freezing level in the atmosphere; but on some occasions, especially in the winter, a shallow layer of subfreezing air is found at the ground. In this case, two or more freezing levels may exist. Some examples are reported on the freezing level chart shown in figure 13-5.

FRONTS

As you have seen, the intensity and type of icing depends to a large degree on the strength of vertical motions. Fronts are sources of significant vertical motions, clouds, and precipitation. Additionally, fronts often produce temperature patterns that favor the development of freezing precipitation, so it is not surprising that they offer extensive icing hazards, especially in the winter.

The presence of ice pellets at the surface is evidence that there is freezing rain at a higher altitude.

Figure 13-5. Freezing level chart. Open circles indicate location of sounding stations where freezing levels are reported in hundreds of feet MSL. Note stations reporting multiple freezing levels. For example, at Boise, Idaho (field elevation 3,858 feet), the surface temperature is below freezing (BF) while there is a warm layer (T>0°C) between 3,000 and 6,300 feet MSL. The solid lines indicate the position of the 4,000, 8,000, and 12,000 feet MSL freezing levels. The dashed line indicates where the freezing level intersects the ground.

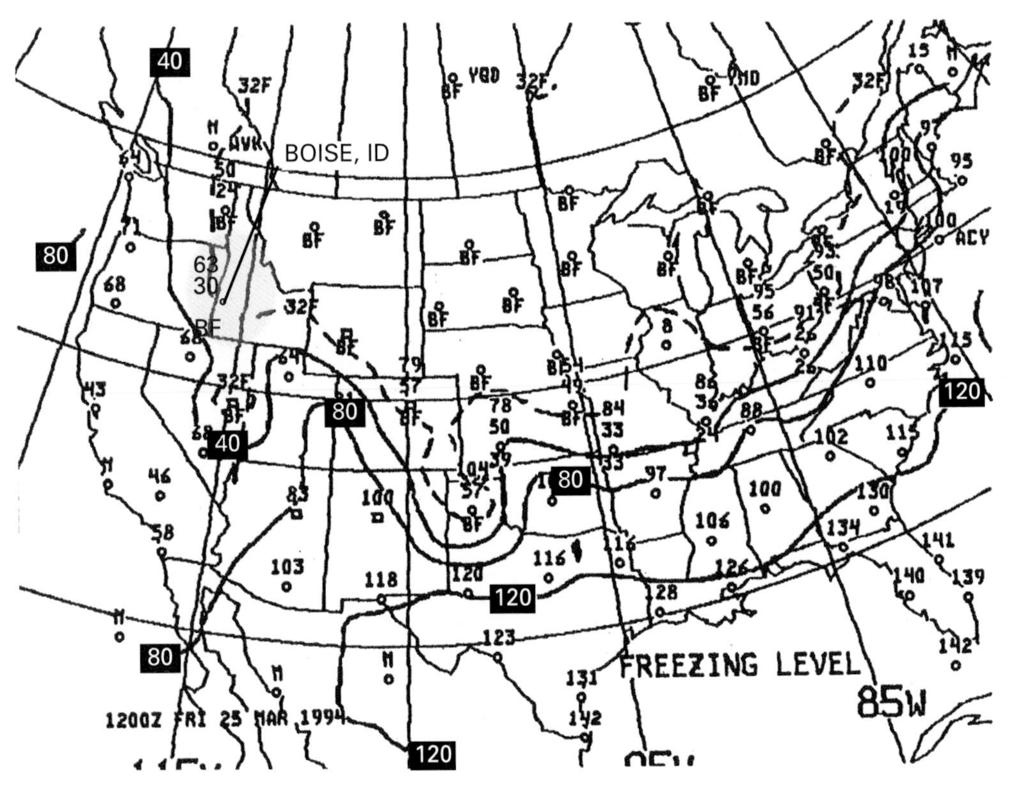

It is estimated that 85% of all reported icing occurs in the vicinity of frontal zones.

In freezing rain, temperatures are always warmer than freezing at some higher altitude. If you decide to climb, move quickly; a delay may leave you with too much ice. If your indicated airspeed is too low during climb, a high angle of attack may cause ice to accumulate on the under surfaces of the wing. This situation can substantially increase drag and dramatically increase the stall speed.

Figure 13-6 shows cross sections through warm and cold fronts with temperature structures conducive to icing. As expected, above the frontal surfaces, the frontal clouds contain layers of clear, mixed, and rime icing. If the air lifted by the front is unstable, the icing patterns reflect the presence of cumuliform clouds; that is, the icing patterns are extensive vertically and limited horizontally to convective cells. Where the air is stable, icing patterns are vertically limited in stratiform clouds, but are often horizontally extensive.

Icing may extend 300 n.m. ahead of a warm front and 100 n.m. behind a cold front. In the frontal zone and in the shallow cold air below each type of front, freezing rain (severe icing) and drizzle (moderate icing) are common hazards. Because of the differences in frontal slopes, freezing precipitation with a warm front or in a stationary front typically extends over a much broader area than with a cold front.

When a wave cyclone occludes, the low is cold throughout the troposphere. Moderate icing is usual in the circular cloud mass that marks the center of the occluded low.

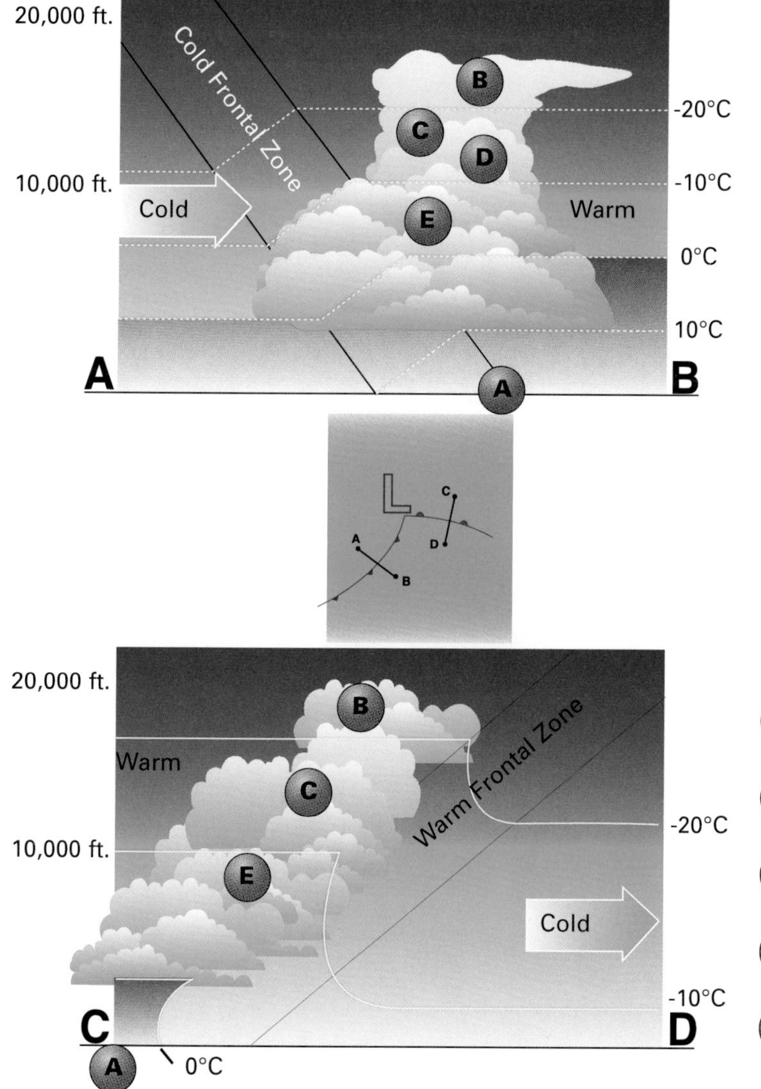

Figure 13-6. Cross sections through idealized cold front (top) and warm front (bottom) showing icing zones. Red shaded regions indicate temperatures above 0°C. The inset shows the cross section lines on a simplified surface chart. If the cold front is preceded by a squall line, icing conditions also are found in that location. Heights of cloud bases may vary significantly across frontal zones with considerable layering of clouds.

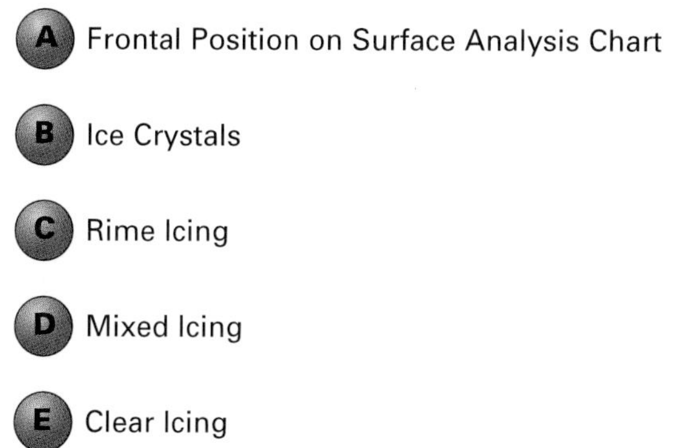

OROGRAPHIC EFFECTS

Mountainous terrain should always be considered a serious icing hazard when clouds are present. The clouds that form in upward motions on the windward sides of the mountains carry liquid water well into subfreezing regions. The worst icing zone is primarily over and on the windward side of mountain ranges. These areas can extend to about 5,000 feet above the peaks. (Figure 13-7)

If the air crossing the mountains becomes unstable, cumulus clouds develop with their deeper icing regions. When fronts cross mountain ranges, the potential for the worst possible icing conditions exist as the effects of fronts, mountains, and instability are combined.

CLIMATOLOGY

In middle latitudes, the icing season typically occurs in the winter, when temperatures are low, the freezing level is close to the ground, and storms and associated frontal zones are more frequent. Early spring and late fall also may have significant icing events. As would be expected, icing occurrences are very low in summer.

Regionally, icing is more frequent in the northern half of the lower 48 states. In winter, icing is common along the East Coast. Frontal activity is prevalent in that area as cold air moving eastward from the land encounters moist air over warm water. Icing conditions are also common downwind of large inland water bodies, such as the Great Lakes when they are ice free. The northwest U.S. provides a favorable environment, as winter storms bring abundant amounts of moisture inland. Icing conditions prevail as this moisture supply is carried across the West Coast and Cascade ranges and the Northern Rockies. Wintertime icing is also frequent over the northern and western parts of the North Pacific and North Atlantic Oceans. In high latitudes over continents, including the Arctic, icing occurrences are minimal in the winter because temperatures are so low. An exception occurs when oceanic air from lower latitudes penetrates the region.

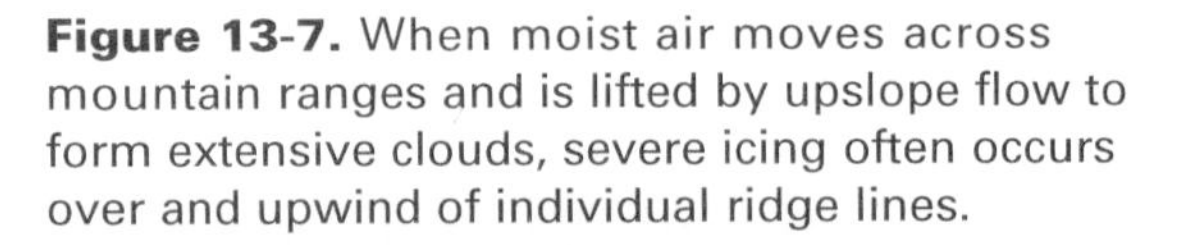

Figure 13-7. When moist air moves across mountain ranges and is lifted by upslope flow to form extensive clouds, severe icing often occurs over and upwind of individual ridge lines.

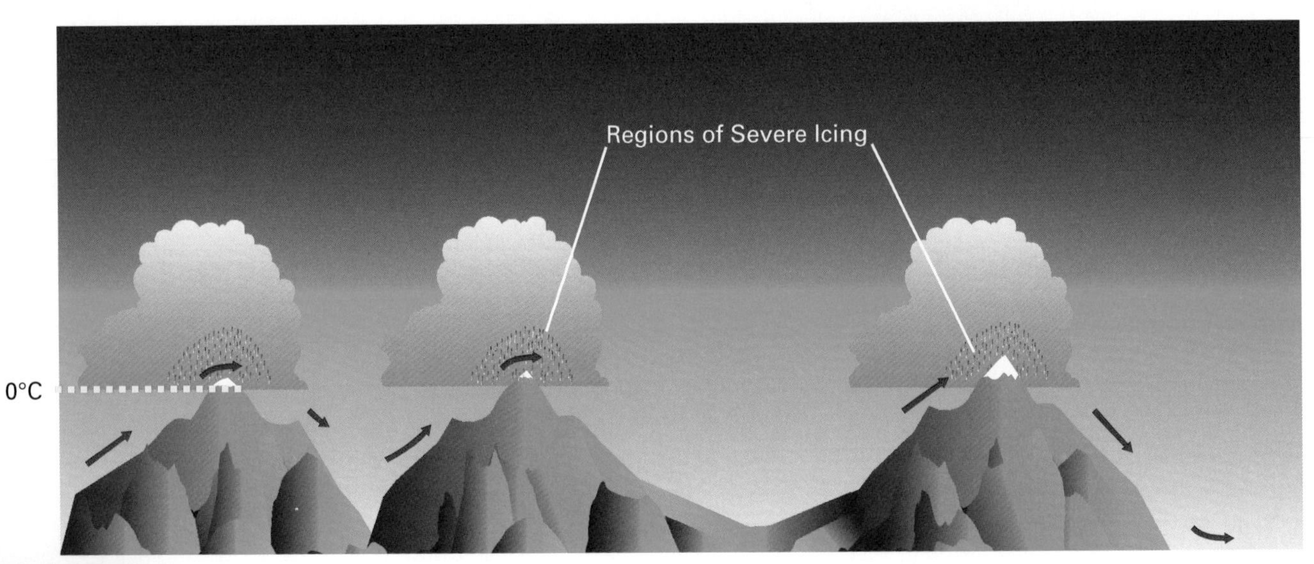

AVOIDANCE

Icing can occur in all phases of flight, given the right temperature and cloud/moisture conditions. Several rules of thumb have been presented throughout this chapter to help you recognize and avoid icing conditions. Some additional advice is presented in figure 13-8 (FAA, 1975; USAF, 1990).

Figure 13-8. Rules of thumb to aid in avoiding or minimizing icing problems.

On the ground:

- Remove all frost, ice, and snow immediately prior to takeoff.
- In cold weather, where possible, avoid taxiing or taking off through mud, water, or slush. If you taxi through any of these, make a preflight check to ensure freedom of controls and cycle retractable landing gear after takeoff.

Inflight:

- When climbing out through an icing layer, climb at an airspeed faster than normal to avoid a stall.
- Use de-icing or anti-icing equipment when accumulations of ice are not too great. When such equipment becomes less than totally effective, change course or altitude to get out of the icing as quickly as possible.
- If your aircraft is not equipped with a heated pitot-static system, be alert for erroneous readings from your altimeter and from your airspeed and vertical speed indicators.
- Avoid abrupt maneuvers when your aircraft is heavily coated with ice since the aircraft has lost some of its aerodynamic efficiency.

Landing:

- When "iced-up," fly your landing approach with power.

SUMMARY

The icing problem does not leave much room for error. This is especially true when it is combined with the additional complications of turbulence, wind shear, and IMC. In this chapter, you have not only learned how structural and induction icing can form, but how the roles of temperature, water content, fronts, and mountains contribute to the production of icing. Furthermore, you have learned that, despite its innocent appearance, frost can have a serious impact on aircraft performance. Finally, on the basis of icing causes and characteristics, a number of practical rules of thumb have been established to help you avoid or at least minimize icing efects.

KEY TERMS

Anti-icing Equipment
Carburetor Icing
Clear Ice
De-icing Equipment
Freezing Level
Freezing Level Chart
Frost
Icing
Icing Environment
Induction Icing
Mixed Ice
Rime Ice
Structural Icing

CHAPTER QUESTIONS

1. Why do large droplets at temperatures just below zero favor the formation of clear ice?

2. Several layers of clouds exist between 2,000 feet MSL and FL200. The freezing level is at 5,000 feet MSL. Use the standard atmosphere to estimate the top of the layer where clear ice would most likely accumulate during a climbout to 15,000 feet MSL.

3. What is it about fronts and mountainous terrain that makes icing conditions so much worse?

4. Is carburetor icing a problem at 10°C? Explain.

5. If it happened that cumulus cloud bases at some location were the same winter and summer, near the freezing level, icing is more likely in the summer. Explain.

6. (True, False) Frost covering the upper surface of an airplane wing usually will cause the airplane to stall at an angle of attack that is lower than normal. Explain.

7. Explain why ice pellets indicate temperatures are above freezing at a higher altitude.

CHAPTER 14

Instrument Meteorological Conditions

Introduction

Instrument Meteorological Conditions (IMC) refers to any state of the atmosphere where ceiling and visibility are below specific minimum values. Low ceilings and visibilities are common occurrences in the meteorological environment in which you fly. Unless the causes and properties of these weather conditions are understood and respected, serious flight problems can result. The purpose of this chapter is to describe the characteristics and primary causes of meteorological phenomena which limit ceiling and visibility.

When you complete this chapter, you will know the technical terminology used to specify current and forecast ceilings and visibilities; you will understand how they develop and the large scale conditions under which they form. Finally, you will learn some useful rules of thumb that will help you deal with instrument meteorological conditions. As you read this chapter, keep in mind that a current instrument rating is required to operate in instrument conditions. But, even experienced instrument-rated pilots will choose not to fly in some types of IMC.

Section A

BACKGROUND

Instrument flight is governed by the Federal Aviation Regulations, which establish the minimum criteria for operating within the National Airspace System in weather conditions less than that required for visual flight. These regulations govern IFR and VFR flights on the basis of weather conditions, aircraft equipment, pilot qualifications, airspace, and flight altitude. It is every pilot's responsibility to know and understand these regulations before beginning any flight. For further information on the regulations, consult FAR Parts 61 and 91.

A wide variety of weather information is available to you to help you make sound go/no-go preflight and in-flight decisions. In order to interpret this information efficiently and effectively, you must be aware of the important terminology and criteria that meteorologists use to define visual and instrument flight.

The counterpart to IMC is visual meteorological conditions (VMC). These two terms are rather broad classifications that are used to describe the state of the ceiling and/or visibility with regard to aviation operations. Specific ceiling and visibility categories were introduced in Chapter 6 and are repeated below for reference purposes. (Figure 14-1)

Figure 14-2 lists other key terms and measurements related to IMC that were defined in Chapter 6. This is very important background information for your understanding of flight hazards caused by low ceilings and visibilities. If you are unsure of any of these items, review them before proceeding with this chapter.

Ceiling	**Runway Visibility (RVV)**
Cloud Amount	**Runway Visual Range (RVR)**
Cloud Height	**Sector Visibility**
Cloud Layer	**Surface Visibility**
Obscuration	**Temperature-Dewpoint Spread**
Partial Obscuration	**Tower Visibility**
Prevailing Visibility	**Vertical Visibility**
Relative Humidity	**Weather Depiction Chart**

Reminder: The weather reporting terminology in regular surface observations for some cloud and visibility items listed above may differ depending on observation type (SA, ASOS, AWOS, METAR). Consult the appendix for details.

Figure 14-2. Key measurements and terms associated with Instrument Meteorological Conditions.

The determination of ceiling and visibility is not a perfect science; measurements are often rough approximations and large variations can occur over short distances and short time spans, especially with IMC. Because of these problems and the serious consequences that IMC can have on flight, a conservative interpretation of the measurements should be used when in doubt.

Keep in mind that the ceiling and visibility information given in SA, AWOS, ASOS, and METAR reports is determined by a ground-based observation. Uncertainties arise simply because of the way visibility is measured. The reported visibility is the greatest horizontal distance over which objects can be seen and identified (daytime) or bright lights can be seen (nighttime). Under the same meteorological conditions, these two definitions can result in different distances being reported for the same meteorological conditions, the only difference being whether it is dark or light. In addition, visibilities often vary from one quadrant to another.

Category	Ceiling (feet AGL)	and/or Visibility (statute miles)
Visual Flight Rules (VFR)	None or > 3,000	and > 5
Marginal Visual Flight Rules (MVFR)	1,000 to 3,000	and/or 3 to 5
Instrument Flight Rules (IFR)	500 to 1,000	and/or 1 to 3
Low Instrument Flight Rules (LIFR)	< 500	and/or < 1

Figure 14-1. Ceiling and visibility categories that define VFR, MVFR, IFR, and LIFR conditions.

Even without these problems, surface observations of visibility are not always a good indication of what you see from the air, especially when clouds are present. Looking down on an airport from your aircraft during a partial or total obscuration, you can often see objects not visible to the ground observer. Another important consideration is slant range visibility or "slant visibility" on final approach. This is the oblique distance at which you can see landing aids, such as runway lights and markings. This value also is not necessarily the same as the visibility reported on the ground. These different visibilities are illustrated in figure 14-3.

Uncertainties in measurements are not limited to visibilities. Surface reports of the heights of cloud bases are subject to increasing estimation errors as the altitude of the cloud base increases. Sometimes the variability in the altitude of very low clouds is indicated with a remark of an SA report, such as "CIG RGD" (ceiling ragged). Because of the differences between surface and pilot observations, surface observations should always be supplemented by current PIREPs whenever possible.

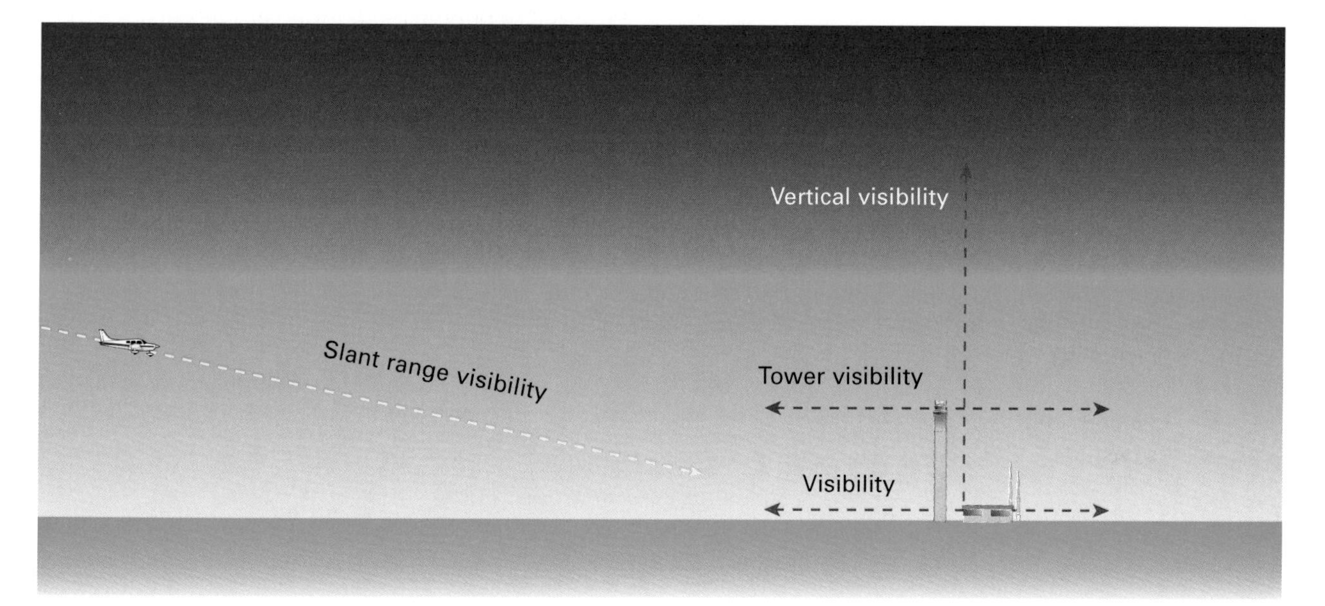

Figure 14-3. Slant range visibility. Other visibilities are shown for comparison. Visibility determined from the control tower is designated "tower visibility." When the surface visibility is also determined from another location, such as the weather station, it is called visibility. Vertical visibility is reported as the height of an indefinite ceiling.

It is good to keep in mind that the visibility given in weather reports is the prevailing horizontal surface visibility. It is a useful guide only if it is properly interpreted.

Section B

CAUSES OF IMC

Visibility is decreased by particles that absorb, scatter, and reflect light. Such particles are always present to some degree in the atmosphere, but there are large variations. Sometimes they are too few and too small to have any significant effect on your ability to see and recognize objects at a distance. On other occasions, high concentrations of particles reduce the visibility to zero. In order to understand how these different conditions arise, it is useful to examine the types and behavior of particles that are found in the atmosphere.

For our purposes, we can separate atmospheric particles into two general groups: those composed of H_2O, such as water droplets and ice crystals; and dry particles, such as those from combustion, wind-borne soil, and volcanoes to name only a few sources. Both types of particles may be very large and fall rapidly (as rain or some volcanic debris), while others may be so small that they remain suspended (as cloud droplets or haze particles). The number and size of these particles influences not only the visibility, but also the color of the sky.

FOG AND LOW STRATUS CLOUDS

All clouds, whether they are composed of ice crystals or water droplets or a mixture of both, have varying influences on visibility. In cirrus clouds, visibilities of over one-half mile are common, while low clouds and CB have visibilities which range from 100 feet down to zero. We have discussed conditions in thunderstorms extensively in previous chapters; at this point, we will concentrate on fog and low stratus clouds.

Technically, fog is a low cloud which has its base within 50 feet of the ground. As we have seen, fog can be an obscuration or a partial obscuration to visibility. In most cases, fog forms in stable air; that is, it is cooled to saturation by contact with the cold ground (radiation fog and advection fog). It is also caused by adiabatic cooling of stable air (upslope fog) or by some combination of contact cooling and adiabatic cooling. A situation where fog forms in unstable air (at least in the lowest layers) is steam fog. In this case, warm water evaporates and saturates a thin layer of cold air, causing the fog.

In some cases, rather than advection fog, stratus clouds (bases greater than 50 feet AGL) form when strong winds cause mechanical mixing of the air. In this case, surface visibility may not be so bad, but slant visibility is poor until the aircraft descends below the cloud base.

When radiation fog is shallow (ground fog), vertical visibility may be very good, while the reported surface visibility (RVV and RVR) is low. You may experience a progressive deterioration in visibility as the aircraft descends. (Figure 14-4)

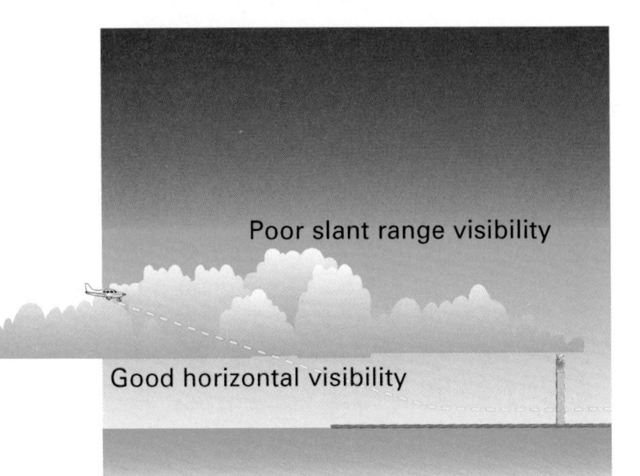

Figure 14-4. Slant range visibility may be better or worse than surface visibility, depending on aircraft altitude and height of the cloud base.

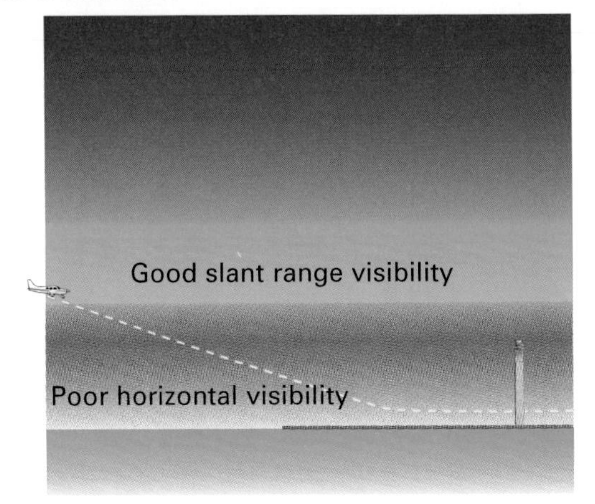

Be especially alert for fog or low stratus clouds when moderate or stronger winds carry stable, moist air up an extended slope. The smaller the temperature-dewpoint spread, the greater the chance of fog. The temperature-dewpoint spread will decrease about 4F° for every 1,000 feet the air is lifted.

In cold climates, ice fog may form. Ice fog is a radiation type fog which, as indicated by its name, is composed of ice crystals. It forms at low temperatures (-20° F or less) and may be quite persistent, especially in cities or industrial areas where many combustion particles are present to act as cloud nuclei. At colder temperatures (-30° F or colder), the sudden addition of moisture and particulates can cause ice fog to form. For example, with very cold temperatures during winter in high latitudes, an airport can rapidly go below VFR minimums with ice fog when an aircraft starts its engines, injecting water vapor and exhaust particles into the atmosphere.

PRECIPITATION

Precipitation can affect ceiling and visibility in a couple of ways. First, drizzle, rain, and snow particles are significantly larger than cloud droplets, so the effect on visibility can be significant. You have seen this in the dark rainshafts associated with heavy convective precipitation. Also recall that if snow or drizzle is light or greater, the visibility will be 5/8 s.m. or less.

If fog occurs with rain or drizzle and the precipitation is forecast to continue, expect little improvement in conditions.

The second way precipitation can affect ceiling and visibility is by saturating layers of the air between the cloud base and the ground. Ragged fractocumulus or fractostratus clouds (sometimes called scud) form below the original cloud base, causing the ceiling to lower over time. Also, precipitation fog may develop when rain saturates the layer near the ground.

Blowing snow occurs when the wind raises snow particles from the surface reducing visibility to less than 7 statute miles. This is a particular problem in polar regions where the snow is dry and fine. Weather extremes are often reached under blizzard conditions. A blizzard exists when low temperatures combine with winds that exceed 30 knots and great amounts of snow, either falling or blowing. The term "ground blizzard" is used when the conditions continue after the snow stops falling.

WEATHER SYSTEMS

Fog and low stratus clouds develop under identifiable larger scale weather conditions. Two of the most common are with warm fronts and when warm, moist air "overruns" a stationary front. These weather patterns typically occur in late fall, winter, and early spring and have similar structures. Air above the front is lifted in a stable upglide motion, producing clouds and precipitation. Precipitation fog and low clouds develop in the cold air near the surface with extensive IMC. You will recall from the previous chapters that icing and wind shear also can be hazards in these conditions. (Figure 14-5)

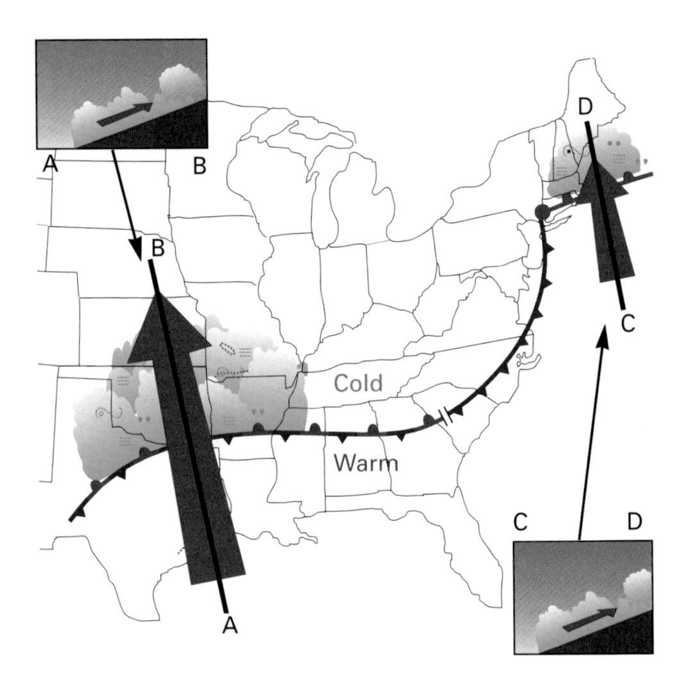

Figure 14-5. Weather map with cross sections showing precipitation falling from the overrunning warm air saturating the shallow cold air near the ground. Low clouds, fog, and, occasionally, icing occurs in a broad band on the cold air side of the front under these conditions.

> Be especially alert for low ceilings and visibilities with low stratus clouds and drizzle whenever low-level moisture overrides a shallow, cold airmass.

Since radiation fog favors clear skies, cold ground, and light winds, it is commonly found in high pressure areas over land in the winter. It is not unusual for fog to form one or two nights after rain associated with a frontal passage dampens the surface. By this time, a high pressure has moved into the area with clearing skies, thus providing light winds and allowing nighttime radiational cooling.

> The temperature-dewpoint spread is a useful index for possible fog formation. Here are two useful rules:
>
> 1. If, at dusk, the temperature-dewpoint spread is 15F° or less, skies are clear, and winds are light, fog is likely the next morning.
>
> 2. Fog is likely when the temperature-dewpoint spread is 5F° or less and is decreasing.

Radiation fog typically dissipates after the sun rises, but there are exceptions. Under light wind conditions, damp, cold valleys are favorite sites for nighttime radiation fog in winter as cold air drains into the valley bottom. If the valley does not receive adequate solar radiation during the following day, the fog will not completely dissipate. If large scale wind and stability conditions remain the same, the fog will worsen with each succeeding night. The central valley of California is a good example. It is noted for its persistent wintertime "valley fog." In some winters, there have been weeks of IMC conditions in the Sacramento and San Joaquin Valleys as the valley fog persisted day and night. It usually takes a major change in the macroscale circulation (such as a frontal passage) to clear out the valley fog.

> Expect little improvement in visibility when fog exists below heavily overcast skies.

Advection fog is common wherever warm, moist air is carried over a cold surface. Over land, this usually is a wintertime phenomenon. Advection fog may contribute to the precipitation fog ahead of a warm front; for example, when the front lies across the East Coast of the United States. Easterly or southeasterly flow carries moist air from the Atlantic over cold land surfaces north of the warm front.

Advection fog also occurs when air over the warm Gulf Stream current blows across the colder waters of the Labrador current to the north. This causes the North Atlantic off the coast of Labrador to be an exceptionally foggy region.

In the warmer months of the year, the coast and coastal waters of California are the location of extensive advection fog and stratus clouds. The cause is northerly and northwesterly winds which carry relatively warm, moist air from the North Pacific over a band of cold waters along the Oregon and California Coast. The cold surface water in that area is the result of upwelling of bottom water. Under typical conditions, stratus clouds persist day and night over the coastal waters. The clouds tend to move inland at night and dissipate back to the coast during the day. This cycle is due to the radiational cooling and warming of land surfaces.

Although visibility below the bases of coastal stratus is often good, the altitude of the cloud bases is typically below 2,000 feet AGL. This can have an important impact on air traffic. For example, at San Francisco International Airport (SFO), a high concentration of traffic arrives from the East Coast at about the time the stratus dissipates in the late morning. A later-than-normal dissipation can severely limit the aircraft arrival rate. This happens because side-by-side approaches to SFO are not allowed when the ceiling is below 2,000 feet AGL. This limitation can cause air traffic delays all the way back to the East Coast.

SMOKE AND HAZE

Smoke is the suspension of combustion particles in the air. You can often identify smoke by a reddish sky as the sun rises or sets, and an orange-colored sky when the sun is well above the horizon. When smoke travels large distances (25 miles or more), large particles fall out and the smoke tends to be more evenly distributed. In this case, the sky takes on a more grayish or bluish appearance and haze, rather than smoke, is often reported.

The impact of smoke on visibility is determined by the amount of smoke produced at the source, the transport of smoke by the wind, the diffusion of the smoke by turbulence, and the distance from the source. In light winds, the most serious visibility reduction is in the vicinity of the source of the smoke. If winds are light, and the atmosphere is stable, visibilities remain low. Unstable conditions cause the smoke to be mixed through deep layers with less impact on visibilities, except near the source.

When smoke is produced by a large, hot fire, it rises to heights where it can be carried great distances in elevated stable layers. A common SA remark from a station near a forest fire is "KOCTY" (smoke over city).

In large industrial areas, smoke provides many condensation nuclei. Fogs tend to be more dense and long lasting in those locations.

Smoke plumes from single sources may be carried far downwind. They are mixed with their environment, depending on their height, the stability, and the windspeed. The height of the plume will not necessarily be the same as the stack height. If the effluent is hot, the plume will initially rise above the stack height until its temperature is equal to that of the environment. In unstable conditions, plumes will be mixed through deep layers. The instability is often visible as thermals cause a plume to take on a looping pattern as it leaves the stack. Under stable conditions, smoke plumes will maintain their shape, except for a tendency to fan out horizontally. A rough rule of thumb applicable for a few miles downwind is that the width of a smoke plume is about 1/20 of its distance from its source. These smoke problems can be complicated near the source (the smoke stack), where heat output in the stack can be large enough to cause significant turbulence in the plume if you happen to fly through it.

Haze is a suspension of extremely small dry particles. Individually, they are invisible to the naked eye, but in sufficient numbers can give the air an opalescent appearance. Haze particles may be composed of a variety of substances, such as dust, salt, or residue from distant fires or volcanoes (after large particles have fallen out). Haze tends to veil the landscape so that colors are subdued. Dark objects tend to be bluish, while bright objects, like the sun or distant lights, have a dirty yellow or reddish hue. When the sun is well above the horizon, haze gives sunlight a peculiar silvery tinge.

When the relative humidity increases beyond 60%, the dry haze particles begin to grow due to the presence of water vapor. The appearance of the haze changes; this "wet haze" layer takes on more of a whitish appearance with decreased visibility. These conditions (haze particles plus high humidities) often occur near the ocean where there is an abundance of moisture and salt particles. Also, a whitish, wet haze is often present early in the morning anywhere temperatures are low and humidities are high. The whitish haze diminishes and visibilities typically increase in the afternoon as temperatures increase.

When smoke or haze is present under overcast skies, there will be little improvement in visibility.

As with smoke, some of the worst haze problems occur in large industrial areas and cities where many air pollution sources add gases and more particulates to any naturally occurring haze particles. In some geographical areas, persistent elevated stable layers often combine with topographical barriers and light winds to trap pollutants. This results in the build-up of air pollution concentrations and further reduces visibility. Most of the primary gaseous pollutants are not visible, but they almost always exist in the presence of smoke and haze in large cities. One exception is nitrogen dioxide (NO_2), caused by the oxidation of nitrogen at high temperatures in internal combustion engines. In high enough concentrations, it has a reddish-brown color. Some well-known locations where air pollution problems are common are Los Angeles (where the term "smog" for smoke and fog was coined), Athens, and Mexico City. Flight into these areas usually means restricted visibilities, especially in the summer.

When a persistent, weak pressure gradient and stable air (for example, in a stationary high) are located over an industrial area, low visibilities due to high concentrations of haze and smoke should be expected.

Besides generally reducing visibilities, haze also contributes to the problem of glare. This often causes visibility to be less looking toward the sun than away from the sun. For example, an observer on the ground may be able to see your aircraft when you can't see the ground because of the glare of the sun from the top of a haze layer. Glare can also be caused by snow and water surfaces and the flat top of a cloud layer.

DUST

Dust refers to fine particles of soil suspended in the air. The actual source of dust or sand may have occurred far away from the point of observation and the dust may be reported as haze. Dust gives a tan or gray tinge to distant objects. The sun's disk becomes pale and colorless, or has a yellow tinge. (Figure 14-6)

Blowing dust is dust raised by the wind restricting visibility to less than 7 statute miles. Visibility is less than 5/8 s.m. in a duststorm, and less than 5/16 s.m. in a severe duststorm. Blowing sand is described similarly to blowing dust, but it is more "localized."

When dust extends to high levels and no frontal passage or precipitation is forecast to occur, low visibilities will persist.

Figure 14-6. A wall of dust marks the leading edge of cold air and strong winds. (Photograph, National Center for Atmospheric Research/University Corporation for Atmospheric Research/National Science Foundation.)

> Be alert for low visibilities due to dust and sand over semiarid and arid regions when winds are strong and the atmosphere is unstable. If the dust layer is deep, it can be carried hundreds of miles from its source.

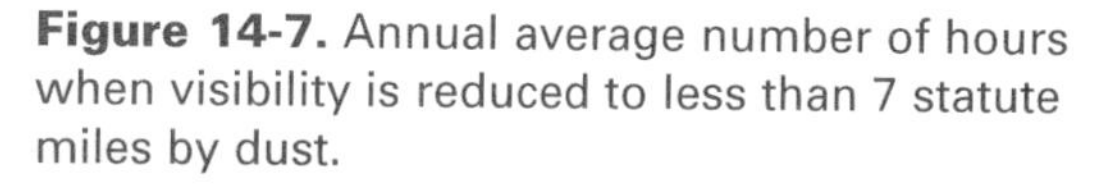

Figure 14-7. Annual average number of hours when visibility is reduced to less than 7 statute miles by dust.

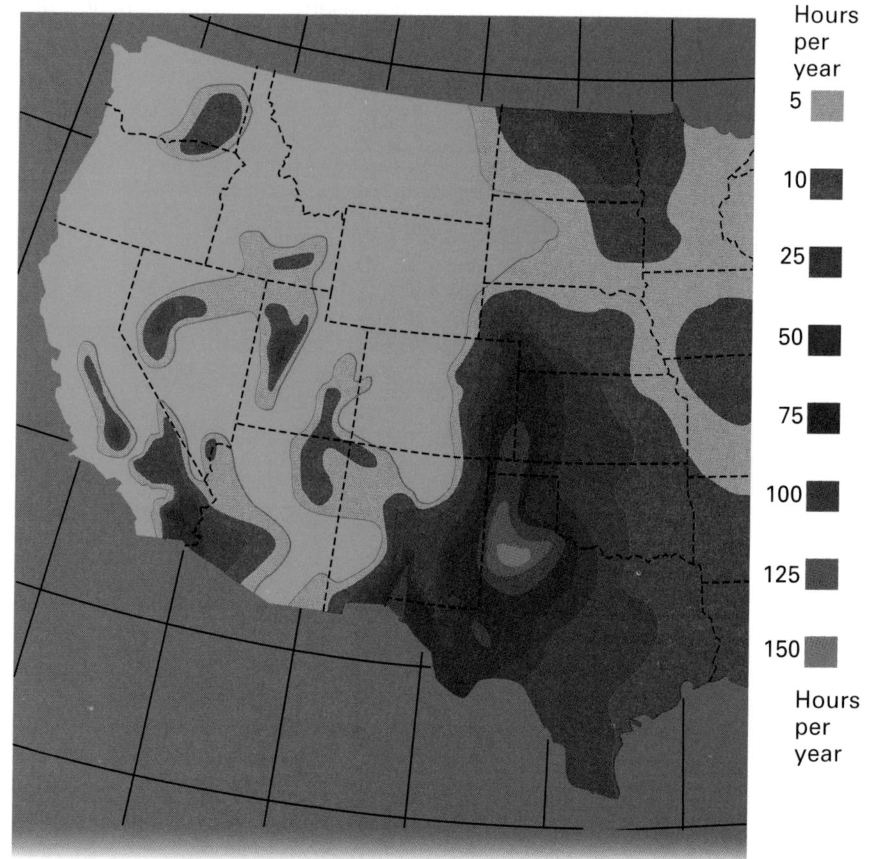

Blowing dust and sand occurs when the soil is loose, the winds are strong, and the atmosphere is unstable. These conditions occur in some locations in the western U.S., such as desert dry lakes and areas of dry land farming. (Figure 14-7)

Figure 14-7 shows that lower visibility due to dust is common in the Texas Panhandle where dry land farming is extensive. Reduced visibilities occur there especially in the spring with southwesterly winds and the passage of cold fronts that have little or no precipitation. In areas along the east slope of the Rockies, chinook winds also cause blowing dust when ground conditions are dry.

CLIMATOLOGY

Limited ceilings and visibilities can occur anywhere, but they occur more frequently in some locations than others, and often have a seasonal preference. A climatology of IFR conditions across the U.S. gives a useful indication of the geographical regions and seasons where the worst problems exist. (Figure 14-8)

Figure 14-8 illustrates the tendency for fewer IFR periods over the west than the east, with exception of the West Coast. The Northwest U.S. lies in the track of wave cyclones that move in from the Pacific. This accounts for the low ceilings and visibilities, especially in the fall and winter. Coastal stratus is common in summer along the California coast as indicated by the higher frequency of IFR conditions near Los Angeles and San Francisco. East of the north-south mountain chains, the frequent occurrence of west winds, with descending dry air, discourages fog production. Even with dust production in these areas in the spring, IFR conditions are relatively infrequent. In the east, air from the Gulf of Mexico and the Atlantic provides adequate moisture to cause a variety of IFR conditions in the winter, including radiation fog, precipitation fog, and advection fog, depending on the location. Conditions also are favorable for the formation of fog and other IMC conditions as lows and their associated fronts move eastward across the Appalachians and the East Coast. This is especially true in the Midwest and the Northeast when these systems are followed by large high pressure regions.

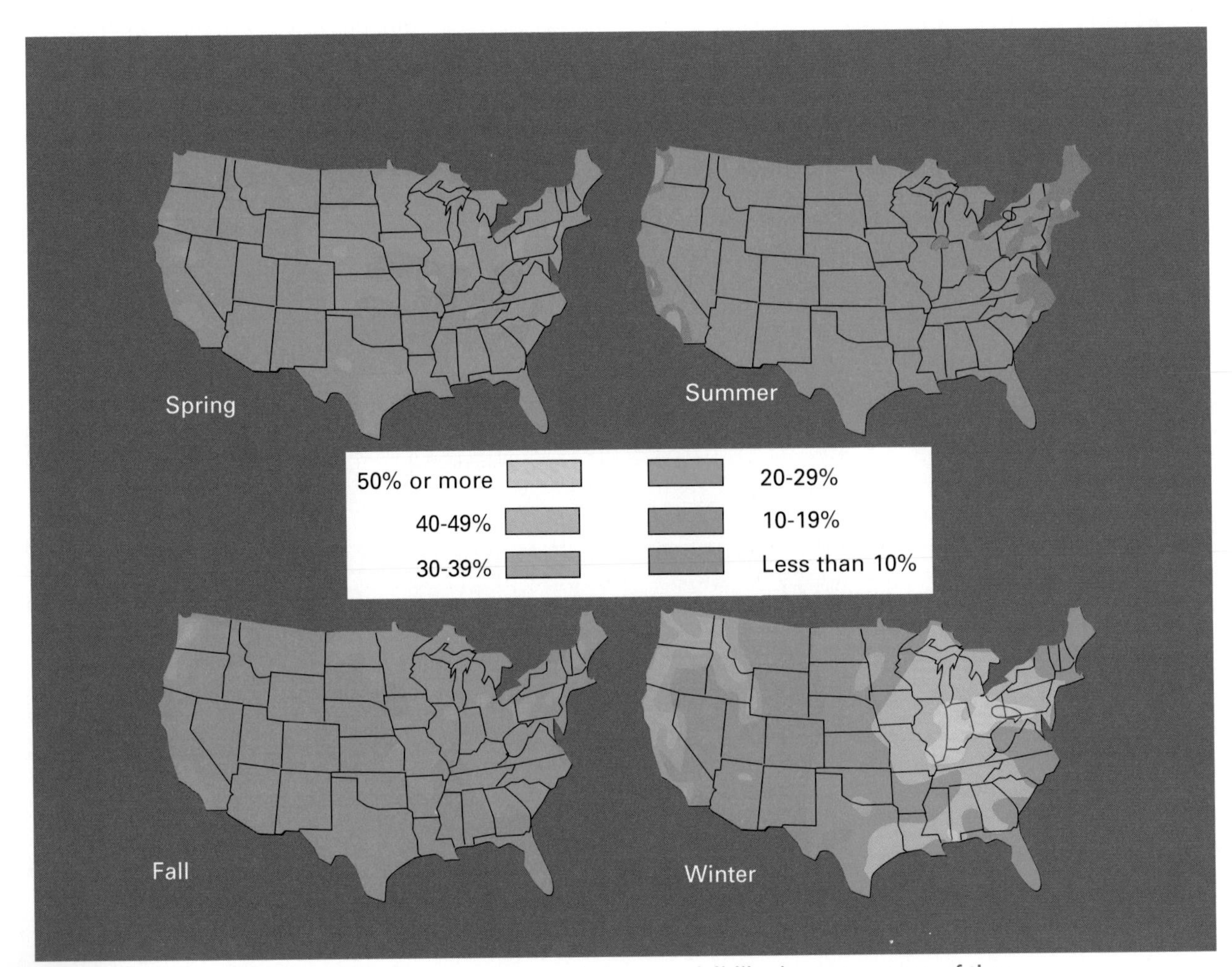

Figure 14-8. Seasonal IFR restrictions to visibility by percentage of time.

SUMMARY

IMC occurs when ceilings and visibilities are reduced by clouds, fog, precipitation, haze, and any other particles produced by natural or anthropogenic sources. These conditions are produced by identifiable large scale weather systems and often in particular geographical areas. Although surface visibility and ceiling observations are made on the basis of specific definitions and procedures, the observations are approximations. Furthermore, surface observations alone give little information with respect to in-flight conditions.

Although imperfect, the system of weather observations is the only one that exists. The wise pilot knows IMC rules and regulations, the technical language of observations and reports of IMC conditions, and the shortcomings of those observations. Finally, when uncertain about conditions, the wise pilot takes a conservative approach to preflight and in-flight decision making with respect to IMC. Remember the advice of the experienced pilot, "There may be rocks in those clouds."

KEY TERMS

Advection Fog
Blizzard
Blowing Dust
Blowing Snow
Dust
Duststorm
Fractocumulus
Fractostratus
Haze
Ice Fog
Instrument Flight Rules (IFR)
Instrument Meteorological Conditions (IMC)
Low Instrument Flight Rules (LIFR)
Marginal Visual Flight Rules (MVFR)
Precipitation Fog
Radiation Fog
Scud
Severe Duststorm
Slant Range Visibility
Smoke
Steam Fog
Upslope Fog
Visual Flight Rules (VFR)
Visual Meteorological Conditions (VMC)

CHAPTER QUESTIONS

1. If low clouds are present the next time you are at the airport, you can do a simple exercise that will illustrate the variability in cloud observations. Be sure to record the time for each observation.

   1. Before you look at any reports, estimate the altitude of the lowest cloud bases.

   2. Record the height reported in the latest SA or other report.

   3. Obtain an estimate of cloud height from the latest local PIREPs and/or your own in-flight observations, both over the airport and 10 miles away.

   4. Discuss, explaining differences in measurements.

2. Construct an exercise similar to question 1, but for horizontal visibility, runway visibility, and slant range visibility.

3. Discuss the meteorological concerns of a VFR descent.

4. You are at an isolated airport in a generally flat area far from the ocean. It is a fall evening, the skies are clear, and winds are light. There is no forecast for your local airport, but the forecast for your destination 100 miles away indicates the chance of dense fog at the time you intend to take off (0600 the next morning). You are carrying a perishable cargo that you must begin loading four hours before takeoff. It is very expensive to load and to unload, so you want to make a decision before you begin to move the cargo. Aside from the previous forecast, you only have local observations. At 0200, it is clear. What should you do?

5. At sunrise on a clear, calm day, a whitish haze has reduced the visibility at the local airport. Ground fog exists along nearby creeks. Do you expect visibility conditions to improve in the next few hours? Explain.

6. It is midnight in winter. Large scale weather conditions show a stagnant high pressure system over your area. Radiation fog occurred under clear skies last night and burned off at 0900 this morning. Fog has formed again this evening. The midnight sky condition is -X 100 OVC. Will it be possible to fly VFR tomorrow morning at 0930? Discuss.

CHAPTER 15
Additional Weather Hazards

Introduction

There are a number of aviation weather hazards that do not fall clearly into any categories of the previous four chapters. The purpose of this chapter is to describe these hazards and to examine their causes and effects. Additionally, information is presented to help you anticipate their occurrence and avoid their negative effects. After you complete this chapter, you will understand such diverse phenomena as atmospheric electricity, stratospheric ozone, volcanic ash, condensation trails, and whiteouts.

ATMOSPHERIC ELECTRICITY
- *Lightning*
  - *Lightning Effects*
- *Static Electricity*

STRATOSPHERIC OZONE

VOLCANIC ASH
- *Volcanic Ash Hazards*
- *Ash Cloud Behavior*
- *Reports and Warnings*

CONDENSATION TRAILS

MISCELLANEOUS HAZARDS
- *Whiteout*
- *Low-Level Inversions*
- *Runway Conditions*

Section A

ATMOSPHERIC ELECTRICITY

The two forms of atmospheric electricity that are of primary concern to pilots are lightning and static discharges.

LIGHTNING

Lightning is defined as any or all of the various forms of visible electric discharge produced by thunderstorms. You will recall from Chapter 9 that the "thunder" of a thunderstorm is, in fact, a result of lightning. The discharge is caused by the separation of electrical charges within the cloud. The cloud is usually a CB, but lightning also has been observed in the vicinity of a volcanic eruption.

Lightning may occur within the cloud, between the cloud and the ground, and between clouds. Discharges have also been observed between a cumulonimbus and the clear air surrounding the storm. Cloud-to-ground lightning strikes are the most dramatic manifestations of this phenomenon, but they are only a fraction of the total number of discharges. The great majority of lightning discharges occur within or between clouds where aircraft are vulnerable targets. (Figure 15-1)

Lightning discharges occur wherever thunderstorms are found. The climatology of thunderstorms in Chapter 9 gives an indirect estimate of where lightning activity should be expected most often across the continental United States.

LIGHTNING EFFECTS

Lightning strikes on aircraft result in a variety of adverse effects. Although most of them are minor;

Figure 15-1. Lightning discharges may be from cloud-to-cloud, cloud-to-ground, within clouds, and, occasionally, from cloud to clear air. (Photograph, National Center for Atmospheric Research/University Corporation for Atmospheric Research/National Science Foundation.)

in some cases, the damage can be severe enough to result in an accident or incident. A lightning flash is extremely bright. Temporary blindness is not an unusual occurrence.

During flight near thunderstorms, avoid looking directly at the storm to reduce the danger of temporary blindness due to lightning. Turn up the cockpit lights to full bright, even during daylight hours, to lessen the temporary blindness from lightning.

Aircraft structural damage is usually restricted to effects such as small holes in the fuselage. Wingtips, engines, and other equipment protruding from the aircraft have taken lightning strikes. Problems include twisted and burned antennas (interfering with communications and navigation equipment) and damaged pitot tubes (causing erroneous airspeed readings).

Lightning is not typically associated with visible damage; but, it may cause significant interference with electrical systems. Electric motors (for example, wing-folding motors) have been known to operate spontaneously after lightning strikes. An increased reliance on digital flight control systems as opposed to analog or mechanical systems has made some aircraft even more vulnerable to the lightning problem. The damage in these situations results in errors in output from electronic processing equipment. Instruments, avionics, radar, and navigational systems can be influenced.

After a lightning strike, all instruments should be considered invalid until their proper operation is verified.

Potentially, one of the most serious effects of a lightning strike is the ignition of vapors in a fuel tank. However, the small number of suspected cases in the past, and improvements in fuel tank designs over the last 30 years or so, make the probability of such occurrences very small, but still not impossible.

All airports require suspension of refueling operations during nearby (within about 5 n.m.) lightning activity.

The potential for aircraft strikes is high in the thunderstorm anvil, even after the main thunderstorm cell has weakened in the dissipation stage. However, the highest frequency of lightning strikes is at lower troposphere.

To reduce the chance of a lightning strike, do not fly within ± 5,000 feet of the freezing level or, alternately, do not fly in the layer where the OAT is between +8°C and -8°C.

Another word of caution: just because the chance of a catastrophic lightning strike is small doesn't mean that the problem should be considered "minor." It can be said generally that the most frequent, and often, the most serious flight problems do not occur when a weather hazard is at its worst, but rather when it is in combination with other hazards. Lightning offers a good example of this rule of thumb. Consider a situation where you find yourself flying in a thunderstorm environment. Conditions are IMC with turbulence and icing. Clearly, your full attention is required to deal with the problems at hand. It is conceivable that a nearby lightning flash could temporarily blind and disorient you just long enough for you to lose control of the aircraft. Be prepared. Avoid thunderstorms.

STATIC ELECTRICITY

When an aircraft flies through clouds, especially thunderstorms, the impact of particles, such as precipitation, dust, or sand may cause an electrical charge to build up on the aircraft. Static electricity refers to the spark or point discharges that occur when the electric charge difference between the aircraft and its surroundings become large enough. The most common effect of static electricity (also called precipitation electricity) is low frequency radio noise. This static can be particularly loud and bothersome in the 200 to 400 kHz frequency range. It can also be heard in the VHF range, although the effect is not nearly as bad as in HF range. Less frequently, a corona discharge, known as St. Elmo's Fire, appears as a bushy halo around some prominent edges or points on the aircraft structure and around windscreens.

Section B

STRATOSPHERIC OZONE

In Chapter 1 the ozone layer in the lower stratosphere was described as a prominent feature of the lower stratosphere. Ozone (O_3), has both good and bad qualities. On the good side, it absorbs damaging UV radiation from the sun. On the bad side, it is not good in an environment where animals, people, and plants are present because it is toxic. In large enough concentrations, it has an acrid smell, it irritates the eyes, and can cause respiratory difficulties. These effects are well known in heavily populated areas, such as Los Angeles, where an ample supply of solar radiation interacts with oxides of nitrogen from automobile exhausts. This process produces "photochemical smog" of which ozone is a primary component.

In the lower atmosphere, the highest concentrations of ozone usually occur in the afternoon. If an elevated stable layer traps air pollutants, ozone concentrations tend to be relatively large and persistent just below the inversion. Under clear sunny skies, this level of potentially high ozone concentrations can be recognized as the top of a haze layer which extends to the surface.

Naturally occurring ozone from the stratosphere may also create a hazard to flights at and above the tropopause. Exposure can occur in two ways. First, an aircraft may simply be so high in the stratosphere that it is close to the maximum concentrations in the ozone layer. The supersonic transport aircraft (SST) and some military aircraft reach such altitudes.

Another way that aircrew exposure to ozone increases is when atmospheric motions bring stratospheric air with high concentrations of ozone down to altitudes near the tropopause where more aircraft fly. These processes go on near jet streams, especially where extratropical cyclones are very strong or are rapidly intensifying. Typically, the tropopause is much lower over the cyclone, and ozone-rich stratospheric air is more likely to be found there. Also, stratospheric air is "injected" into the upper troposphere on the left side of the jet stream axis (looking downwind). (Figure 15-2)

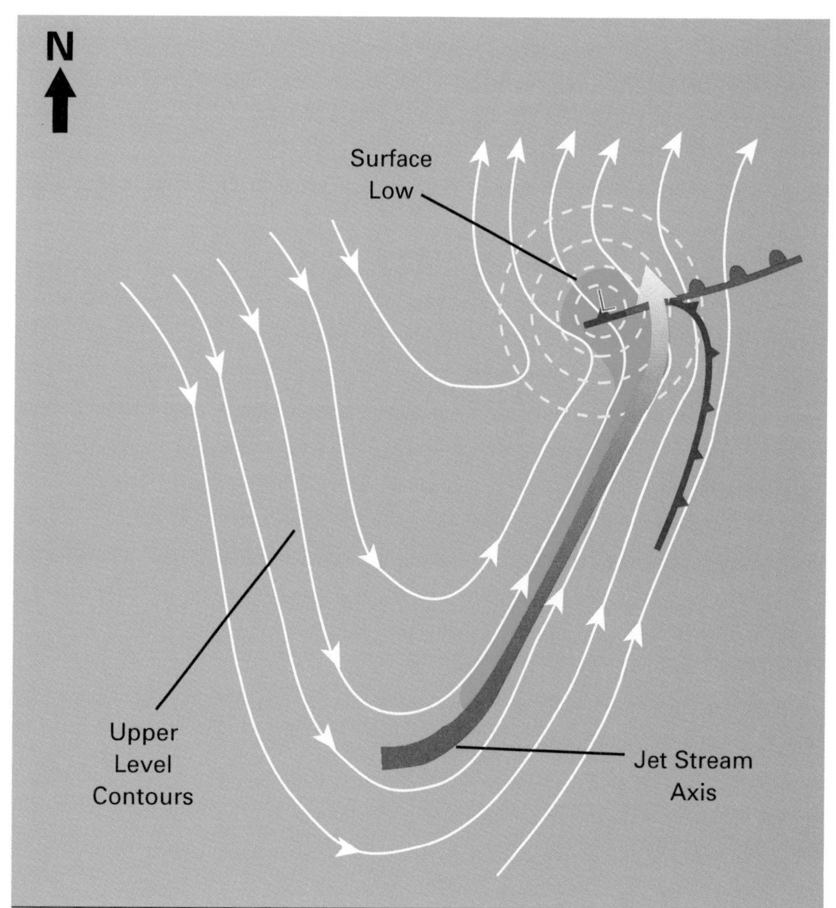

Figure 15-2. Ozone-rich stratospheric air (shaded area) is brought down to the upper troposphere over an intense surface low pressure area and along the left side of a strong jet stream (looking downwind).

When these processes take place at high latitudes, there is an increased probability of bringing high ozone concentrations to common airline cruise altitudes. This is because the average tropopause height decreases toward the poles. Some airlines may restrict flights to lower altitudes when ozone concentrations are estimated to be above a critical level.

There is no question that exposure of passengers and crew to elevated ozone concentrations will occasionally occur within the service ceiling of many corporate and airline aircraft. The amount and impact of the exposure is uncertain. Until there are regular measurements made in the free atmosphere and onboard aircraft, exposure on any given occasion can only be roughly estimated. Attention to the design of aircraft ventilation systems, knowledge of meteorological conditions, and the use of good flight procedures can minimize the potential problem.

Section C

VOLCANIC ASH

Volcanoes are a fact of life around the world. Earthquakes, explosive volcanic eruptions, heavy volcanic ash fallouts, and lava flows may be the most spectacular aspects of volcanic activity on the ground; but in the air, it is the volcanic ash cloud. Volcanic ash, which consists of gases, dust, and ash from a volcanic eruption, can spread around the world and remain in the stratosphere for months or longer. This volcanic material has important effects on the amount of solar radiation received at the earth's surface and, therefore, on the weather and climate. A major influence of volcanic ash clouds is the interruption of flight activities. (Figure 15-3)

VOLCANIC ASH HAZARDS

When an aircraft approaches an ash cloud some distance from a volcano, the cloud isn't always easy to distinguish from ordinary water or ice clouds. However, upon entering the cloud, the situation is distinctly different. Dust and smoke may enter the cabin, often with the odor of an electrical fire. Visible indications of the ash particles

Figure 15-3. Mount St. Helens eruption in Washington state, May 1980.

include lightning, St. Elmo's Fire around the windshield, and a bright orange glow around jet engine inlets. Because the ash is highly abrasive, particle impacts can pit the windscreen and landing lights to the point where they become useless. Depending on the conditions, there may be worse effects. Control surfaces can be damaged and the pitot-static system and ventilation systems can become clogged, causing instruments to malfunction. The ingestion of volcanic ash damages jet engines; it can cause compressor stalls, torching from the tailpipe, and flameouts.

Piston aircraft are less likely than jet aircraft to lose power due to ingestion of volcanic ash, but severe damage is almost certain to occur, especially with a volcanic cloud only a few hours old.

ASH CLOUD BEHAVIOR

Volcanic ash clouds are most dangerous close to the volcano when an eruption has just occurred because the ash particles are large. When the ash cloud is within 30 n.m. of the volcano, it may be identifiable with a nearby NEXRAD or other special radar installation. Depending on the size of the ash cloud, radar sites as far away as 400 n.m. may be able to detect the cloud. However, the presence of precipitating clouds can mask the target.

It is most important to avoid any encounter with a volcanic ash cloud, especially those which are only a few hours old. Make every effort to remain on the upwind side of the volcano.

The effect of the ash cloud on flight activities is not limited to the region of the largest particles. When volcanic material is injected into the stable stratosphere, fallout is slow. The cloud of gases and smaller particles spreads out and is carried away by atmospheric winds. For example, at a nominal speed of 25 knots, a typical ash cloud spreads downwind at 600 n.m. per day. If the cloud is near the jet stream, a more rapid movement may occur. In a recent case, an ash cloud from an eruption in Alaska reached the lower 48 states in a couple of days, causing disruption of air traffic in the upper midwest. (Figure 15-4)

Volcanic clouds may extend to great heights and over hundreds of miles. Pilots should not attempt to fly through or climb out of the cloud.

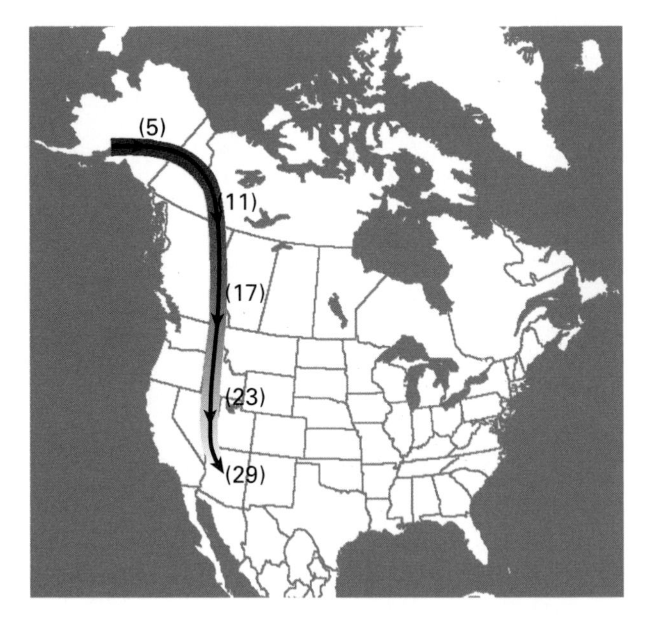

Figure 15-4. Estimated trajectory of the ash cloud from the Mt. Redoubt eruption. Times in hours after the eruption are indicated in parentheses. Note that the ash cloud passed across the major airline routes of the western U.S. and Canada in less than 30 hours.

REPORTS AND WARNINGS

Currently, NWS monitors volcanic eruptions through pilot reports, radar, and satellite observations. Forecasts of the future locations of ash clouds are made with the help of trajectory estimations from computer forecast models at the National Meteorological Center. Subsequent advisories (SIGMETs) are issued by the appropriate weather service forecast office (WSFO), air route traffic control center (ARTCC), and center weather service unit (CWSU) via center weather advisories (CWA).

The most important link in volcano observation and warning programs for aircraft is the pilot. Especially in remote areas, the pilot is often the first to see an eruption and, of course, any aircraft that inadvertently flies into an ash cloud becomes a direct sensor of the cloud location and effects. All pilots are advised to report volcanic activity.

Pilots must depend on reports from air traffic control (ATC) and other pilots to determine the location of the ash cloud and to remain well clear of the area.

If you see a volcanic eruption and have not been previously notified about it, immediately report it to ATC.

Section D

CONDENSATION TRAILS

A condensation trail or contrail is defined as a cloud-like streamer that frequently forms behind an aircraft flying in clear, cold, humid air. Contrails most often develop in the upper troposphere; however, they can occur at any altitude depending on a variety of things such as temperature, humidity, and type of aircraft. Aside from obvious military concerns (aircraft detection), the hazard presented by contrails is the development of a cloud deck with reduced visibility at a flight level where, previously, no cloud existed. (Figure 15-5)

There are two types of contrails; aerodynamic or wing-tip contrails and exhaust contrails. Aerodynamic contrails are formed when the pressure is lowered by air flowing over propellers, wings, and other parts of the aircraft. Adiabatic cooling brings the air to saturation. These contrails are typically thin and short-lived. Aerodynamic trails are often visible in wingtip vortices, especially in aircraft with heavy wing loading during takeoffs and landings in mild, damp weather. Occasionally, small, fast aircraft performing aerobatic maneuvers sustain large wing loads and produce distinctive helix-shaped clouds in wingtip vortex trails.

Exhaust contrails form when hot, moist exhaust gases mix with cold air. A critical condition for this type of contrail is low temperature, depending on the altitude. For example, the temperature must be less than -24°C near sea level and less than -45°C at FL500 for the formation of exhaust contrails.

In comparison to aerodynamic trails, exhaust trails can be more substantial and longer-lived, especially if the air in which they form is very cold and near saturation. In fact, it is not unusual for exhaust trails to form in the same layer that already contains wisps of cirrus clouds.

A major impact of contrails is occasionally seen from the ground in otherwise clear skies in busy airspace. Over a period of a few hours, the sky can become overcast with cirrus clouds as contrails from many aircraft form, spread out, and linger.

Figure 15-5. Contrails form behind aircraft at high altitudes when the air is cold and humid. In some cases, they spread out to form a high-level cloud deck.

The formation of contrails can be prevented by seeking a layer that is drier and/or warmer. Sometimes, a climb from the tropopause into the stratosphere will accomplish this. At other than tropopause level, a climb or descent out of a well-defined layer in which patches of altocumulus or cirrus clouds are present is often sufficient to prevent contrail formation.

Occasionally, a contrail casts a distinct dark shadow on a smooth deck of cirrostratus clouds at a lower altitude. Seen from the ground, this phenomenon is sometimes misinterpreted as a distrail. A dissipation contrail or distrail is actually a streak of clearing that occurs behind an aircraft as it flies near the top of, or just within a thin cloud layer. In this case, the heat added by the aircraft exhaust and/or the mixing of dry air into the cloud layer by the aircraft downwash causes the dissipation of the cloud along the aircraft track. Distrails are much less common than contrails.

Section E

MISCELLANEOUS HAZARDS

In Part III, you were introduced to weather hazards that affect a significant portion of flight operations. The list is not complete. Other hazards exist, many of which are unique to limited geographical areas. It is beyond the scope of this text to cover all of them, but in order to emphasize the importance of the continued study of weather hazards, we end the chapter with some brief descriptions of other potential problems that you may encounter.

WHITEOUT

Snow-covered regions at high latitudes can present some special visibility problems. For example, on some days, the reflection of light by snow-covered surfaces reduces contrast on the ground, making it difficult to identify objects. A related, but more serious problem is whiteout. This is a situation where all depth perception is lost because of a low sun angle and the presence of a cloud layer over a snow surface. The diffusion of light from the sun by the cloud layer causes the light to be reflected from many angles when it reaches the ground. Repeated reflections between the ground and the cloud eliminate all shadows. The horizon cannot be identified and disorientation may occur.

LOW-LEVEL INVERSIONS

Strong low-level inversions are common in snow-covered regions in the winter with severe cases occurring most frequently in the Arctic and Antarctic. An aircraft climbing out under such conditions experiences a marked decrease in climb performance due to higher temperatures at the top of the inversion.

RUNWAY CONDITIONS

A common hazard that is not limited to polar regions is a decrease of braking effectiveness. This occurs when friction between the tires and the surface is reduced. This occurs with takeoffs or landings when the runway is covered with ice, slushy snow, or water. Also, if the ground is covered with volcanic ash which subsequently becomes wet, braking action can be seriously affected. Braking effectiveness is greatly reduced by hydroplaning which occurs when a thin layer of water separates the tire from the runway surface. Heavy rain and/or slow drainage of the runway surface causes these conditions. This hazard is just one more reason why landing in the face of a downburst is inadvisable.

When you operate in conditions where braking effectiveness is poor or nil, be sure the runway length is adequate and the surface wind is favorable.

Hydroplaning is most likely to occur under conditions of standing water, slush, high speed, and smooth runway texture.

At very low temperatures (-40°F), the structure of ice surfaces begin to get "sticky." This condition affects aircraft equipped with snow skis. Longer takeoff runs and more power are required. At -60°F, takeoff in ski planes is nearly impossible.

SUMMARY

This chapter alerted you to a variety of additional aviation weather hazards, some of which are rare and others that are more often nuisances. However, all have the potential of contributing to serious difficulties when they occur with other flight problems. Also, a few of them by themselves can create critical flight conditions (for example, lightning, volcanic ash, whiteout, and hydroplaning). As with all weather hazards, your newly gained knowledge of their causes and of the conditions under which they occur should help you anticipate and avoid the hazards.

KEY TERMS

Aerodynamic Contrail
Condensation Trail
Contrail
Dissipation Contrail
Distrail
Exhaust Contrail
Hydroplaning
Lightning
Ozone
St. Elmo's Fire
Static Electricity
Volcanic Ash
Whiteout

CHAPTER QUESTIONS

1. In whiteout conditions, it has been said that objects appear to "float in the air." What causes this?

2. What causes lightning to occur with volcanic eruptions?

3. Mt. St. Helens, which erupted in southern Washington state, caused much devastation on the ground. However, the ash cloud did not have the same long term effects as ash-clouds from either Mt. Pinutubo in the Philippines or El Chichon in Mexico. Why? (Hint: do some research on the description of the Mt. St. Helens eruption.)

4. An aircraft makes a wintertime flight from Seattle to Boston at FL410. Weather conditions included major low pressure systems and associated fronts on both coasts and a large high pressure center over the central U.S. Draw a diagram showing the aircraft track and the likely upper air patterns at flight level. Show contours and jet streams. Where would increased concentrations of stratospheric ozone most likely be encountered?

5. You are somewhere north of 30°N latitude. Because you are in a mountain valley, surface winds aren't a very reliable indicator of large scale wind patterns. A volcanic eruption has occurred within a few hundred miles of your location in the last 12 hours. A light dusting of ash covers the airport. The pressure has been falling steadily during that period and clouds have increased from the west (the clouds are visible despite the falling ash). Based on this information alone, where is the volcano (distance and direction) relative to your location? Support your answer with a consistent description and sketches of the large scale weather pattern.

Part IV

Applying Weather Knowledge

Part IV
Applying Weather Knowledge

Now that you understand weather-producing processes, the behavior of weather systems, and the flight hazards that weather phenomena generate, you can begin to apply this knowledge as a pilot. Part IV introduces the forecasting process and provides an overview of weather information sources. Successfully using this information to evaluate weather situations for flight is an important skill to acquire. The ability to visualize the weather and create an accurate weather picture is an essential component of an effective self-briefing procedure.

When you complete Part IV, you will have an understanding of the forecasting process and its limitations. After becoming familiar with the various weather information sources available, you must be able to interpret the information obtained in briefings, printed reports, forecasts, graphic weather products, and other sources. Part IV provides the framework for interpretation of this information and guidelines to help you formulate a mental image of the weather and the impact that conditions may have on a specific flight. As a pilot, the development of weather visualization skills will enhance your flight safety and prepare you to successfully utilize the innovative aviation weather products of the future.

(Challenger photograph on previous page, source: Bombardier Business Aircraft)

Chapter 16

Aviation Weather Resources

Introduction

Weather information and forecasts are beneficial in numerous ways. For example, a prediction of warm temperatures or the chance of rain, helps us decide whether to plan a picnic or carry an umbrella for the day. More importantly, forecasts of severe weather such as blizzards, thunderstorms, or hurricanes, help communities prepare to prevent property damage and save lives. As a pilot, weather influences your life in a unique way. Determinations regarding weather conditions must be made before every flight and crucial weather decisions may have to be made in flight. A wide variety of weather resources are available to assist you in this decision making process. In this chapter, we examine weather information sources and the forecasting process, from the compilation of weather observations to the transmission of data to you.

When you complete Chapter 16, you should understand how weather information is compiled and the methods used to produce forecasts. You should also be familiar with the weather information sources available to you and the variety of printed reports and forecasts, as well as graphic weather products that can assist you in developing a comprehensive weather picture before a flight.

Section A

THE AVIATION WEATHER FORECASTING PROCESS

If the initial conditions of a system are known, can an accurate determination be made about that system's state in the future? This is the question that weather forecasters must consider every day, but predicting the future state of a weather system is not an easy task. To fully comprehend and effectively use aviation weather information, you must first understand the process that leads to the development of these resources.

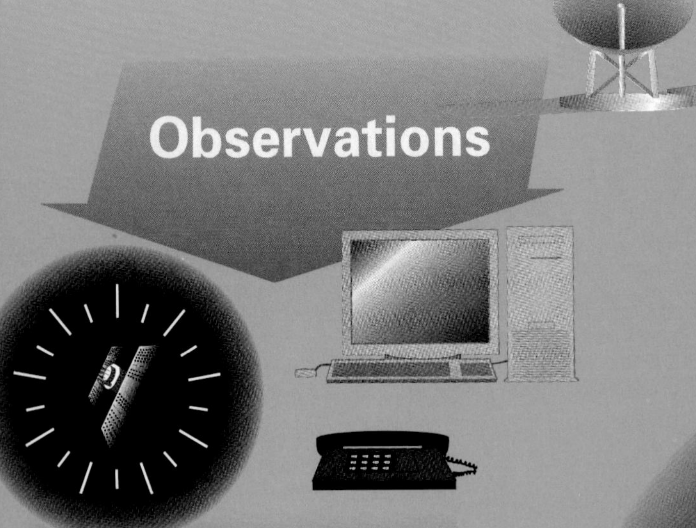

COMPILING WEATHER DATA

In order to develop a forecast for a specific location, the present weather conditions over a large area must be known. Observations of surface weather conditions are provided by a network of over 10,000 observing stations located throughout the world, as well as by most airports. Weather radar systems, such as NEXRAD, obtain additional information about precipitation and wind. Upper-air data are chiefly provided by radiosondes, wind profilers, aircraft, and satellites. The World Meteorological Organization, a United Nations agency consisting of over 130 nations, is responsible for the international exchange of weather data, and ensures that observation procedures are consistent between nations.

After an observation is taken, it is relayed by telephone, computer, or satellite to a communication substation. The combined data obtained from many observations is then sent to the three World Meteorological Centers located in Melbourne, Australia; Moscow, Russia; and Washington, D.C. From these centers the compiled worldwide weather information is transmitted to National Meteorological Centers in each participating country, including the National Meteorological Center in Camp Springs, Maryland, located just outside Washington, D.C.

World Meteorological Centers

- Melbourne, Australia
- Moscow, Russia
- Washington, D.C.

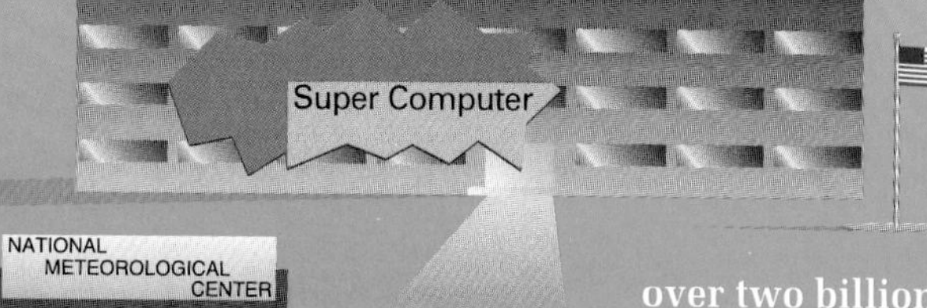

Text and Graphics

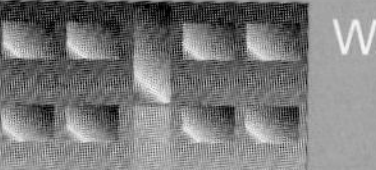

FSS

Private Agency

WEATHER DATA PROCESSING

The National Meteorological Center (NMC) is the focus of United States' weather forecasting. Analyzing data, preparing charts and printed reports, and forecasting the weather on a worldwide and national level are all accomplished at the center. A few of these charts and forecasts are still manually prepared by meteorologists, but most are produced automatically from the output of the NMC's super computer which can perform over two billion operations per second. The computer not only analyzes data but also produces forecasts. Each day, the NMC generates approximately 20,000 different kinds of text and 6,000 graphic images. This information is sent to public and private agencies worldwide, including Weather Service Forecast Offices (WSFO) and Weather Service Offices (WSO), flight service stations, television stations, and airlines.

You

FORECASTING METHODS

Local forecasters use the weather data supplied by the NMC combined with their own knowledge of the weather characteristics of a specific region to create local weather predictions. A variety of methods are used to generate forecasts. Some of these techniques require minimal meteorological information or background. For example, the simplest weather forecast is the persistence forecast. This prediction is based on the concept that future weather will be the same as current weather. For example, if you look outside at this moment and observe a splendid sunny day without a cloud in the sky, you can predict with reasonable assurance that it will be a sunny day an hour from now (provided the sun doesn't set between now and then). The persistence forecast is most accurate for time periods of a few hours, with accuracy deteriorating after that, especially when the weather is bad.

The climatological forecast is based on the average weather (climatology) of a particular region. Long-period predictions about weather conditions are based on weather records from the past. For example, by predicting that there will be no snowstorms in Los Angeles in July, you could consider yourself to be quite a successful forecaster. Climatological forecasts work best where there are only small changes in day-to-day weather conditions.

Meteorological forecasts are predictions which are based on meteorologists' scientific knowledge of atmospheric processes. Numerical weather prediction, which employs mathematical equations to predict the weather, is an essential component of modern meteorological forecasting. The equations represent the physical laws which govern the behavior of the atmosphere. Beginning with observations of current conditions, the equations are solved to determine future conditions. Due to the large number of calculations involved in this task, computers must be used. When a system of equations is programmed into a computer, it is called a numerical prediction model. The NMC super computer uses numerical prediction models to produce forecasts for periods up to two weeks or more for the entire world.

Once the computer forecasts are made, local forecasters use the numerical weather predictions as guidance, and then add detail based on their own scientific knowledge and experience. For example, some forecasters use analogue forecasts. This method depends on the ability of the forecaster to recognize weather patterns that have occurred in the past. Once a pattern is identified, past records of the associated weather conditions can be used to predict future weather. Another simple, but useful, forecast method is extrapolation. This is based on the concept that surface weather systems tend to move in the same direction and at approximately the same speed as they have been moving. A simple calculation using speed and distance allows the forecaster to predict the approximate time that a system will pass through a particular location. These and other local forecast techniques are used to adjust numerical model output for some mesoscale effects which the model doesn't handle well.

FORECASTING ACCURACY AND LIMITATIONS

Both the prediction method and the forecast period determine the accuracy of weather prediction. Forecasts can be quite accurate out to 12 hours. From 12 to 24 hours, predicting the movement of large scale extratropical weather systems and the variations in temperature, precipitation, cloudiness, and air quality associated with these systems are generally well forecast. Usually, forecasters can accurately predict the occurrence of large-scale circulation events such as cold waves and significant storms several days in advance. After five days, however, the ability to predict details deteriorates rapidly. Due to the difficulty in making accurate predictions about specific weather conditions too far in the future, you can appreciate the importance of using the most current weather information while flight planning. It would not be wise to base a go/no-go decision on a forecast that you received 2 days ago. Hopefully you won't be too disappointed when the trip you planned a week ago, based on an excellent long-range forecast, has to be canceled due to a snowstorm. (Figure 16-1)

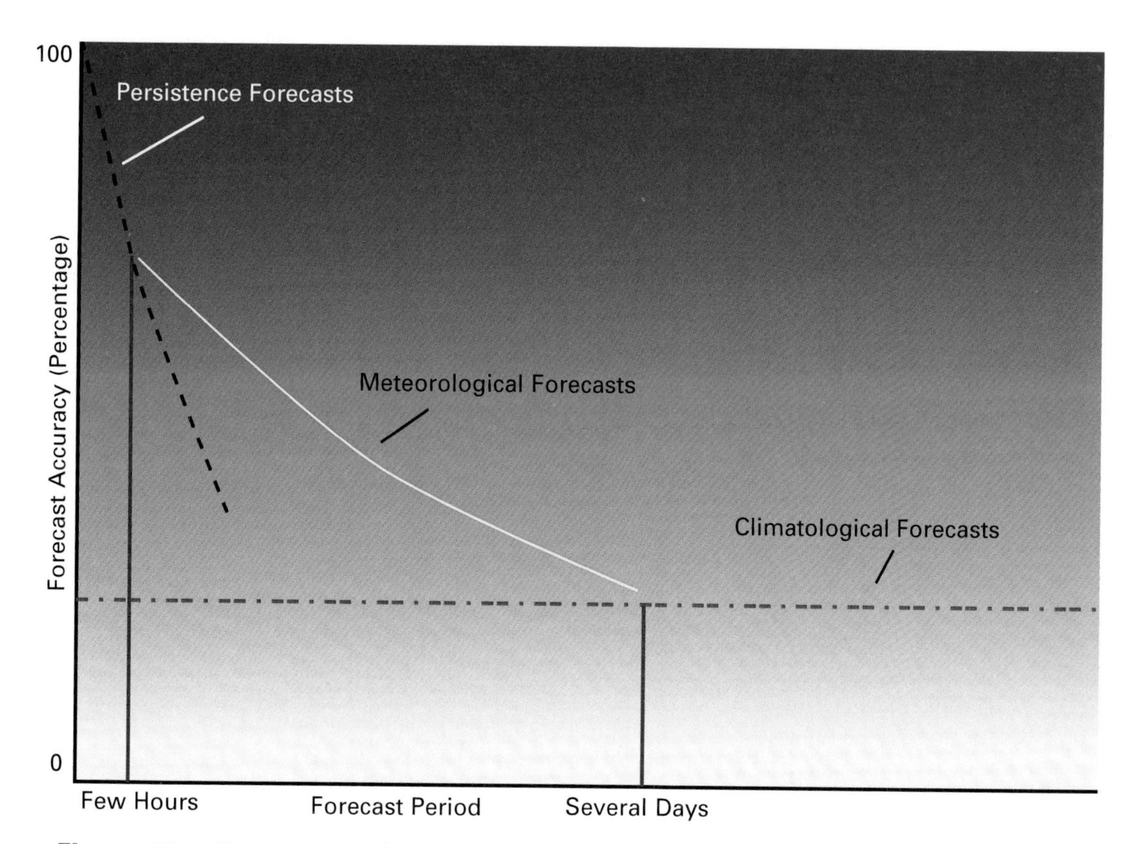

Figure 16-1. The accuracy of weather prediction is a function of the forecast period and the forecasting method. This diagram compares the accuracy of persistence, meteorological, and climatological forecasts from a few hours to a period of several days.

Although forecasting methods and accuracy continue to improve, especially with the aid of computers, there are limits to weather prediction. This is caused to a great degree by the large distance between weather stations and the long time between weather reports. It is difficult for forecasts to be more precise than the current observations despite our understanding of the physics of the problem. In addition, numerical prediction models are based on certain simplifying assumptions about the atmosphere. Occasionally, the assumptions are wrong and the models will not come close to portraying the actual atmospheric state. This is a particular problem for microscale and some mesoscale weather phenomena. Another problem arises because the computer solves equations based on representations of the atmosphere at locations called grid points which may be 20 nautical miles or more apart. Even if it is observed accurately to begin with, a small circulation, such as a thunderstorm, will not be indicated on computer-generated forecast charts. The effects of some surface features, such as water, ice, and mountains on local weather systems are also more difficult for computer models to adequately predict.

THE CHAOS OF WEATHER

It may sound as if forecasting the weather is somewhat chaotic, which is not too far off the mark. Chaos theory describes the difficulty in weather prediction based on "sensitive dependence on initial conditions." The science of chaos began in the early 1960s with a mathematician and atmospheric scientist named Edward Lorenz, and his exploration of computer forecasts.

Edward Lorenz graduated from Dartmouth College in 1938, believing that mathematics would be his chosen profession. World War II interfered with this plan, and he was assigned the duty of weather forecaster in the Army Air Corps. After the war, Lorenz decided to use his mathematical background to pursue mathematical and theoretical problems involving weather systems. Although scientists understood that measurements could never be perfect, the introduction of computer models fostered a renewed enthusiasm for the idea that; given an approximate knowledge of a system's initial conditions and an understanding of physical laws, one can calculate the approximate behavior of the system. In order for this concept to be successful, very small disturbances in a system would have to have little effect on its overall behavior.

Astronomers relied on this assumption in applying mathematical equations to describe the motions of stars, planets, and comets. Precise computer predictions of trajectories of spacecraft and missiles were also based on this idea. So, why couldn't this same concept be used to calculate the movements of weather systems or even the development of an individual cloud? That is the question that Edward Lorenz hoped to answer. Lorenz developed a set of equations that represented certain aspects of weather and programmed them into his Royal McBee computer. The computer output revealed changing patterns that reflected actual weather systems. This seemed to confirm Lorenz's intuition that weather repeated itself, displaying familiar patterns over time.

One day in 1961, Lorenz wanted to reexamine a particular forecast pattern. Instead of beginning the computer run again with the original figures, he typed numbers from the existing printout in the location that he wanted the computer to begin. Assuming that the new run would resemble the earlier printout, Lorenz was shocked to observe that the weather diverged rapidly from the existing pattern. After generating a three-month forecast, all resemblance between the earlier pattern and this new printout had disappeared. What went wrong? The program was the same and the computer was not malfunctioning. The mystery must lie in the numbers that Lorenz had entered. He soon discovered that the numbers were the problem. The computer calculated numbers to six decimal places (.819631), but the printout only indicated three (.819). The numerical forecast changed dramatically due to the very small difference between the initial data. At first the two weather patterns produced by Lorenz's computer closely resembled each other, but the patterns soon began to diverge until all resemblance disappeared. Although each computer run had the same starting point, the small differences in the numbers Lorenz entered created large differences in results. The same effects occur with actual weather systems.

What conclusion about weather forecasting can be drawn from this incident? The outcome provides evidence that small atmospheric disturbances can result in large scale effects on weather systems. This "sensitive dependence on initial conditions" is also known as the "Butterfly Effect." This name stems from the notion that a butterfly flapping its wings in Hong Kong could influence the development of a snowstorm in Denver a few months later. What the Butterfly Effect suggests is that all the conditions affecting a forecast can never be completely known, and as time progresses, a forecast will diverge further and further from reality. Even with improving forecast accuracy due to better methods of observation, a denser network of observing stations, and near perfect computer models, chaos theory may ultimately limit accurate, detailed day-by-day forecasts to about two weeks.

Section B

WEATHER INFORMATION SOURCES

The final step in the forecasting process is disseminating the weather information to you, the user. A wide variety of resources are available to assist you in determining the weather for flight planning.

FAA FLIGHT SERVICE STATIONS

The flight service station (FSS) is the most common source of weather information for pilots. New automated flight service stations (AFSSs), numbering about one per state, are the result of consolidating older, manual stations. More aviation weather briefing services are provided by FAA flight service stations than any other government service. The most common method of obtaining weather information from an FSS is an over-the-phone weather briefing.

If you have not had the opportunity to acquire preliminary weather information prior to a flight, you should request a standard briefing which provides you with the most complete weather picture tailored to your specific flight. A standard briefing normally includes the following items.

1. Adverse conditions (you may wish to cancel the briefing after receiving this information)
2. VFR flight not recommended (if appropriate)
3. Weather synopsis
4. Current weather
5. Forecast weather (enroute and destination)
6. Forecast winds and temperatures aloft
7. Alternate routes (if any)
8. Aeronautical information (NOTAMS)
9. ATC delays
10. Request for PIREPs

An abbreviated briefing enables you to supplement mass disseminated data, update a previous briefing, or request specific information. If the proposed departure time of your flight is six or more hours in the future, an outlook briefing provides a general overview of forecasted weather.

You should request a standard briefing if you have received no preliminary weather information and are departing within the hour. To supplement mass disseminated data, an abbreviated briefing should be requested. The FSS provides an outlook briefing 6 or more hours in advance of your proposed departure time.

Recorded weather information is also available to you through flight service stations. The pilot's automatic telephone weather answering service (PATWAS) is a continuous telephone briefing service provided by some manual FSSs. PATWAS provides a forecast of the local area within a 50 n.m. radius of the station, although selected stations may also include route forecasts. PATWAS information is frequently updated to ensure current and accurate weather data. The transcribed information briefing service (TIBS) is another FSS service which provides continuous telephone recordings of meteorological and/or aeronautical information. TIBS provides route briefings and, depending on user demand, aviation weather observations, terminal forecasts, and wind and temperatures aloft forecasts.

In addition to the services supplied to pilots over the telephone, FSSs furnish weather information to pilots in flight. The enroute flight advisory service (EFAS) is probably the most familiar in-flight service to pilots. To use this service, contact the specific EFAS by using the words "Flight Watch." The frequency for Flight Watch below 18,000 feet MSL is 122.0 MHz. Upon your request, the flight watch specialist can provide aviation weather information, time-critical enroute assistance if you are facing hazardous or unknown weather conditions, and may recommend alternate or diversionary routes. The receipt and rapid dissemination of pilot weather reports is a primary responsibility of EFAS.

You can contact Flight Watch on 122.0 for information regarding current weather along your proposed route of flight.

The transcribed weather broadcast (TWEB) is aired continuously over selected low and medium frequency nondirectional beacons, NDBs (190-535 kHz) and very high frequency omnidirectional ranges, VORs (108.0-117.95 MHz). TWEBs are generally designed to provide in-flight information for flight planning and are based on a route-of-flight format. Broadcast data normally includes a synopsis and route forecast specifically prepared by the National Weather Service (NWS). Adverse conditions, outlooks, and winds and temperatures aloft forecasts are adapted from in-flight advisories, area forecasts and the NMC winds and temperatures aloft forecast. Radar and pilot reports may also be provided.

TWEBs contain in-flight cross-country weather information including winds and temperatures aloft forecasts.

Another in-flight service is the hazardous in-flight weather advisory service (HIWAS). This service provides a continuous broadcast over selected VORs to inform you of hazardous flying conditions such as turbulence, icing, IFR conditions, and high winds. These advisories include SIGMETs, Convective SIGMETs, AIRMETs, severe weather forecast alerts, and center weather advisories.

OTHER FAA WEATHER SOURCES

Additional FAA facilities also furnish weather information to pilots. The FAA terminal controller is responsible for informing arriving and departing aircraft of pertinent weather conditions. At some locations, the NWS shares the duty of reporting visibility observations with the air traffic control tower (ATCT); while at other tower facilities, the controller has the full responsibility for observing, reporting, and classifying weather conditions for the terminal area. Automatic terminal information service (ATIS) is available at most major airports that have operating control towers. This service helps reduce frequency congestion and improves controller efficiency. ATIS is a prerecorded report, broadcast on a dedicated frequency, which includes information regarding current weather and pertinent local airport conditions. The ATIS broadcast is normally recorded every hour but may be updated at any time whenever conditions change. Additionally, at most locations, ATIS can be accessed by telephone.

Another FAA-operated information system is the direct user access terminal service (DUATS). This computer-based program provides NWS and FAA products that are normally used in pilot weather briefings. By using a personal computer and modem, you can access weather information prior to your flight. Flight plans also can be filed and amended through DUATS.

THE NATIONAL WEATHER SERVICE

The National Weather Service (NWS) also provides selected weather services for pilots. Weather Service Forecast Offices (WSFOs) serving a state or large portion of a state, prepare public and aviation regional weather forecasts as well as advisories and warnings of severe weather. Terminal forecasts and the forecasts used in TWEBs are available at WSFOs. Area forecasts and in-flight advisories are also provided at WSFOs in Alaska and Hawaii. Selected WSFOs also furnish formal pilot weather briefings.

In addition, the NWS operates over 200 Weather Service Offices (WSOs) which furnish public forecasts and weather warnings for smaller regions, such as metropolitan areas. You can request specific weather information from all WSOs. Formal pilot weather briefings are available at selected offices.

A major modernization and associated restructuring (MAR) is currently being undertaken by the National Weather Service. The result of this program will be the realignment of the present field offices into Weather Forecast Offices (WFOs). These new offices are designed to improve weather services by taking advantage of updated technology such as the WSR-88D Doppler radar. The WFOs will replace the present WSFO/WSO organization, serve smaller areas than current WSFOs, and will be staffed primarily by meteorologists.

ALTERNATIVE WEATHER SOURCES

A rapidly growing number of alternative weather resources are becoming available to pilots. In addition to weather forecasts provided by local news broadcasts, The Weather Channel, aired on cable television, presents national and international weather information that can help you create a general weather picture for your region prior to flight planning. AM Weather is a TV information source that focuses on aviation weather. This program utilizes professional meteorologists from the National Weather Service and the National Environmental Satellite Data and Information Service (NESDIS), and provides weather information primarily for pilots to make better go/no-go flight decisions. National and regional weather maps, satellite sequences, radar reports, winds and temperatures aloft forecasts, severe weather watch alerts, and in-flight weather advisories all can be obtained from AM Weather. This fifteen-minute weather program is aired Monday through Friday mornings over more than 300 PBS stations.

A variety of private vendors also provide weather information products available through computer services and facsimile. Depending on the specific service, you may access a wide range of data, including coded and plain language observations, forecasts, and advisories. Some businesses provide black and white or color graphic weather products, route briefings specific to your flight, and even flight plan filing. Jepp/Link, JeppFax, CompuServe, Prodigy, and America On Line are several sources for weather information. You can also access weather data from the NWS and other sources through the Internet.

Section C

AVIATION WEATHER DATA

Now that you are familiar with the forecasting process and the information sources available, let's take a closer look at several specific weather products. Visualization is the key to understanding. By forming a mental image of the weather, you can determine how the conditions may affect your flight. Using a combination of weather information products helps you develop a comprehensive weather picture. You must learn to transform the abbreviations, codes, and symbols of printed reports, forecasts, and graphics into a complete picture of the weather. This section provides a general overview of some common weather products and the information they contain. A decoding key for selected products is included in the appendix. In addition, FAA Advisory Circular 00-45 contains descriptions and decoding of National Weather Service materials.

Forming a comprehensive representation of the weather before a flight always means using the latest meteorological information. You may be very skilled at interpreting meteorological conditions, but if you are using old information, the weather picture you formulate could turn out to be grossly inaccurate. The first step is to determine whether the information describes observations taken at a specific time, or a forecast for future weather. "Current" observations may not be so current depending on how often the reports are taken, and the specific reporting station. For example, many smaller airports are not staffed to take late night and early morning observations. Old reports often remain in the current data base until the next report is received. You should verify the time of the report, especially when planning early morning flights. The latest observation may be from the night before. Terminal forecasts for smaller airports may also be delayed until after the station opens. Knowledge of the hours of operation for part-time weather stations is valuable information.

Forecasts may be valid for a specific time or period of time. For example, a winds and temperatures aloft forecast predicts conditions over a 12-hour period, while a low-level significant weather prog portrays conditions as they are expected to be at a specific point in time. A terminal forecast predicts weather over a 24-hour period, but also forecasts events, such as a cold front passage, to occur at specific points in time.

In addition to verifying that the forecast is valid for the time of your flight, you should also be aware of the issue time. Although a forecast may be valid for a 24-hour period, it may be issued every three hours. The latest forecast provides the most accurate prediction. Since weather conditions rapidly change, it may be valuable to wait for an updated forecast.

PRINTED REPORTS AND FORECASTS

You have already been introduced to several weather information products in previous chapters. Some of the reports and charts covered in this chapter will be familiar. The focus here is to provide a foundation for use of these materials to evaluate weather for flight and to provide perspective on how products may be combined to create an accurate weather picture.

A surface aviation observation that is reported and transmitted is called a surface aviation weather report (SA). An SA is normally issued hourly, although these reports are updated if there is a significant change in the observed conditions. An SA may contain as many as 11 separate elements. The valuable weather information you can glean from these reports includes sky condition, visibility, weather and obstructions to vision, sea level pressure, temperature/dewpoint, wind direction, wind speed and character, and the altimeter setting. The remarks section often contains some of the most important information in an observation, including routinely reported data such as runway visual range (RVR), or items the observer considers significant to aviation. (Figure 16-2)

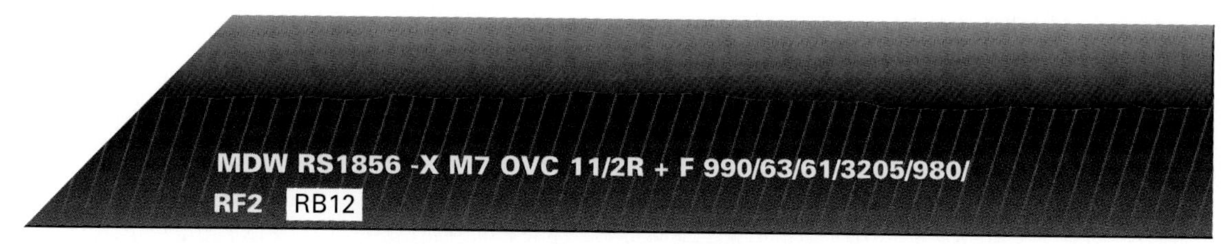

Figure 16-2. The rain at Chicago Midway began 12 minutes past the hour (RB12) according to this SA.

Statistics indicate that most pilots only rarely file pilot reports. Approximately 125,000 hours are flown each day by all sectors of civil aviation, yet the NWS National Aviation Weather Advisory Unit in Kansas City receives and transmits an average of only 2,000 pilot reports daily.

The international weather reporting code (METAR) will be used in the U.S., after January 1, 1996. METAR will alter the format and current coding used in the surface aviation weather report and other printed reports. This new format is illustrated throughout this text and a decoding key is provided in the appendix.

Pilot weather reports (PIREPs) are perhaps the most valuable observations used to anticipate and avoid weather hazards. Observations made from the cockpit are the most timely and the only means of directly observing cloud tops, icing, and turbulence. Often the presence of forecast conditions can be verified by PIREPs. For example, a forecast may predict turbulence in an area from the surface to 18,000 feet MSL, and PIREPs can help you determine if the turbulence exists and at what altitudes. You are encouraged to make pilot reports often, in both good and adverse weather.

AVIATION FORECASTS

The terminal forecast (FT) provides weather conditions expected to occur within a 5 nautical mile radius of the runway complex at an airport. Typically, terminal forecasts are issued 3 times daily and are valid for a 24-hour period. Specific issuance and valid times are scheduled according to time zones. The terminal forecast is one of your most valuable sources for predicting future weather conditions at a specific airport. Predicted sky condition, visibility, weather and obstructions to vision, wind direction and speed, and expected changes in conditions, such as a frontal passage, can be derived from terminal forecasts. (Figure 16-3)

A new, revised terminal aerodrome forecast (TAF) format and code is in effect in most countries and will be in effect in the United States after January 1, 1996. A TAF is a concise statement of the expected weather conditions at an airport for a specified period. The format for these reports can be found in the appendix.

OKC FT 112222 C20 BKN 1610 OCNL 20 SCT. 01Z CFP 40 SCT 3110 etc...

Figure 16-3. This excerpt from a terminal forecast indicates a cold front is expected to pass (CFP) through Oklahoma City around 0100Z with weather improving behind the front.

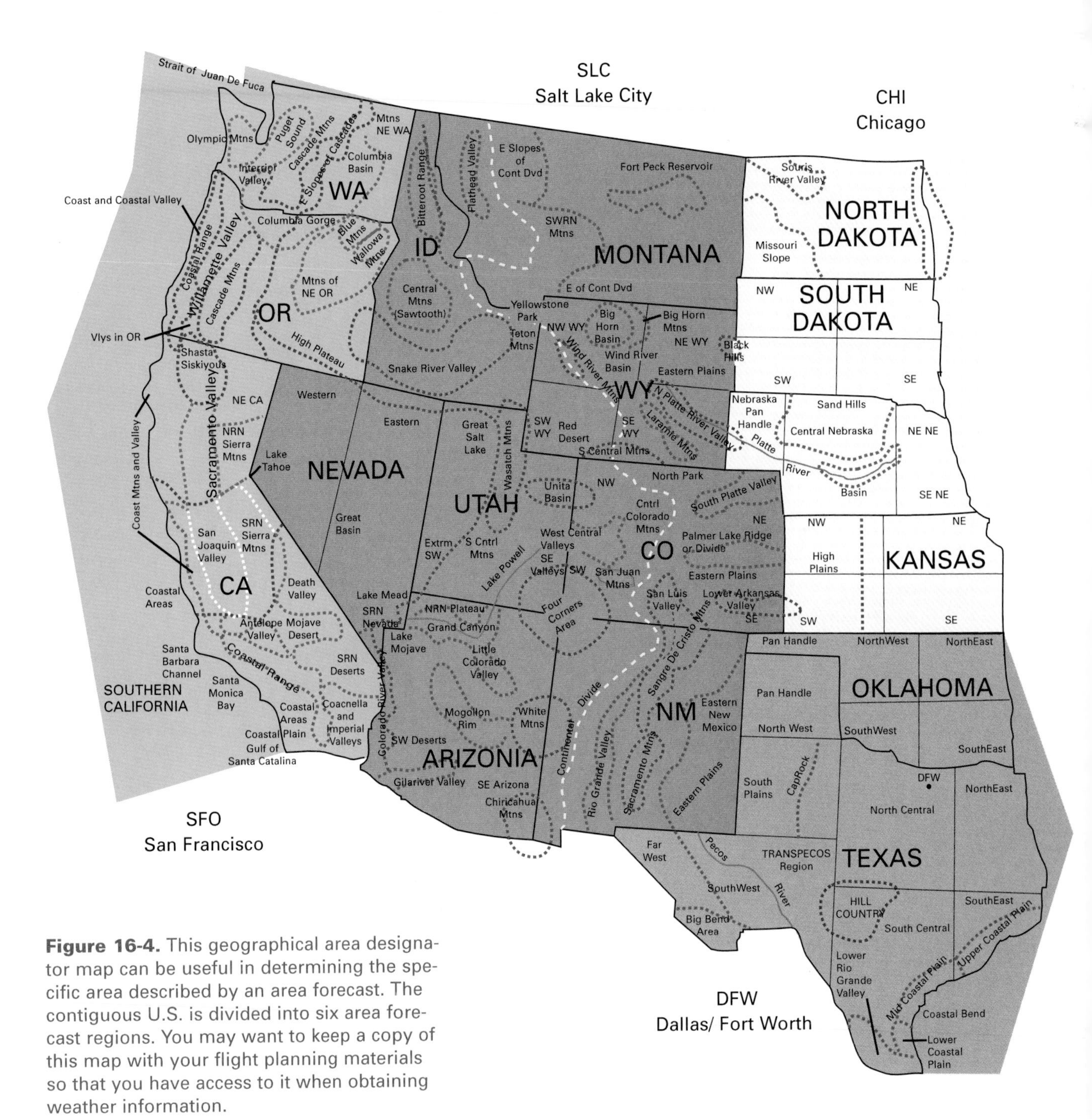

Figure 16-4. This geographical area designator map can be useful in determining the specific area described by an area forecast. The contiguous U.S. is divided into six area forecast regions. You may want to keep a copy of this map with your flight planning materials so that you have access to it when obtaining weather information.

Expected weather conditions over an area the size of several states are described in an area forecast (FA). You can use this forecast to determine enroute weather, and conditions at airports which do not have terminal forecasts. Area forecasts are normally issued three times daily and are valid for 18 hours. A synopsis provides a summary of the location and movement of pressure systems, fronts, and circulation patterns for an 18-hour period. The VFR clouds and weather section of

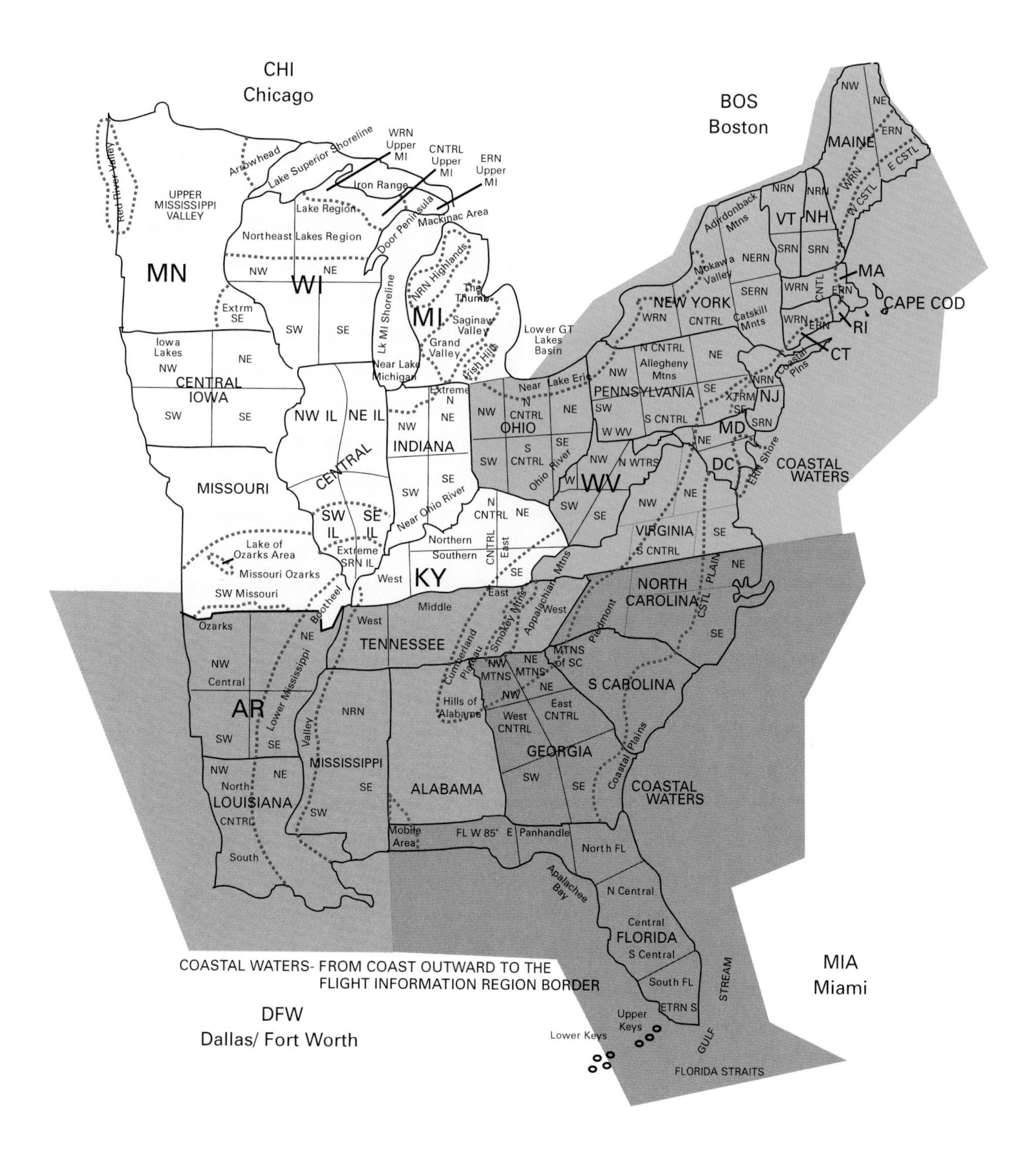

the FA provides expected sky condition, visibility, and weather for 12 hours plus a 6-hour outlook. To use the area forecast most effectively, a geographical area designator map depicting the forecast area is useful in visualizing the regions described in the forecast, and the location and movement of fronts and pressure systems. (Figure 16-4) This map is printed in FAA AC 00-45. Using

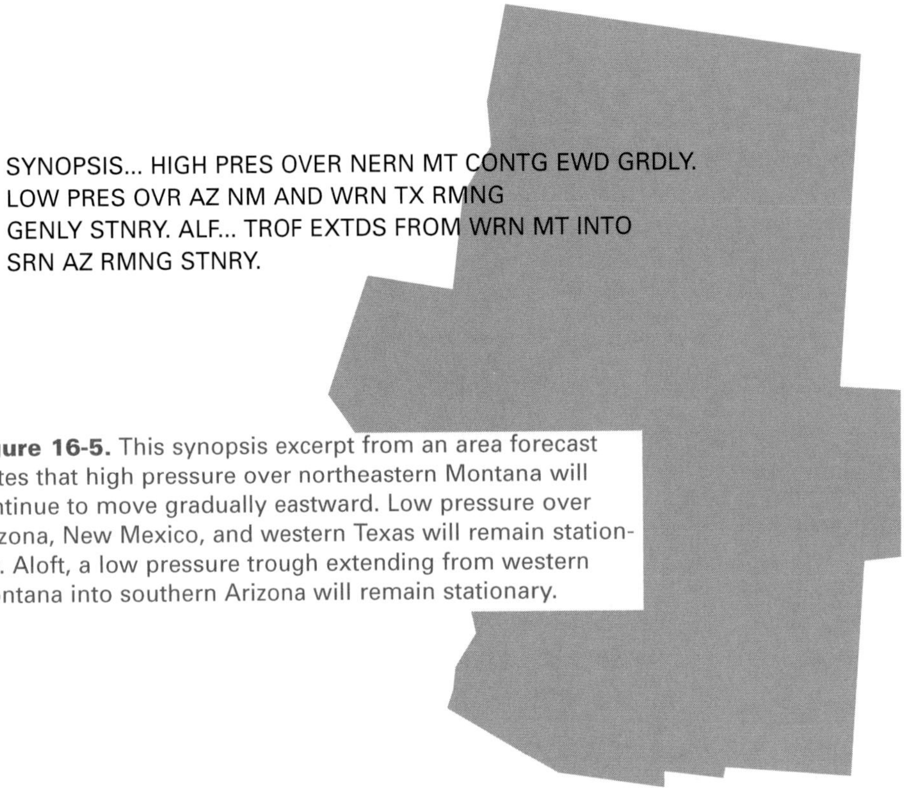

SYNOPSIS... HIGH PRES OVER NERN MT CONTG EWD GRDLY.
LOW PRES OVR AZ NM AND WRN TX RMNG
GENLY STNRY. ALF... TROF EXTDS FROM WRN MT INTO
SRN AZ RMNG STNRY.

Figure 16-5. This synopsis excerpt from an area forecast states that high pressure over northeastern Montana will continue to move gradually eastward. Low pressure over Arizona, New Mexico, and western Texas will remain stationary. Aloft, a low pressure trough extending from western Montana into southern Arizona will remain stationary.

The area forecast covers an area of several states and can be used to determine enroute weather and conditions at the estimated time of arrival if your destination has no FT.

the area forecast in combination with a graphic forecast product, such as the significant weather prognostic chart, helps you develop a more complete image of the forecast weather conditions along your route of flight. (Figure 16-5)

A winds and temperatures aloft forecast (FD) furnishes a prediction of wind direction in relation to true north, wind speed in knots, and the temperature in degrees Celsius for selected altitudes. Depending on the station elevation, the forecast usually includes nine levels of altitude between 3,000 feet MSL and 39,000 feet MSL. Forecasts are based on data taken twice daily, and are also issued twice daily for use during specific time intervals stated on the forecast. FDs are generally used to select flight altitudes, determine aircraft performance, and calculate groundspeed. In addition to flight planning calculations, an FD can add information to the overall weather picture. For example, strong winds aloft indicate a potential for turbulence. (Figure 16-6)

Winds and temperatures aloft contain wind direction in relation to true north, wind speed in knots, and temperature in degrees Celsius for a range of altitudes.

IN-FLIGHT WEATHER ADVISORIES

In-flight aviation weather advisories consist of either an observation and a forecast or just a forecast for the development of potentially hazardous weather. Although identified as in-flight, pertinent advisories are an important part of preflight weather planning as well. These advisories are classified as AIRMETs, SIGMETs, and convective SIGMETs.

Although AIRMETs (WAs) are of operational interest to all aircraft, the weather conditions specified are particularly hazardous to light aircraft having limited capability due to structural integrity or minimal equipment and instrumentation. Pilots with limited experience or qualifications should pay special attention to these advisories. AIRMET bulletins are issued on a scheduled basis every six hours with unscheduled updates and corrections issued as necessary. Each bulletin contains any current AIRMETs in effect, an outlook for weather expected after the AIRMET valid period, and any significant conditions that don't meet AIRMET criteria.

The criteria for issuance of an AIRMET are as follows:

1. Moderate icing
2. Moderate turbulence
3. Sustained surface winds of 30 knots or more
4. Ceiling less than 1,000 feet and/or visibility less than 3 miles affecting over 50 percent of the area at one time
5. Extensive mountain obscurement

Weather conditions that are particularly hazardous to small single-engine aircraft are contained in an AIRMET.

A SIGMET (WS) describes conditions of higher intensity which can pose hazards to all aircraft. A maximum forecast period of four hours is defined for SIGMETs. If the following phenomena are observed or expected to occur, a SIGMET is issued.

1. Severe icing not associated with thunderstorms
2. Severe or extreme turbulence or clear air turbulence not associated with thunderstorms
3. Duststorms, sandstorms, or volcanic ash lowering surface and in-flight visibilities to below three miles
4. Volcanic eruption

FD KWBC 151640
BASED ON 151200Z DATA
VALID 151800Z FOR USE 1700-2100Z TEMPS NEG ABV 24000

FT	3000	6000	9000	**12000**	18000	24000	30000	34000	39000
ALA			2420	2635-08	2535-18	2444-30	245945	246755	246862
AMA		2714	2725+00	2625-04	2531-15	2542-27	265842	256352	256762
DEN			2321-04	**2532-08**	2434-19	2441-31	235347	236056	236262
HLC		1707-01	2113-03	2219-07	2330-17	2435-30	244145	244854	245561
MKC	0507	2006+03	2215-01	2322-06	2338-17	2348-29	236143	237252	238160
STL	2113	2325+07	2332+02	2339-04	2356-16	2373-27	239440	730649	731960

Figure 16-6. According to this FD, the winds aloft over Denver at 12,000 feet MSL are forecast to be from 250° at 32 knots. The temperature at this altitude is -8°C. You can expect turbulence during a flight in this area due to strong winds blowing nearly perpendicular to the mountains just west of Denver.

SIGMETs are issued as warnings of hazardous weather, such as severe icing, which is of operational interest to all aircraft.

A forecaster may issue a convective SIGMET for any convective activity that is potentially hazardous to all categories of aircraft. Bulletins are issued hourly with special advisories issued as required. Convective SIGMETs are warnings for any of the following conditions.

1. Severe thunderstorms
2. Embedded thunderstorms
3. A line of thunderstorms
4. Thunderstorms greater than or equal to VIP level 4 affecting 40% or more of an area at least 3,000 square miles.

Convective SIGMET bulletins are issued for the Eastern (E), Central (C), and Western (W) U.S. based on the location of the majority of observations and forecast conditions. AIRMETs and SIGMETs are issued for the 6 regions corresponding to the FA areas. These "widespread" advisories must be either affecting, or forecast to affect at

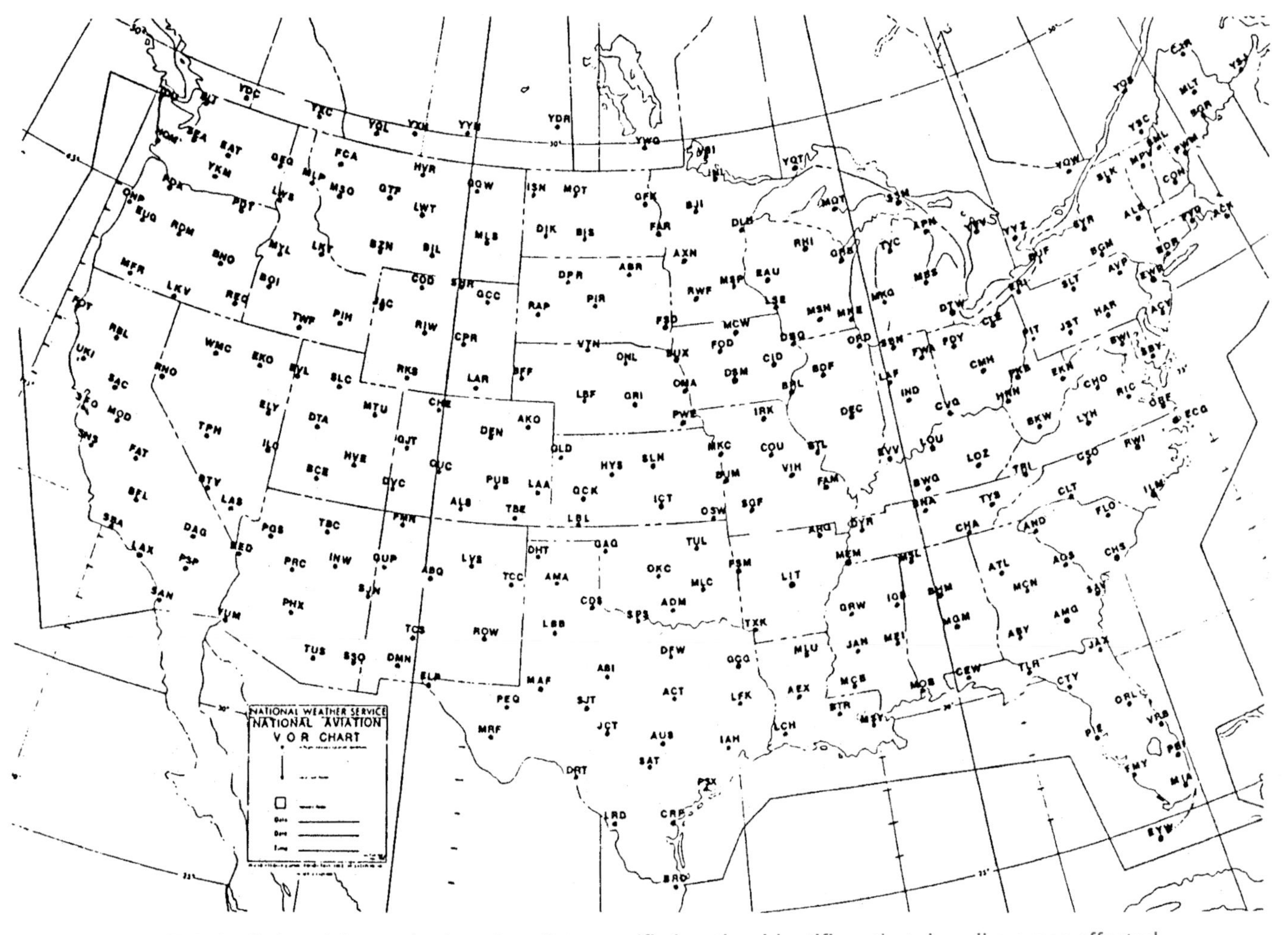

Figure 16-7. This in-flight advisory plotting chart lists specific location identifiers that describe areas affected by AIRMETs, SIGMETs, and Convective SIGMETs. This chart can be a useful tool in the cockpit to locate specific

least 3,000 square miles at any one time. The total area influenced during the forecast period may be extremely large. For example, a 3,000 square mile weather system may be forecast to move across an area totaling 25,000 square miles during the forecast period. The affected areas are described by location identifiers of VORs, airports, or well-known geographic areas. An in-flight advisory plotting chart is valuable in determining the dimensions of the area influenced by the advisory. (Figure 16-7)

An air route traffic control center (ARTCC) may issue a center weather advisory (CWA) as a "nowcast" for adverse weather conditions beginning within the next two hours. A CWA may be issued as a supplement to an existing advisory. If PIREPs indicate conditions meet in-flight advisory criteria, a CWA is the quickest method to alert pilots to hazardous weather. When the criteria have not been met to issue an AIRMET or SIGMET, but conditions are affecting the safe flow of air traffic, a CWA also may be necessary.

An alert severe weather watch (AWW) message is transmitted to alert forecasters, briefers, and pilots that a severe weather watch bulletin (WW) is being issued. A WW is an unscheduled bulletin which defines areas of possible severe thunderstorms or tornado activity.

GRAPHIC WEATHER PRODUCTS

To develop the most complete picture of the impact weather will have on a proposed flight, graphic weather products must be used in combination with printed reports and forecasts. The surface analysis chart is a computer-prepared chart which is normally transmitted every three hours. The valid time of the chart represents the time at which the observations were taken. The most valuable function of the surface analysis chart is to provide a visual means of identifying the location of pressure systems, fronts, and observations from reporting stations. These observations, depicted in the form of station models, furnish an overview of winds, temperatures, and dewpoint at the valid time shown on the chart.

The weather depiction chart which also is issued every three hours is computer prepared from surface aviation observations. The station models portray cloud heights, as well as weather and obstructions to vision. The most valuable feature of this chart is the depiction of areas of IFR, MVFR, and VFR category conditions. Due to the delay between observation and transmission time, comprehensive flight planning must include the use of forecasts, prognostic charts, and the latest pilot, radar, and surface reports to supplement the general information shown on the weather depiction chart. (Figure 16-8)

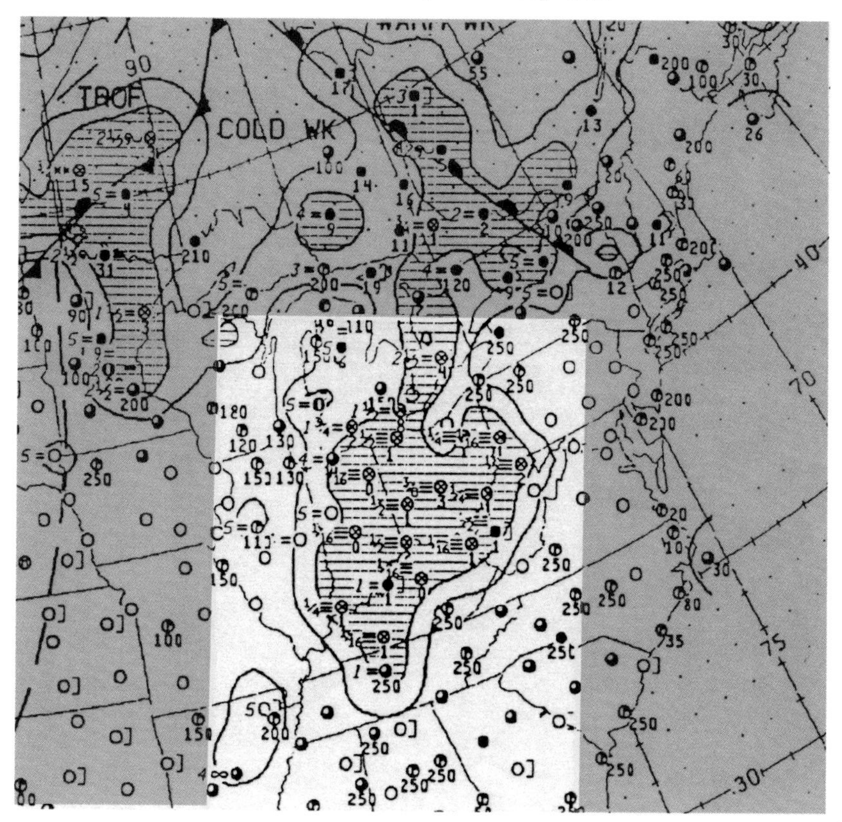

Figure 16-8. In this example from a weather depiction chart, widespread areas of MVFR and IFR conditions are indicated. Areas enclosed in smooth lines are MVFR, while the areas with cross-hatching are IFR.

The weather depiction chart is valuable to determine general weather conditions on which to base flight planning.

To view a graphic representation of the location and intensity of precipitation, you should refer to the radar summary chart. This chart is constructed by computer from regularly scheduled radar observations and is valid at the time of the report. The radar summary chart is normally issued 35 minutes after every hour. The availability of the chart varies depending on the system being used. The type of precipitation echoes, intensity, trend, configuration, coverage, movement, and echo tops and bases are depicted on this chart. (Figure 16-9)

The composite moisture stability chart is a four panel analysis of observed upper air data between the surface and the 500 mb level (18,000 feet MSL). By referring to the stability panel, you can determine the lifted and the K index for the area of your flight. The lifted index represents the stability of the atmosphere. If convective activity is forecast along your route of flight, you can predict the severity of thunderstorms if they occur. The K index examines both the temperature and moisture profile of the environment and is primarily a tool for the meteorologist to determine the probability of airmass thunderstorms. (Figure 16-10)

The freezing level panel is useful in predicting the areas in which icing may occur. This contour analysis portrays an overall view of the lowest observed freezing level for an area. You can anticipate possible icing in clouds or precipitation especially when the temperature is between 0°C and -10°C. The precipitable water panel, which is used to determine water vapor content in the air, is especially valuable to meteorologists concerned with flash flood events. To determine the average air saturation, you can refer to the average relative humidity panel. Clouds and possible precipitation can be expected with a high humidity level.

From constant pressure analysis charts, you can approximate the observed temperature, wind, and temperature/dewpoint spread along a proposed route. The information depicted is at the specified pressure on the chart. The constant pressure chart is issued twice daily for six different pressure levels (850 mb, 700 mb, 500 mb, 300 mb, 250 mb, and 200 mb). A specific chart can be selected based on your proposed flight altitude. Often, the causes of weather and its movement can be determined more clearly by referring to a constant

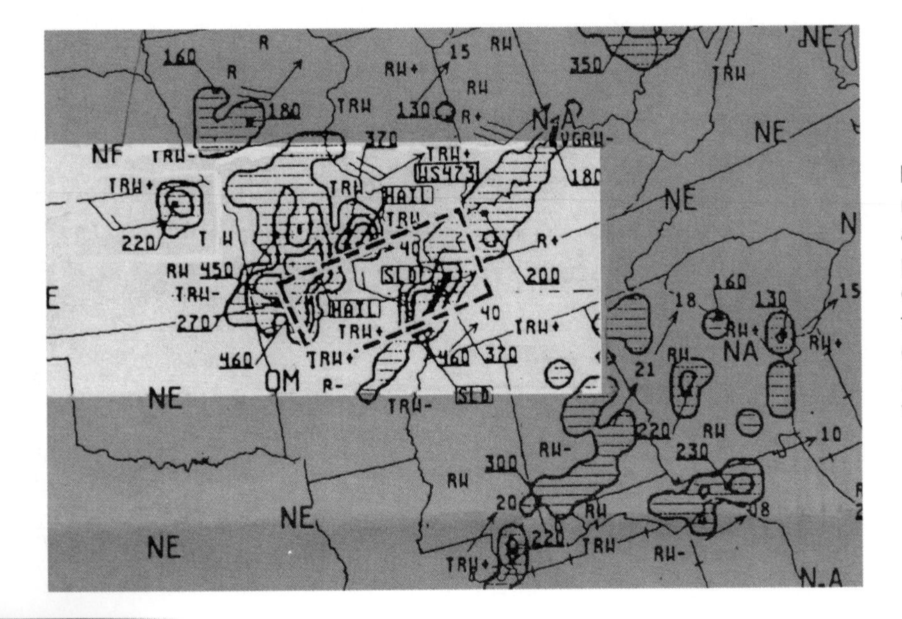

Figure 16-9. This excerpt from a radar sumary chart depicts hazardous weather conditions for a proposed flight in this area. Echoes exceeding VIP level 5, a line of thunderstorms, and hail are indicated. The dashed-line rectangular box denotes a severe weather watch area.

pressure chart rather than a surface map. For example, a large scale wind flow around a low aloft may extend low ceilings and precipitation over a more extensive area than depicted by a synoptic surface chart alone.

So far, the graphic weather products that have been discussed all depict observations taken at a specific time indicated on the chart. However, the significant weather prognostic chart or prog chart portrays a weather forecast that may influence flight planning. Both manually and computer-prepared significant weather progs are designed for flight planning to 24,000 feet MSL (low-level prog) and from 24,000 to 63,000 feet MSL (high-level prog). The charts are normally issued four times a day and depict conditions as they are

A review of the table previously presented in Chapter 3, will assist you in determining pressure altitude for each of the six pressure surfaces depicted in constant pressure analysis charts.

Pressure Millibars	Pressure Altitude Feet	 Meters
850	4,781	1,457
700	9,882	3,012
500	18,289	5,574
300	30,065	9,164
250	33,999	10,363
200	38,662	11,784

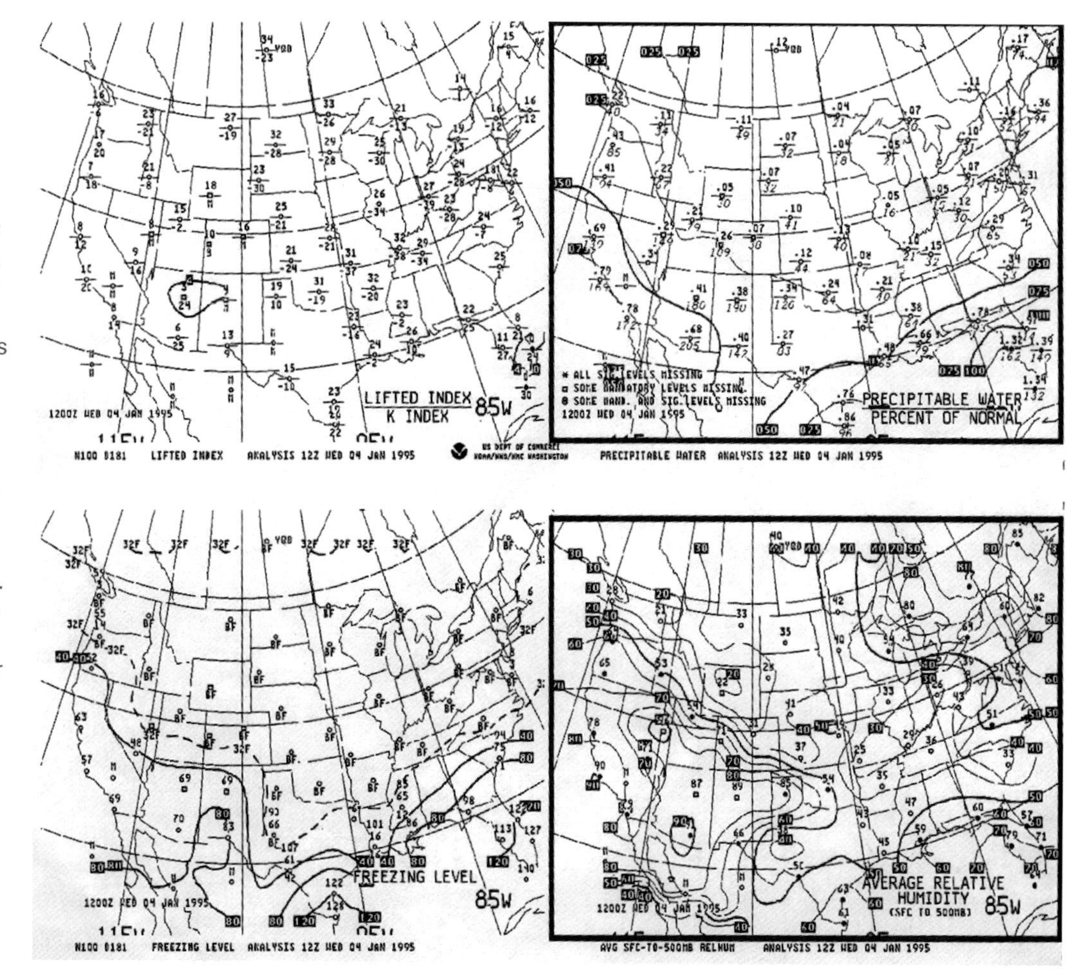

Figure 16-10. The composite moisture stability chart presents an analysis of upper air data. The upper left panel is the stability panel. Freezing level is depicted on the lower left panel and precipitable water is indicated on the upper right panel. The average relative humidity panel is the lower right panel of the chart.

forecast to be at the valid time of the chart. These forecast times are indicated in the lower left corner of each panel. A wide variety of information is depicted on prog charts so it may be helpful to refer to the decoding key in the appendix for symbol interpretation.

The low-level prog chart consists of four panels, with the two lower panels being 12 and 24 hour surface progs, and the two upper panels representing 12 and 24 hour progs of significant weather from the surface to the 400 mb level (24,000 feet MSL). The lower panels use standard symbols to depict fronts, predicted speed and direction of movement of pressure centers, and areas of forecast precipitation. Solid lines enclose areas of expected continuous or intermittent (stable) precipitation, while dash-dot lines indicate showery (unstable) precipitation.

The two upper panels portray forecast areas of IFR, MVFR, and VFR conditions. Smooth lines enclose forecast IFR weather and scalloped lines surround areas of MVFR conditions. Areas of expected moderate or greater turbulence are enclosed by dashed lines with the height of the turbulent layer given by numbers within these lines. Freezing level height contours for the highest freezing level are indicated by solid lines at 4,000 foot intervals. A zig-zag line labeled SFC shows where the freezing level is forecast to be at the surface. Multiple freezing levels are signified by an upper level contour crossing the surface line. This indicates layers of warmer air aloft. If clouds and precipitation are forecast in the area, icing hazards are a possibility. Icing is implied in any clouds and precipitation at altitudes above the freezing level. Low-level significant progs can greatly enhance your ability to visualize the potential weather systems and phenomena along your route of flight. Comparing the features of this chart to printed forecast information, enables you to create a comprehensive forecast picture. (Figure 16-11)

The significant weather prognostic chart can be used to determine areas to avoid, such as forecast locations of icing or turbulence.

The high-level significant weather prognostic chart covers the airspace from 25,000 feet to 60,000 feet pressure altitude. A wide range of

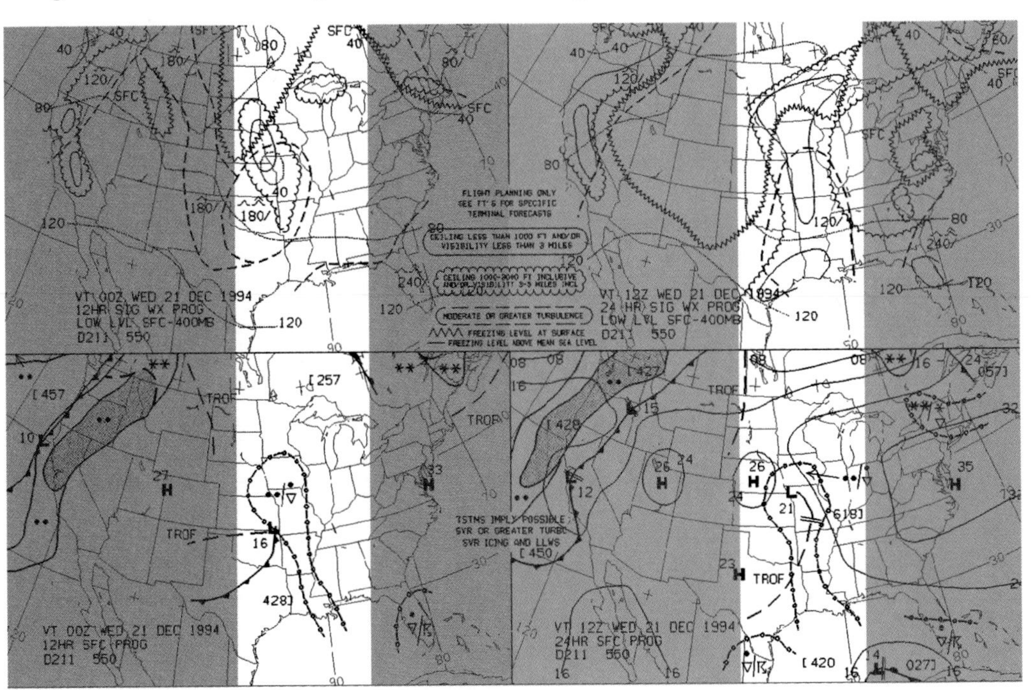

Figure 16-11. By referring to the low-level significant prog chart, you can determine areas of forecast weather to avoid. For example, this prog chart depicts an area of MVFR and IFR weather with showery precipitation located over the midwest extending to the Gulf Coast. This area increases in size during the 12 to 24 hour forecast period.

information can be interpreted from this chart including active thunderstorms, tropical cyclones, surface positions of well-defined convergence zones, positions and movement of frontal systems, and the location of the jet stream. (Figure 16-12)

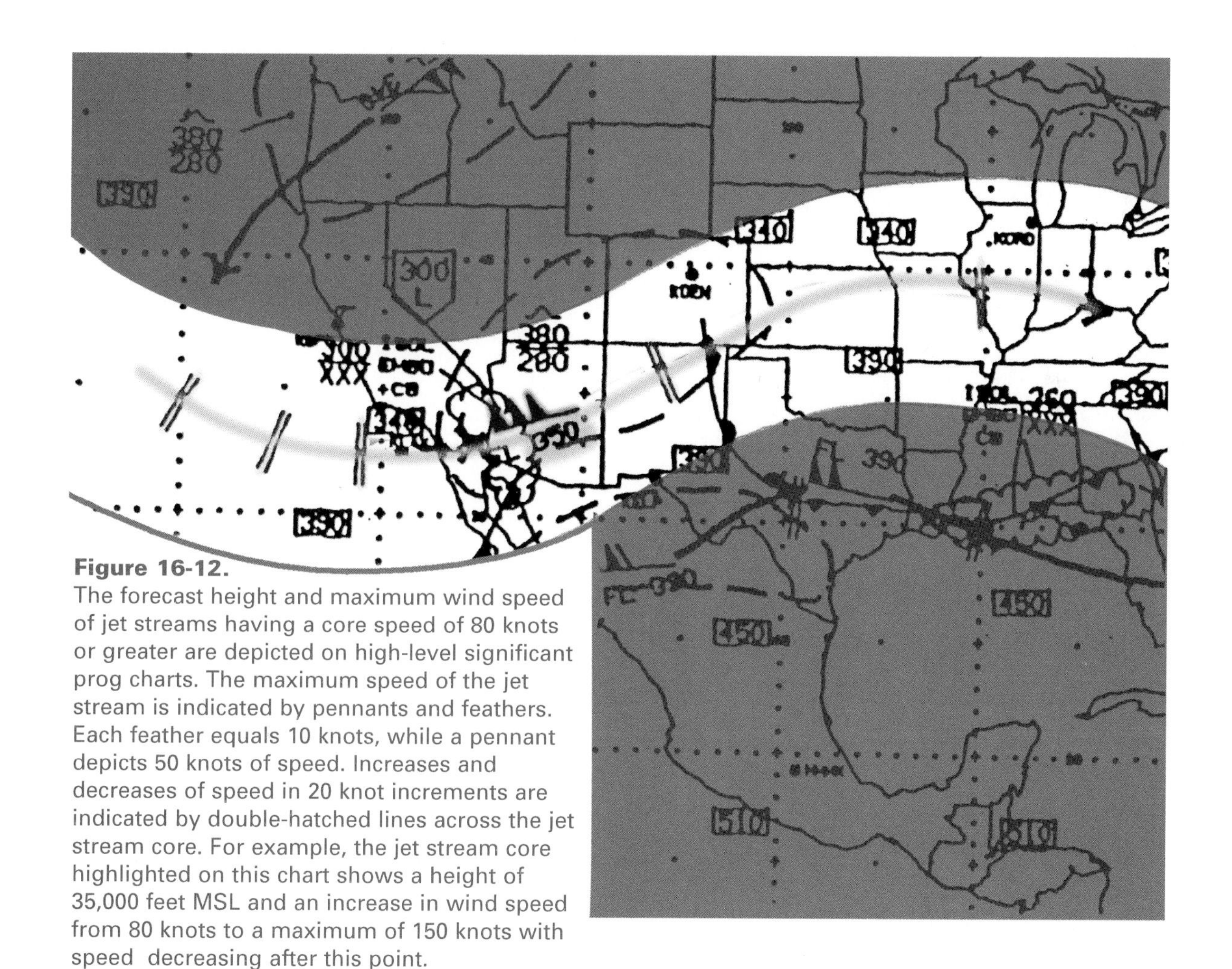

Figure 16-12.
The forecast height and maximum wind speed of jet streams having a core speed of 80 knots or greater are depicted on high-level significant prog charts. The maximum speed of the jet stream is indicated by pennants and feathers. Each feather equals 10 knots, while a pennant depicts 50 knots of speed. Increases and decreases of speed in 20 knot increments are indicated by double-hatched lines across the jet stream core. For example, the jet stream core highlighted on this chart shows a height of 35,000 feet MSL and an increase in wind speed from 80 knots to a maximum of 150 knots with speed decreasing after this point.

SUMMARY

Effectively interpreting weather data is a skill that takes time to develop. An understanding of the procedures used to compile, transmit, and process weather information provides a new perspective on forecasting. Knowledge of the methods used to produce forecasts furnishes insight into the accuracy that can be achieved, and the limitations of weather prediction. This knowledge lays a solid foundation to assist you in interpreting and effectively using weather data for flight planning. Identifying the information sources available and becoming familiar with the specific weather data you can access is just the first step. It takes practice to learn how to extract information that is pertinent to your flight. Applying your knowledge to effectively evaluate weather for flight and then to make a decision based on the weather picture that you have developed, is explored in the final chapter.

KEY TERMS

Abbreviated Briefing
AIRMET (WA)
Alert Severe Weather Watch (AWW)
AM Weather
Area Forecast (FA)
Automated Flight Service Stations (AFSSs)
Automatic Terminal Information Service (ATIS)
Average Relative Humidity Panel
Center Weather Advisory (CWA)
Chaos Theory
Climatological Forecast
Composite Moisture Stability Chart
Constant Pressure Analysis Charts
Convective SIGMET
Direct User Access Terminal Service (DUATS)
Enroute Flight Advisory Service (EFAS)
Flight Service Station (FSS)
Freezing Level Panel
Hazardous In-flight Weather Advisory Service (HIWAS)
High-level Prog Chart
Low-level Prog Chart
METAR
Meteorological Forecasts
Numerical Weather Prediction
Outlook Briefing
Persistence Forecast
Pilot Weather Report (PIREP)
Pilot's Automatic Telephone Weather Answering Service (PATWAS)
Precipitable Water Panel
Radar Summary Chart
Severe Weather Watch Bulletin (WW)
SIGMET (WS)
Significant Weather Prognostic Chart (prog chart)
Stability Panel
Standard Briefing
Surface Analysis Chart
Surface Aviation Weather Report (SA)
Terminal Aerodrome Forecast (TAF)
Terminal Forecast (FT)
Transcribed Information Briefing Service (TIBS)
Transcribed Weather Broadcast (TWEB)
Weather Depiction Chart
Weather Forecast Office (WFO)
Weather Service Forecast Office (WSFO)
Weather Service Office (WSO)
Winds and Temperatures Aloft Forecast (FD)

CHAPTER QUESTIONS

1. Make a weather prediction using the forecast methods of persistence, climatological, and analogue. How accurate are your predictions using each method for different time periods? A few hours? 24 hours? Several days? A week? A month?

2. Prepare a list of all the weather observation and forecast sources for your local flying area. Include the information issue times, mode of acquisition (television, telephone, computer, facsimile), as well as pertinent numbers such as channels and telephone numbers.

3. Prepare a list of a specific weather products that provide observed or forecast information about the following phenomena.
   1. Turbulence
   2. Thunderstorms
   3. IFR conditions
   4. Icing

4. Select a specific printed report or weather chart such as the low-level significant prog chart, and using a decoding key, define every symbol or abbreviation. Try to form a mental image of each weather condition described.

CHAPTER 17

Weather Evaluation For Flight

Introduction

A fact of aviation meteorology today is that you must have a better understanding of weather than your predecessors. There are at least two primary reasons for this. First, approximately 25% of aircraft accidents are weather-related. A weather-wise pilot is much less likely to become a statistic. Second, the rapid growth in the aviation industry and the increasing automation of weather information places a greater burden on you to conduct "self briefings." This chapter provides guidelines to help you develop a self-briefing procedure. An overview of progressive developments in aviation weather resources is also provided. Future weather products designed for the entire aviation community will simplify the weather evaluation process and prove invaluable to flight safety.

When you complete Chapter 17 you will understand the process of combining weather information so you can construct a comprehensive weather picture. You will realize how your weather knowledge and experience affects the self-briefing process. After reading this chapter, you will also understand the importance of visualizing the weather and how future aviation weather products utilize this concept.

Section A

SELF-BRIEFING PROCEDURE

To improve your proficiency in weather evaluation, the development of a system for processing information during flight planning is valuable. We will refer to this system as the self-briefing procedure. A flow diagram outlining this process is shown in figure 17-1.

The self-briefing procedure does not begin an hour before your flight. It begins with appropriate long-term training and study. In addition, a clear understanding of your capabilities as a pilot and your aircraft's limitations is essential. The first step in this process is developing weather awareness. This is achieved by learning as much as you can about weather producing processes and phenomena. Study of this and other textbooks helps provide a solid background in aviation weather. Your flight experience as well as discussions with other pilots regarding their own weather experiences can enhance this awareness. Next, you must have sufficient knowledge of the available and relevant weather products. Chapter 16 furnished you with background information on sources and specific materials that you can use. Before you evaluate the weather, a self analysis is necessary to provide the foundation for making decisions about your flight.

Figure 17-1. The flow diagram illustrates the self-briefing procedure.

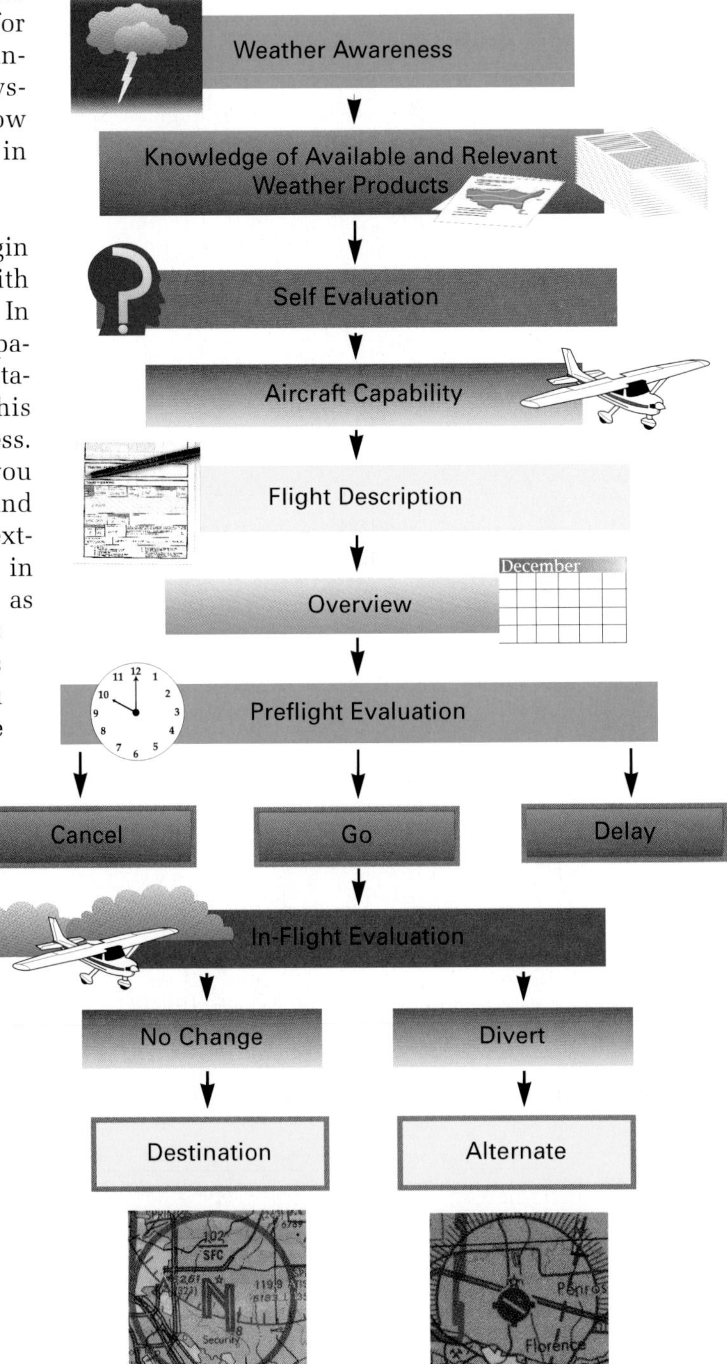

SELF EVALUATION

As a pilot, you must be able to effectively assess your own abilities and limitations when making decisions regarding flight in specific weather conditions. The certificates and ratings you hold provide the most basic criteria for these decisions. Holding an instrument rating and maintaining IFR currency allows you to fly in a wide range of weather conditions. Even if the proposed flight is VFR, the option to file an IFR flight plan, or request a clearance to fly an instrument approach if weather deteriorates, may influence flight planning decisions. In addition to meeting regulatory requirements, you must recognize your own instrument flying skill level and the experience required to maintain that level. For example, your recent instrument experience may consist of approaches practiced only in a simulator without flight operations in actual IFR conditions. It is important to make a realistic evaluation of proficiency before departing on a lengthy flight in IFR weather that may be followed by an instrument approach at your destination.

Whether or not you are an instrument-rated pilot, the amount and recency of your experience in specific flight conditions are still factors that must be considered when making a determination about the weather. Setting personal limitations is an important aspect of flight safety. Lack of extensive flight experience in conditions such as strong gusty winds, crosswinds, turbulence, or low visibility must be considered. For example, if you lack extensive cross-country flying experience in low visibilities, you may want to restrict your flying conditions to fifteen miles visibility or greater.

Another element that affects the weather evaluation process is your ability to interpret weather data properly. You should be thoroughly familiar with the products that you reference and how to decode the material. Having easy access to a decoding key is necessary, especially if an extensive time period has passed since you last used certain information products. Keeping up-to-date on the content and decoding of new weather information products, as well as frequently reviewing material that you have used in the past is essential to maintaining a high level of proficiency in weather data interpretation.

The ability to simply decode weather data, however, is not enough. You must be able to make a competent decision about your flight based on your analysis of the weather situation. A go/no-go decision, a delayed departure, or a change in routing are all options you may need to consider. Your decision must be based on how specific weather will impact your flight. The formation of a mental image of the weather is the key to making the choice. You must develop the skills needed to visualize a complete weather situation, understand its causes and evolution, and anticipate the occurrence of hazards that may affect your flight. You should then be able to make a clear and informed flight decision based on your analysis.

Simply focusing on the more dramatic phenomena such as thunderstorms, microbursts, and icing does not reveal a comprehensive weather picture. Recognizing that a wide variety of weather conditions may pose risks to flight is an important part of weather evaluation. For example, a study by the National Transportation Safety Board reveals that wind, gusts, and turbulence are involved in 20% of all aviation accidents and nearly 80% of weather-related accidents. Most fatal accidents are related to clouds and low ceilings, obscuration to visibility, and precipitation.

AIRCRAFT CAPABILITY

A complete understanding of your aircraft's performance capability and limitations is essential in evaluating the weather's impact. For example, the aircraft's climb performance might not be sufficient to provide safe passage over a mountain ridge, especially in the presence of strong updrafts and downdrafts. Something as simple as a high temperature may prove to be a hazard on a short runway in high terrain. A strong surface wind may exceed your aircraft's maximum crosswind component, or turbulence could lead to structural damage.

The equipment on board your aircraft is another factor to consider when assessing aircraft capabilities. For example, an aircraft equipped with a de-icing system, advanced navigation instruments, or weather radar would be capable of flying in a wider range of weather conditions than an aircraft without this equipment. Cabin pressurization or on-board oxygen may allow you to select an altitude above weather hazards such as icing.

Based on a complete assessment of yourself and your aircraft, you can set specific weather restrictions for your flights. Wind, ceiling, and visibility limitations can be determined as a foundation for flight planning. This allows for more efficient decisions regarding weather.

FLIGHT DESCRIPTION

The next step in the self-briefing procedure is to establish a complete flight description. This allows you to access the weather information that pertains specifically to your flight. You will need to supply the details of your flight to a briefer or to a computerized weather service to obtain data tailored to your flight. The flight plan form can be used as a reference for the items that should be included in the flight description. (Figure 17-2)

When you request a briefing, identify yourself as a pilot and supply the briefer with the following information: type of flight planned (VFR or IFR), aircraft number or pilot's name, aircraft type, departure airport, route of flight, destination, flight altitude(s), estimated time of departure, and estimated time enroute or estimated time of arrival.

MILITARY TRAI...

FLIGHT PLAN FORM

2. AIRCRAFT IDENTIFICATION: N1234A
3. AIRCRAFT TYPE/SPECIAL EQUIPMENT: C-182/A
4. TRUE AIRSPEED: 140 KTS
5. DEPARTURE POINT: DFW
6. DEPARTURE TIME PROPOSED (Z): 1900Z
7. CRUISING ALTITUDE: 4,500
8. ROUTE OF FLIGHT: LOCAL
9. DESTINATION: DFW INTL
10. EST. TIME ENROUTE: HOURS 2 MINUTES 00
12. FUEL ON BOARD: HOURS 5 MINUTES 00
13. ALTERNATE AIRPORT(S): NONE
14. PILOTS NAME, ADDRESS & TELEPHONE NUMBER & AIRCRAFT HOME BASE: JANE BAILEY 219 S. GRANT DALLAS, TX (214) 744-2783 AIR CENTER 1
15. NUMBER ABOARD: 2
16. COLOR OF AIRCRAFT: RED/WHITE

CLOSE VFR FLIGHT PLAN WITH ______ FSS ON ARRIVAL

Figure 17-2. A flight plan form can be used as a reference for the flight description items supplied to a briefer.

Section B

WEATHER EVALUATION PROCESS

Now that the foundation has been laid for self-briefing, the analysis of weather for a specific flight can begin. Developing effective weather-evaluation skills takes time and practice. As you gain experience, the self-briefing procedure becomes easier. It is worthwhile to review pertinent weather material frequently to assist you in interpreting information from printed reports, graphics, and weather briefings. As you proceed through the following exercise, you may want to reference previous chapters in the text to review characteristics of certain weather phenomena. The availability of weather information, and your familiarity with specific products will influence your weather evaluation. Creating a self-briefing checklist is a valuable tool. Maintaining a binder that contains decoding keys and reference maps is helpful in simplifying the interpretation of weather data.

An example is given in the following paragraphs of how to evaluate weather conditions for a planned flight using the guidelines developed for the self-briefing procedure. The goal is to obtain a comprehensive mental image of the weather, recognize the impact that specific conditions may have on your flight, and make a decision regarding the proposed flight based on your evaluation.

This self-briefing exercise requires some creative role playing on your part. Imagine that you are a noninstrument-rated private pilot with 150 hours of flight time. Your weather awareness, and knowledge of available and relevant weather products is excellent since you have just recently finished reading this textbook. You fly frequently in the local area and, although your experience is limited, you strive to maintain a high level of proficiency.

The scenario begins as you are planning a local flight in the area surrounding Dallas, Texas. Your aircraft of choice is a single-engine, high-performance airplane which seats four passengers. The proposed departure time of your flight is 1900Z or 1:00 p.m. local time on Wednesday, December 28. Your plans include operating at altitudes up to 5,000 feet MSL while practicing maneuvers and enjoying the local scenery.

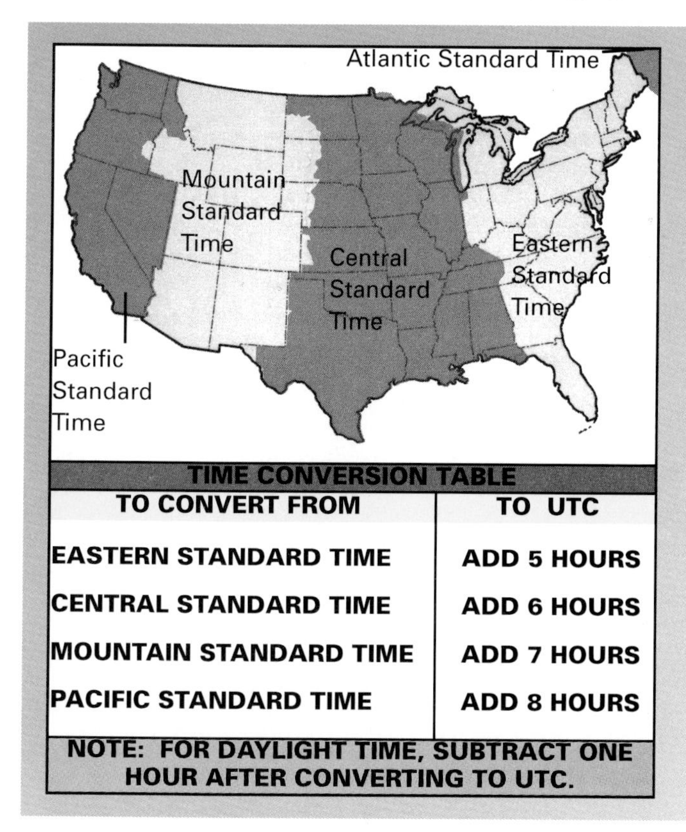

TIME CONVERSION TABLE	
TO CONVERT FROM	TO UTC
EASTERN STANDARD TIME	ADD 5 HOURS
CENTRAL STANDARD TIME	ADD 6 HOURS
MOUNTAIN STANDARD TIME	ADD 7 HOURS
PACIFIC STANDARD TIME	ADD 8 HOURS
NOTE: FOR DAYLIGHT TIME, SUBTRACT ONE HOUR AFTER CONVERTING TO UTC.	

Before you can successfully interpret weather information for your flight, you must be familiar with Coordinated Universal Time (UTC) referred to as Zulu time in aviation. Because a flight may cross several time zones, estimating arrival time at your destination using local time at your departure airport can be confusing. By using the 24-hour clock system and UTC time, the entire world is placed on one time standard. Zulu time is local standard time at longitude 0° which passes through Greenwich, England.

Air traffic control operates on Zulu time, and aviation weather information valid times, forecast periods, and issue times are indicated in Zulu time. In the U.S. when you convert local time to Zulu time you add hours. Converting Zulu time to local time requires subtraction of hours. You can reference the following table to convert times as you interpret weather data in the self-briefing scenario.

OVERVIEW

December

It is worthwhile to begin assessing the general weather situation a few days before the flight. The overview step allows you to develop a picture of macroscale weather patterns that may affect your proposed flight. You may begin this step simply by paying attention to weather forecasts on radio and television broadcasts. These programs can alert you to any major weather systems threatening your proposed flight. The more detailed overview step is initiated about 24 hours in advance of your departure. Your goal at the end of the overview is to answer these questions:

1. Where are the potential areas of adverse weather currently located?
2. How have those areas been moving/developing in the last 24 hours or so?
3. Where will those areas be at the time of your flight?
4. What are the specific flight hazards associated with those weather systems?

We will now put together the overview for your pending flight in the Dallas area. The overview evaluation can be accomplished by using a few select weather products. Large-scale weather systems typically travel 600 miles a day so you need to examine a wide region, and look for systems that are well upstream of your proposed flying location. In this example you have obtained a satellite image, a surface analysis chart, a 300 mb constant pressure analysis chart, and a low-level significant weather prognostic chart.

Your attention to weather broadcasts over the last couple of days has informed you that a large low pressure area exists southwest of Dallas and it is moving northeastward. It is now December 27, the day before your proposed flight. The satellite image in figure 17-3 was taken on the 27th at 1845Z. It is important to visualize the location of your proposed flight on the satellite photo so you can examine cloud patterns associated with weather systems in relationship to your flight. This satellite picture shows a large area of clouds positioned over most of Texas. The lightest (coldest) clouds, indicated by the bright white area are located along the southwestern border of the state. This is most likely the location of the low pressure area mentioned earlier. You will want to investigate this weather system further by referring to the surface analysis chart and 300 mb chart.

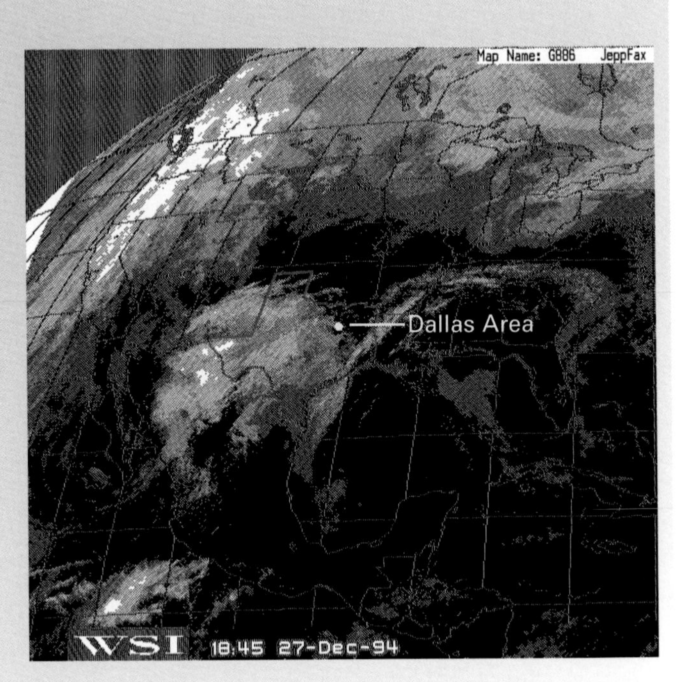

Figure 17-3. Infrared satellite image 27 December at 1845Z.

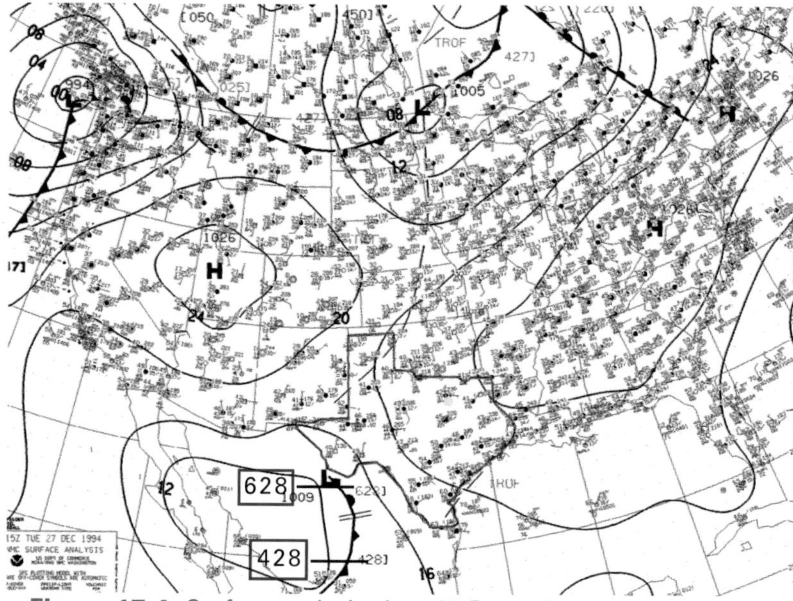

Figure 17-4. Surface analysis chart 27 December at 1500Z.

A surface analysis chart issued at 1500Z on December 27, provides a more detailed picture of the storm seen on the satellite image. (Figure 17-4) An examination of the chart reveals a ridge of high pressure extending into northeast and north-central Texas. However, the low pressure center positioned to the southwest of the state is more significant. You may remember from reading the previous chapters that a surface low is generally a producer of poor flying weather. Low ceilings and precipitation can be expected with this system. The development and movement of the low pressure system is a factor that may impact your flight tomorrow.

Interpreting information about this low may give you some insight into its future development. The three-digit number near the front classifies its type. Decoding this number by using the highlighted tables reveals an occluded front and a cold front. Both are weak with little or no change, and diffuse. The low at the surface is not well defined. This is typical for a frontal system moving across the southwestern United States and Mexico. However, there is clear evidence of rain and overcast skies over south Texas. A look at the 300 mb chart provides some additional information.

Type of Front	
Code Figure	Description
0	Quasi-stationary at surface
1	Quasi-stationary above surface
2	Warm front at surface
3	Warm front above surface
4	Cold front at surface
5	Cold front above surface
6	Occlusion
7	Instability line
8	Intertropical front
9	Coverage line

Intensity of Front	
Code Figure	Description
0	No specification
1	Weak, decreasing
2	Weak, little or no change
3	Weak, increasing
4	Moderate, decreasing
5	Moderate, little or no change
6	Moderate, increasing
7	Strong, decreasing
8	Strong, little or no change
9	Strong, increasing

Character of Front	
Code Figure	Description
0	No specification
1	Frontal area activity, decreasing
2	Frontal area activity, little change
3	Frontal area activity, increasing
4	Intertropical
5	Forming or existence expected
6	Quasi-stationary
7	With waves
8	Diffuse
9	Position Doubtful

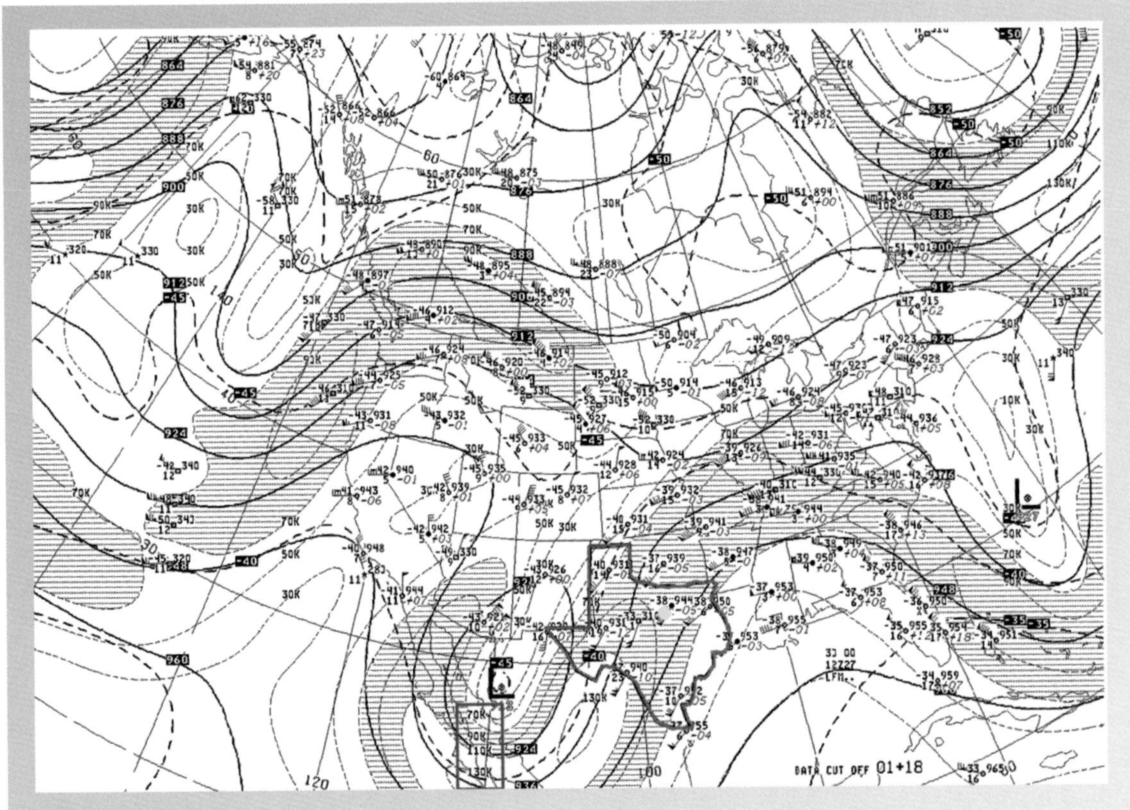

Figure 17-5. 300 mb constant pressure analysis chart 27 December at 1200Z.

Many times surface weather conditions can be more clearly associated with an upper-air pattern than with the features on the surface map. This is the case with the low pressure area over Mexico. An examination of the 300 mb chart valid at 1200Z shows a well developed closed low aloft. Strong jet stream winds (a jet streak) are associated with this low. The hatched lines indicate areas of wind speed from 70 to 110 knots. The clear area inside the hatching depicts wind speeds between 110 knots and 150 knots. This strong low aloft may lead to further development of poor weather conditions near the surface. (Figure 17-5)

The typical low pressure system moves eastward or northeastward at 25 knots. In this case, the low aloft is probably moving more slowly because it is so strong. Based on this knowledge, you can expect that this low pressure area will continue to move toward central Texas by the time of your flight. This situation will most likely produce clouds and precipitation in the Dallas area. Examination of a forecast chart will provide additional information about the potential weather conditions in Dallas tomorrow.

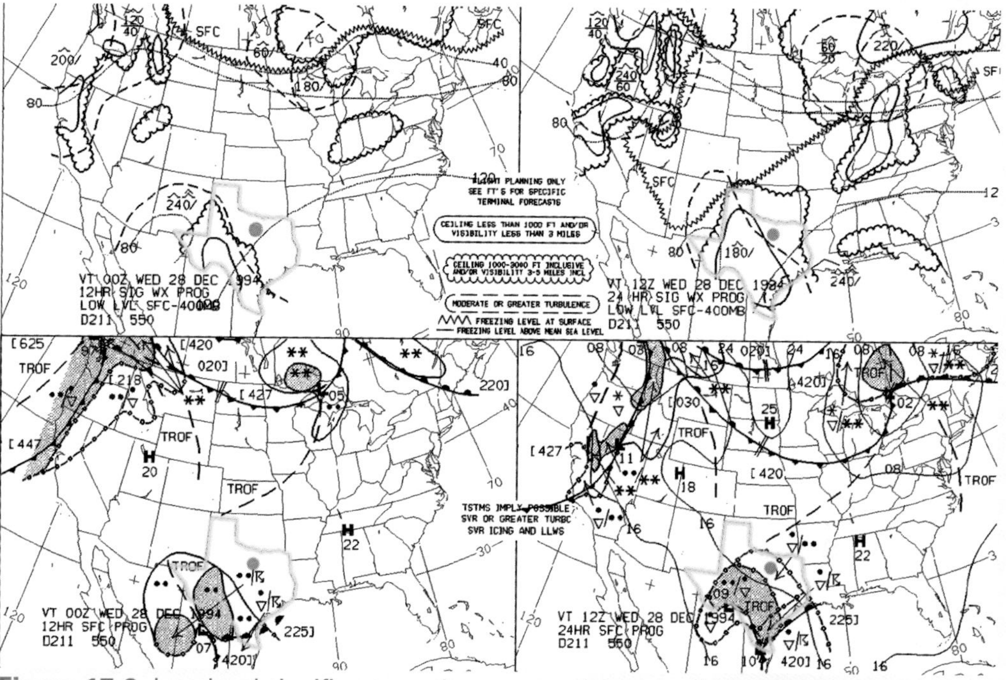

Figure 17-6. Low-level significant weather prognostic chart valid for 0000Z (left) and 1200Z (right) 28 December.

The most recent low-level significant weather prognostic chart at the time of the overview provides you with a more complete picture of forecast weather conditions that may impact your flight. The low-level prog chart provides forecast information valid at 0000Z and 1200Z on December 28. The upper right panel predicts a widespread area of IFR conditions over southern Texas extending into the center of the state. Marginal VFR weather is forecast for almost the entire state by 1200Z.

Your efficient use of weather information requires that you determine the conditions which are significant to your flight. For example, an area of moderate turbulence above 18,000 feet MSL is forecast to cover most of Texas. This area is of no direct relevance to your flight since your proposed altitude is 5,000 feet MSL or lower. Weather information products provide a wide variety of information. It is your responsibility to determine the key information of each product that may impact your flight.

The lower left panel of the low-level surface prog chart predicts a trough over southcentral Texas with areas of continuous precipitation, thunderstorms, and rain showers. This is the same low pressure system that you identified previously on the surface analysis and 300 mb charts. The system continues to track eastward toward the coast. By 1200Z an area of showery precipitation extends into the Dallas area. Thunderstorms are expected along the Gulf Coast where the storm system is the most intense.

An outlook briefing from flight service furnishes additional information at this stage. Your initial assessment of the forecast weather situation provides you with a foundation for posing specific questions to the briefer. The conclusion at the completion of the overview evaluation is that there is a good chance for MVFR or IFR conditions and precipitation over the Dallas area for your flight tomorrow.

PREFLIGHT EVALUATION

The preflight weather evaluation is usually accomplished within a few hours of the proposed departure time due to the importance of procuring the latest information prior to the flight. For your flight in the Dallas area, you obtain an appropriate AIRMET, surface analysis chart, weather depiction chart, lifted index chart, and radar summary chart. Your weather evaluation also includes an examination of the surface aviation weather report, a PIREP, the terminal forecast, and area forecast. A standard briefing from flight service used in combination with select weather products provides you with a comprehensive weather evaluation.

Supplementing a briefing with graphic materials enables you to visualize the weather more effectively. In addition, obtaining weather data before you speak to an FSS briefer provides background for asking questions about the weather situation.

You begin your preflight evaluation the morning of the flight on the 28th. One of the first steps at this stage is to determine if any advisories or warnings are issued for the area at the time of your flight. AIRMET Sierra, issued at 1445Z, is valid until 2100Z. This WA warns of a widespread area of IFR weather over the state of Texas. These conditions are expected to end over southern Texas between 1700Z and 1900Z. The IFR conditions are expected to continue elsewhere beyond 2100Z and spread northeastward through 0300Z. Visualizing the area that is described in the AIRMET is very difficult without the aid of an in-flight advisory plotting chart. By drawing lines between each location identifier specified, you can obtain a clearer picture of the area affected by the advisory. (Figure 17-7)

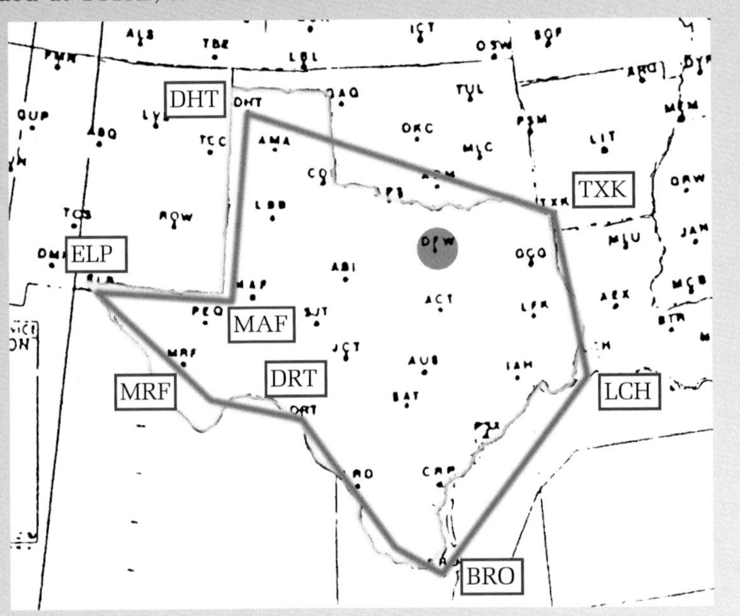

Figure 17-7. AIRMET Sierra 28 December at 1445Z.

The region described in the AIRMET includes the Dallas area, and the IFR conditions are forecast to continue beyond your departure time. Your initial assessment at this time is that the flight may have to be canceled or possibly delayed to a later time. By examining additional weather information, you can paint a more complete picture of the weather and make an informed decision about your flight.

```
281448
DFW WA 281445 AIRMET SIERRA UPDT 2 FOR IFR AND MTN OBSCN VALID
UNTIL 282100 . AIRMET IFR...OK TX AR LA AND CSTL WTRS FROM DHT TO
TXK TO 30SW LCH TO BRO TO 80WNW BRO TO DRT TO 90SSE MRF TO ELP
TO 40W MAF TO DHT OCNL CIG BLO 10/VSBY BLO 3 IN CLDS/PCPN F. CONDS
ENDG SWRN TX 17- 19Z..CONTG ELSW BYD 21Z AND SPRDG NEWD THRU
```

The surface analysis chart, valid at 1500Z, shows that the weather system has tracked easterly and a low pressure area is now centered just off the Gulf Coast. A trough ("TROF") extends over central Texas including the Dallas area. The low-level prog chart that you reviewed in the overview step was very accurate in describing the movement and development of this low pressure system. The low pressure area should continue to produce poor weather conditions over Texas as it draws moisture from the gulf. (Figure 17-8)

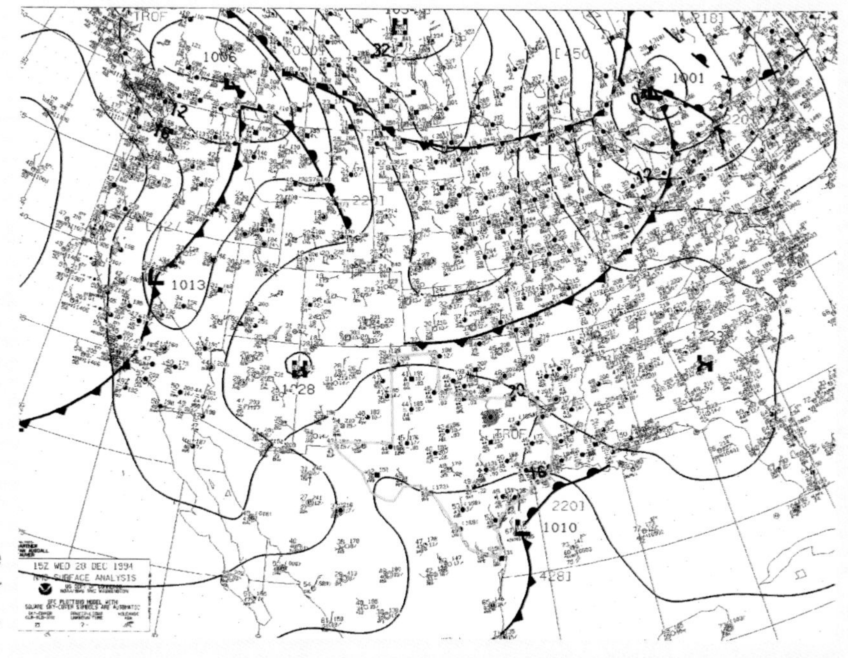

Figure 17-8. Surface analysis chart 28 December at 1500Z.

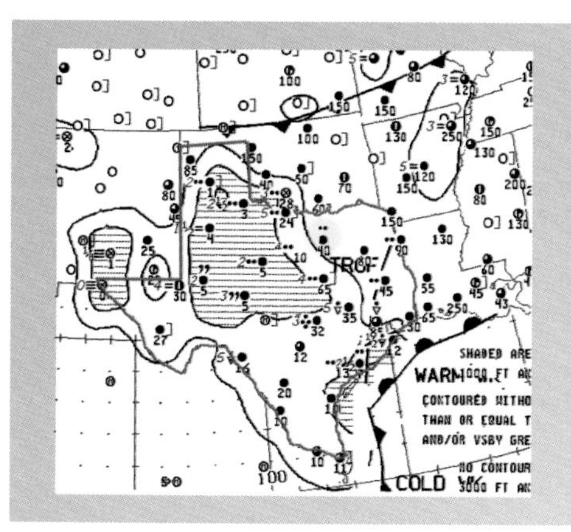

A look at the weather depiction chart helps you visualize the ceiling and visibility conditions associated with this system. This chart, issued at 1600Z indicates marginal VFR and IFR conditions just to the south and west of Dallas. The station model near Dallas depicts an overcast ceiling of 4,000 feet, and the symbol above the model indicates continuous rain. If this system continues to track eastward the weather conditions over Dallas will deteriorate further. (Figure 17-9)

Figure 17-9. Weather depiction chart 28 December 1600Z.

The lifted index chart, based on 1200Z data provides insight into the intensity of potential thunderstorms in the area. Index values ranging from 0 to -2, indicate a low potential for severe thunderstorms. The Dallas area shows a 0 indication on the chart while the southern coast shows a moderate potential for severe thunderstorms (lifted index -4). (Figure 17-10)

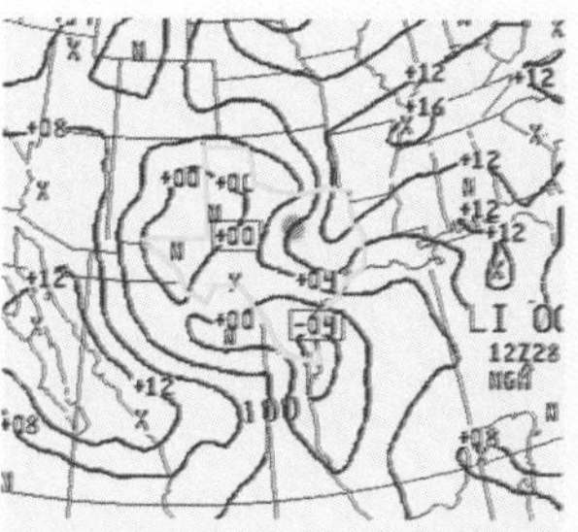

Figure 17-10. Lifted index 28 December at 1200Z.

Your departure time is drawing near and the chances of your flight taking off as planned become increasingly marginal, along with the weather. Within an hour before the flight, you examine the latest weather information available to fully assess the situation.

The radar summary chart provides additional details about the type of precipitation in the area, its intensity, and direction of movement. An examination of the radar summary chart within the hour before the flight reveals light to moderate precipitation over Dallas with a small echo just northwest of the Dallas area that exceeds VIP 5. The tops of the precipitation range from 20,000 feet MSL to 25,000 feet MSL. The area of precipitation is moving north at 15 knots. It is unlikely that any clearing will occur soon near Dallas since the area of precipitation is fairly large and extends to the southeast border of the state. By comparing this chart with the lifted index examined earlier, you can see that the most intense thunderstorms exist in the area where the -4 value was indicated. (Figure 17-11)

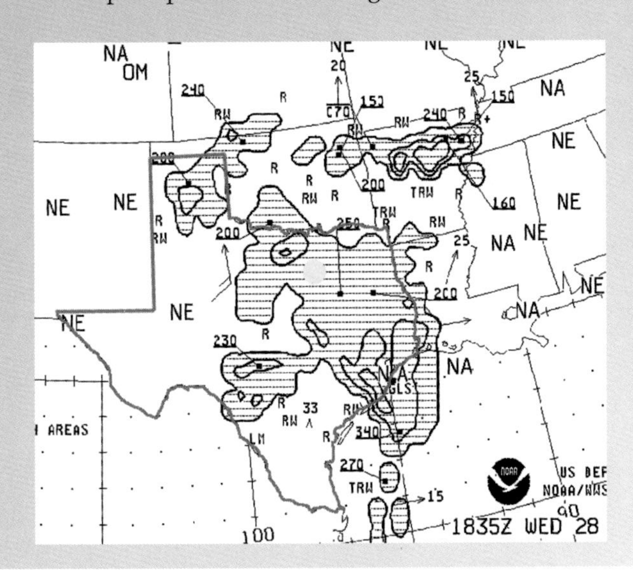

Figure 17-11. Radar summary chart 28 December 1835Z. An arrow shaft indicates direction of area movement and each barb equals ten knots of speed. A half barb represents five knots.

The surface aviation weather report at 1750Z for Dallas/Ft. Worth International Airport (DFW) describes a measured ceiling of 2,800 broken 6,000 overcast with six miles visibility in light rain and fog. (Figure 17-12)

DFW SA 1750 M28 BKN 60 OVC 6R-F 208/48/47/3305/015/BKN V SCT

Figure 17-12. Surface aviation weather report DFW 28 December at 1750Z.

A PIREP at 1735Z over Dallas/Ft. Worth by the pilot of a 727 describes the overcast cloud bases at 3,000 feet MSL extending to 15,000 feet MSL. Since the field elevation is 603 feet MSL, this indicates a ceiling of 2,400 feet AGL, slightly lower than that reported in the surface aviation weather report. (Figure 17-13)

281828
DFW UA /OV DFW/TM 1735/FLDURD/TP B727/SK 030 OVC 150

Figure 17-13. PIREP over DFW 28 December at 1735Z.

Based on the preflight evaluation to this point, you determine that marginal VFR conditions will exist in the Dallas area at your departure time. You may be able to delay your departure until later in the afternoon. An examination of the terminal and area forecasts will help you reach a more informed decision about your flight. Although you may be able to make a go/no-go decision based solely on the terminal forecast, the area forecast allows you to more clearly visualize the forecast movement and development of weather systems in the area.

For this flight, the terminal forecast provides the latest forecast information. The most recent area forecast that can be obtained was issued at 1205Z. The area forecast is valid for your planned flight time, but the information is not as timely as that contained in the terminal forecast. It is important to prioritize data during your weather evaluation. For example, the terminal forecast issued at 1800Z takes precedence since the more recent forecast provides the more accurate prediction. A short wait for an updated forecast may be worthwhile if no recent weather data is available.

According to the most recent terminal forecast for DFW, marginal VFR with occasional IFR conditions are forecast through 2200Z. The weather then deteriorates with IFR conditions through 1600Z on the 29th. Only a slight improvement is predicted at the end of the forecast period. (Figure 17-14)

281749
DFW FT 281818 20 SCT C35 OVC 5R-F 0308 OCNL C20 BKN 2RW-F. 20Z C20
OVC 4R-F 0108 OCNL 2RW-F. 22Z C10 OVC 2R-F 3606. 06Z C6 OVC 1F 3506
HC 1R-F. 16Z C10 OVC 3F 3508..

Figure 17-14. Terminal forecast valid 28 December 1800Z through 29 December 1800Z.

DFW FA 281205 AMD SYNOPSIS AND VFR CLDS/WX SYNOPSIS VALID UNTIL
290500 CLDS/WX VALID UNTIL 282300...OTLK VALID 282300-290500 OK TX
AR TN LA MS AL AND CSTL WTRS . SEE AIRMET SIERRA FOR IFR CONDS AND
MTN OBSCN. TSTMS IMPLY SVR OR GTR TURBC SVR ICG LLWS AND IFR
CONDS. NON MSL HGTS DENOTED BY AGL OR CIG. . SYNOPSIS...11Z LOW
INVOF BRO WITH TROF EXTDG NW ALG A BRO-DRT- MRF LN. BY 05Z LOW
INVOF IAH WITH CDFNT EXTDG S FM THE LOW INTO THE GLFMEX. ANOTHER
CDFNT ALG A BKW-BWG-TUL-DHT LN. ...

The synopsis of the area forecast describes a low in the vicinity of BRO with a trough extending northwest along a line from BRO to DRT to MRF. By 0500Z the low should be positioned in the vicinity of IAH with a cold front extending south from the low into the Gulf of Mexico. Another cold front is located along a line from BKW to BWG to TUL to DHT. If you draw the weather systems described by the synopsis on a map, you can visualize the development and movement of the specific weather conditions. (Figure 17-15)

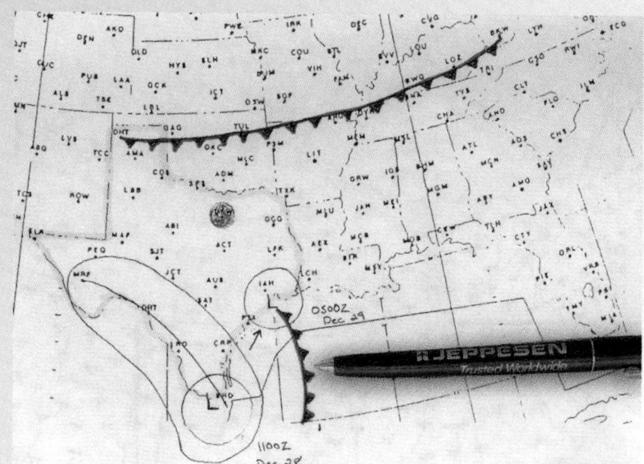

Figure 17-15. Synopsis section of the area forecast amendment issued 28 December 1205Z valid through 29 December 0500Z.

The VFR clouds and weather section of the FA are valid until 2300Z. The outlook portion is valid from 2300Z December 28 to 0500Z December 29. Forecast conditions for northcentral Texas apply to the Dallas area as indicated by the geographical area designator map. You can expect scattered clouds at 400 feet with an overcast ceiling of 1,000 feet. The visibility is forecast to be three to five miles in light rain and fog. After 1300Z widely scattered embedded thunderstorms are predicted with cumulonimbus tops to 35,000 feet. The outlook is for IFR conditions due to low ceilings with rain and fog. (Figure 17-16)

. . N CNTRL TX AGL 4 SCT CIG 10 OVC. VSBY 3-5R-F. AFT 13Z WDLY SCT EMBDD TRW-. CB TOPS 350. OTLK...IFR CIG R F. . NERN TX CIG 70 OVC LYRD 320. VSBY 3-5F. 16-18Z CIG 30 OVC. VSBY 3-5F. WDLY SCT RW-. 20-22Z CIG 15 OVC. VSBY 3-5R-F. WDLY SCT EMBDD TRW-. CB TOPS 350. OTLK...IFR CIG R F. . S CNTRL TX AGL 5 SCT CIG 10-15 OVC LYRD 250. VSBY 3-5F. SCT RW. WDLY SCT EMBDD TRW-. TSTMS PSBLY SVR NR CST. CB TOPS 430. OTLK...VFR SWRN SXNS..MVFR CIG RW ELSW. . SERN TX CIG 10-20 BKN-SCT 40 OVC LYRD 350. VSBY 3-5F. SCT RW. 14Z AGL 5 SCT CIG 10-20 OVC. VSBY 3-5F. SCT RW. WDLY SCT EMBDD TRW-. TSTMS PSBLY SVR NR CST. CB TOPS 430. OTLK...IFR CIG R TRW F. .

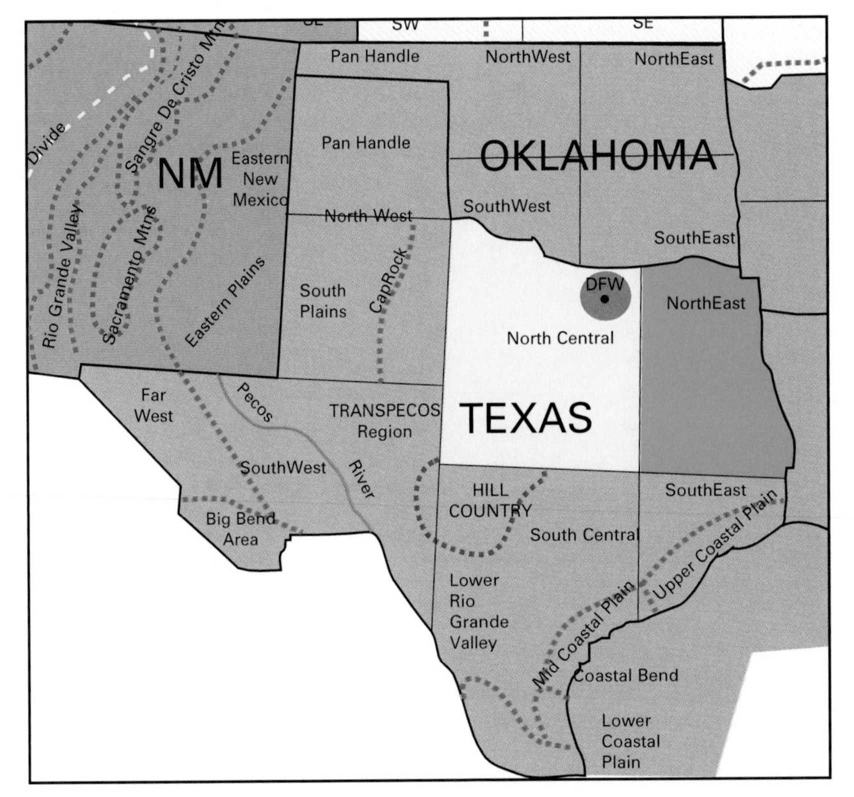

Figure 17-16. VFR clouds and weather section of the area forecast amendment issued 28 December 1205Z valid through 2300Z. Outlook valid 28 December 2300Z through 29 December 0500Z.

To conclude the weather evaluation, you must make a decision based on your analysis. Marginal VFR weather exists in the area of your planned flight at this time and there is a strong probability for IFR conditions in the near future. IFR conditions do exist just to the south of the Dallas area. You are not qualified or rated to fly in IFR weather. Based on your experience level, and your weather knowledge, you make the decision to cancel the flight for today. Although the weather does not appear to be too promising for tomorrow, you have already laid the foundation for the overview step. An additional examination of a low-level significant prog chart and upper-air data will provide you with a more accurate picture of the weather forecast for tomorrow.

An important component of the self-briefing procedure that was not addressed in the preceding scenario is your own weather observation. Simply by scanning the sky, you add to the weather evaluation. You may observe a thunderstorm developing before it shows up on a radar report. Lenticular clouds provide a signpost in the sky alerting you to turbulence. A fog bank may be visible that is not near a weather reporting station, or gusty winds may be apparent before a report has been issued. Your knowledge of the local region's weather can greatly supplement your self briefing. In addition, discussions with other pilots can lead to valuable information about flight conditions. Pilots returning from flights can give you one-on-one pilot reports. Areas of low ceilings, turbulence, and icing can be determined simply by querying your fellow pilots.

IN-FLIGHT EVALUATION

Your evaluation of the weather doesn't end if a "go" decision is made. The dynamic nature of weather makes in-flight weather evaluation essential to safety. The in-flight weather assessment is an on-going process that begins with observation. A visual assessment of your environment is made continuously in flight. In addition to scanning the sky for significant weather, changes in temperature and pressure can alert you to the presence of weather systems. By calculating groundspeed enroute, you also can determine changes in winds aloft.

You must also utilize in-flight weather services such as recorded surface observations and forecasts. Contacting Flight Watch is one of the most effective methods to receive updated weather information tailored to your flight. For example, if you observe developing thunderstorms along your route, a call to Flight Watch may provide you with the latest radar and pilot weather reports for the area. With this information you can make an informed decision to continue along the route to your destination, change your route, or divert to an alternate airport.

Section C

FUTURE AVIATION WEATHER RESOURCES

What does the future hold for aviation weather services and products? The FAA and the National Weather Service are implementing a plan leading to an advanced and improved aviation weather system. This plan is designed to integrate present and pending technologies to develop higher resolution and more accurate observations of atmospheric variables. Increased computer power and modeling capabilities will significantly enhance weather analyses and forecasts. Finally, weather information will be transformed into decision aids specifically for aviation that focus on weather variables impacting flight. These decision aids will be available in pilot-briefing rooms, the cockpit, air traffic control facilities, and airline flight-planning and dispatch centers.

MODERNIZED OBSERVATION SYSTEMS

The focus of modernized weather observation is the production of high-quality data in nearly real time. These data are generated in digital formats that can be readily accessible without compromising quality, precision, accuracy, or resolution. The density of measurements, spatially and temporally, can be increased by 100% over the current observation systems available. The present observing system was conceived to focus on large-scale weather phenomena, such as cyclones, anticyclones, and frontal systems. Although knowledge of the presence of these systems is important in providing you with a general picture of the weather, mesoscale and microscale phenomena are often of the most interest to you as a pilot. Advances in weather observation systems will lead to improvement in detection of smaller-scale weather.

The next generation Doppler weather radar network (NEXRAD) is currently being implemented. The sensitivity of NEXRAD allows measurements of clear-air echoes and air motion. Regions of convergence in the boundary layer that often lead to thunderstorm development can be more clearly detected with this system. Terminal Doppler weather radar (TDWR) is also presently being installed at numerous airports in the United States. The TDWR provides high-resolution Doppler measurements and delivers automated products that describe such weather phenomena as low-level wind shear and microbursts.

Automated surface-observation stations are becoming a major part of the surface weather observing network. Temperature, dewpoint, winds, pressure, precipitation, ceiling, and visibility observations are made automatically. Timeliness is improved over manual stations, since automated systems can often provide data at one-minute intervals. Automated surface observing systems (ASOS) and automated weather observing systems (AWOS) are currently in use. Although highly efficient, the use of the automated systems is a concern to some pilots. Questions exist as to whether these systems can ever fully replace human ability to observe and process information regarding visibility, ceiling, layered clouds, and locations of nearby thunderstorms. The NWS and the FAA hopes to address this issue and supplement automated stations with data from satellites, radar, lightning detectors, and other sources.

The low-level wind shear alert system (LLWAS) is an important complement to terminal Doppler radar. The LLWAS has been installed at more than 100 airports in the U.S. which have a combination of a heavy volume of air traffic and a high frequency of thunderstorms. Each LLWAS consists of an array of anemometers positioned at various locations on the airport. The anemometer readings are transmitted to a central computer where the data is processed. A wind shear advisory is issued if a reading differs from the network average wind by at least 15 knots, or if a divergence is detected in the wind field. LLWAS sites are being enhanced by adding more anemometers and improving data processing and display. Integrating LLWAS with the terminal Doppler radar system provides much improved warning capability.

The radiosonde network has been the primary upper air data source for weather forecasting for many years. New technologies are being investigated to improve this system. Global positioning system (GPS) navigational tracking techniques may be utilized for wind measurements. Radiosondes are also being supplemented by new sounding systems, radar, satellites and commercial aircraft.

One of the most promising new sounding systems is the wind profiler. This system consists of a vertically pointing microwave radar which measures horizontal wind speed and direction between about 1,500 feet AGL and 53,000 feet MSL. A demonstration profiler network is now in place in the central United States to determine optimal profiler spacing for a more extensive network. The principal advantage of the wind profiler is that a sounding may be made every six minutes. The soundings are averaged and used to determine hourly wind speed and direction profiles.

The wind profiler relies on Doppler radar which emits pulses of microwave radiation that are backscattered from a target resulting from the small fluctuation in temperature and moisture created by eddies that move with the wind. As the eddies move away from the receiving antenna, the returning radar pulse will change in frequency. The measured change (Doppler shift) is used to calculate wind speed and direction at various altitudes. The potential of the profilers to detect clear air turbulence is now being explored.

Another development in atmospheric observation is information supplied by aircraft instruments. Over 10,000 measurements of winds and temperatures are being made daily by transport aircraft over the United States. The observations are reported in digital format through Aeronautical Radio, Inc. (ARINC) Communications Addressing and Reporting System (ACARS). The FAA is supporting a program to explore the possibility of humidity sensing by aircraft. In addition, airborne turbulence sensors may also become a part of the airborne data base.

Satellites provide a variety of aviation weather information including images of cloud patterns, accurate cloud-top measurements, temperature soundings, water vapor fields, and precipitation mapping when combined with ground-based radars. The geostationary operational environmental satellite (GOES) provides simultaneous imaging and sounding profiles. The high-resolution imaging produced by the satellite, supplements radar and surface observations. GOES satellites are capable of making low resolution soundings in areas where no soundings were previously available. Satellite imagery does have limitations, for example, satellite observations do not have sufficient resolution in vertical soundings and GOES satellites cannot observe conditions within and below clouds. They are capable, however, of making low resolution soundings in areas where no soundings were previously available.

IMPROVED WEATHER ANALYSIS AND FORECASTING

A widespread agreement in the aviation community is that weather information needs to be simplified and transformed into graphic representations that facilitate decision making. The NWS hopes to transform the new observations into highly effective, user-friendly aviation weather information. The National Weather Service Modernization Committee has concluded that this type of graphic presentation can be derived from a four-dimensional data base that is shared by the aviation community. Data transmitted from modernized observational systems would be stored according to latitude, longitude, height above mean sea level, and time. When this information is combined with improved numerical computer models, three-hour forecasts can be produced operationally. A principal component of this modernization is the aviation gridded forecast system (AGFS). This system converts computer forecasts into more accurate high-resolution presentations of hazardous conditions, including icing, turbulence, and convective activity.

PRESENTING WEATHER INFORMATION EFFECTIVELY

As demonstrated in the weather evaluation process, you must be able to interpret weather information in a wide variety of formats. Currently, this information must then be analyzed and combined to form a mental image of the weather and its impact on your flight. Finally, a decision must be made based on your evaluation. This process could be greatly simplified with more effective presentation of aviation weather information. Obtaining and evaluating weather information in the cockpit can be especially difficult, yet some of the most crucial and timely weather decisions have to be made in flight.

Effective aviation weather products must meet several requirements. The need for pilots and air traffic controllers to visualize the evolving weather situation demands that information be presented in easily comprehended graphic formats. The presentation of weather information must be readily available by computer to pilots, as well as air traffic controllers. An effective means of accessing this information from the cockpit must be developed as well. Route-specific displays of conditions that impact flight can easily enable pilots to examine alternatives and make informed decisions.

The FAA's aviation weather development program is addressing the need for improved weather presentation. Two significant components of this program are the aviation weather products generator and the integrated terminal weather system.

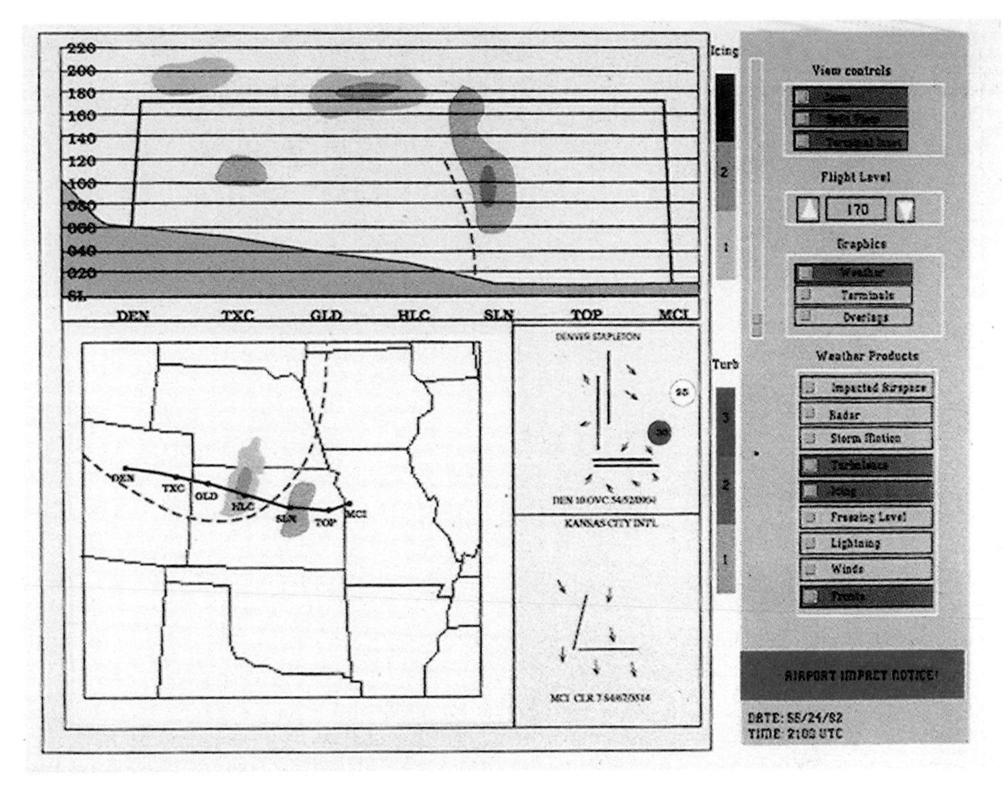

Figure 17-17. This example illustrates the possibilities of the AWPG. The proposed route of flight is displayed as a solid black line. The pilot has requested weather graphics with terminal insets of Denver Stapleton and Kansas City International. The location of specific weather hazards can be requested and are graphically displayed. In this example, areas of turbulence and icing have been requested. The arrows shown on the terminal views show wind direction. Microbursts occurring at Denver are depicted by circles including the magnitude of the wind shear shown inside the circle. Weather display photo source: National Center for Atmospheric Research/University Corporation for Atmospheric Research/National Science Foundation.

The aviation weather products generator (AWPG) will produce new analysis and forecast products tailored for aviation. Pilots, controllers, air traffic managers, and airline dispatchers will benefit from the development of the AWPG. The graphic displays which are generated will provide users with tools to quickly identify aviation weather hazards including icing, turbulence, convective storms, and winds aloft. The AWPG focus is on weather information for the enroute phase of flight. These displays "visualize" the weather for you and require only limited interpretation for maximum information. AWPG products will be available to air traffic control, airline dispatchers, and are designed for use through data links to the cockpit. (Figure 17-17)

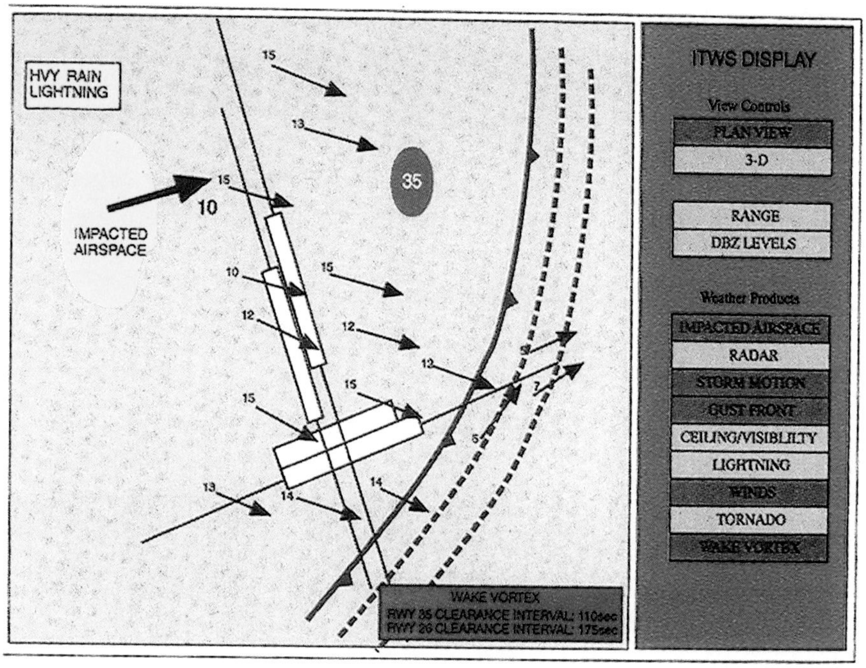

Detailed graphic displays designed for the terminal area will be provided by the integrated terminal weather system (ITWS). ITWS will provide high resolution graphic displays of hazards at airports. Ceilings and visibilities, winds, convective activity, and microbursts will be illustrated on these presentations. (Figure 17-18)

Fig 17-18. This example of a prototype ITWS display indicates an area of heavy rain and lightning (shown in yellow) moving east-northeast at ten knots. A gust front is depicted by dashed blue lines and the associated cold front is shown as a solid blue line with barbs. Wind speeds and directions in the runway complex vicinity are also represented. Wake vortex clearance information can be found in the lower right corner. Weather display photo source: National Center for Atmospheric Research/University Corporation for Atmospheric Research/National Science Foundation.

CONCLUSION

The future holds exciting advancements in aviation weather technology including improved observations, analyses and forecasts, and innovative aviation weather products. Until there are significant improvements in the presentation of aviation weather information, however, the development of a useful mental image of the weather is necessary. This process involves careful analysis of many different products. You are required to assemble the information, like pieces of a puzzle, to construct an accurate picture of the weather. The more background knowledge you have about the process of weather, weather systems, and hazards, the easier the weather evaluation becomes. However, it does take time and practice to efficiently transform weather data into useful decision aids for flying. Your efforts to understand aviation weather will be rewarded with a safer and more satisfying flying experience.

KEY TERMS

Aviation Gridded Forecast System (AGFS)
Aviation Weather Products Generator (AWPG)
Geostationary Operational Environmental Satellite (GOES)
Integrated Terminal Weather System (ITWS)
Low-level Wind Shear Alert System (LLWAS)
Next Generation Doppler Weather Radar Network (NEXRAD)
Terminal Doppler Weather Radar (TDWR)
Wind Profiler

CHAPTER QUESTIONS

1. Develop a checklist that organizes your own self-briefing procedure.

2. Practice several self-briefings for imaginary flights using your checklist.

3. Obtain an area forecast from flight service or a computerized service. Try to sketch the weather systems described in the FA on a map. Now compare your sketch with an appropriate surface analysis chart and low-level prog chart.

4. Create a diagram similar to the future aviation-weather products generator (AWPG) displays. Use weather information for a specific flight to draw areas of hazards along your route of flight.

Appendix A

Standard Atmosphere

Standard Atmosphere

Altitude (Feet)	Temperature (˚C)	Pressure (mb)
0	15.000	1013.25
100	14.802	1009.50
500	14.009	995.07
1,000	13.019	977.16
2,000	11.038	942.13
4,000	7.077	875.13
6,000	3.116	812.04
8,000	-.844	752.71
10,000	-4.803	696.94
12,000	-8.761	644.58
14,000	-12.718	595.46
16,000	-16.675	549.42
18,000	-20.631	506.32
20,000	-24.586	466.00
22,000	-28.541	428.33
24,000	-32.494	393.17
26,000	-36.447	360.40
28,000	-40.399	329.87
30,000	-44.351	301.48
32,000	-48.301	275.11
34,000	-52.251	250.64
36,000	-56.200	227.97
*36,200	-56.500	225.79
38,000	-56.500	207.14
40,000	-56.500	188.23
42,000	-56.500	171.04
44,000	-56.500	155.42
46,000	-56.500	141.24
48,000	-56.500	128.35
50,000	-56.500	116.64
52,000	-56.500	106.00
54,000	-56.500	96.332
56,000	-56.500	87.547
58,000	-56.500	79.565
60,000	-56.500	72.312

* Tropopause

Dewpoint and Humidity Tables

TABLE 1 Dewpoint Temperature (°C)

To find the dewpoint temperature, locate the air temperature (dry-bulb temperature) along the left-hand column. Then, locate the wet-bulb depression along the top of the chart. The intersection of these two values is the dewpoint temperature. For example, an air temperature of 20°C with a wet-bulb depression of 3°C produces a dewpoint temperature of 15°C. Note: dewpoint temperature readings are appropriate for pressures near 29.92 in.Hg. Footnote: The wet-bulb depression is the difference between the air temperature (dry-bulb temperature) and the wet-bulb temperature which can be determined from a psychrometer.

Air (Dry-Bulb) Temperature (°C)	Wet-Bulb Depression (°C)															
	0.5	1.0	1.5	2.0	2.5	3.0	3.5	4.0	4.5	5.0	7.5	10.0	12.5	15.0	17.5	20.0
-20	-25	-33														
-17.5	-21	-27	-38													
-15	-19	-23	-28													
-12.5	-15	-18	-22	-29												
-10	-12	-14	-18	-21	-27	-36										
-7.5	-9	-11	-14	-17	-20	-26	-34									
-5	-7	-8	-10	-13	-16	-19	-24	-31								
-2.5	-4	-6	-7	-9	-11	-14	-17	-22	-28	-41						
0	-1	-3	-4	-6	-8	-10	-12	-15	-19	-24						
2.5	1	0	-1	-3	-4	-6	-8	-10	-13	-16						
5	4	3	2	0	-1	-3	-4	-6	-8	-10	-48					
7.5	6	6	4	3	2	1	-1	-2	-4	-6	-22					
10	9	8	7	6	5	4	2	1	0	-2	-13					
12.5	12	11	10	9	8	7	6	4	3	2	-7	-28				
15	14	13	12	12	11	10	9	8	7	5	-2	-14				
17.5	17	16	15	14	13	12	12	11	10	8	2	-7	-35			
20	19	18	18	17	16	15	14	14	13	12	6	-1	-15			
22.5	22	21	20	20	19	18	17	16	16	15	10	3	-6	-38		
25	24	24	23	22	21	21	20	19	18	18	13	7	0	-14		
27.5	27	26	26	25	24	23	23	22	21	20	16	11	5	-5	-32	
30	29	29	28	27	27	26	25	25	24	23	19	14	9	2	-11	
32.5	32	31	31	30	29	29	28	27	26	26	22	18	13	7	-2	
35	34	34	33	32	32	31	31	30	29	28	25	21	16	11	4	
37.5	37	36	36	35	34	34	33	32	32	31	28	24	20	15	9	0
40	39	39	38	38	37	36	36	35	34	34	30	27	23	18	13	6
42.5	42	41	41	40	40	39	38	38	37	36	33	30	26	22	17	11
45	44	44	43	43	42	42	41	40	40	39	36	33	29	25	21	15
47.5	47	46	46	45	45	44	44	43	42	42	39	35	32	28	24	19
50	49	49	48	48	47	47	46	45	45	44	41	38	35	31	28	23

TABLE 2 Relative Humidity (Percent)

To find the relative humidity, locate the air temperature (dry-bulb temperature) along the left-hand column. Then, locate the wet-bulb depression along the top of the chart. The intersection of these two values is the relative humidity. For example, an air temperature of 20°C with a wet-bulb depression of 3°C produces a relative humidity of 74%. Note: relative humidity readings are appropriate for pressures near 29.92 in.Hg. Footnote: The wet-bulb depression is the difference between the air temperature (dry-bulb temperature) and the wet-bulb temperature which can be determined from a psychrometer.

Air (Dry-Bulb) Temperature (°C)	Wet-Bulb Depression (°C)																	
	0.5	**1.0**	**1.5**	**2.0**	**2.5**	**3.0**	**3.5**	**4.0**	**4.5**	**5.0**	**7.5**	**10.0**	**12.5**	**15.0**	**17.5**	**20.0**	**22.5**	**25.0**
-20	70	41	11															
-17.5	75	51	26	2														
-15	79	58	38	18														
-12.5	82	65	47	30	13													
-10	85	69	54	39	24	10												
-7.5	87	73	60	48	35	22	10											
-5	88	77	66	54	43	32	21	11	1									
-2.5	90	80	70	60	50	42	37	22	12	3								
0	91	82	73	65	56	47	39	31	23	15								
2.5	92	84	76	68	61	53	46	38	31	24								
5	93	86	78	71	65	58	51	45	38	32	1							
7.5	93	87	80	74	68	62	56	50	44	38	11							
10	94	88	82	76	71	65	60	54	49	44	19							
12.5	94	89	84	78	73	68	63	58	53	48	25	4						
15	95	90	85	80	75	70	66	61	57	52	31	12						
17.5	95	90	86	81	77	72	68	64	60	55	36	18	2					
20	95	91	87	82	78	74	70	66	62	58	40	24	8					
22.5	96	92	87	83	80	76	72	68	64	61	44	28	14	1				
25	96	92	88	84	81	77	73	70	66	63	47	32	19	7				
27.5	96	92	89	85	82	78	75	71	68	65	50	36	23	12	1			
30	96	93	89	86	82	79	76	73	70	67	52	39	27	16	6			
32.5	97	93	90	86	83	80	77	74	71	68	54	42	30	20	11	1		
35	97	93	90	87	84	81	78	75	72	69	56	44	33	23	14	6		
37.5	97	94	91	87	85	82	79	76	73	70	58	46	36	26	18	10	3	
40	97	94	91	88	85	82	79	77	74	72	59	48	38	29	21	13	6	
42.5	97	94	91	88	86	83	80	78	75	72	61	50	40	31	23	16	9	2
45	97	94	91	89	86	83	81	78	76	73	62	51	42	33	26	18	12	6
47.5	97	94	92	89	86	84	81	79	76	74	63	53	44	35	28	21	15	9
50	97	95	92	89	87	84	82	79	77	75	64	54	45	37	30	23	17	11

TABLE 3 Dewpoint Temperature (°F)

See table 1 for instructions on finding the dewpoint temperature.

Air (Dry-Bulb) Temperature (°F)	Wet-Bulb Depression (°F)																							
	1	2	3	4	5	6	7	8	9	10	11	12	13	14	15	16	17	18	19	20	25	30	35	40
0	-7	-20																						
5	-1	-9	-24																					
10	5	-2	-10	-27																				
15	11	6	0	-9	-26																			
20	16	12	8	2	-7	-21																		
25	22	19	15	10	5	-3	-15	-51																
30	27	25	21	18	14	8	2	-7	-25															
35	33	30	28	25	21	17	13	7	0	-11														
40	38	35	33	30	28	25	21	18	13	7	-1	-14												
45	43	41	38	36	34	31	28	25	22	18	13	7	-1	-14										
50	48	46	44	42	40	37	34	32	29	26	22	18	13	8	0	-13								
55	53	51	50	48	45	43	41	38	36	33	30	27	24	20	15	9	1	-12						
60	58	57	55	53	51	49	47	45	43	40	38	35	32	29	25	21	17	11	4	-8				
65	63	62	60	59	57	55	53	51	49	47	45	42	40	37	34	31	27	24	19	14				
70	69	67	65	64	62	61	59	57	55	53	51	49	47	44	42	39	36	33	30	26	-11			
75	74	72	71	69	68	66	64	63	61	59	57	55	54	51	49	47	44	42	39	36	15			
80	79	77	76	74	73	72	70	68	67	65	63	62	60	58	56	54	52	50	47	44	28	-7		
85	84	82	81	80	78	77	75	74	72	71	69	68	66	64	62	61	59	57	54	52	39	19		
90	89	87	86	85	83	82	81	79	78	76	75	73	72	70	69	67	65	63	61	59	48	32		
95	94	93	91	90	89	87	86	85	83	81	80	79	78	76	74	73	71	70	68	66	56	43	24	
100	99	98	96	95	94	93	91	90	89	87	86	85	83	82	80	79	77	76	74	72	63	52	37	12
105	104	103	101	100	99	98	96	95	94	93	91	90	89	87	86	84	83	82	80	78	70	61	48	30
110	109	108	106	105	104	103	102	100	99	98	97	95	94	93	91	90	89	87	86	84	77	68	57	43
115	114	113	112	110	109	108	107	106	104	103	102	101	99	98	97	96	94	93	92	90	83	75	65	54
120	119	118	117	115	114	113	112	111	110	108	107	106	105	104	102	101	100	98	97	96	89	81	73	63

TABLE 4 Relative Humidity (Percent)

See Table 2 for instructions on finding the relative humidity.

Air (Dry-Bulb) Temperature (°F)	Wet-Bulb Depression (°F)																							
	1	**2**	**3**	**4**	**5**	**6**	**7**	**8**	**9**	**10**	**11**	**12**	**13**	**14**	**15**	**16**	**17**	**18**	**19**	**20**	**25**	**30**	**35**	**40**
0	67	33	1																					
5	73	46	20																					
10	78	56	34	13																				
15	82	64	46	29	11																			
20	85	70	55	40	26	12																		
25	87	74	62	49	37	25	13	1																
30	89	78	67	56	46	36	26	16	6															
35	91	81	72	63	54	45	36	27	19	10	2													
40	92	83	75	68	60	52	45	37	29	22	15	7												
45	93	86	78	71	64	57	51	44	38	31	25	18	12	6										
50	93	87	80	74	67	61	55	49	43	38	32	27	21	16	10	5								
55	94	88	82	76	70	65	59	54	49	43	38	33	28	23	19	14	9	5						
60	94	89	83	78	73	68	63	58	53	48	43	39	34	30	26	21	17	13	9	5				
65	95	90	85	80	75	70	66	61	56	52	48	44	39	35	31	27	24	20	16	12				
70	95	90	86	81	77	72	68	64	59	55	51	48	44	40	36	33	29	25	22	19	3			
75	96	91	86	82	78	74	70	66	62	58	54	51	47	44	40	37	34	30	27	24	9			
80	96	91	87	83	79	75	72	68	64	61	57	54	50	47	44	41	38	35	32	29	15	3		
85	96	92	88	84	80	76	73	69	66	62	59	56	52	49	46	43	41	38	35	32	20	8		
90	96	92	89	85	81	78	74	71	68	65	61	58	55	52	49	47	44	41	39	36	24	13	3	
95	96	93	89	85	82	79	75	72	69	66	63	60	57	54	51	49	46	43	41	38	27	17	7	1
100	96	93	89	86	83	80	77	73	70	68	65	62	59	56	54	51	49	46	44	41	30	21	12	4
105	97	93	90	87	83	80	77	74	71	69	66	63	60	58	55	53	50	48	46	43	33	23	15	7
110	97	93	90	87	84	81	78	75	73	70	67	65	62	60	57	55	52	50	48	46	36	26	18	11
115	97	94	91	88	85	82	79	76	74	71	68	66	63	61	58	56	54	52	49	47	37	28	21	13
120	97	94	91	88	85	82	80	77	74	72	69	67	65	62	60	58	55	53	51	49	40	31	23	17

Codes for Aviation Weather Products

KEY TO MANUAL AVIATION WEATHER OBSERVATIONS							
STATION DESIGNATOR TYPE AND TIME OF REPORT	SKY CONDITIONS AND CEILING	VISIBILITY, WEATHER, AND OBSTRUCTIONS TO VISION	SEA LEVEL PRESSURE	TEMPERATURE AND DEWPOINT	WIND DIRECTION, SPEED AND CHARACTER	ALTIMETER SETTING	REMARKS AND CODED DATA
MCI SA 0758	15 SCT M25 OVC	1R - F	132	58 / 56	1807	993	R01VR20V40

STATION DESIGNATOR:
3 alphanumeric characters (usually the airport identifier).

TYPE OF REPORT:
SA = Scheduled Record (hourly) Observation.
SP = Special Observation taken between Record Observations to report a significant change in weather.
RS = Record Special, a Record Observation that reports a significant change in weather.
USP = Urgent Special Observation (tornado)

TIME OF REPORT:
Coordinated Universal Time (UTC) using 24-hour clock.
Example: 2255 = 10:55 pm

SKY CONDITION AND CEILING:
Sky condition contractions are for each layer in ascending order. Numbers preceding contractions are base height in hundreds of feet above ground level (AGL). Sky condition contractions are (- = Thin):
CLR = Clear: Less than 0.1 sky cover.
SCT = Scattered layer aloft: 0.1 through 0.5 sky cover.
BKN = Broken layer aloft: 0.6 through 0.9 sky cover (constitutes a ceiling layer).
OVC = Overcast layer aloft: More than 0.9 sky cover, or 1.0 sky cover (constitutes a ceiling layer).
X = Surface-based obscuration (all of sky is hidden by surface-based phenomena; constitutes a ceiling layer).
-X = Surface-based partial obscuration (0.1 or more, but not all, of sky is hidden by surface- based phenomena).
Ceiling designator: A letter preced ing height of a layer; identifies a ceiling and indicates how ceiling was obtained.
M = Measured. **E** = Estimated.
W = Indefinite. Vertical visibility into a surface-based obscuration.
V following height = variable ceiling.

VISIBILITY:
Reported in statute miles and fractions. **V** = Variable.

WEATHER & OBSTRUCTIONS TO VISION:
- = Light. (no sign) = Moderate. **+** = Heavy.

Weather:
T+ Severe thunderstorm
T Thunderstorm
R Rain
RW Rain shower
L Drizzle
ZR Freezing rain
ZL Freezing drizzle
A Hail
IP Ice pellets
IPW Ice pellet shower
S Snow
SW Snow shower
SP Snow pellets
SG Snow grains
IC Ice crystals

Obstructions to Vision:
BD Blowing dust
BN Blowing sand
BS Blowing snow
BY Blowing spray
D Dust
F Fog
GF Ground fog
H Haze
IF Ice fog
K Smoke

SEA-LEVEL PRESSURE:
Pressure in millibars/hectoPascals (hectoPascal is the metric equivalent to millibars; i.e., one millibar equals one hectoPascal). Shown as 3 digits. Leading 9 or 10 and decimal point is omitted. Examples: 150 = 1015.0 950 = 995.0

TEMPERATURE AND DEWPOINT:
Reported in degrees Fahrenheit (°F).

WIND DIRECTION, SPEED & CHARACTER:
Direction in tens of degrees from true north, speed in knots. 0000 = calm. G = gusts. Q = squalls. Peak speed of gusts in the past ten minutes follows G or Q. WSHFT in Remarks = windshift occurred at time indicated. Example: 3627G40 = 360° at 27 peak gusts 40 knots.

ALTIMETER SETTING:
Actual altimeter setting with last three digits transmitted and decimal point omitted. Examples: 005 = 30.05" 992 = 29.92":

REMARKS:
Runway Visibility (RVV) or Runway Visual Range (RVR) Runway visibility (RVV) is the visibility from a particular location along an identified runway and is reported in miles and fractions of miles. Runway visual range (RVR) is the maximum horizontal distance down a specified instrument runway at which a pilot can see to identify standard high intensity runway lights, reported in hundreds of feet. The VV and VR reports are for a 10 minute period preceding observation time. Runway number precedes the reports. V = Variable.

DECODED REPORT:
Kansas City Int'l Airport: Record Observation completed at 0758 UTC. 1500 feet scattered clouds, measured ceiling 2500 feet over cast, visibility 1 mile, light rain, fog, sea level pressure 1013.2 millibars/hectoPascals, temperature 58°F, dewpoint 56°F, wind 180°, 7 knots, altimeter setting 29.93". Runway 01 visual range varying from 2000 to 4000 feet in the past 10 minutes.

KEY TO ASOS (AUTOMATED SURFACE OBSERVING SYSTEM) WEATHER OBSERVATIONS

STATION DESIGNATOR, TYPE OF REPORT, TIME OF REPORT, STATION TYPE	SKY CONDITIONS AND CEILING BELOW 12,000'	VISIBILITY, WEATHER, AND OBSTRUCTIONS TO VISION	SEA-LEVEL PRESSURE/ TEMPERATURE / DEWPOINT / WIND DIRECTION, SPEED AND CHARACTER /ALTIMETER SETTING/	REMARKS AUTOMATED REMARKS GENERATED AUTOMATICALLY IF CONDITIONS EXIST. AUGMENTED REMARKS ADDED IF CONDITIONS EXIST AND CERTIFIED WEATHER OBSERVER IS ATTENDING THE SYSTEM	REMARKS AND CODED DATA
HTM RS 1755 A02A	M19V OVC	1R - F	125 / 36 / 34/ 2116G24 / 990	R29LVR10V50 CIG16V22 TWRVSBY 2 PK WND 2032 / 1732 PRESFR	ZRNO $

STATION DESIGNATOR:
3 alphanumeric characters (usually the airport identifier).

TYPE OF REPORT:
SA = Scheduled Record (hourly) Observation.
SP = Special Observation taken between Record Observations to report a significant change in weather.
RS = Record Special, a Record Observation that reports a significant change in weather.
USP = Urgent Special Observation (tornado)

TIME OF REPORT:
Coordinated Universal Time (UTC) using 24-hr clock.

STATION TYPE:
A02 = Unattended (no observer) ASOS.
A02A = Attended (observer present) ASOS.

SKY CONDITION AND CEILING BELOW 12,000' AGL:
Sky condition contractions are for each layer in ascending order. Numbers preceding contractions are base height in hundreds of feet above ground level (AGL).
CLR BLO 120 = Less than 0.1 sky cover below 12,000'
SCT = Scattered: 0.1 to 0.5 sky cover.
BKN = Broken: 0.6 to 0.9 sky cover.
OVC = Overcast: More than 0.9 sky cover. A letter preceding the height of a base identifies a ceiling layer and indicates how ceiling height was determined.
M = Measured
W = Indefinite
E = Estimated
X = Obscured sky
The letter **V** is added immediately following the height of a base to indicate a variable ceiling: see Remarks.

VISIBILITY:
Reported in statute miles and fractions from **<1/4** through **10+**. **V** = variable: see Remarks.

PRESENT WEATHER:
TORNADO (when augmented).
T = Thunder (when augmented): see Status Remarks.
R = Liquid precipitation that does not freeze (e.g., rain).
P – = Light precipitation in unknown form.
ZR = Liquid precipitation that freezes on impact (e.g., freezing rain): see Status Remarks.
A = Hail (when augmented).
S = Frozen precipitation other than hail (e.g., snow).
+ = Heavy. No sign = Moderate. **–** = Light.

OBSTRUCTIONS TO VISION:
Reported only when visibility is less than 7 statute miles.
F = Fog **H** = Haze
VOLCANIC ASH (when augmented).

SEA-LEVEL PRESSURE:
Pressure in millibars/hectoPascals (hectoPascal is the metric equivalent to millibars; i.e., one millibar equals one hectoPascal). Shown as last 3 digits only without decimal point (e.g., 950 = 995.0).

TEMPERATURE AND DEWPOINT:
Degrees Fahrenheit.

WIND DIRECTION, SPEED AND CHARACTER:
Direction in tens of degrees from **true** north. Voice broadcast in degrees from **magnetic**. Speed in knots. **0000** = calm. **E** = estimated. **G** = gusts. **Q** = squalls. Variable wind, peak wind, wind shift: see Remarks.

ALTIMETER SETTING:
Inches of mercury. Shown as last 3 digits only without decimal point (e.g., 005 = 30.05 inches).

MISSING DATA:
Reported as **M**.

DENSITY ALTITUDE:
Included on voice broadcast only when 1000 or more feet above airport elevation.

REMARKS:
Can Include:
RVR (Runway Visual Range), **VOLCANIC ASH, VIRGA, TWR VSBY** (Tower visibility), **SFC VSBY** (Surface visibility), **VSBY V** (Variable visibility), **CIG V** (Variable ceiling), **WSHFT** (Windshift), **PK WND** (Peak wind), **WND V** (Variable wind direction), **PCPN** (Precipitation amount), **PRESRR** (Pressure rising rapidly), **PRESFR** (Pressure falling rapidly), **PRJMP** (Pressure jump), **B** (Time weather began), **E** (Time weather ended).

STATUS REMARKS:
PWINO = Present weather information not available.
ZRNO = Freezing rain information not available.
TNO = Thunderstorm information not available.
$ = Maintenance check indicator.

DECODED REPORT:
Hometown Municipal Airport, record special observation at 1755 UTC, ASOS with observer. Measured ceiling 1900 feet variable, overcast. Visibility 1 mile, light rain, fog. Sea-level pressure 1012.5 millibars/hectoPascals, temperature 36°F, dew point 34°F, wind from 210° true at 16 knots gusting to 24 knots, altimeter 29.90 inches. Runway 29L visual range 1000 variable to 5000 feet. Ceiling 1600 variable to 2200 feet, tower visibility 2 miles, peak wind 200° true at 32 knots at 1732 UTC, pressure falling rapidly. Freezing rain information not available, maintenance check indicator.

NOTE:
Refer to *ASOS Guide for Pilots* and the *Airman's Information Manual* for more information. Refer to the *Airport/ Facility Directory*, aeronautical charts, and related publications for broadcast, telephone and location data. Check *Notices to Airmen* for ASOS system status.

KEY TO AWOS (AUTOMATED WEATHER OBSERVING SYSTEM) OBSERVATIONS

STATION DESIGNATOR, TYPE OF REPORT, TIME OF REPORT, STATION TYPE	SKY CONDITIONS AND CEILING BELOW 12,000'	VISIBILITY	TEMPERATURE/ DEWPOINT/WIND DIRECTION, SPEED AND CHARACTER/ ALTIMETER SETTING/	REMARKS AUTOMATED REMARKS GENERATED AUTOMATICALLY IF CONDITIONS EXIST. AUGMENTED REMARKS ADDED IF CONDITIONS EXIST AND CERTIFIED WEATHER OBSERVER IS ATTENDING THE SYSTEM
HTM SA 1755 AWOS	M20 OVC	1V	36 / 34 / 2015G25 / 990	P010 / VSBY 1/2V2 WND 17V23 / WEA: R-F

LOCATION IDENTIFIER:
3 alphanumeric characters (usually the airport identifier).

TYPE OF REPORT:
SA = Scheduled record (routine) observation. All observations identified as SA. Most are transmitted at 20-minute intervals (approximately 15, 35, and 55 minutes past each hour).

TIME OF REPORT:
Coordinated Universal Time (UTC) using 24-hour clock.

STATION TYPE:
AWOS = Automated Weather Observing System site. **Note:** In the future, some systems will use "AO" designators.

SKY CONDITION AND CEILING:
Sky condition contractions are for each layer in ascending order. Numbers preceding contractions are base heights in hundreds of feet above ground level (AGL).
CLR BLO 120 = No clouds below 12,000 ft.
SCT = Scattered: 0.1 to 0.5 sky cover.
BKN = Broken: 0.6 to 0.9 sky cover.
OVC = Overcast: More than 0.9 sky cover.
X = Obscured sky **-X** = Partially obscured
A letter preceding the height of a base identifies a ceiling layer and indicates how ceiling height was determined.
M = Measured **W** = Indefinite

VISIBILITY:
Reported in statute miles and fractions. Visibility greater than 10 not reported. **V** = variable: see Automated Remarks

TEMPERATURE AND DEW POINT:
Reported in degrees Fahrenheit.

WIND DIRECTION, SPEED & CHARACTER:
Direction in tens of degrees from true north, except voice broadcast is in degrees magnetic. Speed in knots. **0000** = calm. **G** = gusts. See Automated Remarks for variable direction.

ALTIMETER SETTING:
Inches of mercury. Shown as last 3 digits only without decimal point (e.g., 30.05 inches = 005).

PRESENT WEATHER/OBSTRUCTIONS TO VISION:
Reported only when observer is available. See Augmented Remarks. In the future, some systems will report precipitation, fog, and haze in the body of the observation.

AUTOMATED REMARKS:
Precipitation accumulation reported in hundredths of inches (e.g., P110 = 1.10 inches; P010 = 0.10 inch). **WND V** = variable wind direction. **VSBY V** = variable visibility. **DENSITY ALTITUDE** is included in the voice broadcast when more than 1000 feet above airport elevation.

MISSING DATA:
Reported as "M".

AUGMENTED REMARKS:
"WEA:" indicates manual observer data. Remarks include operationally significant weather conditions within a five mile radius of the airport (e.g., thunderstorms, precipitation, obstructions to vision when visibility is 3 miles or less, fog banks). Standard weather observation contractions are used.

DECODED REPORT:
Hometown Municipal Airport, observation at 1755 UTC, AWOS report. Measured ceiling 2000 feet overcast. Visibility 1 mile variable. Temperature 36 degrees (F), dewpoint 34 degrees (F), wind from 200 degrees true at 15 knots gusting to 25 knots, altimeter setting 29.90 inches. Precipitation accumulation during past hour 0.10 inch. Visibility variable between 1/2 and 2 miles. Wind direction variable from 170 degrees to 230 degrees true. Observer reports light rain (R-) and fog (F).

NOTE:
Refer to the *Airman's Information Manual* for more information. Refer to the *Airport/Facility Directory* , aeronautical charts, and related publications for broadcast, telephone and location data. Check *Notices to Airmen* for AWOS system status.

KEY TO METAR (NEW AVIATION ROUTINE WEATHER REPORT) OBSERVATIONS

TYPE OF REPORT	STATION DESIGNATOR, TIME OF REPORT	WIND	VISIBILITY,	WEATHER AND OBSTRUCTIONS TO VISIBILITY	SKY CONDITIONS	TEMPERATURE / DEWPOINT	ALTIMETER SETTING	REMARKS
METAR	KSEA 1250Z	08032G45KT	1/2SM R32L/1200FT	TSRA	SCT008 OVC012CB	15 / 08	A2995	RMK RETSB24RAB24

TYPE OF REPORT:

There are two types of report - the METAR which is a routine observation report and SPECI which is a Special METAR weather observation. The type of report, METAR or SPECI, will always appear in the report header or lead element of the report.

STATION DESIGNATOR:

The METAR code uses ICAO 4-letter station identifiers. In the contiguous 48 states, the 3-letter domestic station identifier is prefixed with a "K"; i.e., the domestic identifier for Seattle is SEA while the ICAO identifier is KSEA. Elsewhere, the first two letters of the ICAO identifier indicate what region of the world and country (or state) the station is in. For Alaska, all station identifiers start with "PA"; for Hawaii, all station identifiers start with "PH."

TIME:

The time the observation is taken is transmitted as a four digit time group appended with a Z to denote Coordinated Universal Time (UTC). Example: 1250Z.

WIND:

The wind is reported as a five digit group (six digits if speed is over 99 knots). The first three digits is the direction the wind is blowing from in ten's of degrees, or "VRB" if the direction is variable. The next two digits is the speed in knots, or if over 99 knots, the next three digits. If the wind is gusty, it is reported as a "G" after the speed followed by the highest gust reported.

Examples:

13008KT - wind from 130 degrees at 8 knots

08032G45KT - wind from 080 degrees at 32 knots with gusts to 45 knots.

VRB04KT - wind variable in direction at 4 knots

00000KT - wind calm

210103G130KT - wind from 210 degrees at 103 knots with gusts to 130 knots.

If the wind direction is variable by 60 degrees or more and the speed is greater than 6 knots, a variable group consisting of the extremes of the wind direction separated by a "V" will follow the prevailing wind group.

Example:

32012G22KT 280V350

VISIBILITY:

Visibility is reported in statute miles with "SM" appended to it.

Examples:

7SM - seven statute miles

15SM - fifteen statute miles

1/2SM - one half statute mile

Runway Visual Range (RVR), when reported, is in the format: R(runway)/(visual range)FT. The "R" identifies the group followed by the runway heading, a "/", and the visual range in feet (meters in other countries).

Example:

R32L/1200FT - runway 32 left visual range 1200 feet

WEATHER:

The weather as reported in the METAR code represents a significant change in the way weather is currently reported. In METAR, weather is reported in the format:

Intensity, Proximity, Descriptor, Precipitation, Obstructions to visibility, or Other

Intensity - applies only to the first type of precipitation reported. A "-" denotes light, no symbol denotes moderate, and a "+" denotes heavy.

Proximity - applies to and reported only for weather occurring in the vicinity of the airport (between 5 and 10 miles of the center of the airport runway complex). It is denoted by the letters "VC."

Descriptor - these seven descriptors apply to the following precipitation or obstructions to visibility:

TS - thunderstorm　DR - low drifting
SH - shower(s)　MI - shallow
FZ - freezing　BC - patches
BL - blowing

Precipitation - there are eight types of precipitation in the METAR code:

RA - rain　GR - hail (> 1/4")
DZ - drizzle GS - small hail/snow pellets
SN - snow　PE - ice pellets
SG - snow grains　IC - ice crystals

Obstructions to visibility - there are eight types of obstructing phenomena in the METAR code:

FG - fog (vsby < 5/8 mile)
PY -spray
BR - mist (vsby 5/8 - 6 mi)
SA - sand
FU - smoke
DU - dust
HZ - haze
VA - volcanic ash

Note: Fog (FG) is reported only when the visibility is less than five eighths of a mile otherwise mist (BR) is reported.

Other - there are five categories of other weather phenomena which are reported when they occur:

SQ - squall SS - sandstorm
DS - duststormPO - dust/sand whirls
FC - funnel cloud/tornado/waterspout

Examples:

TSRA - thunderstorm with moderate rain
+SN - heavy snow
-RA FG - light rain and fog
BRHZ - mist and haze (vsby > than 5/8 mile)
FZDZ - freezing drizzle
VCSHRA - rain shower in the vicinity

SKY CONDITION:

The sky condition as reported in METAR represents a significant change from the way sky condition is currently reported. In METAR, sky condition is reported in the format:

Amount, Height, (Type), or **Vertical Visibility**

Amount - the amount of sky cover is reported in eighths of sky cover, using the contractions:

SKC - clear (no clouds)
SCT - scattered (1/8 to 4/8's of clouds)
BKN - broken (5/8's to 7/8's of clouds)
OVC - overcast (8/8's of clouds)

Note: A ceiling layer is not designated in the METAR code. For aviation purposes, the ceiling is the lowest broken or overcast layer, or vertical visibility into an obscuration. Also, there is no provision for reporting thin layers in the METAR code.

Height - cloud bases are reported with three digits in hundreds of feet.

(Type) - if towering cumulus clouds (TCU) or cumulonimbus clouds (CB) are present, they are reported after the height which represents their base.

Examples:

SCT025TCU BKN080 BKN250 - scattered towering cumulus at 2,500 feet, broken clouds at 8,000 feet, broken clouds at 25,000 feet.

SCT008 OVC012CB - scattered clouds at 800 feet, overcast cumulonimbus cloud at 1,200 feet

SKC - clear, no clouds

Vertical Visibility - total obscurations are reported in the format "VVhhh" where VV denotes vertical visibility and "hhh" is the vertical visibility in hundreds of feet. There is no provision in the METAR code to report partial obscurations.

Example:

1/8SM FG VV006 - horizontal visibility one eighth of a mile in fog, vertical visibility six hundred feet.

TEMPERATURE/DEWPOINT:

Temperature and dewpoint are reported in a two-digit form in degrees Celsius. Temperatures below zero are prefixed with an "M."

Examples:

15/08 - temperature 15 degrees, dewpoint 8 degrees

00/M02 - temperature zero degrees, dewpoint minus 2 degrees

ALTIMETER:

Altimeter settings are reported in a four-digit format in inches of mercury prefixed with an "A" to denote the units of pressure.

Example:

A2995 - twenty-nine point nine-five inches of mercury

REMARKS:

Remarks are limited to reporting operationally significant weather, the beginning and ending times of certain weather phenomena, and low-level wind shear of significance to aircraft landing and taking off. The contraction "RMK" precedes remarks. The contraction "RE" is used to denote recent weather events. Wind shear information is denoted by "WS" followed by "TKO" for takeoff or "LDG" for landing, and the runway "RW" affected.

Example: RMK RETSB24RAB24

ReMarKs follow, REcent weather event, ThunderStorm Began 24 past the hour, RAin Began 24 past the hour.

DECODED REPORT:

Routine observation report for Seattle, Washington at 1250 UTC. Wind from 080 degrees at 32 knots with gusts to 45 knots. Visibility 1/2 statute mile, runway 32 left visual range 1,200 feet, thunderstorm with moderate rain. There are scattered clouds at 800 feet and overcast cumulonimbus clouds at 1,200 feet. Temperature 15 degrees Celsius, dewpoint 8 degrees Celsius, altimeter setting 29.95 inches of mercury. Remarks: recent weather event, thunderstorm began 24 minutes past the hour, rain began 24 minutes past the hour.

TERMINAL FORECAST (FT)

```
   1       2      3   4    5      6               7
STL FT  251010  C5 X 1/2  S-BS  3325G35  OCNL CO X OS+ BS.
                          8
16Z C30 BKN 3BS 3320 CHC SW-. 22Z 20 SCT 3315. 00Z CLR.
      9
04Z VFR WND..
```

1. **Station Identifier**
2. **Date-time Group** – The date-time group specifies the day of the month and the valid times.
3. **Sky and Ceiling** – The letter "C" always identifies a forecast ceiling layer. Cloud heights are AGL.
4. **Visibility** – The absence of a visibility entry indicates the visibility is more than six statute miles.
5. **Weather and Obstructions to Vision**
6. **Wind** – Omission of the wind entry indicates wind less than six knots. Winds are given in relation to true north.
7. **Remarks**
8. **Expected Changes**
9. **Six-hour Categorical Outlook** – The double period signifies the end of the report.

TERMINAL AERODROME FORECAST (TAF)

TAF
KPIT B.F. 091720Z 1818 22020KT 3SM -SHRA BKN020 FM20 30015G25KT 3SM SHRA OVC015 PROB40 2022 1/2SM TSRA OVC008CB FM23 27008KT 5SM -SHRA BKN020 OVC040 TEMPO 0407 00000KT 1SM -RA FG FM10 22010KT 5SM -SHRA OVC020 BECMG 1315 20010KT P6SM NSW SKC

FORECAST	EXPLANATION
TAF	Message type: TAF-routine and TAF AMD-amended forecast
KPIT	ICAO location indicator
091720Z	Issuance time: ALL times in UTC "Z", 2-digit date and 4-digit time.
1818	Valid period: first 2 digits begins and last 2 ends forecast
22020KT	Wind: first 3 digits mean true-north direction, nearest 10 degrees, (or VaRiaBle;) next 2 digits mean speed and unit, KT, (KMH or MPS); as needed, Gust and 2-digit maximum speed; 00000KT for calm.
3SM	Prevailing visibility: in U.S., Statute Miles & fractions; above 6 miles Plus6SM. (Or, 4-digit minimum visibility in meters and as required, lowest value with direction)
-SHRA	Significant present, forecast and recent weather.
BKN020	Cloud amount, height and type: SKy Clear, SCaTtered, BroKeN, OVerCast; 3-digit height in hundreds of feet; and either Towering CUmulus or CumulonimBus. Or Vertical Visibility for obscured sky and height "VV004", or unknown height "VV///"' More than one layer may be forecast or reported. CLeaR for "clear below 12 thousand feet" at automated observation sites.
FM20	FroM and 2-digit hour: indicates significant change
PROB40 2022	PROBability and 2-digit percent: probable condition during 2-digit beginning and 2-digit ending time period
TEMPO 0407	TEMPOrary: changes expected for less than 1 hour and in total, less than half of 2-digit beginning and 2-digit ending time period
BECMG 1315	BECoMinG: change expected during 2-digit beginning and 2-digit ending time period

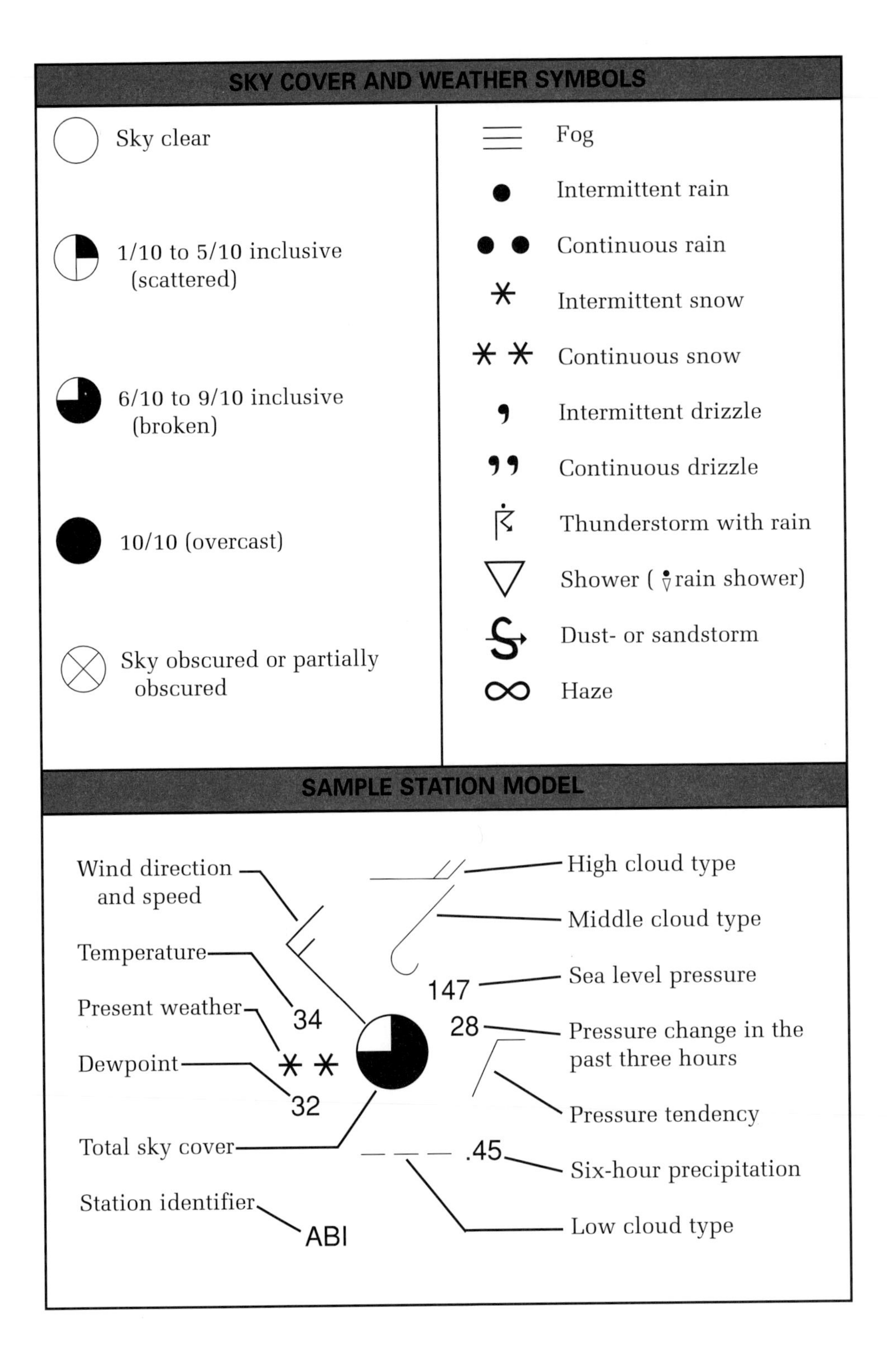

SKY COVER AND WEATHER SYMBOLS
Sky clear
1/10 to 5/10 inclusive (scattered)
6/10 to 9/10 inclusive (broken)
10/10 (overcast)
Sky obscured or partially obscured
Fog
Intermittent rain
Continuous rain
Intermittent snow
Continuous snow
Intermittent drizzle
Continuous drizzle
Thunderstorm with rain
Shower (rain shower)
Dust- or sandstorm
Haze
SAMPLE STATION MODEL
Wind direction and speed
Temperature
Present weather
Dewpoint
Total sky cover
Station identifier
34
32
ABI
147
28
.45
High cloud type
Middle cloud type
Sea level pressure
Pressure change in the past three hours
Pressure tendency
Six-hour precipitation
Low cloud type

RADAR SUMMARY CHART

Symbol	Meaning	Symbol	Meaning
(cloud outline)	Intensity level 1-2 Weak and moderate	240/80	Echo top 24,000' MSL Echo base 8,000' MSL
(second contour)	Intensity level 3-4 (second contour) Strong and very strong	+	Intensity increasing or new echo
		-	Intensity decreasing
(third contour)	Intensity level 5-6 (third contour) Intense and extreme	SLD	Solid, over 8/10 coverage
		LEWP	Line echo wave pattern
		HOOK	Hook echo
(dashed box)	Dashed lines define areas of severe weather	HAIL	Hail
		WS999	Severe thunderstorm watch
(box)	Area of echoes	WT999	Tornado watch
		NE	No echoes
(line)	Line of echoes	NA	Observation not available
→ 20	Cell moving east at 20 knots	OM	Equipment out for maintenance
(barbed arrow)	Line or area is moving east at 20 knots (10 knot barbs)	STC	STC on -all precipitation may not be seen

LOW LEVEL SIGNIFIGANT WEATER PROG

Symbol	Meaning	Symbol	Meaning	Symbol	Meaning
	Showery precipitation (thunderstorms/rain showers) covering half or more of the area		Rain shower		Moderate turbulence
			Snow shower		Severe turbulence
	Continuous precipitation (rain) covering half or more of the area		Thunderstorms		Moderate icing
					Severe icing
	Showery precipitation (snow showers) covering less than half of the area		Freezing rain		Rain
			Tropical storm	*	Snow
,	Intermittent precipitation (drizzle) covering less than half of the area		Hurricane (typhoon)	,	Drizzle

HIGH-LEVEL SIGNIFICANT WEATHER PROG

Symbol	Meaning	Symbol	Meaning
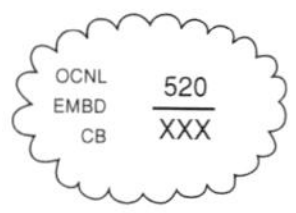 OCNL EMBD CB 520/XXX	Embedded cumulonimbus, 1/8 to 4/8 coverage, tops 52,000 feet, bases below 24,000 feet	 FL420 FL370	Jet stream with maximum speeds of 100 knots at FL 420 at one location and 90 knots at FL 370 at another location
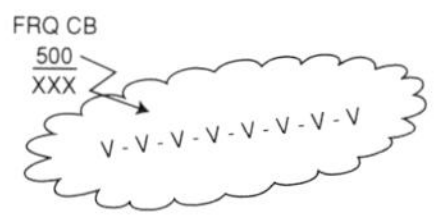 FRQ CB 500/XXX V-V-V-V-V-V-V-V	Forecast severe squall line, CB, coverage 5/8 to 8/8, bases below FL 240 and tops to FL 500	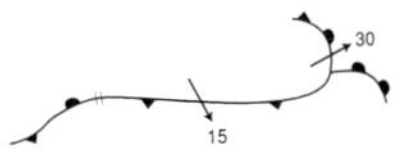 30 15	Forecast surface position, speed (knots), and direction of movement of frontal system
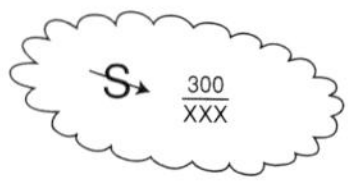 S 300/XXX	Widespread sandstorm or duststorm, bases below FL 240 (i.e., at the surface), tops FL 300	 FRQ CB 500/XXX	Thunderstorm area (5/8 to 8/8 area coverage, bases below FL 240, tops FL 500) associated with a tropical cyclone

Appendix D

Glossary

GLOSSARY OF WEATHER TERMS

Abbreviated briefing: A shortened briefing to supplement mass disseminated data.

Absolute altitude: The altitude of an aircraft above the ground.

Absolute instability: The state of an atmospheric layer when the actual temperature lapse rate exceeds the dry adiabatic lapse rate. An air parcel receiving an initial upward displacement in an absolute unstable layer will accelerate away from its original position.

Absolute zero: The temperature at which all molecular motion ceases.

Acceleration: The change of the speed and/or direction of a mass of air as it moves along its path.

Accretion: The production of a precipitation particle when a supercooled water droplet freezes as it collides with a snow flake or a smaller ice particle. Such particles may become the nucleus of a hailstone.

Adiabatic cooling: Cooling of a gas by expansion.

Adiabatic heating: Warming of a gas by compression.

Advection: The horizontal transport of air or atmospheric properties. In meteorology, sometimes referred to as the horizontal component of *convection.*

Advection fog: Fog resulting from the transport of warm, humid air over a cold surface.

Aerodynamic contrail: As an aircraft moves through moist air the forces created by dynamic flow over the lifting surfaces cause the surrounding atmosphere to reach saturation, to form a cloud like trail. Usually this is generated by high performance aircraft.

Air density: The mass of air per unit volume.

Airmass: An extensive body of air within which the conditions of temperature and moisture in a horizontal plane are essentially uniform.

Airmass thunderstorm: A "nonsevere" or "ordinary" thunderstorm produced by local airmass instability. May produce small hail, wind gusts less than 50 knots. See **severe thunderstorm**.

Airmass wind shear: Wind shear that develops near the ground at night under fair weather conditions in the absence of strong fronts and/or strong surface pressure gradients.

AIRMET (WA): An advisory pertinent to aircraft with limited capabilities, containing information on:

1. moderate icing,
2. moderate turbulence,
3. sustained surface winds of 30 knots or more,
4. ceilings less than 1,000 feet and/or visibility less than 3 miles affecting 50 percent of the area at one time,
5. extensive mountain obscuration.

Albedo: The reflectivity of the earth and its atmosphere.

Altimeter: An instrument which determines the altitude of an object with respect to a fixed level. See **pressure altitude**.

Altimeter setting: The value to which the scale of a *pressure altimeter* is set so as to read *true altitude* at field elevation.

Anemometer: An instrument for measuring *wind speed.*

Aneroid barometer: An instrument for measuring atmospheric pressure, its key component is a partially evacuated cell which changes dimensions in proportion to the change in atmospheric pressure.

Anti-icing equipment: Aircraft equipment used to prevent structural icing.

Anticyclone: An area of high atmospheric pressure which has a closed circulation that is anticyclonic, i.e., as viewed from above, the circulation is clockwise in the Northern Hemisphere, counterclockwise in the Southern Hemisphere.

Anticyclonic flow: In the Northern Hemisphere the clockwise flow of air around an area of high pressure and a counterclockwise flow in the Southern Hemisphere.

Anvil cloud: Popular name given to the top portion of a *cumulonimbus* cloud having an anvil-like form.

Archimedes' Principle: When an object is placed in a fluid (liquid or gas), it will be subject to an upward or downward force depending on whether or not the object weighs more or less than the fluid it displaces.

Arctic airmass: An airmass with characteristics developed mostly in winter over Arctic surfaces of ice and snow. Surface temperatures are basically, but not always, lower than those of polar air.

Atmosphere: The envelope of gases that surrounds the earth.

Atmospheric moisture: The presence of H_2O in any one or all of the states: water vapor, water, or ice.

Atmospheric pressure: The weight of a column of air (per unit area) above the point of measurement.

Attenuation: In radar meteorology, any process which reduces intensity of radar signals.

Aviation turbulence: Bumpiness in flight.

Backing: Change of wind direction in a counterclockwise sense (for example, west to northwest) with respect to either space or time; opposite of veering.

Backscatter: Pertaining to radar, the energy reflected or scattered by a target; an *echo.*

Barometer: An instrument for measuring the pressure of the atmosphere; the two principle types are *mercurial* and *aneroid.*

Billow cloud: A cloud layer having a "herring bone" appearance, these nearly parallel lines of clouds are oriented at right angles to the wind shear.

Black ice: Transparent ice that forms on black pavement, making it difficult to see. It may be caused by the refreezing of melted water or from freezing rain. Also a thin sheet of transparent ice that forms on the surface of water.

Blizzard: A severe weather condition characterized by low temperatures and strong winds bearing a great amount of snow, either falling or picked up from the ground.

Blowing dust: Dust particles picked up locally from the surface and blown about in clouds or sheets.

Blowing sand: Sand picked up locally from the surface and blown about in clouds or sheets.

Blowing snow: Snow picked up from the surface by the wind and carried to a height of 6 feet or more.

Blowing spray: Water particles picked up by the wind from the surface of a large body of water.

Boiling: The process whereby water changes state to vapor throughout a fluid. Occurs when *saturation vapor pressure* equals the total air pressure.

Boiling point: The temperature at which pure water boils at standard pressure, 212°F or 100°C.

Boundary layer: The layer of the earth's atmosphere from surface to approximately 2,000 feet AGL, where surface friction influences are large.

Buoyancy: The property of an object that allows it to float on the surface of a liquid, or ascend through and remain freely suspended in a compressible fluid such as the atmosphere.

Buys Ballot's law: If an observer in the Northern Hemisphere stands with his back to the wind, lower pressure is to his left.

Calm: The absence of wind or of apparent motion of the air.

Cap cloud (also called cloud cap): A standing or stationary cap-like cloud crowning a mountain summit.

Capping stable layer: The elevated stable layer found on top of a convective boundary layer. Usually marks a sharp transition between smooth air above and turbulent air below.

Capture: The process by which small droplets are swept up by faster-falling large droplets. Also called coalescence.

Carburetor icing: Occurs when moist air is drawn into the carburetor and is cooled to a dew-point temperature less than 0°C.

Ceiling: In meteorology in the U.S., (1) the height above the surface of the base of the lowest layer of clouds or *obscuring phenomena* aloft that hides more than half of the sky, or (2) the *vertical visibility* into an *obscuration.*

Ceiling balloon: A small balloon used to determine the height of a cloud base or the extent of vertical visibility.

Ceilometer: A cloud-height measuring system. It projects light on the cloud, detects the reflection by a photoelectric cell, and determines height by triangulation.

Celestial dome: The hemisphere of the sky as observed from a point on the ground.

Celsius temperature scale (abbreviated C): A temperature scale with zero degrees as the melting point of pure ice and 100 degrees as the boiling point of pure water at standard sea level atmospheric pressure.

Centrifugal force: The component of apparent force on a body in curvilinear motion, as observed from that body, that is directed away from the center of curvature or axis of rotation.

Centigrade temperature scale: Same as *Celsius temperature scale.*

Change of state: In meteorology, the transformation of H_2O from one form, i.e., solid (ice), liquid, or gaseous (water vapor), to any other form. See **phase change**.

Chinook: A warm, dry, gusty wind blowing down the eastern slopes of the Rocky Mountains over the adjacent plains in the U.S. and Canada.

Circulation: The organized movement of air. Also called an *eddy.*

Clear air turbulence (abbreviated CAT): Usually, high level (or jet stream) turbulence encountered in air where no clouds are present, may occur in nonconvective clouds.

Clear icing (or clear ice): Generally, the formation of a layer or mass of ice which is relatively transparent because of its homogeneous structure and small number and size of air spaces; used commonly as synonymous with *glaze*, particularly with respect to aircraft icing. Compare with *rime icing.* Factors which favor clear icing are large drop size, such as those found in *cumuliform* clouds, rapid accretion of supercooled water, and slow dissipation of *latent heat* of fusion.

Climatology: The study of the average conditions of the atmosphere.

Climatological forecasts: A forecast based on the average weather (climatology) of a particular region.

Closed low: A low pressure area enclosed in at least one closed isobar or (aloft) one closed contour.

Cloud: A visible body of very fine droplets of water or particles of ice suspended in the atmosphere at altitudes ranging up to several miles above sea level.

Cloud amount: The amount of sky covered by each layer of clouds.

Cloud height: The height of the base of the cloud layer above ground level *(AGL).*

Cloud layer: Refers to clouds with bases at approximately the same level.

Clouds with great vertical development: Cumulus and cumulonimbus clouds.

Cloudy convection: The upward movement of saturated air that is warmer than its surroundings.

Cold air funnel: A weak vortex that occasionally develops behind a cold front with rain showers and nonsevere thunderstorms.

Cold downslope winds: A bora type wind.

Cold airmass: An airmass that is colder than the ground it is passing over.

Cold front: A line or zone along which colder air replaces warmer air.

Cold front occlusion: An occlusion where very cold air behind a cold front lifts the warm front and the cool airmass preceding it.

Comma cloud: A cloud mass shaped like a comma as seen in satellite imagery.

Condensation: Change of state from water vapor to water.

Condensation level: The height at which a rising *parcel* or layer of air would become saturated if lifted adiabatically.

Condensation nuclei: Small particles in the air on which water vapor condenses or sublimates.

Condensation trail (or contrail): A cloud-like streamer frequently observed to form behind aircraft flying in clear, cold, humid air.

Conditionally unstable air: Unsaturated air that will become unstable on the condition it becomes saturated.

Conduction: The transfer of heat by molecular action through a substance or from one substance in contact with another; transfer is always from warmer to colder temperature.

Constant pressure chart: A weather chart that represents conditions on a constant pressure surface; may contain analyses of height, wind, temperature, humidity, and/or other elements.

Contact cooling: The process by which heat is conducted away from the warmer air to the colder earth.

Contour: In meteorology, a line of equal height on a constant pressure chart; analogous to *contours* on a relief map.

Contrail: Contraction for *condensation trail.*

Convection: (1) In general, mass motions within a fluid resulting in transport and mixing of the properties of that fluid. (2) In meteorology, atmospheric motions that are predominantly vertical, resulting in vertical transport and mixing of atmospheric properties; distinguished from *advection.*

Convective cloud: A cloud of vertical development that forms in an unstable environment (stratocumulus, cumulus, cumulonimbus, altocumulus, cirrocumulus).

Convective condensation level (abbreviated CCL): The lowest level at which condensation will occur as a result of *convection* due to surface heating. When condensation occurs at this level, the layer between the surface and the CCL will be thoroughly mixed, temperature *lapse rate* will be dry adiabatic, and *mixing ratio* will be constant.

Convective lifting: In unstable atmospheric conditions, a parcel of air warmer than its surroundings rises.

Convective SIGMET: A weather product that forecasts weather of a convective nature including:
1. severe thunderstorms,
2. embedded thunderstorms,
3. a line of thunderstorms,
4. thunderstorms greater than or equal to VIP level 4 affecting 40 percent or more of an area at least 3,000 square miles.

Convergence: The condition that exists when the distribution of winds within a given area is such that there is a net horizontal inflow of air into the area.

Coriolis force: A deflective force resulting from earth's rotation; it acts to the right of wind direction in the Northern Hemisphere and to the left in the Southern Hemisphere.

Cumuliform: A term descriptive of all convective clouds exhibiting vertical development in contrast to the horizontally extended *stratiform* types.

Cumulus stage: The initial stage of a thunderstorm. The cloud grows from cumulus to towering cumulus and usually lasts 10 or 15 minutes.

Cyclogenesis: Any development or strengthening of cyclonic circulation in the atmosphere.

Cyclone: An area of low atmospheric pressure which has a closed circulation. As viewed from above, the circulation is counterclockwise in the Northern Hemisphere, clockwise in the Southern Hemisphere.

Cyclonic flow: In the Northern Hemisphere the counterclockwise flow of air around an area of low pressure and a clockwise flow in the Southern Hemisphere.

Cyclostrophic wind: A wind in a circular path, in which Coriolis acceleration is negligible as compared to centrifugal forces and pressure gradient acceleration.

Dart leaders: An individual event in a return stroke that the eye cannot distinguish, that occurs following the initial discharge. See **lightning**.

Deepening: A decrease in the central pressure of a pressure system; usually applied to a *low* rather than to a *high*, although technically, it is acceptable in either sense.

De-icing equipment: Aircraft equipment that is actuated to *remove* ice from the structure of the aircraft that has already formed.

Density: Mass per unit volume.

Density altitude: The altitude above mean sea level (MSL) at which a given atmospheric density occurs in the standard atmosphere. See **altitude**.

Deposition: The state where water vapor changes directly to ice.

Dew: Water condensed onto grass and other objects near the ground, the temperatures of which have fallen below the initial dewpoint temperature of the surface air, but is still above freezing. Compare with *frost*.

Dewpoint (or dewpoint temperature): The temperature to which a sample of air must be cooled, while the amount of water vapor and barometric pressure remain constant, in order to attain saturation with respect to water.

Differential heating: The heating of objects that have dissimilar heat capacities. See **heat capacity**.

Direct User Access Terminal Service (DUATS): A computer-based program providing NWS and FAA weather products that are normally used in pilot weather briefings.

Dissipation contrail (distrail): A streak of clearing that occurs behind an aircraft as it flies near the top of, or just within a thin cloud layer.

Dissipating stage: Thirty minutes or so after a single-cell airmass thunderstorm begins, downdrafts spread throughout the lower levels of the cell. Without the necessary source of energy, (heat and moisture), the end of the thunderstorm is near and the clouds take on a strataform appearance.

Diurnal variation: Daily, especially pertaining to a cycle completed within a 24-hour period, and which recurs every 24 hours.

Divergence: The condition that exists when the distribution of winds within a given area is such that there is a net horizontal flow of air outward from the region. The opposite of *convergence.*

Doppler radar: A radar system that has the capability to determine the velocity of a target toward or away from the radar site by measuring the frequency difference between the transmitted and received radiation. When the returning signal is lower frequency than the transmitted frequency then the target is moving away from the site and if the return is higher the cell is moving toward the radar site.

Downburst: A concentrated, severe downdraft that induces an outward burst of damaging winds at the ground.

Downdraft: A downward current of air.

Downslope wind: Wind moving down a slope, the wind can be either a cold downslope wind or a warm downslope wind.

Drainage wind: A shallow, small scale current of cold dense air accelerated down a slope by gravity.

Drag: The resistance of the atmosphere to the relative motion of an aircraft.

Drizzle: A form of *precipitation.* Very small water drops that appear to float with the air currents while falling in an irregular path (unlike *rain,* which falls in a comparatively straight path, and unlike *fog* droplets which remain suspended in the air).

Dry adiabatic lapse rate: The rate of decrease of temperature with height when unsaturated air is lifted adiabatically (3°C/1,000 feet)

Dry adiabatic process: The cooling of an unsaturated parcel of air by expansion and the warming of a parcel of air by compression.

Dry bulb: A name given to an ordinary thermometer used to determine temperature of the air; also used as a contraction for *dry-bulb temperature.* Compare *wet bulb.*

Dry-bulb temperature: The temperature of the air.

Dry line: The moisture boundary, where the moisture content of the air changes rapidly from one side to the other.

Dust: Small soil particles suspended in the atmosphere.

Dust devil: A small, vigorous *whirlwind,* usually of short duration, rendered visible by dust, sand, and debris picked up from the ground.

Duststorm: An unusual, frequently severe weather condition characterized by strong winds and dust-filled air over an extensive area.

D-value: Departure of true altitude from pressure altitude; obtained by algebraically subtracting true altitude from pressure altitude.

Echo: In radar terminology,

1. the energy reflected or scattered by a *target;*
2. the radar scope presentation of the return from a target.

Eddy: An organized movement of air in a circulation of a particular size. The organization is more obvious in the larger scale circulations.

Embedded circulation: A relatively small scale circulation embedded in, and driven by, a larger scale circulation.

Enroute Flight Advisory Service (EFAS): A service provided by the FSS (Flight Watch) for pilots to get current enroute weather by way of the VHF aircraft radio.

Equilibrium level: The altitude where the updraft temperature is equal to its surroundings.

Equinox: Noon on the day when the sun's rays are perpendicular to the earth's surface at the equator.

Evaporation: Change of state from liquid to vapor.

Exhaust contrails: Forms when the water vapor added from an aircraft exhaust is sufficient to saturate the atmosphere.

Extratropical cyclone: A macroscale low-pressure disturbance that develops outside the tropics.

Eye: The roughly circular area of calm or relatively light winds and comparatively fair weather at the center of a well-developed *tropical cyclone.* A *wall cloud* marks the outer boundary of the eye.

Eye wall: The cloudy region embedded with cumulonimbus (CB) clouds immediately adjacent to the eye of an intense tropical cyclone.

Fallstreaks: Ice crystals that descend from cirrus clouds.

Filling: An increase in the central pressure of a pressure system; opposite of *deepening;* more commonly applied to a low rather than a high.

Flight Service Station: The most common source of weather information for pilots. Each state in the U.S. has a single automated flight service station, providing toll free telephone access.

Foehn: A warm, dry, downslope wind, the warmness and dryness being due to adiabatic compression upon descent; characteristic of mountainous regions. See **adiabatic process**, **Chinook**, **Santa Ana**.

Fog: Cloud consisting of numerous minute water droplets and based at the surface; droplets are small enough to be suspended in the earth's atmosphere indefinitely. (Unlike *drizzle*, it does not fall to the surface; differs from cloud only in that a cloud is not based at the surface; distinguished from haze by its wetness and gray color.)

Form drag: Skin friction caused by turbulence induced by the shape of the aircraft.

Freezing: Change of state from liquid to solid.

Freezing drizzle: Drizzle which freezes on contact.

Freezing level: The altitude at which the temperature is 0°C (32°F).

Freezing level chart: A chart depiction of the freezing levels, reported in hundreds of feet.

Freezing rain: Rain that freezes upon contact with the ground or other objects, such as trees, power lines and aircraft.

Frequency: The number of waves that pass some fixed point in a given time interval, measured in cycles per second (cps) or Hertz (Hz).

Friction: The force that resists the relative motion of two bodies in contact.

Front: A transition zone between two adjacent *airmasses* of different densities.

Frontal cyclone: A low pressure area and associated counterclockwise winds (Northern Hemisphere) that develops on the polar front and moves west to east as a macroscale eddy embedded in the prevailing westerlies. Also called a frontal low or wave cyclone.

Frontal lifting: The lifting of a warm airmass over a relative cold airmass.

Frontal wind shear: The change of windspeed or direction per unit distance across a frontal zone.

Frontal zone: A narrow region of transition between two airmasses.

Frost: Ice crystal deposits formed by sublimation when temperature and dewpoint are below freezing.

Funnel cloud: A *tornado* cloud extending downward from the parent cloud but not reaching the ground.

Funneling effect: An increase in winds due to airflow through a narrow mountain pass.

General circulation: The wind system that extends over the entire globe; it is a *macroscale* phenomena with a typical horizontal dimension of 10,000 nautical miles.

Geostrophic balance: The balance of forces that exists when Coriolis force and pressure gradient force are equal in magnitude but in the opposite direction.

Geostrophic wind: The wind that occurs when Coriolis force and pressure gradient are equal and opposite.

G-load: Gust load, the incremental change in vertical acceleration of an aircraft.

Glacier winds: One of the *cold downslope winds.* A shallow layer of cold, dense air that rapidly flows down the surface of a glacier.

Glaze: A coating of ice, generally clear and smooth, formed by freezing of supercooled water on a surface. See **clear icing**.

Global circulation system: The combination of the general and monsoon circulations.

Gradient: In meteorology, a horizontal decrease in value per unit distance of a parameter in the direction of maximum decrease; most commonly used with pressure, temperature, and moisture.

Gravity waves: A small scale wave of air moving in vertical oscillations caused by gravity. Occurring in a stable atmosphere gravity plays the major role in forcing the air parcels to return to their equilibrium level.

Greenhouse effect: The capture of terrestrial radiation by certain atmospheric gases. These gases are commonly called greenhouse gases.

Ground fog: In the United States, a *fog* that is generally less than 20 feet deep.

Gust: A sudden brief increase in wind; according to U.S. weather observing practices, gusts are reported when the variation in wind speed between peaks and lulls is at least 10 knots.

Gust front: The sharp boundary found on the edge of a pool of cold air that is fed by the downdrafts and spreads out below the thunderstorm. A gust front is the key to the long life of a multicell thunderstorm.

Gustnadoes: A tornado-like vortex that sometimes occurs near gust fronts and the edge of a downburst.

Hadley Cell: The tropical cell in a three cell circulation model. This cell exists between 0° and 30° both north and south of the equator.

Hail: A form of *precipitation* composed of balls or irregular lumps of ice, always produced by convective clouds which are nearly always *cumulonimbus.*

Hazardous In-flight Weather Advisory Service (HIWAS): This service provides a continuous broadcast over selected VOR's to inform pilots of hazardous flying conditions.

Haze: Fine dust or salt particles dispersed through a portion of the atmosphere; particles are so small they cannot be felt or individually seen with the naked eye (as compared with the large particle of *dust),* but diminish the visibility; distinguished from fog by its bluish or yellowish tinge.

Heat capacity: The amount of heat energy required to raise the temperature of a substance 1°C.

Heavy snow warning: A warning that snowfall may exceed four inches or more in a 12-hour period or six inches in a 24-hour period.

Height gradient: The rate of change of height per unit of distance on a constant pressure chart. See **figure 3-8**.

High: An area of high barometric pressure, with its attendant system of winds; an *anticyclone*. Also called a high pressure system.

Horizontal pressure gradient force: The force which arises because of a horizontal pressure gradient.

Horizontal wind shear: The change in wind direction and/or speed over a horizontal distance.

Horse latitudes: The areas near 30° both north and south where sinking air creates surface high pressure systems, marked by very low precipitation.

Humidity: Ratio of actual amount of water vapor in the air to the amount at saturation. Expressed as a percent.

Hurricane: A *tropical cyclone* in the Western Hemisphere with winds in excess of 65 knots.

Hurricane eye: The circular, nearly cloud free region approximately 10 to 20 nautical miles in diameter, located in the center of the storm.

Hurricane warning: The warning issued within 24 hours of the arrival of hurricane conditions.

Hurricane watch: Issued when hurricane conditions are expected in a particular area within a day or more.

Hydrological cycle: The movement of moisture from the earth to the atmosphere and back to the earth again.

Hydroplaning: A runway condition that occurs when a thin layer of water separates a tire from the runway surface.

Hydrostatic balance: The balance between the downward-directed gravitational force and an upward-directed pressure gradient force.

Ice crystal process: The process by which cloud particles grow to precipitation size. This can only occur where ice crystals and water droplets co-exist and the temperature is below 0°C.

Ice fog: A type of fog composed of minute suspended particles of ice; occurs at very low temperatures.

Ice pellets: Small, transparent or translucent, round or irregularly shaped pellets of ice. They may be (1) hard grains that rebound on striking a hard surface or (2) pellets of snow encased in ice.

Icing: In general, any deposit of ice forming on an object, such as an aircraft. See **clear icing, rime icing, glaze.**

Incipient stage: The time when frontal cyclone development begins, pressure falls at some point along the original stationary front and counterclockwise circulation is generated.

Indefinite ceiling: A ceiling classification denoting *vertical visibility* into a surface based obscuration.

Indicated air temperature (IAT): Is the temperature of the air as measured by the temperature probe on the outside of the aircraft.

Induction icing: The formation of ice on aircraft air induction ports and air filters.

Infrared (IR): Electromagnetic radiation having wavelengths longer than red light.

Initial lift: One of the two requirements for the production of a thunderstorm, the other is *potential instability*.

Instability: A general term to indicate various states of the atmosphere in which spontaneous *convection* will occur when prescribed criteria are met; indicative of turbulence. See **absolute instability, conditionally unstable air.**

Instrument flight rules (IFR): Rules governing the procedures for conducting instrument flight. Also a term used by pilots and controllers to indicate the type of flight plan, such as IFR.

Instrument meteorological conditions (IMC): Meteorological conditions expressed in terms of visibility, distance from cloud, and ceiling less than the minima specified for visual meteorological conditions.

International standard atmosphere (ISA): An idealized atmosphere with vertical distributions of pressure, temperature, and density prescribed by international agreement.

Intertropical convergence zone: The boundary zone between the trade wind system of the Northern and Southern Hemispheres; it is characterized in maritime climates by showery precipitation with cumulonimbus clouds sometimes extending to great heights.

Inversion: An increase in temperature with height — a reversal of the normal decrease with height in the *troposphere*; may also be applied to other meteorological properties.

Ionosphere: A deep layer of charged particles (ions and free electrons) that extends from the lower mesosphere upward through the thermosphere.

Isobar: A line of equal or constant barometric pressure.

Isotach: A line of equal or constant wind speed.

Isotherm: A line of equal or constant temperature.

Isothermal layers: A layer in the atmosphere where the temperature is the same from the bottom of the layer to the top of the layer.

Jet streak: A portion of the jet stream where windspeeds are greater than in regions up- or downstream. Jet streaks are several hundred to 1,000 miles long.

Jet stream: A narrow band of high speed winds (speeds exceed 60 knots). Normally found near the tropopause.

Jet stream axis: The line of maximum winds (>60 knots) on a constant pressure chart.

Jet stream front: High-level frontal zone marked by a sloping layer below the jet core.

Jet stream cirrus: Associated with an extratropical cyclone, these anticyclonically curved bands of cirrus clouds are usually located just downstream of the upper trough.

Katabatic wind: Any wind blowing down slope.

K index: A stability index that is useful to determine the percentage probability of the occurrence of an airmass thunderstorm.

Kinetic energy: Energy by virtue of motion.

Knot: One nautical mile per hour.

Land breeze: A coastal breeze blowing from land to sea, caused by temperature difference when the sea surface is warmer than the adjacent land. Therefore, it usually blows at night and alternates with a *sea breeze,* which blows in the opposite direction by day.

Lapse rate: The rate of decrease of an atmospheric variable with height; commonly refers to decrease of temperature with height.

Latent heat: The amount of heat absorbed or released during a change of state from ice to water.

Layer: In reference to sky cover, clouds or other obscuring phenomena whose bases are approximately at the same level. The layer may be continuous or composed of detached elements.

Lee wave: Any stationary wave disturbance caused by a barrier in a fluid flow. In the atmosphere when sufficient moisture is present, this wave will be evidenced by lenticular clouds to the lee of mountain barriers; also called *mountain wave* or *standing wave*.

Lee wave region: The upper layer of a two layer lee wave system where smooth wave flow dominates and microscale turbulence occasionally occurs.

Lee wave system: A system marked by two distinct layers. The upper layer is the lee wave region, that begins just above mountain to level, and the lower layer is the lower turbulent zone.

Lifted index: The common approach to the evaluation of potential instability requirement for thunderstorm formation. It is the difference between the observed 500 mb temperature and the temperature the parcel of air would have if lifted from the boundary layer to the 500 mb level.

Lightning: Generally, any and all forms of visible electrical discharge produced by a *thunderstorm*.

Long waves: The wave-like structure in the contour and westerly wind patterns in the mid and upper troposphere. Marked by long, (5,000 miles) slow moving wave troughs located frequently along the east coasts of both Asia and North America.

Low: An area of low barometric pressure, with its attendant system of winds. Also called a *cyclone*.

Lower turbulent zone: The portion of the lee wave system, starting at ground level extending to just above the mountain top, and marked by turbulence.

Low IFR (LIFR): Weather characterized by ceilings lower than 500 feet AGL and/or visibility less than one statute mile.

Low-level wind shear: Wind shear below 2,000 feet AGL along the final approach path or along the takeoff and initial climbout path.

Low-level wind shear alert system (LLWAS): A system installed at many large airports that continually monitors surface winds at remote sites on the airport. A computer evaluates the wind differences from the remote sites to determine if a wind shear problem exists.

Macroscale: Spatial scales of 1,000 n.m. or more.

Mammatus: Bulges or pouches that appear under the anvil of a mature cumulonimbus cloud.

Maneuvering: Input by the pilot or autopilot in response to turbulence, resulting in an excess g-load.

Marginal VFR (MVFR): Weather characterized by ceilings 1,000 to 3,000 feet AGL and/or visibility three to five statute miles.

Mature stage: The most intense stage of a thunderstorm. Begins when the precipitation-induced downdraft reaches the ground. Usually lasts about 20 minutes.

Mechanical turbulence: The turbulence that results when airflow is slowed by surface friction.

Melting: The change of state of a solid to a liquid, as ice to water.

Melting point: The temperature at which pure water begins to melt. At standard pressure, 0°C.

Mercurial barometer: A *barometer* in which pressure is determined by balancing air pressure against the weight of a column of mercury in an evacuated glass tube.

Mesopause: The outer extent of the mesosphere, slightly more than 280,000 feet MSL, the boundary between the mesosphere and thermosphere.

Mesoscale: Spatial scales 1 to 1,000 nautical miles.

Mesoscale convective complex (MCC): The nearly circular clusters of thunderstorms 300 n.m. or more in diameter. MCC develops primarily between the Rockies and the Appalachians during the warmer part of the year.

Mesosphere: The layer of the atmosphere immediately above the stratopause, where the temperature again decreases with height.

METAR: The international weather reporting code that will be introduced in the U.S., after January 1, 1996.

Microburst: A *downburst* with horizontal dimensions of 2.2 nautical miles (4 Km) or less.

Microscale: Spatial scales of 1 n.m. or less.

Mist: A popular expression for drizzle or heavy fog.

Mixed icing: A combination of *clear* and *rime* icing. See **clear ice** and **rime ice**.

Moisture: An all-inclusive term denoting water in any or all of its three states.

Monsoon: A wind that in summer blows from the sea to a continental interior, bringing copious rain, and in winter blows from the interior to the sea, resulting in sustained dry weather.

Mountain breeze: Occurring on a larger scale than downslope winds, it blows down the valley with a return flow, or anti-mountain wind, above the mountain tops.

Mountain wave: An atmospheric gravity wave that forms in the lee of a mountain barrier. See **lee wave**.

Mountain wave turbulence (MWT): Turbulence produced in conjunction with the mountain lee wave.

Multicell thunderstorm: A group of thunderstorm cells at various stages of development. The proximity of the cells allows interaction that prolongs the lifetime of the group beyond that of a single cell.

Negative buoyancy: The tendency of an object, when placed in a fluid, to sink because it is heavier than the fluid it displaces.

Neutral stability: A system is characterized by neutral stability if, when displaced, it accelerates neither toward or away from its original position. The atmosphere displays neutral stability when lapse rate is equal to the dry adiabatic lapse rate.

Nocturnal inversion: A surface-based stable layer that occurs due to nightime radiational cooling.

Numerical weather prediction: Meteorological forecasting using digital computers to solve mathematical equations that describe the physics of the atmosphere; used extensively in weather services throughout the world.

Obscuration: Denotes sky hidden by surface-based *obscuring phenomena* and *vertical visibility* restricted overhead.

Occlusion process: The process by which a cold front overtakes the warm front in a wave cyclone, pushing the warm sector air aloft.

Occluded front: The surface front after a *cold front* overtakes a *warm front* in the occlusion process.

Orographic lifting: The lifting of an airmass when it encounters a barrier, for example, mountains or a hill.

Outlook briefing: A general overview forecast of the weather for a period 6 to 12 hours in advance.

Outflow boundary: The remnant of a gust front that continues to exist long after the thunderstorms that created it dissipate.

Outside air temperature (OAT): The measured or indicated air temperature (IAT) corrected for compression and friction heating, also called **true air temperature**.

Overhang: The anvil of a thunderstorm, under which hail may occur and a turbulent wake may create severe turbulence.

Overshooting tops: In thunderstorms, very strong updrafts that penetrate the otherwise smooth top of the anvil cloud.

Overrunning: When a warm, moist, stable airmass moves over a warm front or a stationary front.

Ozone: An unstable form of oxygen; heaviest concentrations are in the stratosphere; corrosive to some metals; absorbs damaging ultraviolet solar radiation.

Ozone hole: The region of the ozone layer that has a lower than normal concentration of O_3.

Ozone layer: A layer of O_3 found in the lower stratosphere. Characterized by a relatively high concentration of ozone, this layer is responsible for the increase of temperature in the stratosphere.

Parcel: A volume of air, small enough to contain uniform distribution of its meteorological properties, but large enough to remain relatively self-contained and respond to all meteorological processes.

Partial pressure: The gases that make up the atmosphere each exert a partial pressure. When all of the partial pressures are added together, they equal the total atmospheric pressure.

Particulates: Very small liquid or solid particles in the atmosphere. When suspended in the atmosphere, they are called aerosols.

Peak wind: The maximum wind speed since the last hourly observation.

Persistence forecast: A weather prediction based on the assumption that future weather will be the same as current weather.

Phase change: A change of state.

Pilot's telephone weather answering service (PATWAS): A recorded, continuous telephone briefing, forecast for the local area within a 50 nautical mile radius of the station, provided by some manual FSS's.

Plan position indicator (PPI) scope: A radar indicator scope displaying range and azimuth of *targets* in polar coordinates.

Polar airmass: An airmass with characteristics developed over high latitudes, especially within the subpolar highs. Continental polar air (cP) has cold surface temperatures, low moisture content, and, especially in its source regions, has great stability in the lower layers. It is shallow in comparison with *Arctic air*. Maritime polar (mP) initially possesses similar properties to those of continental polar air, but in passing over warmer water it becomes unstable with a higher moisture content. Compare *tropical ear.*

Polar easterlies: Surface winds generated by polar highs north of 60°N latitude.

Polar front: The semi-permanent, semi-continuous *front* separating airmasses of tropical and polar origins.

Polar front jet stream: One of two jet streams which commonly occur in the westerlies. Associated with the polar front.

Polar front model: An idealized representation of events which follow the development of a frontal low. The surface component of the model describes the structure and behavior of fronts and airmasses in the lower atmosphere. The upper air part of the model deals with the associated development of troughs, ridges, and jet streams.

Positive buoyancy: The tendency of an object, when placed in a fluid, to ascend or float because it is lighter than the fluid it displaces.

Potential instability: A layer of air that is not only potentially unstable, it is *conditionally unstable,* and has a high moisture content. Potential instability is one of the two basic requirements for the formation of a thunderstorm, the other is *initial lift.*

Precipitation: Any or all forms of water particles, whether liquid or solid, that fall from the atmosphere and reach the surface. It is a major class of *hydrometer,* distinguished from cloud and *virga* in that it must reach the surface.

Precipitation attenuation: See **attenuation**.

Precipitation fog: Develops when rain saturates the air near the ground.

Precipitation-induced downdraft: The downdrafts that are present inside of the thunderstorm that are induced by rainfall and are much stronger than downdrafts that exist outside of the thunderstorm.

Pressure: See **atmospheric pressure**.

Pressure altimeter: An *aneroid barometer* with a scale graduated in altitude instead of pressure using *standard atmospheric* pressure-height relationships; shows indicated altitude (not necessarily true altitude); may be set to measure altitude (indicated) from any arbitrarily chosen level. See **altimeter setting**, **altitude**.

Pressure altitude: The altitude of a given pressure surface in the standard atmosphere. See **altitude**.

Pressure gradient: The rate of change of pressure per unit distance at a fixed time.

Pressure gradient force: The force that arises because of the pressure of a pressure gradient.

Prevailing visibility: In the U.S., the greatest horizontal visibility which is equaled or exceeded throughout half of the horizon circle; it need not be a continuous half.

Prevailing westerlies: The dominant west-to-east motion of the atmosphere, centered over middle latitudes of both hemispheres.

Prevailing wind: Direction from which the wind blows most frequently.

Primary cycle: The most intense portion of a lee wave, located immediately down wind of the mountain.

Prognostic chart (contracted PROG): A chart of expected or forecast conditions.

Psychrometer: An instrument consisting of a *wet-bulb* and a *dry-bulb* thermometer for measuring wet-bulb and dry bulb temperature; used to determine water vapor content of the air.

Quasi-stationary front (commonly called stationary front): A front which is stationary or nearly so; conventionally, a front which is moving at a speed of less than 5 knots is generally considered to be quasi-stationary.

RADAR (contraction for radio detection and ranging): An electronic instrument used for the detection and ranging of distant objects of such composition that they scatter or reflect radio energy.

Radar beam: The focused energy radiated by radar similar to a flashlight or searchlight beam.

Radar echo: See **echo**.

Radar summary chart: A weather product derived from the national radar network, that graphically displays a summary of radar weather reports.

Radiation: The emission of energy by a medium and transferred, either through free space or another medium, in the form of electromagnetic waves.

Radiation fog: *Fog* characteristically resulting when radiational cooling of the earth's surface lowers the air temperature near the ground to or below its initial dewpoint on calm, clear nights.

Radiosonde: A balloon-borne instrument for measuring pressure, temperature, and humidity aloft. Radiosonde observation — a *sounding* made by the instrument. If the balloon is tracked, winds can also be determined.

Rain: A form of *precipitation;* drops are larger than *drizzle* and fall in relatively straight, although not necessarily vertical, paths as compared to drizzle which falls in irregular paths.

Rain bands: Streaks of rain that spiral into a storm, lines of convergence associated with CB's (cumulonimbus) clouds and shower activity.

Rain shadow: The drier downwind side of a mountain.

Rain shower: See **shower**.

Relative humidity: The ratio of the existing amount of water vapor in the air at a given temperature to the maximum amount that could exist at that temperature; usually expressed in percent.

Return flow: The upper branch of a thermal circulation.

Return stroke: The form of lightening that is visible to the eye, marks the path of the positive charge of the step leader, back into the clouds.

Ridge (also called ridge line): In meteorology, an elongated area of relatively high atmospheric pressure; usually associated with and most clearly identified as an area of maximum anticyclonic curvature of the wind flow *(isobars, contours,* or *streamlines*).

Rime icing (or rime ice): The formation of a white or milky and opaque granular deposit of ice formed by the rapid freezing of supercooled water droplets as they impinge upon an exposed aircraft.

Roll cloud: The dense and horizontal cloud band occasionally found parallel to gust fronts. Also used to describe the rotor clouds associated with mountain lee waves.

Runway visibility (RVV): The *meteorological visibility* along an identified runway determined from a specified point on the runway; may be determined by a *transmissometer* or by an observer.

Runway visual range (RVR): An instrumentally derived horizontal distance a pilot should see down the runway from the approach end; based on the sighting of high intensity runway lights.

St. Elmo's Fire: A luminous brush discharge of electricity from protruding objects, such as masts and yardarms of ships, aircraft, lightning rods, steeples, etc., occurring in stormy weather.

Saturated adiabatic lapse rate: The rate of decrease of temperature with height as saturated air is lifted with no gain or loss of heat from outside sources; varies with temperature, being greatest at low temperatures. See **adiabatic process** and **dry adiabatic lapse rate**.

Saturated adiabatic process: The rate at which saturated air cools as it ascends. It is less than the dry adiabatic lapse rate because adiabatic cooling is offset partially by the release of latent heat. The difference is larger at higher temperatures.

Saturated vapor pressure: The partial pressure of water vapor at saturation.

Saturation: A state of equilibrium where the same amount of molecules are leaving a water surface as are returning and the vapor pressures are balanced.

Scalar: A variable that only has magnitude such as temperature and pressure compared to vector.

Scales of circulations: The typical horizontal dimension size and lifetime of an individual circulation. See **macroscale**, **mesoscale**, and **microscale**.

Sea breeze: A coastal breeze blowing from sea to land. It occurs in the daytime when the land surface is warmer than the sea surface. See **land breeze**.

Sea breeze front: The boundary between the cool, inflowing marine air in the sea breeze and the warmer air over land.

Sea level pressure: The *atmospheric pressure* at *mean sea level.*

Sea smoke: Steam fog, evaporation fog.

Sector visibility: *Meteorological visibility* within a specified sector of the horizon circle.

Sensible heat: Heat that can be felt and measured. Opposite of latent heat.

Severe thunderstorm: A thunderstorm having a much greater intensity, larger size, and longer lifetime than an airmass thunderstorm. Associated weather includes wind gusts of 50 knots or more, and/or hail three-quarters of an inch diameter or larger and/or strong tornadoes.

Shearing gravity waves: Short atmospheric gravity wave disturbances that develop on the edges of stable layers in the presence of vertical shears. Wave amplitudes may grow and overturn causing turbulence.

Shelf cloud: A cloud that indicates the rising air over the gust front. Associated with the updraft of a multicell thunderstorm it is located just above the gust front at low levels.

Short wave trough: Troughs in the mid- and upper troposphere and lower stratosphere that correspond to developing frontal lows. Short wave troughs are smaller in scale than long waves. They move toward the east, averaging 600 nautical miles per day.

Shower: *Precipitation* from a *cumuliform* cloud; characterized by sudden onset and cessation, rapid change of intensity, and usually by rapid change in the appearance of the sky; showery precipitation may be in the form of rain, ice pellets, or snow.

SIGMET (WS): This advisory describes conditions of higher intensity which pose hazards to all aircraft, including:

1. severe icing not associated with thunder storms,
2. severe or extreme turbulence or clear air tur bulence not associated with thunderstorms,
3. duststorms, sandstorms, or volcanic ash lowering surface visibilities to below three miles,
4. volcanic eruptions.

Sleet: See **ice pellets**.

Smog: A mixture of *smoke* and fog.

Snow depth: The depth of the snow actually on the ground.

Snow grains: *Precipitation* of very small, white opaque grains of ice, similar in structure to *snow* crystals. The grains are fairly flat or elongated, with diameters generally less than 0.04 inch (1 mm.).

Snow pellets: Precipitation consisting of white, opaque approximately round (sometimes conical) ice particles having a snow-like structure, and about 0.08 to 0.2 inch in diameter; crisp and easily crushed, differing in this respect from *snow* grains; rebound from a hard surface and often break up.

Solar declination: The latitude where the sun is directly overhead.

Solar elevation angle: The angle of the sun above the horizon measured in degrees.

Solar radiation: The total electromagnetic *radiation* emitted by the sun.

Solstice: Noon the first day of summer and the first day of winter, when the sun has reached its highest and lowest latitudes, respectively.

Sounding: In meteorology, an upper-air observation; a *radiosonde* or *rawinsonde* observation.

Speed of light: Electromagnetic radiation in the wavelength range including infrared, visible, ultraviolet, and X-rays. In a vacuum it is to be about 186,281 miles (300,000 kilometers) per second.

Squall: A sudden increase in wind speed by at least 15 knots to a peak of 20 knots or more and lasting for at least one minute. Essential difference between a gust and a squall is the duration of the peak speed.

Squall line: Any nonfrontal line or narrow band of active thunderstorms.

Stability: A state of the atmosphere in which the vertical distribution of temperature is such that a *parcel* will resist displacement from its initial level. See **instability**.

Standard atmosphere: A hypothetical atmosphere based on climatological averages comprised of numerous physical constants of which the most important are:

1. A surface *temperature* of 59°F (15°C) and a surface pressure of 29.92 inches of mercury (1013.2 millibar) at sea level;
2. A *lapse rate* in the troposphere of 6.5°C per kilometer (approximately 2°C per 1,000 feet);
3. A *tropopause* of 11 kilometers (approximately 36,000 feet) with a temperature of 56.5°C; and
4. An *isothermal* lapse rate in the stratosphere to an altitude of 24 kilometers (approximately 80,000 feet).

Standard briefing: The most complete weather picture, tailored to your specific flight. Usually the briefing includes adverse conditions, a weather synopsis, current weather, forecast weather, forecast winds and temperatures aloft, alternate routes, NOTAM, ATC delays, and request for PIREP.

Standard sea level temperature: A surface temperature of 59°F or 15°C. See **standard atmosphere**.

Stationary front: Same as *quasi-stationary front.*

Station pressure: The actual *atmosphere pressure* at the observing station.

Steam fog: Sea smoke, evaporation fog.

Step leader: The first of a series of events that make up lightning. Nearly invisible to the eye, it is the path that carries electrons from the base of the clouds to the ground, creating an ionized channel for the subsequent discharge.

Stratiform: Descriptive of clouds of extensive horizontal development, as contrasted to vertically developed *cumuliform* clouds; characteristic of stable air.

Stratopause: Occurring at an altitude of about 160,000 feet MSL, the stratopause is the top stratosphere.

Stratosphere: The atmospheric layer above the tropopause, average altitude of base and top, seven and 22 miles respectively; characterized by a slight average increase of temperature from base to top and is very stable; also characterized by low moisture content and absence of clouds.

Structural icing: The formation of ice on the exterior or structure of an aircraft.

Sublimation: Change of state from ice to water vapor. The change from vapor to ice may also be called sublimation; but, to avoid confusion, by convention, it is called deposition.

Subsidence: A slow descending motion of air in the atmosphere over a rather broad area; usually associated with *divergence* and stable air.

Subtropical jet stream: One of two jet streams commonly associated with the westerlies. Located near the 25 to 30° latitude, it reaches its greatest strength in the wintertime and is nonexistent in the summer.

Suction vortex: A small vortex, about thirty feet in diameter, embedded in a tornado funnel.

Superadiabatic lapse rate: A *lapse rate* greater than the *dry adiabatic lapse rate.*

Supercell thunderstorm: A severe thunderstorm that almost always produces one or more of the extremes of convective weather: Very strong horizontal wind gusts, large hail, and/or tornadoes. The supercell can occur anywhere in the mid-latitudes, but by far the favored area is the southern Great Plains of the United States. The supercell is so named because it requires extreme instability and a special combination of boundary layer and high level wind conditions.

Supercooled water droplets: Liquid cloud or precipitation droplets.

Surface air temperature: In meteorology, the temperature of the air measured at 1.5 meters (about 5 feet) above the ground.

Surface friction: The resistive force that arises from the combination of skin friction and turbulence near the earth's surface.

Surface-based inversion: An *inversion* with its base at the surface, often caused by cooling of the air near the surface as a result of *terrestrial radiation*, especially at night.

Surface visibility: Visibility observed from eye-level above the ground.

Sustained speed: The average wind speed over a one- or two-minute period.

Temperature: In general, the degree of hotness or coldness as measured on some definite temperature scale by means of any of various types of thermometers. Also, a measure of the direction heat will flow proportional to the mean kinetic energy of the molecules.

Temperature-dewpoint spread: The difference between the air temperature and the dewpoint.

Terminal forecast (FT): Provides weather conditions expected to occur within a five nautical mile radius of the runway complex at an airport.

Temperature gradient: The change of temperature divided by the distance over which the change occurs.

Temperature inversion: See **inversion**.

Terrestrial radiation: The *radiation* emitted by the earth and its atmosphere.

Thermal circulation: The movement of air resulting from differential heating.

Terminal doppler weather radar (TDWR): Installed at many U.S. airports vulnerable to thunderstorms and microbursts. The use of Doppler radar provides a narrower radar beam and with greater power, a more comprehensive wind shear picture is available for wind shear prediction.

Thermals: A rising bubble of warm air. An element of convection.

Thermal turbulence: Low-level turbulence (LLT) that is produced dry convection (thermals) in the boundary layer.

Thermometer: An instrument for measuring *temperature.*

Thermosphere: The layer of the atmosphere that is directly adjacent to the mesosphere and where the temperature increases with an increase in altitude.

Thunderstorm: In general, a local storm invariably produced by a *cumulonimbus* cloud, that is always accompanied by lightning and thunder.

Tornado: A violently rotating column of air, which appears as a pendant from a cumulonimbus cloud, and nearly always observable as "funnel-shaped." It is the most destructive of all small-scale atmospheric phenomena.

Towering cumulus: A rapidly growing *cumulus cloud*, it is often typical of the cumulus stage of thunderstorm development. Its top may reach 20,000 feet AGL or more and have a width of three to five miles.

Tower visibility: *Prevailing visibility* determined from the control tower.

Trace: When precipitation occurs in amounts too small to be measured, less than .01 inches

Transcribed weather broadcast (TWEB): A continuously broadcast weather information service on selected low and medium frequency nondirectional beacons, and on VHF omni-directional ranges, (VOR). The TWEB includes a synopsis and route forecast and are based on a route of flight format specifically prepared by the NWS.

Transmissometer: An instrument system which shows the transmissivity of light through the atmosphere. Transmissivity may be converted automatically or manually into *visibility* and/or *runway visual range*.

Tropical airmass: An airmass with characteristics developed over low latitudes. Maritime tropical air (mT), the principal type, is produced over the tropical and subtropical seas; very warm and humid. Continental tropical (cT) is produced over subtropical arid regions and is hot and very dry. Compare *polar air*.

Tropical cyclone: A general term for a *cyclone* that originates over tropical oceans. There are three classifications of tropical cyclones according to their intensity:

1. **tropical depression:** winds up to 34 knots;
2. **tropical storm:** winds of 35 to 64 knots;
3. **hurricane or typhoon:** winds of 65 knots or higher.

Tropical storm: See **tropical cyclone**.

Tropopause: The boundary between the *troposphere* and *stratosphere,* usually characterized by an abrupt change of *lapse rate*.

Troposphere: That portion of the *atmosphere* from the earth's surface to approximately 36,000 feet MSL. The troposphere is characterized by decreasing temperature with height, and by appreciable water vapor.

Trough (also called trough line): In meteorology, an elongated area of relatively low atmospheric pressure and maximum cyclonic curvature of the wind flow *(isobars or contours*).

True air temperature: See **outside air temperature (OAT)**.

True altitude: The actual altitude of an aircraft above mean sea level (MSL).

True wind direction: The direction, with respect to true north, from which the wind is blowing.

True north: Geographic center of the North Pole, located at 90° north latitude.

Turbulence: In general, any irregular or disturbed flow in the atmosphere; in aviation, bumpiness in flight.

Turbulence in and near thunderstorms (TNT): That turbulence which occurs within, below, above, and around developing convective clouds and thunderstorms.

Turbulent gusts: The atmospheric wind and vertical motion fluctuations caused by turbulent eddies.

Turbulent wake: The turbulent eddies created near the ground when high surface winds are disrupted by obstacles, such as hangers and other large buildings located near an approach path. Also, the turbulent region downwind of a thunderstorm.

Ultraviolet (UV): Frequencies higher than blue, having relatively short wavelengths.

Unstable: See **instability**.

Updraft: A localized upward current of air.

Upper air temperature: The temperature that is referenced to the height or pressure level where they are measured.

Upper front: A *front* aloft not extending to the earth's surface.

Upslope fog: Fog formed when air flows upward over rising terrain and is, consequently, adiabatically cooled to or below its initial *dewpoint*.

Upslope wind: The deflection of the air by hills or mountains, producing upward motions along the slopes of a mountain or hill.

Valley breeze: A breeze that blows upslope.

Vapor pressure: In meteorology, the partial pressure of water vapor in the atmosphere.

Vector: A variable that has magnitude and direction. For example, wind or pressure gradient.

Vertical motion: Movement of air parcels in an upward or downward direction.

Vertical visibility: The distance one can see upward into a surface based *obscuration;* or the maximum height from which a pilot in flight can recognize the ground through a surface based obscuration.

Vertical wind shear: The change in wind speed and/or direction over a vertical distance. See **wind shear**.

Virga: Water or ice particles falling from a cloud, usually in wisps or streaks, and evaporating before reaching the ground.

Visibility: In U.S. observing practice, a main category of *visibility* which includes the subcategories of prevailing visibility and runway visibility. Meteorological visibility is a measure of horizontal visibility near the earth's surface, based on sighting of objects in the daytime or unfocused lights of moderate intensity at night. Compare *slant visibility, runway visual* range, *vertical visibility*. See **surface visibility**, **tower visibility**, and **sector visibility**.

Visual flight rules (VFR): Rules that govern the procedures for conducting flight under visual conditions. The term "VFR" is also used in the U.S., to indicate weather conditions that are equal to or greater than minimum VFR requirements. In addition, it is used by pilots and controllers to indicate type of flight plan (such as VFR).

Visual meteorological conditions (VMC): Meteorological conditions expressed in terms of visibility, distance from cloud, and ceiling equal to or better than specified minima.

Visual range: See **runway visual range**.

Volcanic ash: In general, particulates and gases from a volcanic eruption.

Vortex: In meteorology, any rotary flow in the atmosphere.

Vortex ring: A microscale circulation cell superimposed on the overall rising motion of a thermal, similar to a smoke ring. It has a relatively narrow core of upward motions surrounded by a broad region of weaker sinking motions.

Wake turbulence: *Turbulence* found to the rear of a solid body in motion relative to a fluid. In aviation terminology, the turbulence caused by a moving aircraft. See **turbulent wake.**

Wall cloud: The well-defined bank of vertically developed clouds having a wall-like appearance which form the outer boundary of the *eye* of a well-developed *tropical cyclone*. Also the portions of the rainfree base of a supercell thunderstorm that is lower in the vicinity of the main updraft. Tornadoes often develop here.

Warm airmass: An airmass characterized by temperatures warmer than the ground over which it is moving.

Warm downslope wind: A warm wind that descends a slope on the lee side of a mountain, often called a *chinook* or foehn, is produced by a warm, stable, updraft airmass moving across a range of mountains at high levels.

Warm front: A front along which warmer air replaces colder air.

Warm front occlusion: Characterized by the warm front remaining on the ground and the cold front moving aloft.

Warm sector: The area covered by warm air at the surface and bounded by the warm *front* and *cold front* of a wave *cyclone*.

Water equivalent: The depth of water that would result from the melting of snow or ice.

Waterspout: A tornado that occurs over water. See **tornado**.

Water vapor: The gaseous form of H_2O.

Wave cyclone: A *cyclone* which forms and moves along a front. The circulation about the cyclone center tends to produce a wavelike deformation of the front.

Wavelength: The distance between two successive, identical wave features, such as two wave crests.

Weather: The instantaneous state of the *atmosphere*.

Weather vane: A *wind vane*.

Wet bulb: Contraction of either *wet-bulb temperature* or *wet-bulb thermometer*.

Wet-bulb temperature: The lowest *temperature* that can be obtained on a *wet-bulb thermometer* by evaporation of water (or ice) from the muslin wick; used in computing *dewpoint* and *relative humidity*.

Wet-bulb thermometer: A thermometer with a muslin covered bulb used to measure wet-bulb temperature.

Whirlwind: A small, rotating column of air; may be visible as a dust devil.

White dew: Frozen dew.

Whiteout: A situation where all depth perception is poor. Caused by a low sun angle, and overcast skies over a snow covered surface.

Wind: Air in motion relative to the surface of the earth; generally used to denote horizontal movement.

Wind direction: The direction from which wind is blowing.

Wind shear: The rate of change of *wind velocity* (direction and/or speed) per unit distance; conventionally expressed as vertical or horizontal wind shear.

Wind vane: An instrument to indicate wind direction.

Wind velocity: A vector which includes *wind direction* and *wind speed*.

References and Recommended Readings

Ahrens, C.D., 1994: *Meteorology Today: an introduction to weather, climate, and the environment. Fifth Edition.* West Publishing Co., St. Paul, MN. 592pp.

Anderson, J.D., 1985: *Introduction to Flight.* Second Edition. McGraw-Hill. New York. 560pp.

Atkinson, B.W., 1981: *Mesoscale Atmospheric Circulations.* Academic Press. 495pp.

Atlas, D. (Ed.), 1990: *Radar in Meteorology.* American Meteorological Society, Boston. 806pp.

Bradbury, T.A., and J.P. Kuettner (Eds.), 1976: *Forecasters Manual for Soaring Flight.* Organisation Scientifique et Technique International du Vol a Voile (OSTIV), Geneva. 119pp.

Buck, R.N.,1988: *Weather Flying.* Third Edition. MacMillan, New York. 311pp.

Byers, H.R., 1974: *General Meteorology.* McGraw-Hill, New York.

Byers, H.R., and R.R.Braham, 1949: *The Thunderstorm.* U.S. Weather Bureau, Washington. 287pp.

Caracena, F., R.L. Holle, C.A. Doswell III, 1990: *Microbursts, A Handbook for Visual Identification.* Second Edition. NOAA, ERL, NSSL. 35pp.

Collins, R.L., 1982: *Thunderstorms and Airplanes.* Delacorte Press/Eleanor Friede. New York. 280pp.

Crossley, A.F. and A.G. Forsdyke, 1960: *Handbook of Aviation Meteorology. M.O. 630 (A.P.3340).* Her Majesty's Stationery Office. London.

Elsberry, R.L. (Ed.), W.M. Frank, G.J. Holland, J.D. Jarrel, R.L. Southern, 1987: *A Global View of Tropical Cyclones.* USNPGS, Monterey. Publication sponsored by Office of Naval Research, Marine Meteorological Program. 185pp.

Evans, J., and M.L. Stone, 1993: *Role of the Aviation Weather System in providing a real-time ATC Volcanic Ash Advisory System. Preprints, Fifth Annual Conference on Aviation Weather Systems*, Vienna, VA. American Meteorological Society. Boston, MA.

FAA, 1975: *Aviation Weather. AC 00-6A.* U.S. Department of Transportation, Federal Aviation Administration and U.S. Department of Commerce, National Oceanic and Atmospheric Administration, National Weather Service, Washington, D.C. 219pp.

FAA, 1983: *Thunderstorms. AC 00-24B.* U.S. Department of Transportation, Federal Aviation Administration, Washington, D.C. 7pp.

FAA, 1988: *Pilot Wind Shear Guide. AC 00-54.* U.S. Department of Transportation, Federal Aviation Administration, Washington, D.C. 56pp.

FAA, 1991: *Wake Turbulence. AC 90-23D.* U.S. Department of Transportation, Federal Aviation Administration, Washington, D.C.

FAA, 1995: *Airman's Information Manual.* U.S. Department of Transportation, Federal Aviation Administration.

FAA, 1995: *Aviation Weather Services. AC 00-45.* U.S. Department of Transportation, Federal Aviation Administration, and U.S. Department of Commerce, National Oceanic and Atmospheric Administration, National Weather Service, Washington, D.C.

Fleagle, R.G., and J.A. Businger, 1980: *An Introduction to Atmospheric Physics.* Second Edition. Academic Press, New York. 346pp.

Fujita, T., 1985: *The Downburst. SMRP Research Paper 210.* The University of Chicago. 154pp.

Fujita, T., 1986: *DFW Microburst. SMRP Research Paper 217.* The University of Chicago. 122pp.

Hansman, R.T., 1989: "The influence of ice accretion physics on the forecast of aircraft icing conditions." *Preprints, Third International Conference on the Aviation Weather System.* Anaheim, CA. American Meteorological Society.

Hurt, H.H., 1965: *Aerodynamics for Naval Aviators. NAVWEPS 00-80T-80.* Office of Chief of Naval Operations, Aviation Training Division. 416pp.

Huschke, R.E. (Ed.), 1959: *Glossary of Meteorology.* American Meteorological Society. Boston. 638pp.

Hutcheon, R.J., J.C. Curtis, A.D. Eubanks, G.L. Hufford, H.L. Kelley, and J.E. Kemper, 1993: *Alaska's Volcanic Ash Warning System. Preprints, Fifth Annual Conference on Aviation Weather Systems,* Vienna, VA. American Meteorological Society. Boston, MA.

Jeppesen Sanderson, 1994: *Instrument Rating Manual.* Jeppesen Sanderson, Inc., Englewood, CO. 500pp.

Jeppesen Sanderson, 1995: *Private Pilot Manual.* Jeppesen Sanderson, Inc., Englewood, CO. 456pp.

Johnson, D.L. (Ed.), 1993: *Terrestrial Environment (Climatic) Criteria Guidelines for use in Aerospace Vehicle Development, 1993 Revision. NASA Technical memorandum 4511.* NASA Office of Management, Scientific and Technical Information Program. Marshall Space Flight Center. 472pp.

Jones, T.N., 1994: "Density Altitude." *FAA Aviation News,* September, 1994. 11-13.

Kessler, E. (Ed.), 1983: *Thunderstorm Morphology and Dynamics.* University of Oklahoma Press, Norman. 411pp.

Kessler, E. (Ed.), 1985: *Thunderstorm Morphology and Dynamics.* University of Oklahoma Press, Norman. 411pp.

Kupcis, E.A., 1989: "The FAA Sponsored Wind Shear Training Aid." *Preprints, Third International Conference on the Aviation Weather System,* American Meteorological Society. 317-322.

Lester, Peter F., 1993: *Turbulence, A New Perspective for Pilots.* Jeppesen Sanderson, Inc., Englewood, CO. 286pp.

Lindsay, C.V. and S.J. Lacy, 1976: Soaring Meteorology for Forecasters. Second Edition. Soaring Society of America.

Mathews, M.D., 1988: "National Weather Advisory Unit Operations and Recent Developments." *Paper AIAA-88-0681,* AIAA 26th Aerospace Sciences Meeting, Reno, NV. 7pp.

Miller, E., 1991: *Volcanic Ash and Aircraft Operations. Preprints, Fourth Annual Conference on Aviation Weather Systems,* Paris. American Meteorological Society. Boston, MA.

NRC, 1994: *Weather for Those Who Fly.* National Weather Service Modernization Committee, Commission on Engineering and Technical Systems, National Research Council. National Academy Press, Washington, D.C. 100pp.

REFERENCES AND RECOMMENDED READING

NTSB, NASA, 1992: Turbulence accident and incident descriptions were abstracted from NTSB Accident Reports and from the NASA Aviation Safety Reporting System (ASRS).

OFCM, 1988: Federal Meteorological Handbook No. 1, Surface Observations. FCM-H1-1988. Washington D.C.

Palmen, E., and C. Newton, 1969: *Atmospheric Circulation Systems.* Academic Press. New York.

Pantley, K., and P.F. Lester, 1990: "Observations of Severe Turbulence near Thunderstorm Tops." *Journal of Applied Meteorology, 29.* 1171-1179.

Peixoto, J., and A. Oert, 1994: *Physics of Climate.* American Institute of Physics. New York.

Purdom, J., J. Weaver, R. Green, 1983: "Analysis of Rapid Interval GOES Data for the 9 July 1982 New Orleans Airliner Crash." *Preprints, Ninth Conference on Aerospace and Aeronautical Meteorology.* American Meteorological Society. 331-334.

Riegel, C.A., 1989: *Atmospheric Dynamics and Thermodynamics.* World Scientific Press.

Rolt, I.T.C., 1966: *The Aeronauts.* Walker and Company, New York. 10019

Schroeder, M.J., and C.C. Buck, 1970: *Fire Weather.* USDA USFS Agricultural Handbook 360.

Scorer, R.S., 1972: *Clouds of the World.* Stackpole Press. Harrisburg PA. 176pp.

Scorer, R.S., 1978: *Environmental Aerodynamics.* Ellis Horwood, Ltd. Chichester, England.

Scorer, R.S., 1978: *Natural Aerodynamics.* Ellis Horwood, Ltd. Chichester. 488pp.

Serebreny, S.M., 1995: *Tailwind.* In press.

Stull, R.B., 1988: *An Introduction to Boundary Layer Meteorology.* Kluwer Academic Publishers. 666pp.

Trollip, S.R. and Richard S. Jensen, 1991: *Human Factors for General Aviation.* Jeppesen Sanderson, Inc. Englewood, CO. 308pp.

U.S. Navy, 1968: *U.S. Naval Flight Surgeon's Manual.* Department of the Navy. 871pp.

USAF, 1969: *Forecasters Guide to Aircraft icing. AWSM 105-39*

USAF, 1983: *Air Navigation. AFM 51-40* (Also *NAVAIR 00-80V-49*). 1987. Reprint with change 1 included. Departments of the Air Force and Navy, Washington, D.C. 379pp.

USAF, 1990: *Weather for Aircrews. AFM 51-12.* Volume 1. Department of the Air Force. Headquarters. Washington, D.C. 137pp.

Wallington, C.E., 1966: *Meteorology for Glider Pilots.* Second Edition. John Murray, Ltd., London. 284pp.

Williams, J., 1992: *The USA Today Weather Book.* Vintage Books, Random House, New York. 212pp.

Wolfson, M. M., 1988: "Characteristics of Downbursts in the Continental United States." *The Lincoln Laboratory Journal, Volume 1, No. 1.* MIT. 49-74.

Index

JEPPFAX: Weather By Fax

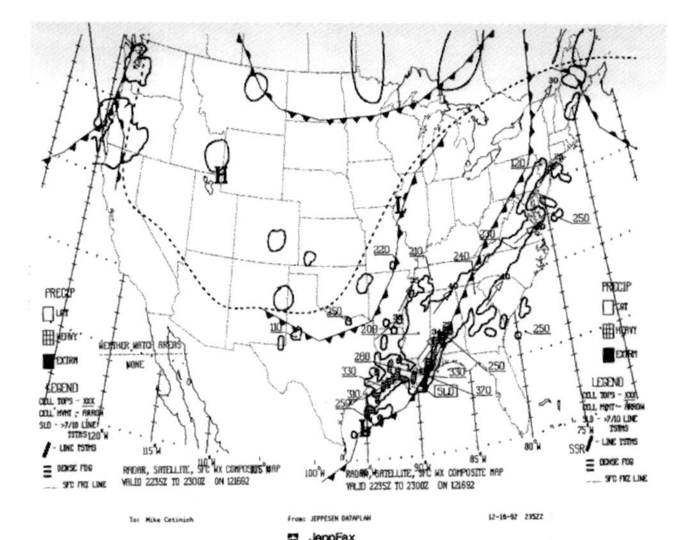

Instant Information, Worldwide
Regardless of your destination or route, JeppFax's interactive weather by fax service gives you instant access to the latest weather maps and briefing. Any time night or day; 24 hours a day, seven days a week, 365 days a year.

The Latest Weather Only a Phone Call Away
All you need is a phone and a fax. No membership fees or subscription required. You choose the weather maps and text briefings that fit your needs: Real-time radar maps • Radar composite maps • Satellite images (infrared and visible • Analysis maps • Surface an low level significant weather forecasts • High level significant weather forecasts • Wind and temperature forecast maps • National Weather Service DIFAX maps • Airport and route briefings specific to your flight.

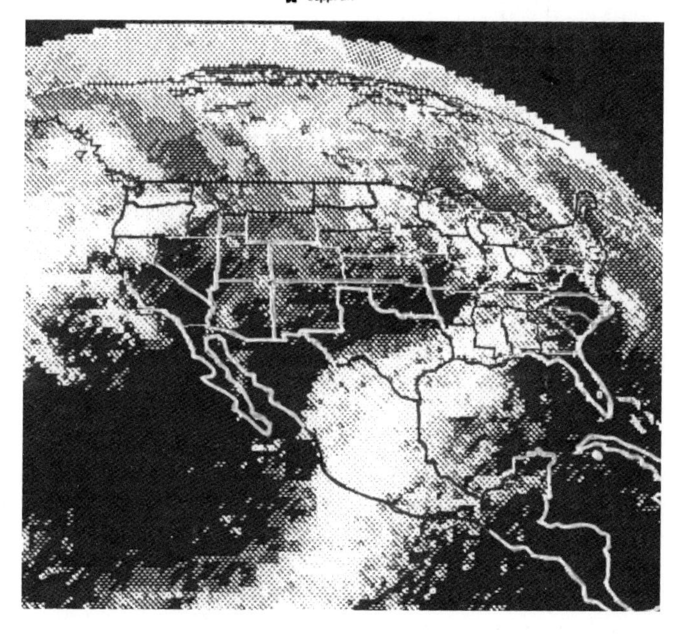

The Price Is Right!
With JeppFax you get more than just weather, more than just price. Your get the added value Jeppesen products and services are known for worldwide.

1 Panel Map or NWS DIFAX Map **$1.25**
1 Panel Satellite or Real Time Radar **$1.65**
2 Panel Map (2 Maps on 1 Page) **$1.95**
2 Panel Map (Incl. 1 Sat or R-T Radar) **$2.35**
2 Panel Map (Include. 2 Sat and/or R-T Radar) **$2.75**
Airport Briefing **$1.00**
All Other Text Briefings **$1.50**

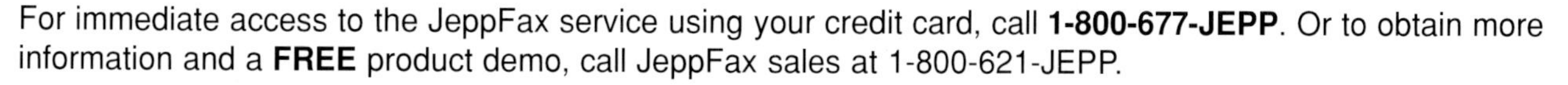

For immediate access to the JeppFax service using your credit card, call **1-800-677-JEPP**. Or to obtain more information and a **FREE** product demo, call JeppFax sales at 1-800-621-JEPP.

Visit Your Jeppesen Dealer or Call 1-800 621-JEPP

Prices Subject to Change

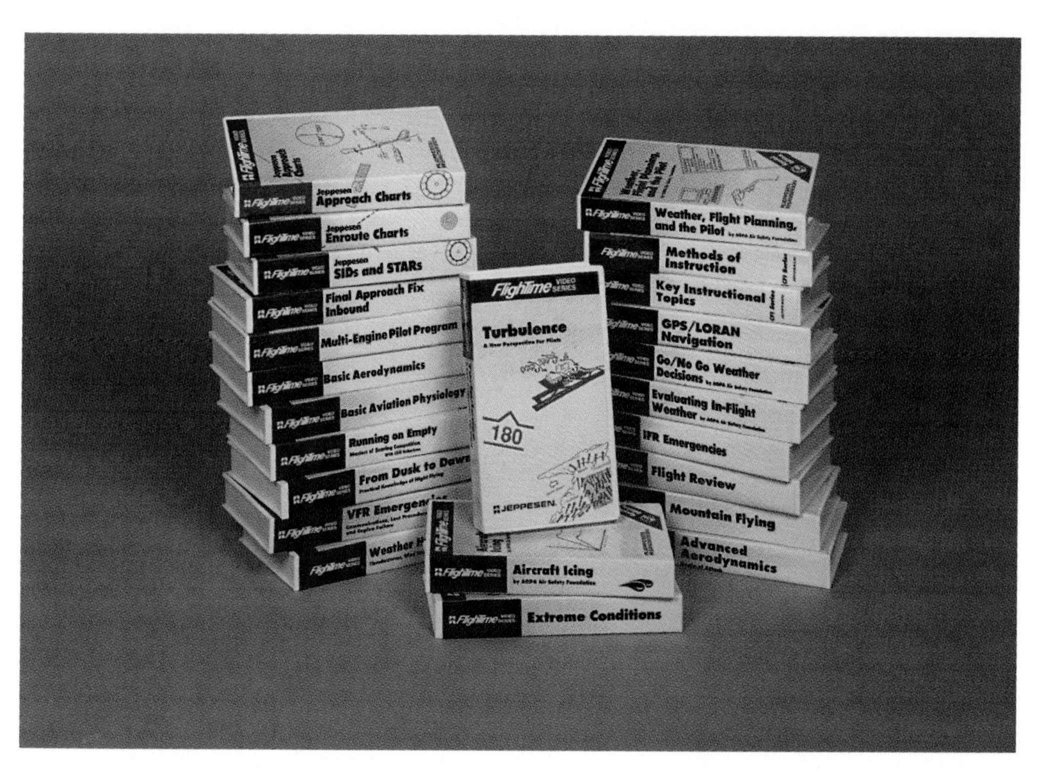

FlighTime video series - Weather Series

Turbulence, A New Perspective Provides the latest, most comprehensive look at turbulence; where it can occur, how to avoid it and what to do if you encounter it. Perfect complement to the text, Turbulence. Available with or without the text.
Item Number JS302162 (VHS) W/Text $48.95
Item Number JS274007 (VHS) W/O Text $29.95

Weather Hazards Illustrates the force of nature behind thunderstorms, wind shear and microbursts. Shows how to improve your ability to avoid the hazards associated with these phenomena.
Item Number JS273070 (VHS) $36.95

Aircraft Icing Emphasizes the varied forms icing can take. Provides techniques to avoid and/or deal with the associated hazards.
Item Number JS273072 (VHS) $29.95

Evaluating Inflight Weather Presents the numerous resources available for gathering and evaluating in-flight weather. Both IFR and VFR situations are demonstrated.
Item Number JS273383 (VHS) $29.95

Go/No-Go Weather Decisions Examines the basic decision making process. Identifies key elements to consider in evaluating weather information to help you make informed weather decisions.
Item Number JS273382 (VHS) $29.95

Extreme Conditions: Hot and Cold Weather Looks at the effects of flying in extreme heat and cold. Includes the impact on aircraft systems and performance. Also addresses how extreme temperatures can affect the pilot.
Item Number JS274005 (VHS) $36.95

Weather, Flight Planning & The Pilot Identifies aviation weather sources and services available. Also covers common errors made in the briefing process.
Item Number JS273381 (VHS) $29.95

Weather Hazards/Turbulence Two volume set, without text. See descriptions above.
Item Number JS302156 (VHS) $64.95

Visit Your Jeppesen Dealer or Call 1-800-621-JEPP

Prices Subject to Change

ADDITIONAL SUPPLIES

Flight Maneuvers Poster Set Excellent student tool provides clear, colorful graphic illustration and text support of maneuvers involved during Traffic Pattern, S-Turns Across a Road, Eights Along a Road, Eights Across a Road and Eights around Pylons.
(Size 22" x 27")
Item Number JS404326 $19.95

Cloud Chart Large, full-scale chart with photographs of 35 typical cloud formations. Each is identified by type and associated weather.
Item Number JS344115 $3.95

Daily Weather Map Reprints of National Oceanic and Atmospheric Administration surface weather maps, high and low temperature and precipitation area and amount amp and height contours of 500-millibar pressure level. Detailed explanation of symbology and map entries.
Item Number JS346152 $ .45

Weather Master Its simple slide rule format makes Jeppesen's exclusive Weather Master highly useful for interpreting aviation weather reports. Meteorology symbols provide ready reference and make it an excellent training tool or memory jogger.
Item Number JS514114 $4.95

FAA Written Exam Study Guides Contain FAA airplane questions arranged in the same sequence as the chapters in our textbooks. Explanations are placed next to each questions and include study references to the page in our textbooks where the topic is covered.

Private Pilot
Item Number JS312400 $12.95

Instrument Rating
Item Number JS312401 $16.95

Commercial Pilot
Item Number JS312402 $14.95

FAR/AIM Manual Comes complete with: Exclusive FAR study guide • FREE Update Summary • Complete Controller Glossary • FAR Parts 1, 61, 91, 135, 141, HMR175 and NTSB 830 • Plus much, much more.
Item Number JS314107 $11.95

Visit Your Jeppesen Dealer or Call 1-800-621-JEPP

Prices Subject to Change